The *Chlamydomonas* Sourcebook

SECOND EDITION

Volume 3: Cell Motility and Behavior

The *Chlamydomonas* Sourcebook

SECOND EDITION

Volume 3: Cell Motility and Behavior

Edited by

George B. Witman, Ph.D.
University of Massachusetts Medical School
Worcester, Massachusetts, USA

Editor in Chief

Elizabeth H. Harris, Ph.D.
Department of Biology
Duke University
Durham, North Carolina, USA

AMSTERDAM • BOSTON • HEIDELBERG • LONDON • NEW YORK
OXFORD • PARIS • SAN DIEGO • SAN FRANCISCO • SYDNEY • TOKYO

Academic Press is an imprint of Elsevier

Cover image: Model for the structure of the 96-nm repeat in the *Chlamydomonas* flagellar axoneme as revealed by cryoelectron tomography. The proximal end is at the left. The protofilaments of the doublet microtubules are gray, the outer dynein arm motor domains are in shades of blue, the outer arm-outer arm dynein linkers are yellow, the inner arms are in shades of green, the dynein regulatory complex is dark green, the outer arm-inner arm dynein linkers are orange, and the radial spokes are purple. See Chapter 9. (Modified from Nicastro et al., 2006, *Science* **313**, 944–948 and kindly provided by Daniela Nicastro, Department of Biology, Brandeis University.)

Academic Press is an imprint of Elsevier

The Boulevard, Langford Lane, Kidlington, Oxford OX5 8GB
30 Corporate Drive, Suite 400, Burlington, MA 01803, USA
525 B Street, Suite 1900, San Diego, CA 92101-4495, USA

First edition 1989
Second edition 2009

Set ISBN: 978-0-12-370873-1

Volume 1: ISBN: 978-0-12-370874-8
Volume 2: ISBN: 978-0-12-370875-5
Volume 3: ISBN: 978-0-12-370876-2

For information on all Academic Press publications
visit our website at www.elsevierdirect.com

Typeset by Charon Tec Ltd., A Macmillan Company. (www.macmillansolutions.com)

Printed and bound by CPI Group (UK) Ltd, Croydon, CR0 4YY

Transferred to Digital Print 2011

Working together to grow
libraries in developing countries

www.elsevier.com | www.bookaid.org | www.sabre.org

ELSEVIER   BOOK AID International   Sabre Foundation

Contents of Volume 3

Preface

Over the past two or three decades, *Chlamydomonas* has become the premier model organism for studying eukaryotic flagella and cilia (essentially identical cell organelles) and the basal bodies from which they arise. The ascendancy of *Chlamydomonas* has been due in large part to its advantages for biochemical, genetic, and molecular and cell biological approaches, which can be uniquely combined to address fundamental questions about these structures, and to a critical mass of devoted and highly interactive investigators studying all aspects of flagella and basal body structure, assembly, and function. As a result, it is fair to say that we now know more about the flagella and basal bodies of *Chlamydomonas* than of any other organism. Fortunately, because flagella and basal bodies have been highly conserved throughout evolution, information obtained from *Chlamydomonas* has yielded profound insights into the basic biology of these organelles and provides a solid foundation for extending these insights in less tractable systems, such as mammalian cells. Therefore, when Lib Harris asked me if I would be interested in editing a volume on motility for the new *Chlamydomonas* Sourcebook, I enthusiastically accepted.

My vision was that this single volume should concisely summarize the accumulated knowledge for all major areas of *Chlamydomonas* cell motility and behavior, as well as contain summary tables and figures that would provide a readily accessible reference source for investigators who wished to quickly look up basic information on any one of the many relevant proteins and mutations, which are difficult for even a *Chlamydomonas* aficionado to remember. This goal, of course, could not have been realized without the cooperation of the many experts who contributed to this volume. It is a testimony to the extraordinary cohesiveness and sense of cooperation within the *Chlamydomonas* motility community that everyone who was invited to contribute agreed to do so and ultimately delivered an outstanding manuscript. Five of the chapters represent collaborations between two or more principal investigators from different laboratories. All of the reviews are highly objective, and authors have generally taken care to present all sides of controversial issues. Overlap between chapters has been kept to the minimum necessary to ensure that each chapter is self-contained and can be understood without reference to other chapters.

Most studies on *Chlamydomonas* cell motility and behavior have been concerned with processes related to flagella and basal bodies, and so this is necessarily the primary focus of this volume. The lengths of the chapters are roughly proportional to how much is known about each subject. For example, the longest chapter covers the enormous amount of information available on the axonemal dyneins, which have been studied in more detail than any other flagellar component, and which have served as important models for the dyneins – including cytoplasmic dynein – of other organisms. The shortest chapter deals with mitosis and cytokinesis, about which much less is known in *Chlamydomonas*, but where *Chlamydomonas* has tremendous potential to contribute more. I hope that the many opportunities for further studies posed in this chapter, and indeed in all the chapters, will stimulate young investigators to use *Chlamydomonas* to address the important outstanding questions in this field.

I am especially pleased to include, as the first chapter, a historical perspective by Joel Rosenbaum, who was a pioneer in the study of *Chlamydomonas* flagella and who mentored, as graduate students or postdocs, many of the authors of chapters in this volume. Joel himself could have written authoritatively on many of the specialized subjects represented by the other chapters, but it seemed most appropriate for him to provide a firsthand account of many of the discoveries, personalities, and collaborations that have shaped this field from its beginning. Joel has been, and continues to be, a central figure in the field and a major stimulus for its progress.

Motility and behavior of course mean movement, and many of the processes discussed in this volume are best represented by movies. Over the years, some spectacular movies of *Chlamydomonas* have been made, but access to these movies is often limited – some are not published, and others are available only as short sequences provided as supplementary material to published articles. Many of these movies are referred to in this volume, and to my delight, the publisher has provided a companion website (http://www.elsevierdirect.com/companions/9780123708731) where, for the first time, all of these movies have been assembled in one place, and often in longer versions than have been available previously. I hope the reader will take the time to view these movies, because they show, with a piquancy that words alone cannot convey, why so many of us continue to be excited and enthralled by *Chlamydomonas*.

Acknowledgments

I am extremely grateful to Lib Harris for her wise counsel, thoughtful suggestions, and careful reviewing from the early planning through the final editing of this volume. Her input on both the science and writing was extremely helpful, and this volume has been much improved as a result of her active and interested role. It has been a real pleasure working with her.

I am equally grateful to all the authors of the chapters, who made time in their busy schedules to contribute outstanding reviews and patiently addressed my many questions and requests. Their enthusiastic cooperation made my task, as an editor, much easier and ensured that our shared vision for this volume would be realized. I also thank Bill Dentler, Dennis Diener, Susan Dutcher, Peter Hegemann, Ritsu Kamiya, Steve King, Karl Lechtreck, Telsa Mittelmeier, Lynne Quarmby, Josh Rappoport, and Joel Rosenbaum for providing beautiful images for a possible cover montage; in the end, the art directors at Elsevier decided to use only a single image, kindly provided by Daniela Nicastro.

Also, thanks to the staff at Elsevier, who worked with us to achieve uniformity in the layout and appearance of all three volumes and to keep production on schedule.

I am much obliged to the members of my laboratory, who understood why I was sequestered in my office for long periods of time writing and editing and less available for scientific discussions than I wished to be. I hope that the information and insights contained in these chapters will do much to make up for this.

Finally, special thanks to my wife, Rita, who endured many lonely weekends while I was working on this volume and put up with me spending most of two vacations editing and corresponding from my laptop while traveling. This volume could not have been completed without her patience and understanding. I know Rita is as glad as I am to see it finally completed!

George Witman
August, 2008

List of Contributors

Peter Berthold Institute for Biology, Experimental Biophysics, Humboldt University, Invalidenstr. 42, 10115 Berlin, Germany

Robert A. Bloodgood Department of Cell Biology, University of Virginia School of Medicine, P.O. Box 800732, Charlottesville, VA 22908, USA

Douglas G. Cole Department of Microbiology, Molecular Biology and Biochemistry, University of Idaho, Moscow, ID 83844, USA

William Dentler Department of Molecular Biosciences, 1200 Sunnyside Avenue, University of Kansas, Lawrence, KS 66045, USA

Susan K. Dutcher Department of Genetics, Washington University School of Medicine, 660 S. Euclid Avenue, St. Louis, MO 63110, USA

Ursula W. Goodenough Department of Biology, Washington University, St. Louis, MO 63130, USA

Peter Hegemann Institute for Biology, Experimental Biophysics, Humboldt University, Invalidenstr. 42, 10115 Berlin, Germany

Ritsu Kamiya Department of Biological Sciences, Graduate School of Science, University of Tokyo, 7-3-1 Hongo, Bunkyo-ku, Tokyo 113-0033, Japan

Stephen M. King Department of Molecular, Microbial and Structural Biology, University of Connecticut Health Center, 263 Farmington Avenue, Farmington, CT 06030, USA

Paul A. Lefebvre Department of Plant Biology, University of Minnesota, 1445 Gortner Avenue, St. Paul, MN 55108, USA

Wallace F. Marshall Department of Biochemistry and Biophysics, UCSF, Genentech Hall, 600 16th Street, San Francisco, CA 94159, USA

David R. Mitchell Department of Cell and Developmental Biology, SUNY Upstate Medical University, 750 E. Adams Street, Syracuse, NY 13210, USA

Daniela Nicastro Department of Biology, Brandeis University, Rosenstiel Center, 415 South Street, Waltham, MA 02454, USA

Gregory J. Pazour Program in Molecular Medicine, University of Massachusetts Medical School, 373 Plantation Street, Worcester, MA 01605, USA

Mary E. Porter Department of Genetics, Cell Biology and Development, University of Minnesota, 6-160 Jackson Hall, 321 Church Street SE, Minneapolis, MN 55455, USA

Lynne M. Quarmby Department of Molecular Biology and Biochemistry, Simon Fraser University, Burnaby, BC, Canada V5A 1S6

Joel Rosenbaum Department of Molecular, Cellular, Development Biology, Yale University, Kline Biology Tower, P.O. Box 208103, New Haven, CT 06520, USA

Winfield S. Sale Department of Cell Biology, Emory University School of Medicine, 465 Whitehead Biomedical Research Building, Atlanta, GA 30322, USA

Elizabeth F. Smith Department of Biological Sciences, Dartmouth College, Hanover, NH 03755, USA

William J. Snell Department of Cell Biology, University of Texas Southwestern Medical Center, 5323 Harry Hines Boulevard, Dallas, TX 75390, USA

Maureen Wirschell Department of Cell Biology, Emory University School of Medicine, 465 Whitehead Biomedical Research Building, Atlanta, GA 30322, USA

George B. Witman Department of Cell Biology, University of Massachusetts Medical School, 55 Lake Avenue North, Worcester, MA 01655, USA

Pinfen Yang Department of Biology, Marquette University, 530 N 15th Street, Milwaukee, WI 53233, USA

Conventions Used

We assume that the reader is familiar with the basic vocabulary of cell and molecular biology. Abbreviations and acronyms specific to particular areas of research or to *Chlamydomonas* are defined in the list that follows. *Chlamydomonas* is assumed to refer to *C. reinhardtii* unless otherwise specified. The reference genome sequence is the DOE Joint Genome Institute (JGI) version 3.0 (http://www.genome.jgi-psf.org/Chlre3/Chlre3.home.html; Merchant et al., 2007). Supplementary material can be found on the Elsevier companion web site, http://www.elsevierdirect.com/companions/9780123708731.

Nomenclature for genes and proteins follows the guidelines, contributed by S.K. Dutcher and E.H. Harris, in the *Genetic Nomenclature Guide* published by *Trends in Genetics* in 1998, and also summarized at http://www. chlamy.org/nomenclature.html. In brief, genes encoded in the nucleus are designated by uppercase italic letters, often followed by an arabic number to distinguish different loci with the same name (*ARG7*). Three-letter names are preferred. Unless otherwise named, proteins are designated by the relevant uppercase gene symbol, but not italic (e.g., ARG7). Mutant alleles are designated in lowercase italics; different mutant alleles at the same locus are distinguished by a number separated from the gene symbol by a hyphen (e.g., *arg7-1*).

Nomenclature for organelle-encoded genes and their products is based on the system used for cyanobacteria and for plants: gene symbols consist of three letters in lowercase followed by one uppercase letter or number, in italics (*petG*, *rps12*). When a protein has no more familiar name, it may be expressed as the same letters in mixed case, not italics (PetG).

Abbreviations

AC	adenylyl cyclase
AKAP	A-kinase anchoring protein
BAC	bacterial artificial chromosome
BBS	Bardet–Biedl syndrome
CALK	*Chlamydomonas* aurora-like kinase
CAM	calmodulin
CaM kinase	calcium/calmodulin-regulated kinase
cAMP	cyclic AMP
CDK	cyclin-dependent kinase
cGMP	cyclic GMP
cNMP	cyclic nucleotide monophosphate
ChR	channelrhodopsin
CK1	casein kinase 1
CR	chlamyrhodopsin
db-cAMP	dibutyryl cAMP
DIC	differential interference contrast
DRC	dynein regulatory complex
EDC	1-ethyl-3-(3-dimethylaminopropyl)carbodiimide, a "zero-length" protein crosslinker
EM	electron microscopy
ER	endoplasmic reticulum
EST	expressed sequence tag
GAK	gliding-associated kinase
GPI	glycosylphosphatidylinositol
HC	heavy chain
HRGP	hydroxyproline-rich glycoprotein
IC	intermediate chain
IFT	intraflagellar transport
Ins(1,4,5)P$_3$	inositol 1,4,5-triphosphate
JME	juvenile myoclonic epilepsy
KAP	kinesin-2-associated protein, also known as kinesin accessory protein
KCBP	kinesin-like calmodulin-binding protein
LC	light chain

LIC	light-intermediate chain
LRC	length regulatory complex
LRR	leucine-rich repeat
MT	mating type
NaPPi	sodium pyrophosphate
NBBC	nucleus-basal body connector
NDK	nucleoside diphosphate kinase
NEPHGE	non-equilibrium pH gradient electrophoresis
ODA-DC	outer dynein arm-docking complex
OID	outer arm–inner arm dynein linker
OMIM	Online Mendelian Inheritance in Man; http://www.ncbi.nlm.nih.gov/sites/entrez?db=OMIM
OOD	outer arm–outer arm dynein linker
PAGE	polyacrylamide gel electrophoresis
PCD	primary ciliary dyskinesia
PCR	polymerase chain reaction
PKA	protein kinase A, also known as cAMP-dependent protein kinase
PKD	polycystic kidney disease
PKG	cGMP-dependent protein kinase
PLC	phospholipase C
PP1	protein phosphatase 1
PP2A	protein phosphatase 2A
PtdIns(4,5)P$_2$	phosphatidylinositol (4,5)-bisphosphate
PTK	protein tyrosine kinase
RSP	radial spoke protein
SDS	sodium dodecyl sulfate
SOFA	site of flagellar autotomy
TEM	transmission electron microscopy

List of Tables

List of Figures

Contents of Volume 1

by Elizabeth H. Harris

Contents of Volume 2

edited by David B. Stern

CHAPTER 1

A Stroll through Time with *Chlamydomonas*

Joel Rosenbaum
Department of Molecular, Cellular and Developmental Biology
Yale University, New Haven, Connecticut, USA

I first met *Chlamydomonas* in 1966 at The University of Chicago. I had been working with the protistan *Ochromonas* because it was able to take up amino acids and I was interested in doing pulse-labeling studies to determine the flagellar growth zone. We were lucky to be able to show that 65–70% of the tritiated leucine-labeled flagellar proteins added on at the tip of the *Ochromonas* flagellum. The days of gene cloning and transformation of *Chlamydomonas* enabled us to show that all the labeled tubulin added on at the tip, but that result came almost 25 years later! In spite of the success with *Ochromonas*, my postdoctoral mentor, Frank Child, was smart enough to realize that the future of studies on the flagellum lay with an organism that had "genetics" and that meant *Chlamydomonas*. I then read Ralph Lewin's book, published in 1962, on the biochemistry and physiology of algae, and that was really my first introduction to the research possibilities of *Chlamydomonas*. It was often said that Lewin's book gave birth to many Ph.D. theses, and if there was a father to the *Chlamydomonas* field before Sager and Levine began their research, then it certainly was Ralph Lewin.

It was not until I took my first position at Yale University in 1967 that I seriously turned to *Chlamydomonas*, when I met David Ringo, a young postdoctoral fellow from Kentucky. He had done his Ph.D. thesis work at the University of Texas and was just getting two lovely manuscripts together for publication in the *Journal of Cell Biology* (on *Chlamydomonas* flagellar ultrastructure) and in the *Journal of Ultrastructure Research*. It was in the former publication that he showed the first picture of intraflagellar transport (IFT) particles between the flagellar membrane and axoneme, something that we were to rediscover 25 years later. Ringo had come to

1

work at Yale with a botanist, Ben Bouck, who worked on brown algae. By chance, Ringo came up to my laboratory in the building next door (probably to talk to my neighbor Joe Gall), and found out that we were studying flagellar biochemistry and assembly. He suggested that we work with *Chlamydomonas*, because of the ease of its culture, the ability to synchronize the cell cycle, and especially its genetics. We became good friends, and eventually he switched his postdoctoral fellowship to my laboratory, and we started studying flagellar regeneration in *Chlamydomonas*. Ringo, by this time, had become a good friend of Ursula Goodenough, who was working on her Ph.D. at Harvard on *Chlamydomonas* under the guidance of Keith Porter and Paul Levine (whom she was later to marry); Levine's laboratory was deeply engaged in the genetics of *Chlamydomonas* chloroplast function and formation. Levine, following the seminal work of Ruth Sager at Hunter College in New York City, was then the guru of *Chlamydomonas*, and he laid the basis for much of the genetics that we can do today with this organism. We are greatly indebted to him for his perseverance with *Chlamydomonas* and developing the tools to study it, as well as for his isolation of many chloroplast mutants, and for making contributions to chloroplast biology in general.

Ringo and I had a lot of fun studying flagellar regeneration in single deflagellated cells. For these studies, a paralyzed mutant of *Chlamydomonas* was essential for our work, as we photographed (with a flash attachment on the microscope) a single cell for at least three hours and could not afford to have the cell move. We would pick out the cell we wanted from a population that had been totally or partially deflagellated, immobilize it by pressing down a bit on the cover slip supported by vaseline so that the organism was trapped between the slide and cover slip, and then proceed to photograph the regenerating flagella. Later, the 35-mm film negatives were projected and we measured flagellar lengths. It was in these studies that Ringo first noticed that detachment of only one flagellum caused the intact one to shorten, sometimes all the way, while the detached flagellum began to grow. The shortening flagellum usually waited for the growing one, and when they were at the same intermediate length they grew out together. We called these "the long-zero" experiments, and after we published our work on *Chlamydomonas* flagellar regeneration, the figures appeared in many textbooks. Interestingly, this phenomenon of simultaneous shortening and growing of different flagella on the same organism is still unexplained, although recently, with the study of IFT and increased insight into the molecular basis for the growth of super-long flagella in the *lf* mutants by Lefebvre (see Chapter 5), there have been some attempts to do so.

By this time, 1968–1970, there were two other laboratories beginning to study flagellar assembly in *Chlamydomonas*. One was that of Sir John Randall, the biophysicist who was knighted for his work on the development of radar during World War II (one of the saviors of Great Britain) and who had decided to enter the field of biophysics in his laboratory at King's College in London. He was director of this laboratory during the initial studies on the high-resolution double helical structure of DNA by Rosalind Franklin, Maurice Wilkins, John Kendrew, and the group at Cambridge (Watson and Crick). Randall, casting about for a good biological problem and an organism in which to study it, decided to work on the assembly of the ciliary/flagellar apparatus, i.e., the basal body/centriole and cilium, and chose the organism *Stentor* to begin his work. *Stentor* is a very large organism, 2–3 mm long, and it already had been shown that its oral apparatus, containing almost 10,000 basal bodies/centrioles, could easily be removed by sucrose treatment, and each cell would synchronously regenerate 10,000 new centrioles and cilia in 5–7 hours. He published a superb paper in the *Journal of Biophysical and Biochemical Cytology* (precursor to the *Journal of Cell Biology*) with Sylvia Fitton-Jackson on the ultrastructure of *Stentor*, with a focus on the oral apparatus. However, as Randall soon realized, culturing large numbers of these organisms was difficult, and if he were to carry out biochemical experiments, he needed large numbers of cells. This, of course, is less of a problem today with the various micromethods and chip technologies that can be carried out for nucleic acid and protein analyses with material from just a few of these cells. Moreover, classical genetic analysis was impossible. Randall gave up working on *Stentor* and started, instead, to work on *Chlamydomonas*, recognizing that one could do genetic analyses with these sexual haploid cells and also study flagellar regeneration. His first paper on *Chlamydomonas*, published in *Nature* in 1964, showed that a non-motile mutant had a disorganized central pair of microtubules, and that this was due to mutation of a single gene. Subsequently, Randall and his *Chlamydomonas* laboratory were the first to show that mating of a cell with super long flagella to a cell with wild-type, normal-length flagella, resulted, in about 30 minutes, in the shortening of the extra-long flagella to wild-type length in the quadriflagellated dikaryon (see Chapter 5). These were some of the first studies that indicated that components of the flagella were moving into the intact flagella from the cytoplasm, that the flagellar parts were probably turning over, and that one could complement mutants *in vivo* by dikaryon rescue. They also carried out some of the first detailed analyses of flagellar regeneration kinetics in *Chlamydomonas*. Although we had, at The University of Chicago, carried out flagellar regeneration studies with *Euglena*, *Ochromonas*, and eventually the ciliate *Tetrahymena*, it was Randall who realized the genetic and biochemical potential of *Chlamydomonas* for studies on flagellar assembly.

Randall's studies also influenced a researcher at The Rockefeller Institute, David Luck, who had carried out seminal studies on mitochondrial formation in *Neurospora*, to begin work on *Chlamydomonas* flagellar assembly. I was invited to Rockefeller around 1970 by George Palade and Phil Siekevitz, who were studying chloroplast formation in *Chlamydomonas*, to give a seminar on our recent work on flagellar regeneration in *Chlamydomonas*. It was there that I met Luck for the first time. I spent a full day at Rockefeller giving a talk and telling them what we were doing. However, no one told me that David Luck already had started working on the very same problem of flagellar assembly in *Chlamydomonas* that Ringo and I were working on at Yale!! Needless to say, we became competitors. Luck continued to work on *Chlamydomonas* flagella until the early 1990s. My last research "competition" with Luck was in 1992 after he and his colleagues at Rockefeller had reported that there was a piece of circular DNA coding for flagellar genes in *Chlamydomonas*, and that this DNA might reside in the two basal bodies. The latter observation on basal body DNA, by Hall and Luck, was published as a full paper, with cover, in the journal *Cell*. Without going into details, both of these observations proved to be wrong for a variety of reasons. The work showing this most conclusively was carried out by Karl Johnson, a postdoctoral student working in my laboratory and now at Haverford College. Using gold-labeled antibodies to DNA, Johnson showed by electron microscopy (EM) of thin sections of *Chlamydomonas* that one could easily quantify gold labeling of the DNA in the nucleus, chloroplasts, and mitochondria, whereas dozens of sections of the basal bodies were unlabeled. Johnson's paper was published as a "Matters Arising" in *Cell*. This, and a number of other papers, firmly put the cap on further research on the presence of DNA in basal bodies/centrioles. The Hall and Luck paper was never formally retracted, and it remains today a historical curiosity in the literature, and one which I assign to Yale seniors in my seminar "Landmark Papers in Cell Biology" to determine what was done wrong. This flagellar basal body nucleic acid story may not be the same for RNA, however.

By the time Ringo, John Moulder (a bright Ph.D. student in the laboratory) and I had published our paper on flagellar regeneration in *Chlamydomonas* in the *Journal of Cell Biology*, George Witman, a new graduate student from UC Riverside, was well into his Ph.D. thesis work on the isolation, fractionation, and analysis of the parts of the *Chlamydomonas* flagellum. He was assisted by Kathryn Carlson, a superb technician who could have easily been a Ph.D. student. I still regard Witman's research as a landmark in *Chlamydomonas* flagella research as it was the first really thorough attempt to separate the flagellum into some of its component parts, other than, of course, the beautiful earlier work of Gibbons and his colleagues on the isolation of ciliary dynein and its reassembly onto the axonemes of *Tetrahymena* cilia *in vitro*. It required a lot of thin sectioning and EM, and with my new NIH grant I purchased my first diamond knife and automated microtome, having used glass knives up until that time. Witman showed, for the first time, the rather insoluble protofilaments shared by the A and B tubules of the outer doublets, on which Stephens and Linck later made major contributions, the pointed structures inside the B tubules of doublets 1, 5, and 6, and some lovely negative stains of isolated flagellar mastigonemes. Witman's work provided the foundation for almost all future studies on the biochemistry of the *Chlamydomonas* flagellum. The subsequent work of Luck and his colleagues Gianni Piperno and Bessie Huang using 2-D gel analysis of various mutant flagella lacking specific structures contributed greatly to the biochemical analysis of flagella. I am happy to say that Witman and I have remained good friends and colleagues since that time and have published on *Chlamydomonas* together for 35 years, and that he still is making major contributions to the field of flagellar assembly and function in *Chlamydomonas*.

It was in this period of the seventies that Ursula Goodenough began her major research effort on gametogenesis and mating in *Chlamydomonas* in her new position in the Biology Department at Washington University in St. Louis. A series of important papers from her laboratory were published in the *Journal of Cell Biology* on this topic, usually firmly based on the ultrastructural approach in which she was expert. There were many new findings, not the least of which was the description of the fibers that attach the basal body to the cell membrane, and which demarcate the flagellar membrane from the cell membrane (each membrane domain is of different composition). These transition fibers had been noted earlier by Gibbons, but Goodenough's work explored them in greater depth. They assume new importance today as the site of the flagellar "gates" where the IFT particles enter with their flagellar axonemal and membrane protein cargoes and where flagellar targeting sequences in the cargo are probably recognized (see Chapters 2, 4, and 10).

By this time my own laboratory was into what I call its renaissance period, 1975–1985. I think every laboratory goes through a time when all the right people come together and a lot of good work gets done. George Witman, Ringo, and Moulder had left (Witman to continue with *Chlamydomonas* in my footsteps at the University of Chicago with Hewson Swift as a postdoctoral mentor, where he published an excellent paper on flagellar tip growth using EM autoradiography, and later with Gary Borisy at the University of Wisconsin, before going to a position as an assistant professor at Princeton). But with their leaving came excellent students, namely, William Snell, William Dentler, Paul "Pete" Lefebvre, Robert Bloodgood, Carolyn Silflow, Lester "Skip" Binder, Leah Haimo, Bruce Telzer, and Roger Sloboda. Part of this influx was due to my taking a position in the Fertilization and Gamete Physiology (FERGAP) course at the MBL at Woods Hole where Charles Metz wanted a flagella person. I met several students there who were later to come to my laboratory. Lefebvre was working at The Rockefeller University, took the FERGAP course, and switched to Yale (he took his Ph.D. exams over again at Yale!); Sloboda was just getting his Ph.D. at Rensselaer Polytechnic Institute and, after the FERGAP course, came to my laboratory to work as a postdoctoral. Binder, from

Oakland University, was already a graduate student in my laboratory when he took the course. We got a lot done and drank a huge amount of beer.

Perhaps the most important person to come to the laboratory during this period was Carolyn Silflow, whom I met while on a seminar trip to the University of Georgia. She was a good molecular biologist, working in Joe Key's laboratory, who came to my laboratory as a postdoctoral fellow; she proceeded to clone and sequence one of the first tubulin genes anywhere, and the work was published in *Cell*. But more importantly she was the first real nucleic acid molecular biologist in the laboratory, at a time when gene cloning was just beginning, and she set us on a course that has not changed since. Her presence was also important to Pete Lefebvre, to whom she is now married. Dentler was an excellent microscopist and filled in EM results on many different projects in the laboratory. He was the first to show the flagellar microtubule tip structures, both on the A tubules of the outer doublets and on the central pair microtubules (see Chapter 10). Snell (together with Dentler, Haimo, and Binder) developed a method for the isolation of the basal bodies of *Chlamydomonas*; this was published in *Science* and served as the basis for the proteomic analysis of *Chlamydomonas* basal bodies by Wallace Marshall and colleagues more than 30 years later. Snell also worked on the mating reaction in *Chlamydomonas* by isolating and analyzing flagellar membranes from gametes. His Ph.D. resulted in two *Journal of Cell Biology* papers, both of which were published under his own name. After a postdoctoral in Saul Roseman's laboratory at Johns Hopkins, Snell returned to *Chlamydomonas* in his position at the University of Texas Southwest Medical Center in Dallas, where he continues to make major contributions to the signaling pathway involved in mating (see Chapter 12).

One of the brightest students was Pete Lefebvre, who decided to analyze RNA and protein synthesis requirements for flagellar assembly in *Chlamydomonas*, using new and modern techniques of two-dimensional gel autoradiography. Lefebvre, on a bet from me, also was the first to develop methods for causing the flagellum to shorten, something I did not think he would be able to do. We bet 50 dollars on it; I don't think I have ever paid him, but there is still time. He demonstrated the importance of calcium ions in the shortening process and also used isobutylmethylxanthine (IBMX), a phosphodiesterase inhibitor (we still do not know why it works) to shorten the flagella. He was the very first person to experimentally shorten the flagella of any organism.

Leah Haimo, who did her undergraduate work at Washington University, did her Ph.D. thesis in the laboratory at this time on the use of *Chlamydomonas* flagellar dynein to decorate brain microtubules assembled *in vitro*. She later extended this work to show that the slant of the dynein could be used to show the polarity of microtubules *in situ* … very similar to the decoration of actin by heavy meromyosin (HMM). It was only later that others demonstrated that the dynein could be attached to surfaces and used to move these microtubules *in vitro*. Haimo was another student who met her husband-to-be, Bruce Telzer, in the laboratory. Leah is now a professor at the University of California, Riverside, and Bruce is on the faculty of nearby Pomona College.

It was also during this time in the early and mid-eighties that Lefebvre and McKeithan, the latter an M.D./Ph.D. student in the laboratory, were able to demonstrate by two-dimensional gel electrophoresis that the alpha tubulin of *Chlamydomonas* flagella ran as two distinct spots. A graduate student, Steve L'Hernault, then demonstrated that one of the spots was acetylated alpha tubulin, and so began an entire field of research concentrated on post-translational modifications of tubulin. He showed that almost all of the flagellar alpha tubulin was acetylated. Shortly afterward, the *Chlamydomonas* laboratory of Gianni Piperno developed a monoclonal antibody to the acetylated lysine epitope, and the ready availability of this antibody allowed many laboratories throughout the world to show that all "stable" microtubules in a great variety of cells are acetylated. There were dozens of reports given, and even separate sessions at the American Society for Cell Biology on this post-translational modification, until Keith Kozminski, working with *Chlamydomonas* in my laboratory, and Jacek Gaertig, working with *Tetrahymena* in Marty Gorovsky's laboratory at the University of Rochester, showed that knocking down (or out) this post-translational modification in cilia and flagella, and indeed on microtubules throughout the cell, resulted in no observable phenotype. I suspect because of these reports, the cottage industry in microtubule acetylation research soon slowed down, and is barely visible today. The function of this post-translational modification is still not entirely known, although it may be involved in motor movement on microtubules.

During the early eighties I was director of the Physiology Course at the MBL in Woods Hole. I brought in Lefebvre and Silflow to teach *Chlamydomonas* genetics in the course. A student in the course at that time was Dennis Diener, and after receiving his Ph.D. at the University of Kansas, working on cell division and cilia regeneration in *Stentor* with Brower Burchill, he came to my laboratory as a postdoctoral student. He has been a mainstay in my laboratory ever since, publishing key papers on flagellar radial spoke polypeptides and their assembly and helping new students with everything from electron microscopy to molecular biology and genetics, and, not of the least importance, their writing of papers. I owe a great deal of the success of my laboratory to him.

By the mid-eighties the *Chlamydomonas* community was beginning to grow and there was an especially productive group starting to work in Geneva, Switzerland, with Jean-David Rochaix, a former postdoctoral at Yale. Rochaix was an expert molecular biologist and put his head to the problem of chloroplast transcription. Among his many contributions, in addition to helping to keep the *Chlamydomonas* chloroplast community alive and growing, were his pioneering efforts to transform genes into *Chlamydomonas*. After much work from many laboratories, homologous integration into the *Chlamydomonas* chloroplast genome – and indeed, into *any* plant genome – was achieved in 1988 by John Boynton, Nick Gillham, and their colleagues at Duke University (Volume 1, Chapter 8). This was extremely important because by then the *Chlamydomonas* community was slowly but surely being replaced by the yeast molecular biology community, principally because the molecular biology of yeast had so many advantages, including facile genetics, easy gene transformation, maintenance of plasmids in the cytoplasm, and most importantly, gene entry into the genome by homologous recombination, allowing replacement of the endogenous gene and enabling gene knockouts and directed mutagenesis. However, efforts to achieve similar transformation of the *Chlamydomonas* nuclear genome had not yet been successful, and this inability threatened to reduce the size of the community interested in working with the cell, and also threatened the ability to obtain research funding from government agencies. Clearly, either we got nuclear transformation to work in *Chlamydomonas* or we were going to be bypassed by researchers working with yeast and other model genetic systems. It was here where the *Chlamydomonas* group, always cooperative, especially in a crisis, started to work intensively together to achieve transformation as a goal. The key workers included the principal molecular biologists and geneticists in the community: Lefebvre and Silflow at the University of Minnesota in St. Paul; Karen Kindle, a new plant molecular biologist at Cornell University who was working on *Chlamydomonas*; Ursula Goodenough at Washington University; Jean Rochaix and his colleagues; Susan Dutcher, a former yeast geneticist now working with *Chlamydomonas*; and Donald Weeks, then at the Philadelphia Cancer Research Institute. There were meetings directed solely toward the progress being made on transforming *Chlamydomonas*. By now, the International Conference on the Cell and Molecular Biology of *Chlamydomonas* was meeting every two years, and it also presented an opportunity to exchange results on this very important problem. Ultimately, it was Paul Lefebvre and his colleague Emilio Fernández from Spain in cooperation with Karen Kindle who made the first major breakthroughs on nuclear transformation when they successfully transformed into *Chlamydomonas* the nitrate reductase gene to complement a mutation in the nitrogen-reducing pathway. I regard this research, quite simply, as the work that saved the field of *Chlamydomonas* research. Carried out in the late eighties and early nineties, it was soon followed by the transformation of other *Chlamydomonas* mutants with wild-type genes, e.g., radial spoke genes to complement radial spoke-paralyzed mutants by Dennis Diener in my laboratory, and various flagellar motility genes to complement paralyzed mutants. There were still some problems with the transformation: the genes entered the genome randomly and not by homologous recombination (and this is still a major problem); and many heterologous genes, e.g., from yeast and humans, could not be used to transform *Chlamydomonas* because of differences in codon usage/bias. Still, this was a major advance in the molecular technology available to *Chlamydomonas* researchers, giving the laboratories using *Chlamydomonas* the lift they needed, and rapidly becoming a routine procedure in the laboratories of most of the researchers working with this cell. The *Chlamydomonas* research community would have withered without this advance; instead it thrived throughout the nineties and continues to expand to this day. We tend to use the technical advances made by others and take them for granted, and yet those advances are the linchpins of our research.

Research on the cell biology of cilia and flagella was always important in the *Chlamydomonas* field, because the cells are phototrophic and the screens for flagellar mutants are, therefore, quite easy, and the mutations are not lethal. Indeed, the largest group of cilia/flagella researchers in the world now works with *Chlamydomonas*, and the ability to transform *Chlamydomonas* ensured this attention.

Additional interest in the flagella of *Chlamydomonas* came about quite by accident. Keith Kozminski, a graduate student in my laboratory, decided to investigate an interesting finding made by Robert Bloodgood, a postdoctoral in my laboratory in the seventies, that latex beads would attach to the flagellar surface and move rapidly both proximally and distally (see Chapter 11 and videos at http://www.elsevierdirect.com/companions/9780123708731). It was a spectacular observation, but its molecular basis had gone without explanation for 15 years. Kozminski decided to look at this phenomenon and try to find something out about the motors, obviously underneath the flagellar membrane, that were moving the beads. Instead, while observing the bead movement using new, high-resolution differential interference contrast (DIC) equipment in the adjacent laboratory of Paul Forscher, he observed something that was also spectacular: the rapid movement, both proximally and distally at 2–4 μm/s, of rather large particles under the flagellar membrane (see video "Chlamy_IFT_Kozminski" at http://www.elsevierdirect.com/companions/9780123708731). He called the process IFT for intraflagellar transport, although at the time there was no information that it was transporting anything. IFT just appeared to be an excellent mechanism to support the tip assembly of the flagella that had been demonstrated years earlier. It was soon shown that a specialized kinesin (kinesin-2) and a special type of cytoplasmic dynein (cytoplasmic dynein 1b) were powering the particle movement in the anterograde and retrograde directions, respectively. The kinesin-2 was especially important because there was a temperature-sensitive mutant for this kinesin available, and placing the mutants at the nonpermissive temperature not only stopped the IFT (the very first indication that the kinesin was involved in IFT) but also caused the flagella to resorb.

Moreover, if one deflagellated the cells at the non-permissive tempera-ture, the flagella did not reassemble. This, then, was hard evidence, using a kinesin mutant, that IFT was important for both flagellar assembly and maintenance.

The excitement of finding this new microtubule motor-driven sub-membrane motility in the flagella of *Chlamydomonas* led us to submit a report to *Science* and then to *Nature* in 1992. Both journals, in their wis-dom and boundless enthusiasm for publishing something new, and even exciting, turned it down without review. It was published in 1993 in the *Proceedings of the National Academy of Sciences*, submitted by Hewson Swift, who had been my second postdoctoral mentor at Chicago. Once again, *Chlamydomonas* was to be responsible for initiating an entire new field, but this time, unlike acetylation of tubulin, it would expand to well beyond our initial expectations and continues to this day to permeate all aspects of research on cilia and flagella, including the heretofore unknown relationship of cilia to many different pathologies dependent on the sensory function of cilia.

The next major advance in the new *Chlamydomonas* flagellar IFT field came from the work of Douglas Cole. Cole had been doing postdoctoral work on kinesin purification in the John Scholey laboratory at UC Davis, and decided to try his hand at purification of the *Chlamydomonas* flagellar IFT kinesin in my laboratory. Working with many liters of cells, he iso-lated flagella and purified, to a high degree of homogeneity, the enzyme, kinesin-2, or heterotrimeric kinesin. He was able to pinpoint this kinesin as the important one (there are several flagellar kinesins) because of the temperature-sensitive *Chlamydomonas* mutant in which this kinesin was defective at nonpermissive temperature. When the cells were grown at nonpermissive temperature, the kinesin disappeared from western blots of flagellar gels (as first demonstrated by Walther and Hall at Rockefeller), as did a large number of flagellar polypeptides sedimenting together at about 16S on sucrose gradients. These were clearly the IFT anterograde motor and the proteins moved by them as observed in DIC, because the particle movement stopped at nonpermissive temperature. Cole separated out the 16S proteins on two-dimensional gels, cloned the genes from the sequences obtained, and made antibodies to many of what turned out to be a group of 16–18 IFT-particle polypeptides in two distinct supramolecular complexes, called complex A and complex B. The antibodies stained the polypep-tides in the flagella but principally around the basal body – their primary location. One of their specific locations near the basal body/centriole was shown by EM gold labeling to be on the transition fibers mentioned earlier.

The cloning of the genes for the IFT polypeptides was, perhaps, the most important step taken by Cole, because those sequences led directly to genes in the world's gene data base that indicated where the gene homologues existed. Perhaps one of Kozminski's and Cole's most important findings was that mutations in the IFT polypeptides themselves, or the motors that moved them, did not allow flagellar assembly. By this time, others in the *Chlamydomonas* community began to enter the IFT field. Witman and Greg Pazour at the University of Massachusetts Medical School and Mary Porter at the University of Minnesota showed that cytoplasmic dynein 1b was the retrograde motor for IFT, and they isolated dynein retrograde IFT mutants, including at least one temperature-sensitive mutant in the heavy chain of the dynein. The anterograde and retrograde IFT temperature-sensitive mutants were to prove invaluable in later studies done by several *Chlamydomonas* researchers.

The entire mushrooming area of cilia-dependent pathologies, almost all involving the sensory function of nonmotile cilia, called primary cilia, really began with Cole's studies described above. The *Chlamydomonas* IFT polypeptide of molecular mass 88 kD, for example, turned out to be the same polypeptide which, when mutant in the mouse, gave rise to the pathology of polycystic kidney disease (PKD). Pazour, then a postdoctoral with Witman and now a faculty member at the University of Massachusetts Medical School, found that IFT88 was essential for the assembly of *Chlamydomonas* flagella, and then showed that the IFT88-defective mice with PKD had very short or no cilia on the epithelial cells of the kidney tubules, which normally have a single long primary cilium. Thus, IFT88 is essential not only for flagella formation in *Chlamydomonas* but also for cilia formation in the mouse, and a mutation in IFT88 in the mouse results in PKD. From these reports by Cole and Pazour, two young members of the *Chlamydomonas* community, came the now respected "Ciliary Hypothesis of PKD," which states that the PKD-encoded gene products on the ciliary surface are ion channels whose proper function keeps the ciliated cells from dividing. The disease results when the cilia are not present, e.g., due to a defect in IFT, or if the cilia are normal and the PKD gene products themselves are mutant. Since these initial discoveries on the importance of cilia in PKD, researchers around the world have localized a variety of gene products on the membranes of nonmotile primary cilia, as well as on motile cilia, including many other genes responsible for a variety of cystic diseases. The list of cilia-related diseases now includes primary ciliary dyskinesia (defective airway cilia), situs inversus (defective cilia of the embryonic node), hydrocephalus (defective cilia in the brain ventricles), Bardet-Biedl syndrome (BBS) (obesity, bone defects, diabetes, etc.) and cystic diseases of the liver and pancreas. Many workers entering the field now, especially those who do not read the *Chlamydomonas* literature, do not realize that the entire "ciliopathy" field started with basic cell biological studies on *Chlamydomonas* flagellar assembly (see Chapter 15).

More recently, since 2002, *Chlamydomonas* has become the cell of choice for high-resolution structural studies of molecular motors, i.e., flagellar dynein. Because of the complexity of dynein motors, being composed of many polypeptides, the availability of mutants in which one or more of these polypeptides are missing has proven invaluable to studies of the motor structure. Quick-freeze/deep-etch procedures developed by John Heuser, and successfully applied to *Chlamydomonas* flagella by Heuser and Goodenough and published in the mid-eighties, still represent one of the best high-resolution views we have of the inner workings of the flagellar axoneme and the associated dynein motors. Other researchers have now followed suit with studies using the cryo-electron microscope in which native cilia and flagella can be observed, analyzing the tomographic results with specially designed computer programs. Researchers such as Daniela Nicastro in the United States, Takashi Ishikawa in Switzerland, and Masahide Kikkawa and Kazuhiro Oiwa in Japan are using *Chlamydomonas* for their high-resolution ultrastructural analyses of flagella and flagellar dynein using cryo-EM. Pietro Lupetti and his colleagues Gaia Pigino and Caterina Mencarelli in Italy have initiated studies on the high-resolution structure of the IFT particles and their motors in *Chlamydomonas* flagella, as well as studies on the isolated IFT particles.

Many laboratories have used the ability to synchronize the cell cycle of *Chlamydomonas* by placing them on a light/dark cycle to study the control of the eukaryotic cell division cycle. This reached a new level of sophistication with the arrival on the *Chlamydomonas* scene of Jim Umen, a postdoctoral in the Goodenough laboratory and now at the Salk Institute. Thanks to the work in the Umen laboratory, *Chlamydomonas* is one of the best division-synchronized cells available in cell biology today. Indeed, one can obtain, using Umen's procedures, over 95% synchrony of *Chlamydomonas* cell division by paying close attention to cell size, light, and temperature. This will be of immeasurable value to researchers studying the control of the cell cycle, and also to flagella assembly/disassembly workers, since it is becoming apparent that the control of flagellar length is also involved in cell cycle control. Once again a technological development lays the platform for research done by many.

A research area on *Chlamydomonas* that has come into its own in the past 15 years is one that I regard as one of the most important topics in general cell biology today: How is organelle size control (length in the case of the flagellum) regulated? Following in the footsteps of Sir John Randall in the sixties, Pete Lefebvre first took this problem in hand by identifying the genes involved in the assembly of the super long flagella in the "long-flagella" *lf* mutants, and the proteins for which they were coding. These are spectacular mutants with flagella that can reach 20–30 μm long (two to three times the length of wild-type flagella). One is not surprised that Lefebvre and his colleagues found specific protein kinases involved as key gene products in this length-control process (see Chapter 5). Almost certainly these genes will, in turn, affect the IFT system since Marshall has shown that the IFT is carrying out the microtubule assembly/disassembly (turnover) at the flagellar tip. I know of no other cell organelle in which such elegant studies on size control have been done, and the ease of doing both flagellar biochemistry and genetics with *Chlamydomonas* promises to make this an increasing area of interest among *Chlamydomonas* researchers and to those in the general cell biology community.

The modern era of gene and protein biotechnology has ensured the place of *Chlamydomonas* among the best model genetic organisms available for use by the world's research community in cell biology. Thanks to the hard work of several of the *Chlamydomonas* laboratories in the past decade, particularly those of Pete Lefebvre, Ursula Goodenough, and Susan Dutcher, the genome of *Chlamydomonas* has been completely sequenced and the genes responsible for coding flagellar proteins have been determined by comparison of the genomes of *Chlamydomonas* with those of non-flagellated organisms. The proteomic analysis of *Chlamydomonas* flagella has been completed by the Pazour and Witman laboratories with analysis of total axonemal and membrane polypeptides. Because the transcription of flagellar mRNAs is upregulated following flagellar detachment, Marshall and colleagues at UCSF have been able to use RNA chip technology to define all those mRNAs that are transcribed in *Chlamydomonas* during flagellar regeneration.

So, all the polypeptide components of the flagella are available to us now, and all that is required is to find out which ones go where and how they function in flagellar motility as well as in the sensory functions involved in the ciliopathies. Is this the end of our travel with *Chlamydomonas*? Probably not, there being sufficient questions remaining to keep us busy for decades. For example, studies on *Chlamydomonas* flagella by Huang and Rosenbaum (unpublished) have shown, that, in addition to being an organelle of motility and sensory function, the flagellum is also an organelle of secretion. The function of the secreted vesicles remains to be determined, so this is an area that has much potential for future investigations.

If I were to pick one research area, using *Chlamydomonas*, that has not received the attention it should have, it would be the study of the upregulation of flagellar gene transcription that occurs immediately following flagellar detachment. It was the initial reason for my starting to work with flagella-regenerating systems in protists, because my Ph.D. background had been in translation. Over 350 mRNAs, most coding for flagellar polypeptides, are rapidly transcribed after the flagella are detached. Once flagellar regeneration is underway, this mRNA synthesis decreases to constitutive levels. Many of our papers in the late seventies and eighties were directed toward this topic, and researchers such as Lefebvre, Silflow, Ellen Baker, and Jeff Schloss did their research in my laboratory in this area. Although our research attention over the past 15 years has been diverted by IFT, flagellar assembly, and the sensory function of cilia and flagella, I regard flagellar gene upregulation upon deflagellation and downregulation once flagella reach full length as an important project that still has to be investigated. The questions can be stated quite simply: What cytoplasmic factors signal the transcription apparatus to upregulate so many genes (over 300) so quickly (within minutes) upon flagellar detachment, and exactly what transcription mechanism is being used to carry out all this rapid mRNA synthesis from so many genes? Moreover, what is downregulating the transcription once the flagella approach full length? We already know that tubulin mRNA synthesis increases almost 10 fold as soon as the flagella are detached, and then rapidly drops down to constitutive levels. *Chlamydomonas*, with its ease of deflagellation, transformability, and synchronous flagellar regeneration kinetics, represents one of the best systems I can think of to investigate the rapid upregulation of hundreds of genes for proteins required for organelle assembly, in any organelle assembly system in cell biology today. It is for this reason that my last hurrah, if that is indeed what it will be, is to go back to working on this transcriptional problem. My starting points will be an observation made by William Snell that a specific protein kinase, in the cell cytoplasm of *Chlamydomonas*, is phosphorylated immediately after one detaches or shortens the flagella, and one made by former postdoctoral student Laura Keller, now at Florida State University, that nuclei isolated from cells with regenerating flagella have upregulated transcription in an *in vitro* system. I figure that these starting points are as good as any.

Editor's note: Joel Rosenbaum began work on *Chlamydomonas* in 1967, and he continues, to this day, over 40 years later, to find excitement in studies of the flagella of this organism. Many of his former students continue to work on *Chlamydomonas*, and indeed he trained many of the authors of chapters in this volume. His work has been supported throughout this time by the National Institute of General Medical Sciences of the National Institutes of Health, from which he most recently received a MERIT Award. In 2006 he was awarded the E. B. Wilson Medal, the highest honor given for research by the American Society for Cell Biology, for his studies on *Chlamydomonas* flagella.

CHAPTER 2

Basal Bodies and Associated Structures

Susan K. Dutcher

Department of Genetics, Washington University School of Medicine,
St. Louis, Missouri, USA

CHAPTER CONTENTS

I. INTRODUCTION

Many features of *Chlamydomonas* basal bodies reflect the organization of these structures in most organisms. These characteristics include the presence of nine triplet microtubules, which consist of a 13-protofilament tubule (the A tubule) and two partial tubules with 11 protofilaments each (the B and C tubules); together these three tubules are referred to as the "microtubule blade." A second attribute is the orthogonal position of the two basal bodies. In an interphase *Chlamydomonas* cell, the two basal bodies are found at right angles to one another at the anterior end of the cell (Ringo, 1967) (Figure 2.1A). This arrangement may allow the two flagella to beat effectively in opposite directions. In association with the two mature basal bodies are two probasal bodies. In contrast to the 260-nm length of the mature basal body, probasal bodies have triplet microtubules that are only 80-nm long. At the proximal end of both basal bodies and probasal bodies is a ring of amorphous material that is ~20 nm in depth (Figure 2.1B-1, C-1). Distal to this region is the cartwheel, which in cross-sectional images appears as a hub with a series of radiating spokes from the center of the basal body to the triplet microtubules where they connect to the A tubule by triangular projections. This cartwheel structure is found in most basal bodies and often in the probasal bodies. Along the spokes of the cartwheel are bulges with unknown function (O'Toole et al., 2003). The cartwheel has a depth of about 40 nm. In longitudinal sections, the cartwheel can be seen as a series of plates or tiers (Cavalier-Smith, 1974). In observations of other green algae, Beech et al. (1991) observed that the number of tiers was related to the cell cycle age of the basal bodies with the older basal body having fewer tiers. The number of tiers in *Chlamydomonas* basal bodies varies from two to seven tiers with three and four tiers being most common (Geimer and Melkonian, 2004). This would suggest that an older basal body generally has two or three (46% of the basal bodies) and the younger basal bodies have four to seven tiers (54% of the basal bodies).

The cylinder of the triplet microtubules extends for an additional 200 nm and is referred to as the basal body shaft. Immunoelectron microscopy has suggested that gamma-tubulin may be present in the interior of the shaft (Silflow et al., 2001). At the end of the basal body proper is the transition region in which the C tubule of the triplet microtubules ends and doublet microtubules begin. In the shaft, a thin fiber that has a diameter of about 10 nm is attached to the microtubule doublets. It links the A tubules of doublets 7, 8, 9, 1, and 2. It passes through the lumen as it moves from

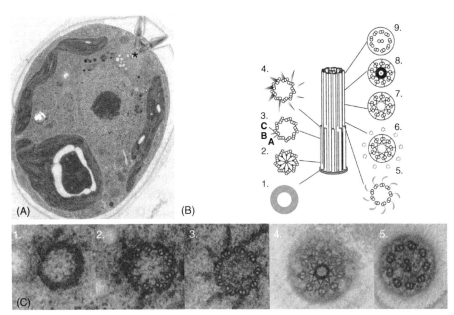

FIGURE 2.1 *(A) Electron micrograph of a* Chlamydomonas *cell preserved by cryofixation methods to illustrate the position of the pair of basal bodies (indicated by ★) in the cell at the anterior end. (Image supplied by Dr. Eileen O'Toole, University of Colorado.) (B) Diagram of basal body and transition zone structures. The distance from the base to the tip is approximately 350 nm. (1) Ring of amorphous material. (2) Triplet microtubules, also known as microtubule blades, with pinwheel. (3) The A, B, and C tubules of one triplet microtubule are indicated. The central shaft is devoid of the pinwheel, but gamma-tubulin has been localized to this region. (4) The transition fibers are shaded and begin at the distal end of the basal body. They extend and remain attached through the doublet microtubules of the transition zone. (5) Transition fibers at the proximal end of the transition zone. (6) Pores of the ciliary necklace; they are the termination of the transition fibers. They surround the transition zone and hold the basal body onto the membrane. The stellate fibers of the proximal transition zone have two distinct regions: (7) the proximal stellate fibers and (8) the distal stellate fiber system with osmiophilic ring and small inner fiber. (9) Proximal end of flagellum prior to appearance of axonemal substructures. (From Dutcher, 2003a with permission) (C) Electron micrographic images. (1) Corresponds to diagram 1 of panel B. (2) Corresponds to diagram 2 of panel B. (3) Corresponds to diagram 4 of panel B. (4) Corresponds to diagram 8 of panel B. (5) Corresponds to diagram 9 of panel B. (From Dutcher, 2001 with permission.)*

doublet 7 to 2 and has a prominent knob. Geimer and Melkonian (2004) refer to this fiber as the acorn. A second set of filaments bisects the acorn. These fibers extend from triplets 4 and 5. These fibers were not observed by tomographic analysis (O'Toole et al., 2003), and no proteins have been associated with the fibers to date.

This region that transitions from triplet to doublet microtubules is also associated with filaments known as the transition fibers. The fibers are striated and triangular at their attachment to the basal bodies (Ringo, 1967; O'Toole et al., 2003). They bend and form a cage around the distal regions of the basal body that attaches the basal bodies to the plasma membrane (Figure 2.1C-3). The ends of the fibers appear similar to the Y-shaped connectors seen in *Spermatozopsis similis* (Lechtreck et al., 1999) (Figure 2.2A). Lastly, there are knob or pore-like structures termed "the ciliary necklace," which serve as anchor points for the basal body (Weiss et al., 1977). These are found at the ends of the transition fibers (Figure 2.2C).

In this transition region, the interior of the cylinder contains two distinct stellate fiber systems (Figures 2.1B-7, B-8, C-4 and 2.2). Each system has nine points around a central ring and each of the nine points attaches to an A tubule. The proximal stellate fiber has osmiophilic staining in the center; the points and the ring show similar staining. The distal stellate fibers lack the center staining, and the ring stains much more intensely than the points (Ringo, 1967; O'Toole et al., 2003). A 10-nm disk of amorphous material separates the two stellate fiber systems (O'Toole et al., 2003). Stellate fibers are unique to the green algae and are not found in all basal bodies. The stellate fibers form the osmiophilic H-shaped structure observed in longitudinal sections (Figure 2.2F).

Structural features of the wild-type *Chlamydomonas* basal body may be seen in the tomographic reconstructions by O'Toole et al. (2003; see videos of electron microscopic tomograms at   http://www.elsevierdirect.com/companions/9780123708731).

The mechanism for the formation of doublet and triplet microtubules as opposed to singlet microtubules is not understood. Biochemical solubilization of flagella with high concentrations of urea from *Chlamydomonas* as well as sea urchins reveal a set of proteins that may form the connection between the A and the B tubules as well as the B and C tubules (Ikeda et al., 2003).

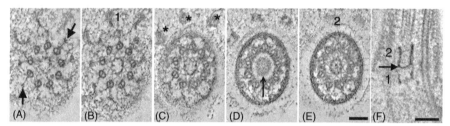

FIGURE 2.2 *Selected tomographic images of the transition zone and fibers in wild-type Chlamydomonas. (A) The proximal region at the end of the transition fibers (between diagram 5 and 6 in Figure 2.1B) has doublet microtubules and Y-shaped connecters (arrows). (B and C) The first stellate fiber array consists of a nine-pointed star with a central hub formed from electron-dense triangular points. The distal tips of the transition fibers and Y-shaped connecters form knobs or pores that attach to the plasma membrane (★). (D) The first stellate array is replaced by a central, amorphous disk (arrow). (E) The second stellate array at the distal end of the transition zone consists of a nine-pointed star with an elaborate center. (F) The two stellate fiber arrays in longitudinal section look like an osmiophilic H. The crossbar (arrow) is the material in panel D; "1" and "2" indicate levels of sections shown in panels B and E, respectively. (From O'Toole et al., 2003 with permission.)*

II. FIBERS ATTACHED TO THE BASAL BODIES

A. Rootlet microtubules

Melkonian (1977) showed that the pattern of a specific class of microtubules, the rootlet microtubules, is a valuable tool for the classification of algae. Microtubular rootlets are composed of four bundles of microtubules that originate from the basal body region and extend under the plasma membrane. In *Chlamydomonas*, there are two bundles with two microtubules and two bundles with four microtubules (Figure 2.3A, B). By electron microscopy, these microtubules are tightly opposed to one another compared to the cytoplasmic microtubules; they are arranged in a pattern of three microtubules in a row over a single microtubule (Figure 2.3C). Most green algae show the 4-*X*-4-*X* pattern, where *X* is the number of microtubules in the other rootlet. By light microscopy, rootlet microtubules can be distinguished from cytoplasmic microtubules by their stability (Holmes and Dutcher, 1989) and by the presence of acetylated alpha-tubulin, which is detected by an antibody to this posttranslational modification on lysine 40 of alpha-tubulin (Piperno and Fuller, 1985; LeDizet and Piperno, 1986; Holmes and Dutcher, 1989). Acetylated alpha-tubulin is also present in the basal bodies and flagella.

The rootlet microtubules play a role in positioning other organelles. The four-membered microtubular rootlet from the daughter basal body ends in the vicinity of the eyespot, which is in the chloroplast (Holmes and Dutcher, 1989). Gruber and Rosario (1974) suggested that the four microtubules of the rootlet splay and each microtubule associates individually with the eyespot, but this splaying was not observed in 70 consecutive serial sections of two cells (K. McDonald and S.K. Dutcher, unpublished results). The two-membered microtubular rootlet from the daughter basal body is associated with contractile vacuoles and with the mating structure (Weiss, 1984).

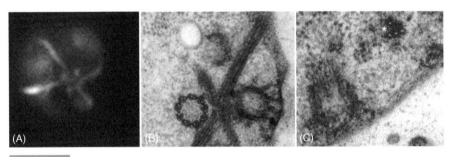

FIGURE 2.3 *Rootlet microtubules. (A) The cross-shaped arrangement of microtubular rootlets as visualized by immunofluorescence microscopy with a monoclonal antibody to acetylated alpha-tubulin. The mature basal bodies at 3 and 9 o'clock and the probasal bodies are also stained and appear as dots. (Taken from Preble et al., 2001 with permission.) (B) Electron micrograph of microtubular rootlets and mature basal bodies. One probasal body is seen at the top of the image between the two- and four-membered microtubular rootlets. (Image from T.H. Giddings, Jr.) (C) Electron micrograph of the three-over-one arrangement of the four-membered microtubular rootlet (★). (Image from T.H. Giddings, Jr.)*

Rootlet microtubules are found associated with specific triplet microtubules of the basal body. The four-membered microtubular rootlets terminate near triplets 2 and 3 and the two-membered microtubular rootlets are near triplet 9; together they form a distinct cross-shaped structure (Figure 2.3B). The rootlet microtubules associate with the cleavage furrow during mitosis. Early electron microscopy identified microtubules, termed "phycoplast microtubules," that were present in the cleavage furrow following the completion of telophase (Johnson and Porter, 1968). The phycoplast is formed by the four-membered rootlet microtubules present in interphase. Unlike the cytoplasmic microtubules, these microtubules do not disassemble at the transition from interphase to mitosis but remain associated with the

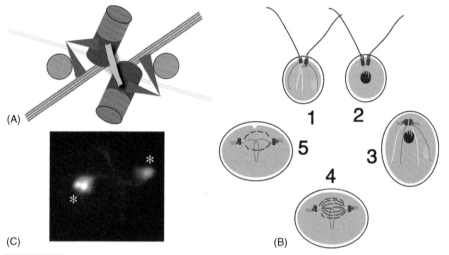

FIGURE 2.4 *The fiber systems attached to the* Chlamydomonas *basal body complex. (A) The mature basal bodies are shown in red, the transition zones in peach, and the probasal bodies in pink. The microtubular rootlets have four microtubules (orange) or two microtubules (yellow) and attach at specific triplet microtubules of the basal body. The distal (solid) and proximal (striped) striated fibers are shown in light blue (here the latter are partially hidden by the former). They connect the two mature basal bodies at the two ends. The lateral fibers are shown in green. They connect the mature basal body to its daughter probasal body across the rootlet microtubules. (B) Changes in the fiber systems during the cell cycle. (1) During interphase the basal bodies and transition zones are continuous with the flagella. The rootlet microtubules are adjacent to the plasma membrane. One of the four-membered rootlet microtubules lies adjacent to the eyespot (rose). (2) Another view of interphase cells illustrates that the basal bodies are connected to the nucleus (dark blue) and to each other by centrin fibers (light blue). (3) At preprophase, the flagella are lost. The probasal bodies elongate. The distal and proximal striated fibers are lost. (4) The two-membered rootlet microtubules shorten. The centrioles (without transition zones) are found at the poles of the spindle. The four-membered rootlet microtubules arc over the spindle. The eyespot is disassembled. (5) Cytokinesis is initiated at one end of the cell. This will be followed by extension of the two-membered rootlet microtubules and reassembly of the striated fibers, as well as assembly of new rootlet microtubules and of a new eyespot in association with one of the new four-membered microtubular rootlets. (Image from Dutcher, 2004.) (C) The rootlet microtubules arching across the spindle as visualized by antibody to acetylated alpha-tubulin. The poles of the spindle are visualized by antibody to centrin (*). (Image from Jessica Esparza, Washington University.)*

duplicated pairs of basal bodies and arch over the spindle, curving at the midpoint (Figure 2.4).

Mutations that affect the assembly of basal body microtubules cause profound defects in the orientation and placement of the rootlet microtubules. These include mutations in *BLD2*, *UNI3*, and *BLD10* (Ehler et al., 1995; Dutcher and Trabuco, 1998; Preble et al., 2001; Matsuura et al., 2004). In these mutants, the cross-shaped pattern of rootlet microtubules is lost and the bundles of microtubules begin to fray as revealed by an increased number of structures that are identified by antibodies to acetylated alpha-tubulin (Ehler et al., 1995). These mutants show defects in the placement of the cleavage furrow with respect to the nucleus (Ehler et al., 1995; Dutcher and Trabuco, 1998; Preble et al., 2001; Matsuura et al., 2004). However, these frayed rootlet microtubules remain associated with the cleavage furrow.

SF-assemblin is a 30-kD protein that forms striated fibers that do not contract. The long SF-assemblin fibers are associated with the four microtubular rootlets at their proximal ends near the basal bodies (Goodenough and Weiss, 1978) and may play a role in the stabilization of the rootlet microtubules (Lechtreck and Silflow, 1997).

B. Distal striated fibers and nucleus-basal body connectors

Centrin is a 20-kD EF-hand protein that forms calcium-sensitive, contractile fibers that are found in at least three distinct locations in the cell (Salisbury et al., 1988). In interphase cells, it is present in the distal striated fiber that holds the older and younger basal bodies to each other. The distal striated fiber attaches to microtubule blades 9, 1, and 2 of the basal bodies (Hoops and Witman, 1983) and appears as a bundle of oriented fibers. The distal striated fiber is connected to the basal bodies by an additional fiber system (O'Toole et al., 2003; Dutcher, Li, and O'Toole, submitted for publication). Centrin is also present in another set of fibers that attach at microtubule blades 7 and 8 and connect the basal bodies to the nucleus; these fibers are termed the nucleus-basal body connectors. They also are referred to as the rhizoplast. Centrin is also present in the transition zone in the stellate fibers and in the inner dynein arms of the flagellar axoneme (see Chapter 6). Several lines of evidence suggest that centrin fibers contract *in vivo* with changes in calcium concentration (Salisbury et al., 1987).

C. Proximal fibers

During interphase, the mature basal bodies are connected to each other by the centrin-containing distal striated fiber as well as a proximal striated fiber (Figure 2.4). The composition of the proximal fiber is unknown. The mature basal body and the probasal body are connected to each other by a 6-nm fiber (Gould, 1975), but again its composition is unknown. In mammalian cells, the attachment of centrioles at the proximal end is mediated by a protein called rootletin (Bahe et al., 2005). It is a coiled-coil protein and is regulated by Nek2, a NIMA-like kinase. Its phosphorylation allows for the separation of the centriole pair. FA2 is a NIMA-like kinase that is localized to the proximal fibers in *Chlamydomonas* (see Chapter 3). FA2 mutants have defects in cell cycle progression and in flagellar autotomy (Finst et al., 1998).

D. Y-shaped connectors

As mentioned above, the doublet microtubules in the transition zone connect to the plasma membrane via Y-shaped fibers, which emerge from the distal end of the transition fibers (Figure 2.2A, B) (O'Toole et al., 2003; Geimer and Melkonian, 2004). Similar fibers have been described in a wide range of ciliated cells, including those from metazoans. The protein P210 was identified in the wall-less alga *S. similis*, and antibodies to it localize to the Y-shaped connectors (Lechtreck et al., 1999). The protein is also present in nascent basal bodies and basal body precursors (Lechtreck and Grunow, 1999; Lechtreck and Bornens, 2001). The antibody to *S. similis* P210 similarly stains the transition zone of some *Chlamydomonas* species (Schoppmeier and Lechtreck, 2002). The gene has not been identified in *C. reinhardtii*, to date, but is likely to have an orthologue that is difficult to detect by bioinformatics due to a high proline and alanine content.

III. GLOBAL APPROACHES TO THE IDENTIFICATION OF BASAL BODY PROTEINS

Intact basal bodies have been isolated by a variety of methods. These preparations differ in the presence of flagella, rootlet microtubules, and nuclei. In preparations that removed noticeable contamination of flagella and nuclei, over 150 polypeptides can be resolved; alpha- and beta-tubulins represent the major components by mass (Figure 2.5). Mass spectrometry of isolated basal bodies identified 195 proteins of which 65 are likely to be contaminants from the mitochondria and chloroplast (Keller et al., 2005). An additional 35 could be flagellar contaminants as they are upregulated by deflagellation. Intraflagellar transport proteins (IFT) (see Chapter 4) have been localized to the basal body region by light microscopy (Vashishtha et al., 1996; Mueller et al., 2005), and IFT52 has been localized to the transition zone fibers by immunoelectron microscopy (Deane et al., 2001). However, no IFT proteins or motors were detected in isolated basal bodies by mass spectrometry, which suggests that the association is lost during the isolation protocol. Gamma-tubulin, which has been localized to the basal body core (Silflow et al., 1999), was also not present in these preparations. Table 2.1A presents a list of possible basal body components that are supported by their presence in human centrosomes or by comparative genomics.

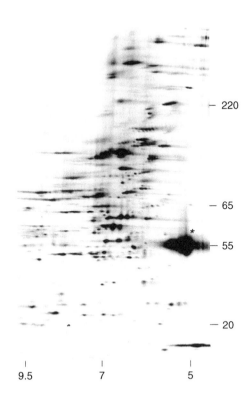

FIGURE 2.5

Two-dimensional gel electrophoresis of isolated basal bodies from wild-type Chlamydomonas. *Apparent molecular weight in kD is shown on the y-axis and the apparent pI is shown on the x-axis. Alpha- and beta-tubulins are indicted by ★. They are estimated to make up 65% of the protein by mass.*

Comparative genomic techniques represent an alternative approach to identify basal body proteins. Since many flagellar and basal body proteins show high levels of similarity with proteins in other ciliated organisms, but no similarity to proteins in nonciliated organisms, it is possible to use BLAST searches to find homologues present in ciliated organisms but missing from nonciliated organisms (Avidor-Reiss et al., 2004; Li et al., 2004; Merchant et al., 2007). A comparison between *Chlamydomonas* and human identifies over 4000 homologues. About 700 of these are missing from *Arabidopsis*, a seed plant that lacks basal bodies and cilia (Figure 2.6). Only 300 of these are reciprocal best matches between *Chlamydomonas* and human. Nineteen of these were also obtained in the proteomic approach (Keller et al., 2005) (Table 2.1A). This dataset was used to identify the human disease gene, Bardet–Biedl syndrome 5 (*BBS5*) on human chromosome 2 (Li et al., 2004; see section IX).

IV. MUTATIONS THAT ALTER BASAL BODIES

A. *BLD2* locus

The *bld2-1* allele was isolated in a screen for cells that are unable to mate. Because flagella are required for the initial stages of partner recognition, numerous aflagellate strains were identified with this approach.

Table 2.1A	Genes and proteins likely to be involved in basal body assembly or function identified by proteomics or comparative genomics, with homologues in other organisms but without mutations in *Chlamydomonas*

Gene	Protein accession #	Found in human centrosomes	Found exclusively in ciliated organisms	Upregulated by deflagellation in *Chlamydomonas* in at least two studies[a]	References
POC3/ CEP290	EDP06028	✓	×	NT[b]	Andersen et al. (2003), Avidor-Reiss et al. (2004), Keller et al. (2005)
NPH4	EDP03489	×	✓	No	Li et al. (2004), Keller et al. (2005)
SAS4	EDP08157	✓	×	NT	Pelletier et al. (2006), Dutcher (2007)
SFI1	EDP01952	×	×	NT	Keller et al. (2005)
POC7/ UNC119/ HRG4	EDO97256	×	✓	No	Li et al. (2004), Keller et al. (2005)
POC1	EDP07349	✓	✓	NT	Avidor-Reiss et al. (2004), Keller et al. (2005)
POC2	EDP05664	×	✓	No	Li et al. (2004), Keller et al. (2005)
POC5	EDP06067	✓	×	NT	Andersen et al. (2003), Keller et al. (2005)
POC6	EDP01597	×	×	NT	Keller et al. (2005)
POC8	EDP08590	✓	×	NT	Andersen et al. (2003), Keller et al. (2005)
POC9	EDO99018	×	✓	NT	Li et al. (2004), Keller et al. (2005)
POC11	EDP03148	✓	×	NT	Andersen et al. (2003), Keller et al. (2005)
POC12	EDO97995	×	✓	No	Li et al. (2004), Keller et al. (2005)
POC13	–	×	✓	NT	Keller et al. (2005)
POC14	EDO96857	✓	×	NT	Andersen et al. (2003), Keller et al. (2005)
POC15	–	×	✓	NT	Li et al. (2004), Keller et al. (2005)
POC16	EDO99093	×	✓	NT	Li et al. (2004), Keller et al. (2005)
POC17	EDP06728	✓	×	NT	Andersen et al. (2003), Keller et al. (2005)
POC18	EDP06797	✓	✓	NT	Andersen et al. (2003), Keller et al. (2005)

[a]*Genes encoding proteins located in the basal body but not the flagellum would not be expected to be upregulated by deflagellation.*
[b]*NT, not tested.*

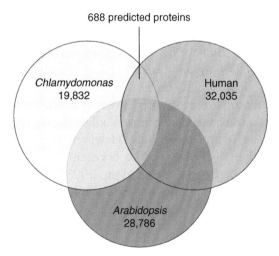

688 predicted proteins

Chlamydomonas
19,832

Human
32,035

Arabidopsis
28,786

FIGURE 2.6 *Comparative computational method to identify flagellar and basal body polypeptides. A genome wide BlastP screen of the predicted proteins from* Chlamydomonas, Homo sapiens, *and* Arabidopsis *was used to find 688 polypeptides of which 320 are reciprocal best matches that are present in ciliated organisms* (Chlamydomonas *and humans) and missing in flowering plants* (Arabidopsis). *(Redrawn from Li et al., 2004.)*

The *bld2* strain was shown to have aberrant basal bodies (Goodenough and St. Clair, 1975). Instead of assembling triplet microtubules, most cells had singlet microtubules and the cylinder was often only 40–50 nm in length instead of 200 nm. The *BLD2* locus encodes epsilon-tubulin, a tubulin isoform that is found only in ciliated organisms with triplet microtubules (Chang and Stearns, 2000; Dutcher et al., 2002; Dutcher, 2003). The mutant allele has a stop codon at amino acid 9. Epsilon-tubulin localizes to the transition fibers in *Chlamydomonas* (Dutcher et al., 2002) and to the subdistal fibers of mammalian centrioles (Chang et al., 2003). It does not appear to be a component of the microtubule cylinder. Yeast two-hybrid experiments with mouse epsilon-tubulin failed to identify another tubulin partner (Dutcher, Li, and O'Toole, submitted for publication). The mechanism by which epsilon-tubulin stabilizes the triplet microtubules remains unclear. Extragenic suppressors (*rgn1*) restore the ability to assemble flagella, and incompletely restore the ability to assembly triplet microtubules (Preble et al., 2001). The *RGN1* gene has not been identified.

B. *UNI3* locus

The *uni3-1* allele is a deletion that removes another tubulin isoform, delta-tubulin. This tubulin is also found only in organisms with triplet microtubules (Dutcher and Trabuco, 1998; Chang and Stearns, 2000; Garreau de Loubresse et al., 2001). The mutant flagellar phenotype is heterogeneous.

One-half of the cells are aflagellate, about one-quarter of the cells have a single flagellum, and the remaining cells have two flagella; this probably is due to a defect in the ability of the basal bodies to mature properly (see section VIII.C). The cells with two flagella can have flagella of different lengths (Dutcher and Trabuco, 1998; Garreau de Loubresse et al., 2001).

Electron microscopic tomography revealed several phenotypes of the basal bodies. First, the basal bodies have doublet rather than triplet microtubules along most of the basal body. However, there is a short stretch of triplet microtubules just before the beginning of the transition zone (Dutcher and Trabuco, 1998; O'Toole et al., 2003). Second, the distal striated fiber fails to assemble. This suggests a role for the C tubule in the attachment of the distal striated fiber. Third, there are additional stellate fibers formed in the cylinder shaft more proximally than observed in wild-type cells. In fact, multiple stellate fibers are observed. This suggests that the presence of doublet microtubules may be the signal for the assembly of the stellate fiber system (O'Toole et al., 2003).

Suppressors of the *uni3-1* allele fall into at least four genes. One of these is the *TUA2* locus, which is one of the two genes encoding alpha-tubulin. Alleles in this gene allow for partial assembly of triplet microtubules and independently confer supersensitivity to colchicine, a microtubule-destabilizing drug (Fromherz et al., 2004). Alterations in alpha-tubulin may allow the basal body to mature in the absence of delta-tubulin. The mutational changes in the alpha-tubulin suppressors (D205N and A208T) are identical to alterations in alpha-tubulin found in twisting mutants in *Arabidopsis* that cause left- or right-handed spiraling of the roots (Ishida et al., 2007).

C. *BLD10* locus

The *bld10-1* strain, like *bld2-1*, is an aflagellate strain that does not assemble microtubule blades, and it is unclear if the amorphous material is present (Matsuura et al., 2004). This mutant represents the most severe basal body assembly defect in *Chlamydomonas*, to date. BLD10 is a 170-kD coiled-coil protein that has an orthologue in humans. Immunoelectron microscopy suggests that the BLD10 protein localizes to the cartwheel in the proximal end of the basal body. Like *bld2* mutants, *bld10-1* has disorganized rootlet microtubules and centrin fibers. Unlike *bld2*, disorganized spindles are observed in this strain (Matsuura et al., 2004). Truncation of the last third of the C-terminus of the BLD10 protein results in basal bodies with eight-fold rather than nine-fold symmetry. The spokes of the cartwheel are reduced in length (Hiraki et al., 2007).

D. *SAS6* locus

The *sas-6* gene was first identified in *Caenorhabditis elegans* as an embryonic lethal with spindle assembly defects. This gene along with five other genes acts in a linear pathway in *C. elegans* to assemble centrioles; *sas-6* is required for the formation of an internal cylinder (Pelletier et al., 2006). The *Chlamydomonas bld12-1* mutation is a deletion of ~40 kb and 12 genes, including the *Chlamydomonas* homologue of *sas-6*. The *bld12-1* mutant has a flagellar phenotype that is rescued by *Chlamydomonas SAS6*; another mutant (*bld12-2*) with a similar flagellar phenotype has a single base substitution in the *SAS6* gene that is predicted to cause abnormal splicing resulting in a premature stop codon that would produce a truncated SAS6 protein (Nakazawa et al., 2007). About 10% of *bld12-1* cells assemble flagella and there are abnormal numbers of doublet microtubules in about 10% of the examined flagella. The basal bodies also have abnormal numbers of triplet microtubules. The SAS6 protein localizes to the cartwheel. It appears that the defect in the *sas6* mutant in *Chlamydomonas* is much less severe than the defects observed in *C. elegans* (Pelletier et al., 2006).

E. Transition zone mutants

Mutants with altered basal bodies or transition zones have been reported. The four independently isolated *uni1* strains (*uni1-1* to *uni1-4*) have wild-type basal bodies, but assemble additional stellate fiber material on the older of the two basal bodies. The majority of cells have a single flagellum. The younger basal body fails to assemble any transition zone material (McVittie, 1972; Huang et al., 1982). Mutations in the *UNI2* gene also result in cells that predominantly have a single flagellum on the older basal body and, like *uni1* mutants, have defects in the transition zone (Piasecki et al., 2008).

F. Other possible mutants

McVittie reported three aflagellate or stumpy mutants with aberrant basal bodies by electron microscopy (McVittie, 1972; Dutcher, 1986). Little is known about the molecular nature of these mutations or genes. The strains with the isolation names NG6 and NG30 have defects in the number or stability of the microtubule blades. The basal body phenotype of NG6 resembles the *hennin* mutant in mice that opens up the B and C tubules; the *hennin* gene encodes an Arl/Arf family member (Caspary et al., 2007).

An alternative approach to identify basal body mutants uses the behavior of dikaryons, which are recently mated cells (see Volume 1, Chapter 5). In a dikaryon formed between most paralyzed flagellar (*pf*) mutants and a wild-type strain, paralysis will be relieved by the incorporation of the wild-type gene product into the pair of flagella from the *pf* parent (Luck et al., 1977). However, when a *uni1* strain is mated to a wild-type strain, the dikaryon remains with three flagella. Deflagellation and regrowth again allows only three flagella to form. Rescue of this phenotype is either hampered by the lack of a pool of wild-type proteins or by a necessity for a point in the cell cycle other than G1 (Dutcher, 1986). Similar behaviors were observed for *fla9-1*, a temperature-sensitive flagellar assembly mutant (Adams et al., 1982; Dutcher, 1986); for *pf10*, an abnormal motility mutant (Dutcher et al., 1988); and for *uni3-1*. This phenotype provides a possible screen for additional basal body mutants.

V. MUTATIONS IN BASAL BODY SEGREGATION

Basal bodies and their associated probasal bodies must segregate properly at each mitotic division. To accomplish this, the older mature basal body with its newly elongated probasal body and the recently matured younger basal body with its newly elongated probasal body must move to opposite mitotic poles. During interphase, the mature basal bodies are connected to each other by the distal striated fiber, by the proximal striated fiber, and by a 6-nm fiber (Gould, 1975). Thus, segregation must require the breaking of these fibers between the two mature basal bodies to allow their separation, while retaining connections between the mature and daughter basal bodies. The separation of the two mature basal bodies is likely to require a cell cycle-regulated proteolysis or disassembly of these fibers.

It is expected that defects in basal body segregation would lead to abnormal numbers of flagella. Three loci were found in initial screens for mutant strains with a variable number of flagella. These are *VFL1*, *VFL2*, and *VFL3*. In each of these mutants, there is a distribution of cells with zero to seven flagella. The *vfl2* mutation is a missense mutation in the centrin gene and reduces the amount of centrin to about 25% of the wild-type level. Both *vfl1* and *vfl2* mutations are likely to affect basal body segregation via the distal striated fiber. In the *vfl2* mutant strain, cells in the population have from zero to many basal bodies as monitored by antibodies (Marshall et al., 2001). As mentioned above, the *uni3-1* mutant is defective in delta-tubulin and fails to assemble the distal striated fiber. Yeast two-hybrid screens with mouse delta-tubulin identified the *STP2* gene, which encodes a 61-kD protein that is conserved in humans. RNA interference against this gene produces strains with very short flagella (<200 nm) and lacking a distal striated fiber. STP2 may serve as another localization signal along with VFL1 for placement and assembly of the distal striated fiber (Dutcher, Li, and O'Toole, submitted for publication). No mutations are known presently that affect the proximal striated fiber.

The *vfl3* mutation is reported to affect the fibers that hold the mature and daughter basal bodies together. The *vfl3* mutant strain is also lacking the distal striated fibers (Wright et al., 1983). This loss results in loss of the normal rotational orientation of the flagella (Hoops et al., 1984). The distal striated fiber must play a role in maintaining the correct orientation of the basal bodies to ensure the correct polarity of the flagellum.

Basal body proteins identified by mutants are included in Table 2.1B.

VI. DNA IN THE BASAL BODY

Experiments using *in situ* hybridization by Hall and Luck suggested that basal bodies might contain DNA (Hall et al., 1989) that has a circular genetic map (Ramanis and Luck, 1986). However, several experiments have refuted these conclusions. Antibodies to DNA do not recognize the basal bodies (Johnson and Rosenbaum, 1991), and although the original results suggested that the DNA must be present in at least two copies in haploid cells, copy number tests did not support this result (Johnson and Dutcher, 1991). *In situ* experiments with additional controls showed that the DNA in question was localized to the nuclear periphery rather than to the basal bodies (Hall and Luck, 1995). Ultimately, the circular genetic element was demonstrated to be the result of the incorrect mapping of the *PF29* marker and high chiasma interference (Holmes et al., 1993).

VII. FUNCTION OF BASAL BODIES

A. Templating

One major function of basal bodies is the templating of flagella. The analysis of the *bld2-2 rgn1-1* strain provides evidence for this function. In this double mutant, basal bodies with singlet, doublet, and triplet microtubules as well as missing microtubules are observed by electron microscopy (Preble et al., 2001). If the basal bodies are missing a microtubule blade, then the axoneme is missing a doublet microtubule (Figure 2.7). Basal bodies with a singlet microtubule will assemble a flagellar axoneme with a singlet microtubule. The capacity for direct templating of the axonemal microtubules resides in the basal body.

B. Docking of IFT proteins

Several lines of evidence suggest that the transition fibers are the site of docking for IFT motors and proteins (see Chapter 4). First, immunofluorescence microscopy with antibodies to motors as well as IFT proteins shows accumulation around this region. Second, immunoelectron microscopy with antibodies to IFT52 shows localization to the transition fibers (Deane et al., 2001). Finally, the presence of transition fibers is correlated with the ability to assemble a flagellum. For example, the *uni1-1* mutant has one basal body with transition fibers and a flagellum, and one basal body without transition fibers and without a flagellum (Huang et al., 1982). Similarly, in the pair of mammalian centrioles, only the centriole with subdistal fibers, which appear to be orthologous to the transition fibers of *Chlamydomonas*, can assemble a primary cilium (Rieder and Borisy, 1982; Vorobjev and Chenstov, 1982; Chretien et al., 1997).

Table 2.1B Genes and proteins likely to be involved in basal body assembly or function identified by genetics/biochemistry

Gene	Protein accession #	Protein/motif	Localization	Mutant phenotype	Linkage group	References
BLD2	EDP05985	Epsilon-tubulin	Transition zone	Aflagellate, cleavage furrow placement, lack of triplet microtubule blades	III	Goodenough and St. Clair (1975), Ehler et al. (1995), Dutcher et al. (2002)
BLD10	BAD00740	Coiled-coil	Pinwheel	Aflagellate, cleavage furrow placement, disorganized spindle, lack of basal body	X	Matsuura et al. (2004), Hiraki et al. (2007)
BBS5	AAS89977		Transition zone	Aflagellate	VI	Li et al. (2004)
FA2	AAL86904	NIMA-like kinase	Proximal fibers	Unable to deflagellate	VII	Mahjoub et al. (2004)
FLA9	–	–	–	Aflagellate at 32°C, flagellar assembly defect not rescued in dikaryons	XIX	Adams et al. (1982), Dutcher (1986)
NG6	–	–	Microtubule blade defect	Aflagellate	–	McVittie (1972)
NG30	–	–	Microtubule blade defect	Aflagellate	–	McVittie (1972)
P210	–	Coiled-coil	Y-shaped connectors in transition zone	–	–	Lechtreck et al. (1999)
RGN1	–	–	–	Suppressor of *bld2-1*	I	Preble et al. (2001)
SAS6	BAF94334		Basal body cartwheel		XII/XIII	Pelletier et al. (2006), Dutcher (2007), Nakazawa et al. (2007)

SFA1	CAC69239	SF-assemblin	Fibers on microtubule rootlets	–	VI	Lechtreck and Silflow (1997)
STP2	EDO96773	LamA motif	Transition zone	Very short flagella	VI	Dutcher and Li, submitted for publication
TUG1	AAA82610, AAB71841	Gamma-tubulin	Proximal shaft of basal body	NA	VI	Silflow et al. (1999)
UNI1	–	–	–	Uniflagellate	XIX	Huang et al. (1982)
UNI2	ABV25961	Coiled-coil	Basal bodies and probasal bodies	Aflagellate and uniflagellate	IX	Dutcher and Trabuco (1998), Piasecki et al. (2008)
UNI3	AAB71840	Delta-tubulin	Basal body	Aflagellate and uniflagellate	III	Dutcher and Trabuco (1998)
VFL1	AAD52203	Leucine-rich repeats	Distal shaft of basal body	Basal body segregation and variable numbers of flagella	VIII	Silflow et al. (2001)
VFL2	P05434	Centrin	Distal striated fibers, nucleus-basal body connector	Basal body segregation, variable number of flagella, chromosome loss	XI	Wright et al. (1985), Salisbury et al. (1988), Wright et al. (1989)
VFL3	AAQ95706	–	–	Basal body segregation, variable numbers of flagella, misorientation of basal bodies and flagella	VI	Wright et al. (1983), Hoops et al. (1984)

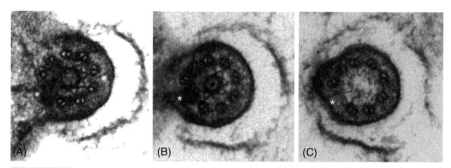

FIGURE 2.7 *The basal body plays a role in templating the flagellar axoneme. Serial section electron micrographs of the basal body and flagellar axoneme. (A) First stellate fiber region missing one of the doublet microtubules. (B) Second stellate fiber region missing one of the doublet microtubules. (C) Flagellar axoneme missing one of the doublet microtubules. The missing structure is indicted by ★ in each panel. (Images from Preble et al., 2000.)*

C. Cellular asymmetry

Most biflagellate algae do not have two identical flagella, and the differences in the flagella correlate with the parental/daughter history of the basal bodies (Beech et al., 1991). This parental difference in the basal bodies is likely to be responsible for the asymmetric positioning of the eyespot. The eyespot in wild-type cells is always positioned near the daughter basal body (Holmes and Dutcher, 1989). As a result of the asymmetric, but different placement of mating structures in the *plus* and *minus* mating-type cells, fusion of the gametes always results in dikaryons with parallel pairs of flagella and the eyespots on the same side of the newly mated cell (Figure 2.8). In addition, the basal bodies as they enter mitosis must retain their same orientation. The two mature basal bodies are attached at specific microtubule blades. As mentioned above, they must break the attachment between mature basal bodies and form attachments with the new daughter basal bodies. This event requires that the basal bodies rotate so that the microtubules that faced the old basal body now face the daughter basal body. Since the basal bodies can be marked by the presence of a contractile vacuole, their behavior can be monitored by the behavior of the contractile vacuole. Both show a clockwise orientation. Furthermore, the spindle pole also retains a fixed orientation. The older basal body at each pole remains closer to the cleavage furrow and the younger one will be near the site for the new eyespot. These asymmetries are likely due to the asymmetries of the basal body pair (Holmes and Dutcher, 1989).

D. Mitosis and cytokinesis

Prior to mitosis, the flagella and the transition zones are lost by the process of resorption or deflagellation; this is followed by a 90° rotation of the cell within the mother cell wall (Kater, 1929; Cavalier-Smith, 1974).

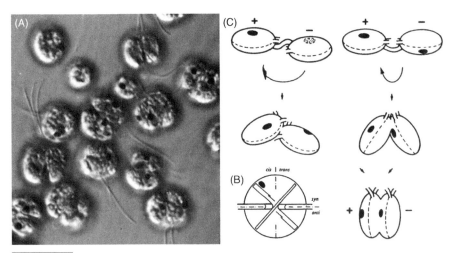

FIGURE 2.8 *Asymmetry of mating in* Chlamydomonas *may be related to basal body maturity. (A) A field of 15-minute-old dikaryons taken with differential interference contrast microscopy. Relative to the position of the eyespot (dark ovals), the mating occurs along the* **syn** *side of one gamete and the* **anti** *side of the other gamete. The outcome of mating is that the flagella from the mature basal bodies (*trans*) are oriented together and the flagella from the daughter basal bodies (*cis*) are oriented together. This arrangement allows the cells to continue to orient towards a light source. (B)* Syn *is defined as the half of the cell with the eyespot and* anti *is defined as the half of the cell lacking the eyespot. The mating-type plus structure is formed on the* **anti** *side. (C) During mating, flagella can adhere in a* cis-*to-*cis *position (right) or in a* cis-*to-*trans *position (left). In either case, the cells fuse with the eyespots on the same side. In the diagram, the left gamete is mating-type plus and the right gamete is mating-type minus. (Image from Holmes and Dutcher, 1989, reproduced with permission of The Company of Biologists.)*

Blocking this rotation by high light does not affect other events (Holmes and Dutcher, 1989). The events of mitosis occur rapidly and have been estimated to take about 15 minutes. During mitosis, the basal bodies remain in the cell perpendicular to the plasma membrane and remain attached to the membrane by the transition zone fibers (Cavalier-Smith, 1974). The basal bodies are found near clusters of nuclear pores as mitosis begins. These pores may help to form the large fenestrae (300–500 nm) at the poles of the spindle. Spindle microtubules terminate in this region. The rootlet microtubules arch over the nuclear envelope and spindle and bend near the metaphase plate (Doonan and Grief, 1987: Gaffal, 1988) (Figure 2.4). Johnson and Porter (1968) referred to these microtubules as the metaphase band. Chromosomes can be found aligned at metaphase. At telophase, the metaphase band has evolved into the phycoplast, which helps to recruit actin to the cleavage furrow (Ehler and Dutcher, 1998). The cleavage furrow appears on one side of the cell at the time of telophase in wild-type cells. In *bld2* cells, the cleavage furrow appears to form in metaphase rather than telophase. The basal bodies play roles in the phycoplast (Ehler et al., 1995) and in the formation of the spindle microtubules (Matsuura et al., 2004). Several mutations affect the segregation of chromosomes as monitored by chromosome loss (Miller et al., 2005; Zamora and Marshall, 2005). Mutations in the two motor subunits of kinesin-2 (FLA8 and FLA10) produce ~3.5-fold increase in the rate of chromosome loss while a mutation in centrin increases the rate of loss by ~6.5-fold (Zamora and Marshall, 2005). Mitosis and cytokinesis are further discussed in Chapter 14.

E. Meiosis

The role of basal bodies in meiosis has remained controversial. Cavalier-Smith (1974) observed the loss of transition fibers and the transition zone soon after the formation of zygotes, but reported that basal bodies are lost in the meiotic cells. However, Triemer and Brown (1976) reported that basal bodies are visible at prophase of the first meiotic division. Thus, it is likely that meiosis is associated with basal bodies. It is not known how the basal body number is resolved, as the diploid cell will have four mature basal bodies and four probasal bodies from the two parental strains. The *bld2-1* mutation has a recessive meiotic defect, which suggests strongly that basal bodies play a role in meiosis (Preble et al., 2001).

VIII. DUPLICATION OF BASAL BODIES

A. Basal body replication

Basal bodies and centrioles usually duplicate once per cell cycle. In mammalian cells, duplication occurs in conjunction with S phase. It is followed by the separation of the centrioles at the beginning of G2/M phase. Each pole of the spindle contains one new or daughter centriole and one old or mother centriole (Kochanski and Borisy, 1990). Unlike in mammalian cells, probasal bodies in *Chlamydomonas* are present in early G1 phase. The basal bodies elongate to form two full-length basal bodies. The timing of this elongation event is unknown and may occur between S phase and preprophase. Several groups have observed that elongation is preceded by the lose of the distal striated fiber to allow the old basal bodies to lie parallel to one another rather than lying perpendicularly to one another (Gould, 1975).

Electron microscopy has been used to follow the events of basal body initiation and elongation. The formation of the probasal body begins with an annulus of amorphous material within 1 hour of cytokinesis (Gould, 1975); this is likely to be the same ring of amorphous material observed in mature basal bodies (O'Toole et al., 2003). Johnson and Porter (1968) reported that the first event of basal body duplication is the formation of a ring of nine singlet microtubules. The B and C microtubules are added in an irregular manner. However, Gould (1975) never observed singlet microtubules. Thus, the pathway remains somewhat unclear.

B. Copy number control

Duplication of basal bodies and centrioles generally occurs near existing basal bodies; this result suggests that the existing basal bodies may have a copy number control function. *Chlamydomonas* cells mutant for the gene *VFL2* (centrin) show a randomization of basal body segregation, often producing daughter cells that lack intact basal bodies. Such cells lacking basal bodies are able to assemble new basal bodies *de novo* (Marshall et al., 2001). If there are no probasal bodies, then these results suggest that basal bodies can duplicate without a template. This *de novo* mechanism of basal body formation in *Chlamydomonas* is also observed in mammalian cells. The *de novo* mechanism may require more time than the templated mechanism. This lag may suggest that the initial events take more time to complete or that the presence of a basal body represses the formation of multiple basal bodies. Moreover, the number of new basal bodies made by the *de novo* pathway is random, which shows a loss of number control when preexisting basal bodies are not present. Control of basal body number is important for normal spindle assembly and integrity.

C. Maturation

Centrioles in mammalian cells are distinguishable from each other in the G1 phase of the cell cycle. The older or mother centriole is distinguished by the presence of subdistal fibers at the distal end and has the ability to assemble a primary cilium (Vorobjev and Chenstov, 1982). Ninein, ODF2, CEP110, dynactin, and epsilon-tubulin are associated with the mother centriole, and immunoelectron microscopy has shown that ninein, CEP110, and epsilon-tubulin localize to the subdistal fibers (reviewed in Dutcher, 2003). The acquisition of these epitopes and the ability to assemble a cilium is termed maturation. In each cell cycle, the daughter centriole will undergo maturation in G2 phase to become a mature centriole and be able to assemble a cilium in the next cell cycle. All basal bodies in the respiratory ciliated epithelium are associated with ODF2 (Ishikawa et al., 2005; Cao et al., 2006). Thus, basal bodies in a ciliated epithelium must undergo maturation in a manner not mediated by the cell cycle. In *Chlamydomonas* the younger basal body must undergo maturation in the G1 phase of the cell cycle so that it can assemble a flagellum. The two mature basal bodies in *Chlamydomonas* are associated with epsilon-tubulin.

Several mutations affect the ability of the basal bodies to undergo maturation in *Chlamydomonas*. The *uni1* mutant strains assemble only one flagellum rather than two flagella in more than 95% of the cells. The older basal body assembles the flagellum and the younger basal body gains the ability to assemble a flagellum after one cycle. The gene product of the *UNI1* locus is not known at the present time, but it is likely to be important to the maturation of the daughter in the first cell cycle.

The *uni2* mutant shows defects in maturation. The distal regions of the transition zone are elongated. The UNI2 protein exists in a phosphorylated and unphosphorylated form; the phosphorylated form accumulates at the end of mitosis with the assembly of the transition zone and new flagella (Piasecki et al., 2008).

The *UNI3* gene, which encodes delta-tubulin, is also important for maturation. Pedigree experiments suggest that the phenotype of the parental strains can predict the number of flagella that are assembled in the daughter cells (Dutcher and Trabuco, 1998). Cells of *uni1* almost always give rise to two uniflagellate daughters. In *uni3-1* cells, every cell gives rise to one aflagellate daughter cell. The number of flagella on the other daughter cell is dependent on the phenotype of the starting mother cell. An aflagellate mother cell produces a uniflagellate cell. A uniflagellate mother cell produces a biflagellate cell, which itself produces a biflagellate cell. It has been postulated that the cell cycle age is related to the phenotypic difference.

IX. GENES WITH CONNECTIONS TO HUMAN DISEASE

Bardet–Biedl syndrome (BBS) is a rare autosomal recessive human disease that was first recognized in the late 1800s. It is associated with renal disease, similar to nephronophthisis, central obesity, polydactyly, retinal degeneration, anosmia, hypogenitalia, and hypertrophy of the heart. At present there are 13 genes that have been identified in human BBS patients. *BBS6* is the only one without a clear homologue in *Chlamydomonas*. *BBS5* was identified using the conservation of ciliary genes from humans to *Chlamydomonas*, and the absence of these genes in nonciliated organisms such as *Arabidopsis* or yeast (Figure 2.6). The *BBS5* gene was mapped in a family from Newfoundland. The region of interest was 14 Mb and contained 230 genes. Only two of the 230 genes were present in the comparative cilia database. One of these genes was unknown and contained a splice site mutation in the affected members of the family. Affected members of three additional families from Kuwait carry mutations that result in premature stop codons in this gene (Li et al. 2004). Antibodies raised to the mouse protein showed localization around the basal bodies in mouse ependymal cells.

DIP13/NA14 is found associated with basal bodies and flagellar axonemes, as well as cytoplasmic microtubules (Pfannenschmid et al., 2003). It interacts with spastin, which is an AAA-ATPase that plays roles in severing microtubules (Errico et al., 2004). NA14 is an autoantigen associated with Sjögren's syndrome, but there is no evidence that it is causally involved in the disease. However, spastin is associated with hereditary spastic paraplegia and degeneration of axons (Ramos-Morales et al., 1998).

Other proteins found by comparative genomics are likely to play roles in the flagella (see Chapter 15). These include Tubby superfamily protein

(TUSP), fibrocystin, UNC119, which is also known as HRG4, and seahorse and qilin, which were found to cause embryonic lethality and cystic kidney disease in zebrafish (Li et al., 2004).

X. CONCLUSION

Basal bodies and flagella are ancient organelles that show amazing evolutionary conservation as evidenced by our identification of the flagellar basal body proteome (Li et al., 2004) and the proteome analyses of flagella and basal bodies (Keller et al., 2005; Pazour et al., 2005). Studies in *Chlamydomonas* will provide the framework for dissecting docking and transport of flagellar proteins in a simple, genetically tractable, cell-autonomous system. In turn, processes in *Chlamydomonas* can be applied to studies of ciliary diseases and signaling events mediated by sensory cilia in diverse metazoan cell types.

REFERENCES

Adams, G.M., Huang, B., and Luck, D.J. (1982). Temperature-sensitive, assembly-defective flagella mutants of *Chlamydomonas reinhardtii*. *Genetics* **100**, 579–586.

Andersen, J.S., Wilkinson, C.J., Mayor, T., Mortensen, P., Nigg, E.A., and Mann, M. (2003). Proteomic characterization of the human centrosome by protein correlation profiling. *Nature* **426**, 570–574.

Avidor-Reiss, T., Maer, A.M., Koundakjian, E., Polyanovsky, A., Keil, T., Subramaniam, S., and Zuker, C.S. (2004). Decoding cilia function: defining specialized genes required for compartmentalized cilia biogenesis. *Cell* **117**, 527–539.

Bahe, S., Stierhof, Y.D., Wilkinson, C.J., Leiss, F., and Nigg, E.A. (2005). Rootletin forms centriole-associated filaments and functions in centrosome cohesion. *J. Cell Biol.* **171**, 27–33.

Beech, P.L., Heimann, K., and Melkonian, M. (1991). Development of the flagellar apparatus during the cell cycle in unicellular alga. *Protoplasma* **164**, 23–27.

Cao, W., Gerton, G.L., and Moss, S.B. (2006). Proteomic profiling of accessory structures from the mouse sperm flagellum. *Mol. Cell. Proteomics* **5**, 801–810.

Caspary, T., Larkins, C.E., and Anderson, K.V. (2007). The graded response to Sonic Hedgehog depends on cilia architecture. *Dev. Cell* **12**, 767–778.

Cavalier-Smith, T. (1974). Basal body and flagellar development during the vegetative cell cycle and the sexual cycle of *Chlamydomonas reinhardii*. *J. Cell Sci.* **16**, 529–556.

Chang, P. and Stearns, T. (2000). Delta-tubulin and epsilon-tubulin: two new human centrosomal tubulins reveal new aspects of centrosome structure and function. *Nat. Cell Biol.* **2**, 30–35.

Chang, P., Giddings, T.H., Jr., Winey, M., and Stearns, T. (2003). Epsilon-tubulin is required for centriole duplication and microtubule organization. *Nat. Cell Biol.* **5**, 71–76.

Chretien, D., Buendia, B., Fuller, S.D., and Karsenti, E. (1997). Reconstruction of the centrosome cycle from cryoelectron micrographs. *J. Struct. Biol.* **120**, 117–133.

Deane, J.A., Cole, D.G., Seeley, E.S., Diener, D.R., and Rosenbaum, J.L. (2001). Localization of intraflagellar transport protein IFT52 identifies basal body transitional fibers as the docking site for IFT particles. *Curr. Biol.* **11**, 1586–1590.

Doonan, J.H. and Grief, J. (1987). Microtubule cycle in *Chlamydomonas reinhardtii*: an immunofluorescence study. *Cell Motil. Cytoskeleton* **7**, 381–392.

Dutcher, S.K. (1986). Genetic properties of linkage group XIX in *Chlamydomonas reinhardtii. Basic Life Sci.* **40**, 303–325.

Dutcher, S.K. (2001). The tubulin fraternity: alpha to eta. *Curr. Opin. Cell Biol.* **13**, 49–54.

Dutcher, S.K. (2003a). Elucidation of basal body and centriole functions in *Chlamydomonas reinhardtii. Traffic* **4**, 443–451.

Dutcher, S.K. (2003b). Long-lost relatives reappear: identification of new members of the tubulin superfamily. *Curr. Opin. Microbiol.* **6**, 634–640.

Dutcher, S.K. (2004). Dissection of basal body and centriole function in the unicellular green alga *Chlamydomonas reinhardtii*. In: *Centrosomes in Development and Disease* (E.A. Nigg, Ed.), pp. 71–88. Wiley-VCH, Weinheim.

Dutcher, S.K. (2007). Finding treasures in frozen cells: new centriole intermediates. *Bioessays* **29**, 630–634.

Dutcher, S.K. and Trabuco, E.C. (1998). The *UNI3* gene is required for assembly of basal bodies of *Chlamydomonas* and encodes delta-tubulin, a new member of the tubulin superfamily. *Mol. Biol. Cell* **9**, 1293–1308.

Dutcher, S.K., Gibbons, W., and Inwood, W.B. (1988). A genetic analysis of suppressors of the *PF10* mutation in *Chlamydomonas reinhardtii. Genetics* **120**, 965–976.

Dutcher, S.K., Morrissette, N.S., Preble, A.M., Rackley, C., and Stanga, J. (2002). Epsilon-tubulin is an essential component of the centriole. *Mol. Biol. Cell* **13**, 3859–3869.

Dutcher, S.K., Li, L., and O'Toole, E., submitted. STP2 plays a role in centrin attachment and microtubule number.

Ehler, L.L. and Dutcher, S.K. (1998). Pharmacological and genetic evidence for a role of rootlet and phycoplast microtubules in the positioning and assembly of cleavage furrows in *Chlamydomonas reinhardtii. Cell Motil. Cytoskeleton* **40**, 193–207.

Ehler, L.L., Holmes, J.A., and Dutcher, S.K. (1995). Loss of spatial control of the mitotic spindle apparatus in a *Chlamydomonas reinhardtii* mutant strain lacking basal bodies. *Genetics* **141**, 945–960.

Errico, A., Claudiani, P., D'Addio, M., and Rugarli, E.I. (2004). Spastin interacts with the centrosomal protein NA14, and is enriched in the spindle pole, the midbody and the distal axon. *Hum. Mol. Genet.* **13**, 2121–2132.

Finst, R.J., Kim, P.J., and Quarmby, L.M. (1998). Genetics of the deflagellation pathway in *Chlamydomonas. Genetics* **149**, 927–936.

Fromherz, S., Giddings, T.H., Jr., Gomez-Ospina, N., and Dutcher, S.K. (2004). Mutations in alpha-tubulin promote basal body maturation and flagellar assembly in the absence of delta-tubulin. *J. Cell Sci.* **117**, 303–314.

Gaffal, K.P. (1988). The basal body-root complex of *Chlamydomonas reinhardtii* during mitosis. *Protoplasma* **143**, 118–129.

Garreau de Loubresse, N., Ruiz, F., Beisson, J., and Klotz, C. (2001). Role of delta-tubulin and the C-tubule in assembly of *Paramecium* basal bodies. *BMC Cell Biol.* **2**, 4.

Geimer, S. and Melkonian, M. (2004). The ultrastructure of the *Chlamydomonas reinhardtii* basal apparatus: identification of an early marker of radial asymmetry inherent in the basal body. *J. Cell Sci.* **117**, 2663–2674.

Goodenough, U.W. and St. Clair, H.S. (1975). *BALD-2*: a mutation affecting the formation of doublet and triplet sets of microtubules in *Chlamydomonas reinhardtii*. *J. Cell Biol.* **66**, 480–491.

Goodenough, U.W. and Weiss, R.L. (1978). Interrelationships between microtubules, a striated fiber, and the gametic mating structure of *Chlamydomonas reinhardi*. *J. Cell Biol.* **76**, 430–438.

Gould, R.R. (1975). The basal bodies of *Chlamydomonas reinhardtii*. Formation from probasal bodies, isolation, and partial characterization. *J. Cell Biol.* **65**, 65–74.

Gruber, H.E. and Rosario, R. (1974). Variation in eyespot ultrastructure in *Chlamydomonas reinhardtii (ac-31)*. *J. Cell Sci.* **15**, 481–494.

Hall, J.L. and Luck, D.J. (1995). Basal body-associated DNA: *in situ* studies in *Chlamydomonas reinhardtii*. *Proc. Natl. Acad. Sci. U.S.A.* **92**, 5129–5133.

Hall, J.L., Ramanis, Z., and Luck, D.J. (1989). Basal body/centriolar DNA: molecular genetic studies in *Chlamydomonas*. *Cell* **59**, 121–132.

Hiraki, M., Nakazawa, Y., Kamiya, R., and Hirono, M. (2007). Bld10p constitutes the cartwheel-spoke tip and stabilizes the 9-fold symmetry of the centriole. *Curr. Biol.* **17**, 1778–1783.

Holmes, J.A. and Dutcher, S.K. (1989). Cellular asymmetry in *Chlamydomonas reinhardtii*. *J. Cell Sci.* **94**, 273–285.

Holmes, J.A., Johnson, D.E., and Dutcher, S.K. (1993). Linkage group XIX of *Chlamydomonas reinhardtii* has a linear map. *Genetics* **133**, 865–874.

Hoops, H.J. and Witman, G.B. (1983). Outer doublet heterogeneity reveals structural polarity related to beat direction in *Chlamydomonas* flagella. *J. Cell Biol.* **97**, 902–908.

Hoops, H.J., Wright, R.L., Jarvik, J.W., and Witman, G.B. (1984). Flagellar waveform and rotational orientation in a *Chlamydomonas* mutant lacking normal striated fibers. *J. Cell Biol.* **98**, 818–824.

Huang, B., Ramanis, Z., Dutcher, S.K., and Luck, D.J. (1982). Uniflagellar mutants of *Chlamydomonas*: evidence for the role of basal bodies in transmission of positional information. *Cell* **29**, 745–753.

Ikeda, K., Brown, J.A., Yagi, T., Norrander, J.M., Hirono, M., Eccleston, E., Kamiya, R., and Linck, R.W. (2003). Rib72, a conserved protein associated with the ribbon compartment of flagellar A-microtubules and potentially involved in the linkage between outer doublet microtubules. *J. Biol. Chem.* **278**, 7725–7734.

Ishida, T., Kaneko, Y., Iwano, M., and Hashimoto, T. (2007). Helical microtubule arrays in a collection of twisting tubulin mutants of *Arabidopsis thaliana*. *Proc. Natl. Acad. Sci. U.S.A.* **104**, 8544–8549.

Ishikawa, H., Kubo, A., Tsukita, S., and Tsukita, S. (2005). Odf2-deficient mother centrioles lack distal/subdistal appendages and the ability to generate primary cilia. *Nat. Cell Biol.* **7**, 517–524.

Johnson, D.E. and Dutcher, S.K. (1991). Molecular studies of linkage group XIX of *Chlamydomonas reinhardtii*: evidence against a basal body location. *J. Cell Biol.* **113**, 339–346.

Johnson, K.A. and Rosenbaum, J.L. (1991). Basal bodies and DNA. *Trends Cell Biol.* **1**, 145–149.

Johnson, U.G. and Porter, K.R. (1968). Fine structure of cell division in *Chlamydomonas reinhardi*. Basal bodies and microtubules. *J. Cell Biol.* **38**, 403–425.

Kater, J.M. (1929). Morphology and division of *Chlamydomonas* with reference to the phylogeny of the flagellar neuromotor system. *Univ. Calif. Publ. Zool.* **33**, 125–168.

Keller, L.C., Romijn, E.P., Zamora, I., Yates, J.R., III, and Marshall, W.F. (2005). Proteomic analysis of isolated *Chlamydomonas* centrioles reveals orthologs of ciliary-disease genes. *Curr. Biol.* **15**, 1090–1098.

Kochanski, R.S. and Borisy, G.G. (1990). Mode of centriole duplication and distribution. *J. Cell Biol.* **110**, 1599–1605.

Lechtreck, K.F. and Bornens, M. (2001). Basal body replication in green algae – when and where does it start? *Eur. J. Cell Biol.* **80**, 631–641.

Lechtreck, K.F. and Grunow, A. (1999). Evidence for a direct role of nascent basal bodies during spindle pole initiation in the green alga *Spermatozopsis similis*. *Protist* **150**, 163–181.

Lechtreck, K.F. and Silflow, C.D. (1997). SF-assemblin in *Chlamydomonas*: sequence conservation and localization during the cell cycle. *Cell Motil. Cytoskeleton* **36**, 190–201.

Lechtreck, K.F., Teltenkotter, A., and Grunow, A. (1999). A 210 kDa protein is located in a membrane–microtubule linker at the distal end of mature and nascent basal bodies. *J. Cell Sci.* **112**, 1633–1644.

LeDizet, M. and Piperno, G. (1986). Cytoplasmic microtubules containing acetylated alpha-tubulin in *Chlamydomonas reinhardtii*: spatial arrangement and properties. *J. Cell Biol.* **103**, 13–22.

Li, J.B., Gerdes, J.M., Haycraft, C.J., Fan, Y., Teslovich, T.M., May-Simera, H., Li, H., Blacque, O.E., Li, L., Leitch, C.C., Lewis, R.A., Green, J.S., Parfrey, P.S., Leroux, M.R., Davidson, W.S., Beales, P.L., Guay-Woodford, L.M., Yoder, B.K., Stormo, G.D., Katsanis, N., and Dutcher, S.K. (2004). Comparative genomics identifies a flagellar and basal body proteome that includes the *BBS5* human disease gene. *Cell* **117**, 541–552.

Luck, D., Piperno, G., Ramanis, Z., and Huang, B. (1977). Flagellar mutants of *Chlamydomonas*: studies of radial spoke-defective strains by dikaryon and revertant analysis. *Proc. Natl. Acad. Sci. U. S. A.* **74**, 3456–3460.

Mahjoub, M.R., Qasim Rasi, M., and Quarmby, L.M. (2004). A NIMA-related kinase, Fa2p, localizes to a novel site in the proximal cilia of *Chlamydomonas* and mouse kidney cells. *Mol. Biol. Cell* **15**, 5172–5186.

Marshall, W.F., Vucica, Y., and Rosenbaum, J.L. (2001). Kinetics and regulation of *de novo* centriole assembly. Implications for the mechanism of centriole duplication. *Curr. Biol.* **11**, 308–317.

Matsuura, K., Lefebvre, P.A., Kamiya, R., and Hirono, M. (2004). Bld10p, a novel protein essential for basal body assembly in *Chlamydomonas*: localization to the cartwheel, the first ninefold symmetrical structure appearing during assembly. *J. Cell Biol.* **165**, 663–671.

McVittie, A. (1972). Flagellum mutants of *Chlamydomonas reinhardii*. *J. Gen. Microbiol.* **71**, 525–540.

Melkonian, M. (1977). The flagellar root system of zoospores of the green alga *Chlorosarcinopsis* (Chlorosarcinales) as compared with *Chlamydomonas* (Volvocales). *Plant Syst. Evol.* **128**, 79–88.

Merchant, S.S., Prochnik, S.E., Vallon, O., Harris, E.H. *et al.* (2007). The *Chlamydomonas* genome reveals the evolution of key animal and plant functions. *Science* **318**, 245–250.

Miller, M.S., Esparza, J.M., Lippa, A.M., Lux, III , F.G., Cole, D.G., and Dutcher, S.K. (2005). Mutant kinesin-2 motor subunits increase chromosome loss. *Mol. Biol. Cell* **16**, 3810–3820.

Mueller, J., Perrone, C.A., Bower, R., Cole, D.G., and Porter, M.E. (2005). The FLA3 KAP subunit is required for localization of kinesin-2 to the site of flagellar assembly and processive anterograde intraflagellar transport. *Mol. Biol. Cell* **16**, 1341–1354.

Nakazawa, Y., Hiraki, M., Kamiya, R., and Hirono, M. (2007). SAS-6 is a cartwheel protein that establishes the 9-fold symmetry of the centriole. *Curr. Biol.* **17**, 2169–2174.

O'Toole, E.T., Giddings, T.H., McIntosh, J.R., and Dutcher, S.K. (2003). Three-dimensional organization of basal bodies from wild-type and delta-tubulin deletion strains of *Chlamydomonas reinhardtii*. *Mol. Biol. Cell* **14**, 2999–3012.

Pazour, G.J., Agrin, N., Leszyk, J., and Witman, G.B. (2005). Proteomic analysis of a eukaryotic cilium. *J. Cell Biol.* **170**, 103–113.

Pelletier, L., O'Toole, E., Schwager, A., Hyman, A.A., and Muller-Reichert, T. (2006). Centriole assembly in *Caenorhabditis elegans*. *Nature* **444**, 619–623.

Pfannenschmid, F., Wimmer, V.C., Rios, R.M., Geimer, S., Krockel, U., Leiherer, A., Haller, K., Nemcova, Y., and Mages, W. (2003). *Chlamydomonas* DIP13 and human NA14: a new class of proteins associated with microtubule structures is involved in cell division. *J. Cell Sci.* **116**, 1449–1462.

Piasecki, B.P., Lavoie, M., Tam, L.W., Lefebvre, P.A., and Silflow, C.D. (2008). The Uni2 phosphoprotein is a cell cycle regulated component of the basal body maturation pathway in *Chlamydomonas reinhardtii*. *Mol. Biol. Cell* **19**, 262–273.

Piperno, G. and Fuller, M.T. (1985). Monoclonal antibodies specific for an acetylated form of alpha-tubulin recognize the antigen in cilia and flagella from a variety of organisms. *J. Cell Biol.* **101**, 2085–2094.

Preble, A.M., Giddings, T.M. Jr., and Dutcher, S.K. (2000). Basal bodies and centrioles: their function and structure. *Curr. Top. Dev. Biol.* **49**, 207–233.

Preble, A.M., Giddings, T.H. Jr., and Dutcher, S.K. (2001). Extragenic bypass suppressors of mutations in the essential gene *BLD2* promote assembly of basal bodies with abnormal microtubules in *Chlamydomonas reinhardtii*. *Genetics* **157**, 163–181.

Ramanis, Z. and Luck, D.J. (1986). Loci affecting flagellar assembly and function map to an unusual linkage group in *Chlamydomonas reinhardtii*. *Proc. Natl. Acad. Sci. U.S.A.* **83**, 423–426.

Ramos-Morales, F., Infante, C., Fedriani, C., Bornens, M., and Rios, R.M. (1998). NA14 is a novel nuclear autoantigen with a coiled-coil domain. *J. Biol. Chem.* **273**, 1634–1639.

Rieder, C.L. and Borisy, G.G. (1982). The centrosome cycle in PtK2 cells; asymmetric distribution and structural changes in the pericentriolar material. *Biol. Cell* **44**, 117–132.

Ringo, D.L. (1967). Flagellar motion and fine structure of the flagellar apparatus in *Chlamydomonas*. *J. Cell Biol.* **33**, 543–571.

Salisbury, J.L., Sanders, M.A., and Harpst, L. (1987). Flagellar root contraction and nuclear movement during flagellar regeneration in *Chlamydomonas reinhardtii*. *J. Cell Biol.* **105**, 1799–1805.

Salisbury, J.L., Baron, A.T., and Sanders, M.A. (1988). The centrin-based cytoskeleton of *Chlamydomonas reinhardtii*: distribution in interphase and mitotic cells. *J. Cell Biol.* **107**, 635–641.

Schoppmeier, J. and Lechtreck, K.F. (2002). Localization of p210-related proteins in green flagellates and analysis of flagellar assembly in the green alga *Dunaliella bioculata* with monoclonal anti-p210. *Protoplasma* **220**, 29–38.

Silflow, C.D., Liu, B., LaVoie, M., Richardson, E.A., and Palevitz, B.A. (1999). Gamma-tubulin in *Chlamydomonas*: characterization of the gene and localization of the gene product in cells. *Cell Motil. Cytoskeleton* **42**, 285–297.

Silflow, C.D., LaVoie, M., Tam, L.W., Tousey, S., Sanders, M., Wu, W., Borodovsky, M., and Lefebvre, P.A. (2001). The Vfl1 protein in *Chlamydomonas* localizes in a rotationally asymmetric pattern at the distal ends of the basal bodies. *J. Cell Biol.* **153**, 63–74.

Triemer, R.E., and Brown, R.M. (1976). The ultrastructure of meiosis in *Chlamydomonas reinhardtii*. *Br. Phycol. J.* **12**, 23–44.

Vashishtha, M., Walther, Z., and Hall, J.L. (1996). The kinesin-homologous protein encoded by the *Chlamydomonas FLA10* gene is associated with basal bodies and centrioles. *J. Cell Sci.* **109**, 541–549.

Vorobjev, I.A. and Chenstov, Y.S. (1982). Centrioles in the cell cycle. I. Epithelial cells. *J. Cell Biol.* **93**, 938–949.

Weiss, R.L. (1984). Ultrastructure of the flagellar roots in *Chlamydomonas* gametes. *J. Cell Sci.* **67**, 133–143.

Weiss, R.L., Goodenough, D.A., and Goodenough, U.W. (1977). Membrane particle arrays associated with the basal body and with contractile vacuole secretion in *Chlamydomonas*. *J. Cell Biol.* **72**, 133–143.

Wright, R.L., Chojnacki, B., and Jarvik, J.W. (1983). Abnormal basal-body number, location, and orientation in a striated fiber-defective mutant of *Chlamydomonas reinhardtii*. *J. Cell Biol.* **96**, 1697–1707.

Wright, R.L., Salisbury, J., and Jarvik, J.W. (1985). A nucleus–basal body connector in *Chlamydomonas reinhardtii* that may function in basal body localization or segregation. *J. Cell Biol.* **101**, 1903–1912.

Wright, R.L., Adler, S.A., Spanier, J.G., and Jarvik, J.W. (1989). Nucleus–basal body connector in *Chlamydomonas*: evidence for a role in basal body segregation and against essential roles in mitosis or in determining cell polarity. *Cell Motil. Cytoskeleton* **14**, 516–526.

Zamora, I. and Marshall, W.F. (2005). A mutation in the centriole-associated protein centrin causes genomic instability via increased chromosome loss in *Chlamydomonas reinhardtii*. *BMC Biol.* **3**, 15.

CHAPTER 3

Deflagellation

Lynne M. Quarmby

Department of Molecular Biology and Biochemistry,
Simon Fraser University, Burnaby, British Columbia, Canada

CHAPTER CONTENTS

I. INTRODUCTION

The terms deflagellation, flagellar excision, and flagellar autotomy are synonymous and refer to the shedding of the flagella into the environment (see Figure 3.1). A wide range of chemical and physical stimuli can induce deflagellation by activating the precise severing of the nine outer doublet microtubules at the distal end of the flagellar transition zone, a location known as the SOFA (site of flagellar autotomy; Blum, 1971; Lewin and Lee, 1985; Mahjoub et al., 2004). Microtubule severing is accompanied by resealing of the plasma membrane over the doublet microtubules of the transition zone, thereby retaining the integrity of the cytoplasm – the deflagellated cell can regrow new flagella in ~90 minutes, if transferred to a stress-free environment. This chapter provides a brief review of how scientists have exploited this behavior, a discussion of its potential value to

Two ways to lose a flagellum. (A) Premitotic flagellar resorption. The first cell division is completed within 15 minutes of loss of the flagella. (B) Deflagellation induced by dibucaine. The flagella are shed milliseconds after treatment with the local anesthetic. Images collected by Moe Mahjoub in the author's laboratory.

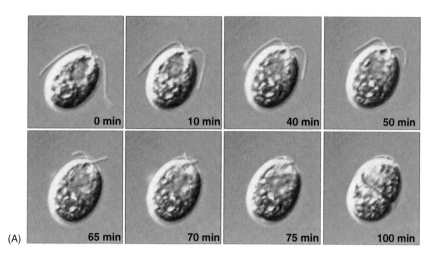

(A)

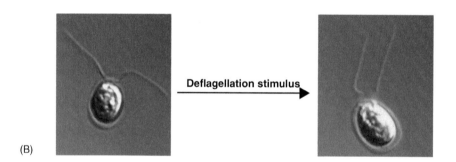

(B)

the cell, and a review of what is known about the mechanism of deflagellation. It also touches on the relationship between deflagellation and the regulation of flagellar assembly and disassembly, leading to an emerging link between deflagellation and cell cycle regulation.

II. DEFLAGELLATION: KEY TOOL OF FLAGELLAR RESEARCH

Most of our knowledge about the structure, function, and assembly of flagella can be directly attributed to the fact that insightful pioneering scientists noticed the deflagellation behavior and recognized its potential scientific value. For example, Rosenbaum and Child (1967) noted that the synchronous regeneration of flagella by a population of deflagellated cells provides a unique opportunity to study the genesis of an organelle. Similarly, use of the deflagellation behavior for the purification of large quantities of flagella (e.g., Witman et al., 1972; Witman, 1986) has been pivotal for many of the discoveries discussed in this volume.

Deflagellation triggers an upregulation of tubulin gene expression (e.g., Weeks et al., 1977; Lefebvre et al., 1978). The accumulation of tubulin transcripts is a consequence of both activation of transcription and stabilization of the message (reviewed by Lefebvre and Rosenbaum, 1986). Numerous other flagellar genes are also upregulated in response to deflagellation (Lefebvre et al., 1980; Remillard and Witman, 1982; Li et al., 2004; Pazour et al., 2005; Stolc et al., 2005). A few deflagellation-activated DNA response elements have been identified (Davies and Grossman, 1994; Periz and Keller, 1997; Kang and Mitchell, 1998) but these do not control transcription for all of the flagellar genes, nor has a program of flagellar gene induction yet been described.

The signal that activates upregulation of the flagellar genes upon deflagellation is an unsolved problem that is ripe for new exploration. It is not known whether the signal for gene expression is generated by the deflagellation event or by a branch of the signaling pathway that triggers deflagellation. In response to environmental stresses that normally trigger deflagellation, mutants that are unable to deflagellate nevertheless upregulate flagellar genes, albeit to a lesser extent than wild-type cells (Cheshire et al., 1994; Evans and Keller 1997a). This suggests that the signal that triggers deflagellation bifurcates and independently triggers upregulation of flagellar genes. However, the situation is not so simple. These same deflagellation-defective mutants resorb their flagella in response to deflagellation-inducing stimuli, suggesting that it may indeed be the absence of flagella that somehow signals upregulation of flagellar genes (Parker and Quarmby, 2003).

Some evidence suggests that calcium is the initial signal in the flagellar regeneration pathway. As discussed in detail below, it is well established that calcium signaling plays an essential role in deflagellation; it would be parsimonious if this same signal activated the program of gene expression required for flagellar assembly. While it has been proposed that calcium activates flagellar gene expression, the signal involved is distinct from the calcium which triggers deflagellation (Cheshire and Keller, 1991; Evans et al., 1997a). Much work needs to be done in order to definitively establish the signaling pathway that activates gene expression, and whether calcium is involved. Technological advances in *Chlamydomonas* research over the past decade make this an attractive problem to revisit.

A second important scientific value of the deflagellation behavior is the ease of postdeflagellation purification of flagella and flagellar fractions (axoneme vs. membrane plus matrix). There is no other subcellular organelle that can be as simply prepared in large quantities of high purity as flagella, especially the flagella of *Chlamydomonas*. Pazour et al. (2005) leveraged the power of this technique and provided the scientific community with a high quality flagellar proteome. Use of the deflagellation behavior as a means to a highly pure subcellular fraction also allows integration of biochemical, physiological, structural, and genetic information. Examples of elegant studies using combinations of analytical approaches are abundantly represented in the chapters of this volume.

III. WHY DO CELLS DEFLAGELLATE?

While we scientists are grateful that Mother Nature has provided the deflagellation behavior, we cannot but wonder what might be in it for the cell. In evolutionary history, has there been positive selection for the deflagellation behavior? If so, what survival advantages does it offer? Is this strictly a laboratory phenomenon, or do cells deflagellate in response to stimuli normally encountered in their environment?

The cell body of *Chlamydomonas* is encased in a complex cell wall (see Volume 1, Chapter 2). In the wall there are two holes (with collars) through which the flagella project. The flagella themselves are not encased in the wall and the flagellar membrane is directly exposed to the environment. Deflagellation thus provides a rapid way to reduce the area of exposed permeable surface. Based on this idea, it has been suggested that deflagellation might provide survival value to free-living ciliated cells, such as *Chlamydomonas*, when exposed to an unfavorable physiochemical environment (Lewin et al., 1982). Indeed, this idea is the reason that deflagellation-defective mutants are named *fa*, for flagellar autotomy, after the process of autotomy whereby amphibians shed and replace a damaged limb (Lewin and Burrascano, 1983).

The hypothesis predicts that deflagellation-defective mutants will have decreased survivorship when exposed to deflagellation-inducing environments. My laboratory has subjected *Chlamydomonas* to a variety of deflagellation-inducing stresses (we have studied pH shock most thoroughly) and we have never observed reduced survivorship in the mutants (unpublished observations). Nevertheless, it is possible that this result is a consequence of the fact that we are only testing the noxious stimuli used to induce deflagellation in the laboratory. What (if anything) induces *Chlamydomonas* to shed its flagella in its natural environment?

Most of the chemical stimuli used to induce deflagellation in the laboratory will kill the cells if they are not transferred to fresh, noxious-stimulus-free, media within minutes. In other words, if we are looking for the natural environment equivalents of the laboratory stimuli, they would need to involve transient exposure, not chronic exposure (i.e., "acid rain" does not provide the same stimulus as "pH shock"). To my knowledge, the only observations of a deflagellation-inducing event relevant to the evolutionary history of *Chlamydomonas* can be seen in the spectacular videos captured by Pickett-Heaps and Pickett-Heaps (1995, and see video *"Chlamydomonas* for dinner?" at http://www.elsevierdirect.com/companions/ 9780123708731, courtesy of Jeremy and Julianne Pickett-Heaps and

Cytographics [Cytographics.com]). This footage shows *Chlamydomonas* cells in culture with a heliozoan, a relatively large protist with a spherical cell body with numerous stiff radial spines, known as axopods. The video images reveal a great deal of particle mobility in the axopods, reminiscent of *Chlamydomonas* intraflagellar transport (IFT) (see Chapter 4). The movies reveal *Chlamydomonas* cells stuck on axopods and reeled in towards the cell body of the heliozoan, where they are ingested by phagocytosis. As has been observed by many laboratory researchers, the flagella of *Chlamydomonas* are particularly sticky – they stick to glass slides and they stick to the axopods of heliozoans. The exciting anecdotal observation is that some *Chlamydomonas* cells manage to deflagellate while being reeled in along an axopod: The heliozoan has flagella for dinner while the *Chlamydomonas* cell floats off and will, presumably, regenerate its flagella. Does this mean that the deflagellation behavior evolved because it provided selective advantage in predator-rich environments? If deflagellation evolved as a defense against predation, then deflagellation-defective mutants should show reduced survivorship when cocultured with predators. To my knowledge, the experiment has not been done.

One argument against the evolution of deflagellation as a defensive behavior is the fact that a wide variety of ciliated or flagellated cells can deflagellate, and it is difficult to imagine a commonality in the lifestyles of these cells. Cells reported to undergo deflagellation (deciliation) include those of sea urchin embryos (Auclair and Siegel, 1966), mollusk gills (Stephens, 1975), and many different types of vertebrate tissues (Stalheim and Gallagher, 1977; Willoughby et al., 1992; Friedrich and Korsching, 1997; Mohammed et al., 1999). Surely all of these cells do not have "escape from cellular predation by deflagellation" recently enough in their evolutionary history to have retained the trait for this reason.

Either different lineages of ciliated/flagellated cells independently evolved deflagellation, or the roots of this trait are ancient. Because the deflagellation behavior is found in both bikonts and unikonts, the most parsimonious explanation is that the ciliated ancestral eukaryote deflagellated. Based on the fundamental conservation of the ciliary structure, it is likely that the molecular mechanism of flagellar excision is conserved, involving orthologous proteins and processes. The ubiquity of a "break-point" in cilia and flagella (Blum, 1971; Quarmby, 2004) supports the idea that a sensitive junction between the cilium-proper and the ciliary transition zone is part and parcel of building a cilium (Parker and Quarmby, 2003).

One corollary is that the break-point did not evolve in response to selection for deflagellation *per se*, but rather for a more fundamental function. One possibility is that controlled disassembly of the outer doublets at this site could facilitate resorption of cilia prior to mitosis. The central idea is that the calcium-sensitive "break-point" serves an adaptive role other than deflagellation and that deflagellation is the result of hyperactivation of this disassembly process. In this scenario, all cilia have the potential to be shed by deflagellation, and some lineages have evolved signaling pathways that allow the cell to elicit this behavior in response to various environmental triggers. These ideas provide a conceptual framework for understanding how the wildly different environments of, say, an infected epithelial cell and a free-living *Chlamydomonas* have provided selective pressure for the evolution of a pathway to trigger deflagellation. Therefore, although the basic mechanism of deflagellation is likely to be highly conserved, the deflagellation is predicted to have different adaptive values for different cells.

Consistent with these ideas, all ciliated cells can be induced to shed their cilia in the laboratory, but for some cells, the behavior is part of their natural life. Sperm deflagellation associated with fertilization is common (but not universal), presumably releasing the centriole for the upcoming meiosis. Some species shed the flagellum before penetration of the egg, but for most the separation of basal body from axoneme occurs after entry into the egg (Paweletz et al., 1987). Sperm also deflagellate in response to environmental toxins, including fluoride (Chinoy and Narayana, 1994). Deciliation of the ventral surfaces of *Paramecium* occurs during mating and is presumed to facilitate close apposition of the membranes prior to fusion (Vivier and Andre, 1961 as cited in Blum, 1971). The ciliated cells of both the respiratory tract and the oviduct shed their cilia in response to infection (Stalheim and Gallagher, 1977; Willoughby et al., 1992).

The next section explores our knowledge of how various agents induce *Chlamydomonas* to deflagellate in the laboratory. The recurring theme of these studies is that, one way or another, all deflagellation-inducing stimuli produce a calcium signal. Section V reviews our knowledge of what happens downstream of the calcium signal, leading to the severing of the axoneme. Finally, section VI presents an emerging story about the relationship between deflagellation and regulation of cell cycle progression, a relationship which could provide the evolutionary explanation for the existence of a calcium-sensitive breakage point in flagella.

IV. SIGNALING PATHWAYS TO DEFLAGELLATION

Scientists have used many different stimuli to trigger deflagellation. These include chemicals (alcian blue, alcohol, chloral hydrate, dibucaine, mastoparan), pH shock, temperature shock, and mechanical shear. It is unknown whether any of these treatments activates pathways that trigger deflagellation in the natural environment, but there are some indications that this is so. Most compelling is the existence of mutants that are specifically defective in the signaling pathway activated by pH shock (see below). I will briefly review our knowledge of the responses to physical and chemical stimuli and then discuss the pathway activated by pH shock.

Although the flagella of wild-type *Chlamydomonas* cells are resistant to substantial shear force (e.g., cells retain their flagella during treatment with a bench top vortex), the more severe shear forces provided by a homogenizer induce deflagellation (Rosenbaum and Child, 1967). This treatment, which requires calcium in the medium, appears to trigger normal calcium-induced axonemal severing and flagellar shedding. The cells survive the treatment and regenerate their flagella. If there is no calcium in the medium, or if the cells are *fa* mutants, then they do not survive mechanical deflagellation (Cheshire et al., 1994), possibly because the flagella are broken at random places and cannot reseal efficiently if the outer doublet microtubules are broken at unequal lengths. Using wild-type cells with calcium present, it is likely that the very high shear forces that are required to induce deflagellation perturb membrane, perhaps most dramatically at the base of the flagella, allowing an influx of calcium.

Temperatures over 40°C induce deflagellation, but nothing is known about the mechanisms involved (Lewin et al., 1982). More interesting are the observations of deflagellation by temperature-sensitive flagellar assembly (*fla*) mutants. These strains were originally isolated in a screen for temperature-sensitive mutants defective in motility and named "drop-down" or *dd* mutants (Huang et al., 1977). Many were later renamed *fla* to reflect the observation that their loss of motility was a consequence of a defect in flagellar assembly (Adams et al., 1982). While strains such as *fla10* have clear flagellar assembly defects (see Chapter 4), other strains such as *fla2* have only subtle assembly defects, but they deflagellate on transfer to the restrictive temperature (33°C; Parker and Quarmby, 2003). These data indicate that temperature may trigger deflagellation via a specific signaling pathway. For example, if the *fla2* strain carries a loss of function mutation, then perhaps elevated temperature triggers deflagellation via the inactivation of specific proteins, possibly proteins that play roles in both deflagellation and flagellar assembly. The *FLA2* gene has not yet been cloned, but based on what we know about deflagellation, it might encode a protein involved in calcium homeostasis.

It is likely that several of the chemical agents that trigger deflagellation act via effects on membrane proteins. Chloral hydrate (key ingredient in the infamous Mickey Finn, a knockout drink) is a lipophilic compound with reported effects on several different membrane proteins (reviewed in Quarmby, 2004). Its mode of action in deflagellation is unknown and it is rarely used by *Chlamydomonas* researchers. More commonly used for flagellar preparations is the local anesthetic, dibucaine (Witman et al., 1978). Dibucaine is also lipophilic and, like chloral hydrate, has diverse cellular targets, which are also primarily, if not exclusively, membrane proteins (reviewed by Butterworth and Striachartz, 1990). The most commonly used dibucaine deflagellation protocols kill the cells, but methodology exists for deciliation of *Tetrahymena* using conditions that support cell survival and ciliary regeneration (Thompson et al., 1974). This suggests that the deflagellation-inducing activity of dibucaine is via effects on specific membrane proteins (or lipids) – affecting either the activity of transporters/ion channels (e.g., Catterall, 1992; Anteneodo et al., 1994) or by triggering Ca^{2+} release from calcium-binding proteins and lipids.

The wasp venom mastoparan is an amphipathic tetradecapeptide that, like chloral hydrate and dibucaine, can disrupt membrane structure. Its best-known pharmacology is as a potent activator of heterotrimeric G proteins (Higashijima et al., 1990). Early experiments suggested that mastoparan-triggered deflagellation of *Chlamydomonas* is mediated by phospholipase C (Quarmby et al., 1992). It is likely that the concentrations of mastoparan used in these experiments were directly activating flagellar severing by making the membrane leaky to calcium (Munnik et al., 1998). Nevertheless, there is evidence suggesting that mastoparan can trigger deflagellation in *Chlamydomonas* without causing gross membrane disruption. Induction of deflagellation by detergent permeabilization of the membrane requires greater than 1 μM extracellular calcium (Sanders and Salisbury, 1989). In contrast, mastoparan-induced deflagellation requires only nanomolar levels of extracellular calcium, implicating signal-activated release from internal calcium stores as the source of calcium for the activation of flagellar severing (Quarmby and Hartzell, 1994).

Alcian blue (a histological stain for glucosaminoglycans) is a cationic phthalocyanin which can activate receptor-operated calcium flux in *Paramecium* and *Tetrahymena* (Tamm, 1994; Francis and Hennessey, 1995; Hennessey et al., 1995). Because Alcian blue is most commonly applied to dead cells, little is known about its physiological effects. It is likely that, like chloral hydrate, dibucaine, and mastoparan, there will be numerous cellular targets, and at this point we have no idea which might be activating the deflagellation pathway. Intriguingly, when applied at threshold concentrations, Alcian blue preferentially induces shedding of the flagellum that is *cis*, relative to the eyespot (Evans and Keller, 1997b). Similarly, we have observed that conditional flagellar assembly mutants preferentially shed their *cis* flagellum in response to a temperature shift (J.D.K. Parker and L.M. Quarmby, unpublished observation). These observations suggest that lower calcium might be required to induce severing of the *cis* relative to the *trans* flagellum.

In the early years of flagellar research, when scientists sought to define conditions for the efficient isolation of intact cilia, there were many complex ethanol-based protocols (Child and Mazia, 1956; Child, 1959; Watson and Hopkins, 1961). We now know that ethanol, without the aid of pH or temperature shock, triggers nonlethal deflagellation (Lewin and Burrascano, 1983). Efficient ethanol-induced deflagellation requires extracellular calcium in amounts comparable to what is required for pH shock-induced deflagellation (50% of cells deflagellate after 10 seconds in 1 mM Ca^{2+}) and much more than what is required to induce deflagellation by detergent permeabilization (Huber et al., 1986; Sanders and Salisbury, 1989; Quarmby and Hartzell, 1994). These data suggest that ethanol triggers deflagellation by stimulation of calcium influx. This could be mediated by any one of the numerous cellular effects of ethanol, including phospholipid metabolism, redox state, and direct potentiation of some ion channels (Baker and Kramer, 1999; Feinberg-Zadek and Treistman, 2007). It is not known whether ethanol triggers the same signaling pathway that is activated in response to pH shock, the signaling pathway about which we know the most.

In one standard protocol for the preparation of flagella for biochemical work, cells are briefly exposed to a solution of pH 4 (Witman et al., 1972). If the cells are returned to neutral media within a minute or two, they survive and regrow their flagella. It is now well established that pH shock works via intracellular acidification and activation of a calcium influx (see Quarmby, 2004 and Figure 3.2). Deflagellation is only induced when weak acids are used; strong acids do not work (Hartzell et al., 1993). This is because in their protonated forms, most weak organic acids are highly membrane permeant. Once in the cytoplasm where their total concentration is lower, the acids dissociate and the cytoplasm becomes acidified,

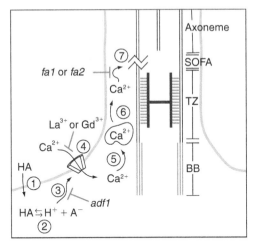

FIGURE 3.2 *Schematic diagram of the signaling pathway for acid-induced deflagellation. (1) Protonated forms of many weak acids are freely permeable across the cell membrane. (2) Inside the cell the protonated acid dissociates, causing acidification of the cytosol. (3) Acidification causes activation of a calcium-influx pathway. This step is blocked in adf mutants. (4) The calcium channel (or pump) is permeant to strontium and is blocked by lanthanum or gadolinium. (5) The calcium (or strontium) then activates release of an internal source of calcium, possibly from a calcium- or IP$_3$-sensitive store that remains to be defined. (6) The calcium released from an internal store binds a calcium-sensing protein (possibly centrin). (7) The severing machinery is activated and the outer doublets are broken at the site of axonemal severing (SOFA). This step is blocked in fa1 and fa2 mutants. FA1 and FA2 both localize to the SOFA. The microtubule-severing protein responsible for actual severing of the outer doublets remains to be identified, but the AAA-ATPase katanin remains a candidate. TZ, transition zone; BB, basal body. Drawing by John Glover.*

triggering deflagellation. How does a drop in cytosolic pH produce the calcium signal that triggers deflagellation?

This question was initially difficult to tackle because calcium buffers are ineffective at low pH (protons compete with calcium for binding sites). The discovery that intracellular acidification activates the pathway allowed the possibility of triggering deflagellation at higher pH using highly membrane-permeant organic acids (such as benzoate). This approach permitted the effective use of calcium buffers, such as BAPTA, to control the concentration of extracellular calcium during a pH shock protocol (Tsien, 1980; Quarmby and Hartzell, 1994). These experiments demonstrated that pH shock-induced deflagellation requires extracellular calcium in the range of 1 mM. Using a calcium-45 flux assay, it was demonstrated that, as predicted, pH shock activates calcium influx (Quarmby and Hartzell, 1994). When pharmacological inhibitors are used to block acid-induced calcium influx, deflagellation is also blocked (Quarmby, 1996).

The essential role of acid-induced calcium influx in signaling the deflagellation response to pH shock is further illustrated by the *adf1* (**a**cid-**d**eflagellation) mutants. These cells are defective in pH shock-induced deflagellation, but are wild type for the calcium-activated machinery of flagellar excision (when the plasma membrane is permeabilized with non-ionic detergent in the presence of calcium, *adf1* cells shed their flagella; Finst et al., 1998). It turns out that *adf1* mutant strains are defective in acid-induced calcium influx, assessed by calcium-45 flux (Quarmby, 1996) and by calcium imaging (Wheeler et al., 2008). The *ADF1* gene has not yet been cloned, but it is predicted to encode either the acid-sensor or the calcium-permeant channel/transporter, which may of course be one and the same protein. Does the calcium influx mediated by the *ADF1* gene product directly activate the machinery of deflagellation, or is the signaling pathway more complex than this?

The calcium which flows into the cell in response to intracellular acidification may not directly activate the machinery of flagellar excision. Evans et al. (1997) found that strontium supports pH shock-induced deflagellation, but does not support deflagellation of detergent-permeabilized cells (even at concentrations 10,000-fold higher than the concentration of calcium that activates severing under the same conditions). These data suggest that the calcium which enters the cell in response to acidification does not directly trigger axonemal severing, but rather, causes a secondary calcium signal generated by release of calcium from intracellular stores (see Figure 3.2). Presumably, strontium enters the cell through an acid-activated influx pathway and triggers calcium release from internal stores, but does not directly activate the severing machinery. There is some evidence that an I(4,5)P$_3$-sensitive store might be involved in the acid-induced deflagellation pathway.

pH shock stimulates inositol phospholipid metabolism (Quarmby et al., 1992). In the classic pathway, activated phospholipase C (PLC) hydrolyses

PtdIns$(4,5)$P$_2$ into Ins$(1,4,5)$P$_3$ and diacylglycerol (reviewed by Clapham, 1995). The Ins$(1,4,5)$P$_3$ then binds to the IP$_3$-receptor, a ligand-gated calcium channel in the membrane of the endoplasmic reticulum, thereby triggering release of calcium from the ER into the cytosol. pH shock of *Chlamydomonas* cells causes activation of PLC, evidenced by an accumulation of Ins$(1,4,5)$P$_3$ and a corresponding decrease in PtdIns$(4,5)$P$_2$ (Quarmby et al., 1992). The predicted accumulation of diacylglycerol is not observed, but this is likely due to an activation of diacylglycerol kinase because an accumulation of phosphatidic acid was measured (Quarmby et al., 1992). We know that the activation of PLC is a response to intracellular acidification and not a consequence of deflagellation because PLC is activated as robustly in the deflagellation-defective mutant *fa1-1* as it is in wild-type cells (Quarmby et al., 1992). Does activation of PLC play an essential role in the deflagellation pathway or is it an independent response to acidification? While there is some evidence in support of the idea that PLC activation is part of the deflagellation pathway, none of it is compelling (reviewed in Quarmby, 2004). The role of PLC activation in mediating pH shock-induced deflagellation remains an open question.

Another protein that has been reported to have a role in excision is CALK (*Chlamydomonas* aurora protein kinase; AAF97501), a member of the aurora family of protein kinases. CALK is in a dephosphorylated state in flagellated cells, but becomes phosphorylated as a result of pH shock; this also occurs during pH shock of *fa1* and *fa2* mutants (see below), indicating that the phosphorylation of CALK could be dependent on events that occur upstream of FA1 and FA2 in the deflagellation pathway (Pan et al., 2004). Alternatively, this phosphorylation event may be an independent response to cytosolic acidification. In an RNAi strain in which CALK was reduced to less than 5% of control levels, pH shock-induced deflagellation was partially inhibited. However, it was not determined whether the deflagellation defect in this strain was a consequence of the reduced levels of CALK or due to the insertional mutation that resulted from construction of the strain. Further studies will be required to determine whether or not CALK has a role in flagellar excision.

In summary, little is known about the signaling pathways activated by most of the agents used to stimulate deflagellation in the laboratory. pH shock is the exception. We know that the pH shock protocol causes intracellular acidification, which in turn stimulates an influx of calcium leading to the severing of the flagella. To date, *ADF1* is the only gene involved in conveying the signal from stress to deflagellation that has been identified (Finst et al., 1998). The simplest hypothesis is that *ADF1* encodes an acid-sensitive calcium channel, but the gene has not yet been cloned. While it remains an open question whether the calcium that enters through the *ADF1* pathway directly activates the machinery of flagellar shedding, as described in the next section, it is clear that calcium is the trigger for the physical separation of the flagellum from the cell body.

V. THE MECHANISM OF FLAGELLAR SEVERING

When the flagellum physically separates from the cell body, two things happen: the axonemal outer doublet microtubules are severed and the flagellar membrane pinches off, leaving the cell body sealed as the flagellum drops off (see Figure 3.3). Electron micrographs of *Tetrahymena* and *Chlamydomonas* cells caught in the act of deflagellation suggest that there is complete (or close to complete) closure of the membrane over the stub by the time the flagellum is shed (Satir et al., 1976; Lewin and Lee, 1985). We know little

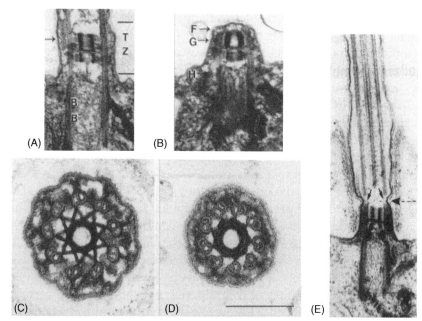

FIGURE 3.3 *Transmission electron micrographs of* Chlamydomonas *flagellar transition zones. (A) Longitudinal section reveals two of the nine triplets of the basal body (BB), extending into the outer doublets of the transition zone (TZ) and then the flagellum. Note the electron-dense "H"-shaped structure inside the outer doublets of the TZ. (B) Similar section as shown in panel A, but after deflagellation. Note that the top of the electron-dense "H" has narrowed to an "A"-shaped structure and that the membrane has sealed over the top of the flagellar stub. (C) Cross-section through the transition zone at approximately the location indicated by the arrow in panel A. Note the stellate array of fibers with tips attached to the A-tubules of the outer doublets. Centrin is a component of these stellate fibers. (D) Similar section as in panel C, but under conditions of high calcium. Note that the centrin-containing stellate fibers have contracted – panels C and D are shown at the same magnification (bar is 0.2 μm). This section may have been from a cell whose flagella had been shed, such as shown in panel B. (E) Longitudinal section of a* Chlamydomonas *flagellum caught in the act of deflagellating. Note that the outer doublet microtubules have been severed just above the TZ and the membrane is beginning to pinch off between the cell body and the flagellum that is being shed. Panels A–D are reproduced from Sanders and Salisbury, 1989,* The Journal of Cell Biology, *108, 1751–1760, Copyright 1989 The Rockefeller University Press. Panel E is from Lewin and Lee, 1985 and reproduced with permission.*

about this pinching off of the membrane, not even whether it is active or passive. For example, coincident with severing of the outer doublets, the transition zone contracts and, aided by mild external shear forces, fission may occur passively, much as mammalian cytokinesis mutants can successfully divide if the two daughter cells can crawl apart. In contrast, the mechanism of axonemal severing is active and triggered by calcium.

Calcium has long been implicated as an important signal in the deflagellation pathway (Rosenbaum and Carlson, 1969; Goldstein, 1974; Thompson et al., 1974; Dunlap, 1977; Huber et al., 1986; Sanders and Salisbury, 1989, 1994; Quarmby et al., 1992) but the first direct evidence that calcium is the trigger of axonemal severing came from *in vitro* experiments. *Chlamydomonas* cells, stripped of their cell walls, were lysed in the presence of millimolar concentrations of the calcium buffer EGTA. Under these conditions, the axoneme remained attached to its basal body, and the two basal bodies remained connected by the distal striated fibers (see Chapter 2). The magic happened when the purified axoneme–basal body structures were exposed to $1\,\mu M$ calcium: the axoneme severed near the base and was released from its respective basal body (Lohret et al., 1998). Thus, all of the components that are downstream of calcium in the deflagellation pathway are integral to the axonemal structure itself.

What is the molecular target of the calcium signal? One favored candidate is the calcium-binding protein centrin (also known as caltractin). The centrins are a subfamily of the 20-kD EF-hand calmodulin-like protein family. They are ubiquitously associated with basal bodies/centrioles, centrosomes, mitotic spindle poles, and yeast spindle pole bodies (Salisbury, 1995). In *Chlamydomonas*, centrin is found in three places: the striated fibers that connect the two basal bodies; the nucleus-basal body connector, a series of fibers that connect the basal body to the nucleus; and most relevant to deflagellation, the stellate fibers internal to the distal flagellar transition zone (Salisbury et al., 1988; Sanders and Salisbury, 1989; Taillon et al., 1992) (see Chapter 2). It is likely that all three sets of centrin fibers contract in response to calcium signals (Wright et al., 1985; Salisbury et al., 1987).

Can calcium-induced contraction of the centrin-containing stellate fibers within the flagellar transition zone generate sufficient shear force to sever the outer doublet microtubules? Coincident with deflagellation, there is a contraction of these fibers, which are located proximal to the site of severing (Figure 3.3). This contraction yields a 12% reduction in the diameter of the transition zone, but because the fibers attach to the A tubule of the outer doublets, there is also a presumed torsional stress and a 42% reduction in the distance between the A tubule and the inner cylinder of the transition zone. The doublet angle relative to the radius of the transition zone changes from 9.6° to 22.1° during deflagellation (Sanders and Salisbury, 1989, 1994). It has been proposed that the combination of torsional and transverse shear resulting from calcium-induced contraction of the stellate fibers causes severing of the outer doublet microtubules. In a genetically amenable organism like *Chlamydomonas*, one would imagine that a centrin mutant would resolve this issue, but in this instance, results with the mutant have been controversial.

The *vfl2-1* strain of *Chlamydomonas* carries a missense mutation in the centrin gene (AY454158), which converts the glutamic acid to a lysine at position 101, the first amino acid of the E-helix of the third EF-hand (Taillon et al., 1992). *vfl2* cells have structural defects in the flagellar transition zone, where the highly organized stellate fibers of the wild-type cell are replaced by poorly organized electron-dense material (Wright et al., 1989; Jarvik and Suhan, 1991). The contraction of the distal transition cylinder that is normally observed in response to deflagellation stimuli does not occur in *vfl2* cells (Jarvik and Suhan, 1991). If contraction of the stellate fibers mediates axonemal severing, then *vfl2* cells should be deflagellation defective. Therein lays the controversy.

There are reports that *vfl2* cells deflagellate normally (Wright et al., 1989; Jarvik and Suhan, 1991) and also that they do not (Sanders and Salisbury, 1989, 1994). Sanders and Salisbury (1989, 1994) report that external shear force, supplied by pipetting or vortexing, is required in conjunction with dibucaine, acid, or detergent in order to induce *vfl2* cells to deflagellate. In my laboratory we found that this effect is subtle and contrasts strongly with the *fa* deflagellation-defective strains, which can be subject to substantial external shear in the context of deflagellation-inducing agents without shedding their flagella (Finst et al., 1998; Lohret et al., 1998). It may be that *vfl2* cells require an external shear force in order to deflagellate, a force that is not required by cells with normal centrin. But given the modest amount of external shear that is required, it is clear that the outer doublet microtubules of *vfl2* cells are severed in the absence of contraction of the centrin-containing stellate fibers. Therefore, torsional and transverse shear generated by contraction of the stellate fibers cannot be the cause of outer doublet microtubule severing. Contraction may, however, facilitate the separation of the flagella from the cell body in the absence of external shear forces.

Although the calcium-stimulated contraction of the centrin-containing stellate fibers does not cause severing of the outer doublet microtubules, is centrin still a candidate for the calcium sensor of the deflagellation pathway? The *vfl2* mutant expresses a stable protein that, although it fails to form an organized array of stellate fibers, does localize to the flagellar transition zone (Taillon et al., 1992). Given that no other candidate Ca^{2+}-binding protein has been implicated in deflagellation, the possibility remains that centrin is the calcium sensor of deflagellation.

If contraction of the stellate fibers does not provide the force necessary to sever the outer doublet microtubules, what does? There is no definitive answer to this question yet, but it is likely that microtubule severing proteins are involved. Breaking a microtubule involves disrupting tubulin–tubulin interactions around the circumference of the tubule or, in the case of the outer doublet microtubules, both the A and B tubules. Few proteins can accomplish this feat, and most are in the katanin family (Quarmby and Lohret, 1999).

In animal cells, katanin is found as a heterodimer of p80 and p60 subunits. Although the precise mechanism by which katanin severs microtubules is not known, the available data support a model in which monomeric p60 subunits exchange bound ADP for ATP, oligomerize, and assemble as a hexameric ring on the wall of a microtubule (Hartman et al., 1998; Hartman and Vale, 1999; McNally, 2000). Hydrolysis of ATP leads to a conformational change that exerts a force on the microtubule, leading to disruption of tubulin–tubulin interactions, and disassembly of the katanin hexamer (Hartman and Vale, 1999; McNally, 2000).

The first question was whether katanin could sever the axoneme, which, as opposed to a simple microtubule, is comprised of highly stable doublet microtubules and numerous accessory proteins. Lohret et al. (1998) demonstrated that exogenous katanin can sever isolated *Chlamydomonas* axonemes. In further support of the idea that katanin might mediate axonemal severing during deflagellation, katanin p60 antibodies block calcium-induced severing of isolated axonemal-basal body complexes. Finally, using a pan-specific antibody to the p60 subunit, *Chlamydomonas* katanin p60 is found in the distal transition zone (Lohret et al., 1998, 1999). While these data are suggestive, they are not compelling. A katanin mutant could provide a more definitive answer.

Repeated failure to isolate katanin mutants led to speculation that katanin is an essential gene. This idea is attractive because in other cells katanin is known to play a role in cell cycle progression (e.g., Buster et al., 2002). In contrast to direct mutagenesis, RNAi approaches allow production of hypomorph strains, which may reveal defects in nonessential processes, such as deflagellation. We screened over ~1500 stable katanin RNAi transformants for defects in deflagellation or cell size (as a proxy for cell cycle defects). Only two clones had distinct phenotypes: both were defective in flagella formation and cell size (M.Q. Rasi and L.M. Quarmby, unpublished observation). Therefore, we could not test for a role in deflagellation because the cells did not make flagella! Nevertheless, we pursued characterization of these strains.

Transformation of the *Chlamydomonas* nuclear genome with recombinant DNA plasmids involves random insertion of the construct within the genome (see Volume 1). Consequently, in the above RNAi experiments, the mutant phenotypes could have resulted from either the intended RNAi knockdown or from disruption of a nontarget gene. Therefore, it was important to determine the site of integration of the RNAi plasmid. In one clone, the insertion disrupted the gene for IFT88/Polaris (AAG37228), which is known to be essential for flagellar assembly (see Chapter 4). In the other mutant strain, the gene encoding a novel cyclic-GMP-dependent protein kinase (PKG2; EDO980015) was disrupted. These data, and the link between cilia and cell cycle regulation (Quarmby and Parker, 2005), suggested that cells lacking flagella would offer a permissive state for the knockdown of katanin, i.e., perhaps katanin is only essential when it is necessary to disassemble a cilium in order to enter the cell cycle. If so, then PKG2, like IFT88, may play an essential role in the assembly of flagella. This and other predictions of this model are currently being tested, but the question of whether katanin mediates axonemal severing during deflagellation remains open.

The nuclear genome of *Chlamydomonas* encodes at least three putative microtubule-severing proteins (Katanin1, KAT1, AAF12877; Katanin2, KAT2, EDP07782; and the related protein, Spastin, EDP09260). Understanding which, if any, of these mediates deflagellation awaits future experimentation. In the meantime, it is intriguing to note that the *Chlamydomonas* ortholog of the p80 subunit of katanin is essential for the formation of the central pair microtubules of the flagella (Dymek et al., 2004). The central pair microtubules are normally nucleated near the site where outer doublet microtubules are severed during deflagellation.

We know that calcium is the signal for axonemal severing and that the axoneme must be severed in order for deflagellation to occur. It may seem ironic that in this genetic model organism mutant screens have yet to identify either the calcium sensor or the microtubule-severing activity. Deflagellation is a nonessential behavior and the only large-scale screen for deflagellation-defective mutants completed to date (Finst et al., 1998) was not designed to uncover genes with additional, possibly essential roles in the cell (see Table 3.1 for a list of mutations that affect deflagellation). A functional relationship between deflagellation and cell cycle progression would explain the difficulty in identifying key components of the deflagellation pathway, and this possibility is discussed in section VI.

In the only published large-scale screen for deflagellation-defective mutants, more than 26,000 mutagenized haploid clones of *Chlamydomonas* were screened for defects in pH shock-induced deflagellation (Finst et al., 1998). Several mutant strains were identified, but after linkage analysis it was determined that only three genes had been uncovered: *ADF1*, *FA1* (AF246990), and *FA2* (AF479588). As described above, strains carrying mutant alleles of *ADF1* are wild type for axonemal microtubule severing, but are defective in the acid-activated calcium-influx pathway (Quarmby and Hartzell, 1994b; Quarmby, 1996). In contrast, the *fa* strains are wild type for calcium influx, but are defective in axonemal severing in response to the calcium signal (Finst et al., 1998).

The first strain carrying a mutation in *FA1* (now known as *fa1-1*) was identified by Lewin and Burrascano (1983); several more alleles were

Table 3.1 Mutants with defects in deflagellation

Mutation	Gene	Protein	Phenotype	Key references
fa1 flagellar autotomy 1	*FA1*	FA1, # AAF66419, novel 171-kD protein	– Flagellar autotomy defective, fails to sever outer doublets in response to calcium – Slow to regrow flagella after mitosis and slow to resorb flagella	Finst et al. (1998), Finst et al. (2000)
fa2 flagellar autotomy 2	*FA2*	FA2, # AAL86904, a NIMA-related kinase	– Flagellar autotomy defective, fails to sever outer doublets in response to calcium – Large cell size due to delay in entering mitosis – Slow to regrow flagella after mitosis and slow to resorb flagella	Finst et al. (1998), Mahjoub et al. (2002, 2004)
adf1 acid deflagellation 1	*ADF1*	Unknown	– Defective in acid-activated calcium influx: does not deflagellate in response to pH shock but does deflagellate in response to other stimuli	Finst et al. (1998), Quarmby (1996)
vfl2-1 variable flagella 2	*VFL2*	Centrin, # EDO98562, calcium-binding protein capable of forming contractile fibers	– Variable number of flagella – Subtle (controversial) defect in deflagellation	Wright et al. (1989), Sanders and Salisbury (1989, 1994), Jarvik and Suhan (1991), Taillon et al. (1992)
fla2-1 flagellar assembly 2	*FLA2*	Unknown	– Reduced rate of retrograde IFT – "Spontaneous" deflagellation at restrictive temperature	Adams et al. (1982), Iomini et al. (2001), Parker and Quarmby (2003)
fla3, fla8, fla10 flagellar assembly	Various	IFT proteins (see Chapter 4)	– Many temperature-sensitive *fla* mutants deflagellate at the restrictive temperature	Parker and Quarmby (2003); see Chapter 4 for additional references

uncovered in the Finst et al. (1998) screen, and the gene was finally cloned (Finst et al., 2000). *fa1* mutants do not shed their flagella in response to any known stimulus (Lewin and Burrascano, 1983; Quarmby and Hartzell, 1994b; Finst et al., 1998). The outer doublet microtubules of *fa1-1* cells do not sever in response to pH shock (Sanders and Salisbury, 1989) and purified axonemal-basal body preparations from *fa1* cells do not sever in response to calcium addition (Finst et al., 1998). The *FA1* gene product, known as FA1 (AAF66419), appears to play an essential role in calcium-stimulated axonemal severing.

It is interesting to note that the inward contraction and change in pitch in the outer doublet microtubules of the flagellar transition zone is attenuated in *fa1-1* cells (Sanders and Salisbury, 1989). This observation may indicate that it is the structural integrity of the outer doublet microtubules that retains the stellate fibers in their beautiful extended configuration, rather than contraction of the fibers playing a causal role in deflagellation, as previously proposed. The *fa1* cells express wild-type centrin and display beautiful stellate fibers, yet in the absence of outer doublet severing, the fibers do not fully contract (Sanders and Salisbury, 1989) even though the acid-stimulated increase in intracellular calcium would be sufficient to activate contraction in the *fa1* cells (Quarmby, 1996). There are two possible explanations for failure of the stellate fibers to contract in *fa1* cells: either they are unable to contract if the outer doublets are not severed or FA1 is important for contraction of the stellate fibers. One further observation indicates that it is unlikely that the stellate fibers play a role in deflagellation: many different types of cells deflagellate, yet few have transition zones with a stellate array of contractile fibers. If facilitating contraction of the stellate fibers is not what FA1 does during deflagellation, what does it do?

The sequence of FA1 provides few hints for the cellular roles of this protein. The 171-kD protein is predicted to have coiled-coil and Ca/CAM-binding domains (Finst et al., 2000). To date, the public databases contain no proteins with highly significant sequence identity to FA1. Because FA1 is essential for deflagellation, and given the high degree of conservation of cilia and of the deflagellation behavior, it is likely that functional orthologs of FA1 exist in species with ciliated cells. Perhaps FA1 is performing scaffolding, rather than catalytic functions, thereby being less constrained in the evolution of its amino acid sequence.

Although FA1 has yet to be localized by indirect immunofluorescence, biochemical fractionation suggests that it is tightly associated with the basal body-flagellar transition zone (Finst et al., 2000). Experiments with dikaryons support the idea that FA1 is a component of a stable complex that only turns over when new flagella are assembled. All of the *fa* mutants identified to date are recessive to wild type in stable diploids, yet temporary dikaryons of *fa/FA* fail to excise the two flagella derived from the *fa* gamete (Finst et al., 1998). Further support for the idea that FA1 is a component of a stable basal body–transition zone complex comes from the observation that calcium-induced axonemal severing occurs in isolated wild-type axonemal-basal body complexes, indicating that the entire machinery for axonemal microtubule severing is a component of a detergent-resistant complex (Lohret et al., 1998).

The phenotype of *fa2* mutants, as with *fa1* mutants, suggests that the *FA2* gene product plays an essential role in deflagellation, downstream of the calcium signal. It was, therefore, a surprise to discover that *FA2* encodes a kinase (AAL86904) in the Nek family of cell cycle kinases (Mahjoub et al., 2002; Quarmby and Mahjoub, 2005). On closer inspection it was discovered that, in addition to their deflagellation defects, *fa2* mutants also have a subtle cell cycle progression defect (Mahjoub et al., 2002). This aspect of FA2 is discussed in section VI; here the focus is on the deflagellation role of FA2. It is perplexing that a kinase is an essential component of calcium-induced axonemal severing, a process that does not require ATP *in vitro* (Lohret et al., 1998). The obvious explanation would be that the kinase activity of FA2 is not required for deflagellation, but in fact a kinase-dead form of the protein localizes correctly, but does not rescue the deflagellation defect of *fa2* mutants. Indirect immunofluorescence studies revealed that FA2 localizes to the SOFA and remains associated with the proximal end of the flagella after deflagellation (Mahjoub et al., 2004). This is distinct from FA1, which remains with the basal body–transition zone fraction after calcium-induced severing of axonemal-basal body complexes (Finst et al., 2000). Perhaps the kinase activity is necessary for priming one of the other components of the complex or for assembly of the complex and is no longer required once the SOFA is assembled. Identifying the protein targets of the kinase activity of FA2 is a high priority for future studies.

VI. DEFLAGELLATION AND THE CELL CYCLE

The relationship between deflagellation and the regulation of cell cycle progression is neither obvious nor clear. Nevertheless, there is growing evidence that a link of some form exists. There is a fundamental connection between cilia and the cell cycle in that the basal bodies that nucleate flagella during interphase are the same organelles that, working under the alias of centrioles, provide the foci for the poles of the mitotic spindle, and play important roles in positioning of the spindle and the cleavage furrow (Ehler et al., 1995, and see Chapters 2 and 14). This focuses attention on the reciprocal regulatory relationships that govern ciliary function and cell cycle progression (see Quarmby and Parker, 2005). There are two lines of evidence that link deflagellation to cell cycle.

Deflagellation is a rapid event that occurs in response to environmental stress. Most cells also lose their flagella prior to mitosis, most commonly by disassembling and resorbing the flagella into the cell body. *Chlamydomonas* flagella are resorbed during preprophase (see Figure 3.1). Superficially, deflagellation and disassembly are dramatically different processes. However, the underlying mechanism and regulatory pathways appear to share common elements, and this provides the first link between deflagellation and cell cycle control.

It is likely that the premitotic disassembly of flagella involves either an inhibition of anterograde IFT, an increase in the rate of retrograde IFT, or both (see Chapter 4). At the permissive temperature (20°C), the flagellar assembly mutant *fla2* has wild-type rates of anterograde IFT, but retrograde IFT is slow (Iomini et al., 2001). At the restrictive temperature (33°C), *fla2* cells, but not wild-type cells, are induced to deflagellate (Parker and Quarmby, 2003). Several other *fla* mutants show this same facile deflagellation phenotype, indicating that the correspondence of the two phenotypes is not associated with a single gene product. It is important to note that this propensity for deflagellation at the restrictive temperature is occurring at the same time that these strains are undergoing flagellar disassembly as a consequence of defects in IFT. Strikingly, deflagellation-defective mutants in the *fa* phenotypic class have flagellar assembly defects: both *fa1* and *fa2* mutants exhibit slow assembly of their flagella after exit from the cell cycle (Mahjoub et al., 2002; Mahjoub and Quarmby, unpublished observation). One final observation further links deflagellation and disassembly: when *fa* mutants are exposed to pH shock or other stimuli that trigger deflagellation of wild-type cells, they do not deflagellate, instead they resorb their flagella, and then build new flagella (Parker and Quarmby, 2003). The observations that IFT mutants have deflagellation defects, deflagellation mutants have flagellar assembly defects, and deflagellation stimuli induce resorption in deflagellation-defective mutants all support the idea that the mechanisms for resorbing flagella share common elements with deflagellation (Parker and Quarmby, 2003).

The second link between deflagellation and cell cycle control comes from the phenotype of *fa2* mutants. As described above, FA2 plays an essential role in calcium-activated severing of the outer doublet microtubules (Finst et al., 1998). In addition to this deflagellation defect, *fa2* mutants are slow to grow new flagella after exit from mitosis, and they are delayed at the G2/M transition of the cell cycle (Mahjoub et al., 2002). The Nek family, of which FA2 is a member, has a deep evolutionary history that appears to be associated with the alternating roles of centrioles as basal bodies and spindle foci (reviewed by Quarmby and Mahjoub, 2005). Intriguingly, the cell cycle role of FA2 is not mediated by its kinase activity because a kinase-dead FA2 rescues the G2/M delay while leaving the cells deflagellation-defective (Mahjoub et al., 2004). This may indicate that the role of FA2 during deflagellation is independent of its role during cell cycle progression. However, the

G2/M delay of *fa2* mutants is coincident with premitotic flagellar disassembly, and the connection could lie in the shared components of deflagellation and the disassembly by resorbing of flagella, as discussed above. For example, perhaps FA2 regulates both processes via effects on microtubule severing activity.

The link between deflagellation and cell cycle progression provides an explanation for the paucity of genes uncovered in the original screen for deflagellation-defective mutants (Finst et al., 1998): Genes whose products have roles in cell cycle progression are unlikely to be uncovered by a mutagenesis protocol which generates primarily null alleles (i.e., insertional mutagenesis). It will be exciting to see whether a new screen, using mutagens that produce a high frequency of missense mutations, will uncover alleles of new deflagellation genes with conditional cell cycle phenotypes.

VII. SUMMARY AND THOUGHTS FOR FUTURE DIRECTIONS

Deflagellation is an important tool for scientists interested in the biology of flagella. Although it is not yet clear whether deflagellation evolved as an adaptive behavior in its own right, the fact that the products of three different genes are essential for pH shock-induced deflagellation indicates that something about the process is important to the cell. Although few laboratories are currently investigating the mechanism of deflagellation, the emerging links with cell cycle control will undoubtedly increase interest in this cellular process.

A conditional screen for deflagellation-defective cells could uncover important genes involved in both deflagellation and cell cycle progression. The original screen for deflagellation mutants failed to identify either the calcium sensor or the microtubule-severing protein, which are hypothesized to be centrin and katanin, both of which are implicated in cell cycle control. What other key players have been missed?

Continued work on FA1, FA2, and the products of the *FLA2* and *ADF1* genes will also be important. What are the substrates of the kinase FA2? Why are *fa2* cells delayed at G2/M? How do FA2 and FA1 work together to mediate axonemal severing? What does *ADF1* encode? With respect to the calcium signaling pathway, it will be important to visualize the temporal and spatial characteristics of the signal(s). Related to this, what control mechanisms govern expression of genes involved in flagellar assembly? Answers to these questions will provide insight into the mechanism of deflagellation, but more importantly, work in this area will illuminate the relationships between deflagellation, flagellar resorption, ciliogenesis, and the cell cycle.

REFERENCES

Adams, G.M.W., Huang, B., and Luck, D.J.L. (1982). Temperature-sensitive, assembly-defective flagella mutants of *Chlamydomonas reinhardtii*. *Genetics* **100**, 579–586.

Anteneodo, C., Rodahl, A.M., Miering, E., Heynen, M.L., Senisterra, G.A., and Lepock, J.R. (1994). Interaction of dibucaine with the transmembrane domain of the Ca^{2+}-ATPse of sarcoplasmic reticulum. *Biochemistry* **33**, 12283–12290.

Auclair, W. and Siegel, B.W. (1966). Cilia regeneration in the sea urchin embryo: evidence for a pool of ciliary proteins. *Science* **154**, 913–915.

Baker, R.C. and Kramer, R.E. (1999). Cytotoxicity of short-chain alcohols. *Annu. Rev. Pharmacol. Toxicol.* **39**, 127–150.

Blum, J.J. (1971). Existence of a breaking point in cilia and flagella. *J. Theor. Biol.* **33**, 257–263.

Buster, D., McNally, K., and McNally, F.J. (2002). Katanin inhibition prevents the redistribution of γ-tubulin at mitosis. *J. Cell Sci.* **115**, 1083–1092.

Butterworth, J.F. and Striachartz, G.R. (1990). Molecular mechanisms of local anesthesia: a review. *Anesthesiology* **72**, 711–734.

Catterall, W.A. (1992). Cellular and molecular biology of voltage-gated sodium channels. *Physiol. Rev.* **72**, S15–S48.

Cheshire, J.L. and Keller, L.R. (1991). Uncoupling of *Chlamydomonas* flagellar gene expression and outgrowth from flagellar excision by manipulation of Ca^{2+}. *J. Cell Biol.* **115**, 1651–1659.

Cheshire, J.L., Evans, J.H., and Keller, L.R. (1994). Ca^{2+} signalling in the *Chlamydomonas* flagellar regeneration system: cellular and molecular responses. *J. Cell Sci.* **107**, 2491–2498.

Child, F.M. (1959). The characterization of the cilia of *Tetrahymena pyriformis*. *Exp. Cell Res.* **18**, 258–267.

Child, F.M. and Mazia, D. (1956). A method for the isolation of the parts of ciliates. *Experientia* **12**, 161.

Chinoy, N.J. and Narayana, M.V. (1994). In vitro fluoride toxicity in human spermatozoa. *Reprod. Toxicol.* **8**, 155–159.

Clapham, D.E. (1995). Calcium signalling. *Cell* **80**, 259–268.

Davies, J.P. and Grossman, A.R. (1994). Sequences controlling transcription of the *Chlamydomonas reinhardtii* beta-tubulin gene after deflagellation and during the cell cycle. *Mol. Cell. Biol.* **14**, 5165–5174.

Dunlap, K. (1977). Localization of calcium channels in *Paramecium*. *J. Physiol.* **271**, 119–133.

Dymek, E.E., Lefebvre, P.A., and Smith, E.F. (2004). Pf15p is the *Chlamydomonas* homologue of the Katanin p80 subunit and is required for assembly of flagellar central microtubules. *Eukaryotic Cell* **3**, 870–879.

Ehler, L.L., Holmes, J.A., and Dutcher, S.K. (1995). Loss of spatial control of the mitotic spindle apparatus in a *Chlamydomonas reinhardtii* mutant strain lacking basal bodies. *Genetics* **141**, 945–960.

Evans, J.H. and Keller, L.R. (1997a). Calcium influx signals normal flagellar RNA induction following acid shock of *Chlamydomonas reinhardtii*. *Plant Mol. Biol.* **33**, 467–481.

Evans, J.H. and Keller, L.R. (1997b). Receptor-mediated calcium influx in *Chlamydomonas reinhardtii*. *J. Euk. Microbiol.* **44**, 237–245.

Evans, J.H., Smith, J.L., and Keller, L.R. (1997). Ion selectivity in the *Chlamydomonas reinhardtii* flagellar regeneration system. *Exp. Cell Res.* **230**, 94–102.

Feinberg-Zadek, P.L. and Treistman, S.N. (2007). Beta-subunits are important modulators of the acute response to alcohol in human BK channels. *Alcohol Clin. Exp. Res.* **31**, 737–744.

Finst, R.J., Kim, P.J., and Quarmby, L.M. (1998). Genetics of the deflagellation pathway in *Chlamydomonas*. *Genetics* **149**, 927–936.

Finst, R.J., Kim, P.J., Griffis, E.R., and Quarmby, L.M. (2000). Fa1p is a 171 kDa protein essential for axonemal microtubule severing in *Chlamydomonas*. *J. Cell Sci.* **113**, 1963–1971.

Francis, J.T. and Hennessey, T.M. (1995). Chemorepellents in *Paramecium* and *Tetrahymena*. *J. Euk. Microbiol.* **42**, 78–83.

Friedrich, R.W. and Korsching, S.I. (1997). Combinatorial and chemotopic odorant coding in the zebrafish olfactory bulb visualized by optical imaging. *Neuron* **18**, 737–752.

Goldstein, S.F. (1974). Isolated, reactivated and laser-irradiated cilia and flagella. In: *Cilia and Flagella* (M.A. Sleigh, Ed.), pp. 111–130. Academic Press, New York.

Hartman, J.J. and Vale, R.D. (1999). Microtubule disassembly by ATP-dependent oligomerization of the AA enzyme katanin. *Science* **286**, 782–785.

Hartman, J.J., Mahr, J., McNally, K., Okawa, K., Akihiro, I., Thomas, S., Cheesman, S., Heuser, J., Vale, R.D., and McNally, F.J. (1998). Katanin, a microtubule-severing protein, is a novel AAA ATPase that targets to the centrosome using a WD40-containing subunit. *Cell* **93**, 277–287.

Hartzell, L.B., Hartzell, H.C., and Quarmby, L.M. (1993). Mechanisms of flagellar excision I. The role of intracellular acidification. *Exp. Cell Res.* **208**, 148–153.

Hennessey, T.M., Kim, M.Y., and Satir, B.H. (1995). Lysozyme acts as a chemorepellant and secretagogue in *Paramecium* by activating a novel receptor operated Ca^{2+} conductance. *J. Membr. Biol.* **148**, 13–25.

Higashijima, T., Burnier, J., and Ross, E.M. (1990). Regulation of G_i and G_o by mastoparan, related amphiphilic peptides, and hydrophobic amines. *J. Biol. Chem.* **265**, 14176–14186.

Huang, B., Rifkin, M.R., and Luck, D.J.L. (1977). Temperature-sensitive mutations affecting flagellar assembly and function in *Chlamydomonas reinhardtii*. *J. Cell Biol.* **72**, 67–85.

Huber, M.E., Wright, W.G., and Lewin, R.A. (1986). Divalent cations and flagellar autotomy in *Chlamydomonas reinhardtii* (Volvocales, Chlorophyta). *Phycologia* **25**, 408–411.

Iomini, C., Babaev-Khaimov, V., Sassaroli, M., and Piperno, G. (2001). Protein particles in *Chlamydomonas* flagella undergo a transport cycle consisting of four phases. *J. Cell Biol.* **153**, 13–24.

Jarvik, J.W. and Suhan, J.P. (1991). The role of the flagellar transition region: inferences from the analysis of a *Chlamydomonas* mutant with defective transition region structures. *J. Cell Sci.* **99**, 731–740.

Kang, Y. and Mitchell, D.R. (1998). An intronic enhancer is required for deflagellation-induced transcriptional regulation of a *Chlamydomonas reinhardtii* dynein gene. *Mol. Biol. Cell* **9**, 3085–3094.

Klink, V.P. and Wolniak, S.M. (2001). Centrin is necessary for the formation of the motile apparatus in spermatids of *Marsilea*. *Mol. Biol. Cell* **12**, 761–776.

Lefebvre, P.A. and Rosenbaum, J.L. (1986). Regulation of the synthesis and assembly of ciliary and flagellar proteins during regeneration. *Annu. Rev. Cell Biol.* **2**, 517–546.

Lefebvre, P.A., Nordstrom, S.A., Moulder, J.E., and Rosenbaum, J.L. (1978). Flagellar elongation and shortening in *Chlamydomonas*. IV. Effects of flagellar detachment, regeneration, and resorption on the induction of flagellar protein synthesis. *J. Cell Biol.* **78**, 8–27.

Lefebvre, P.A., Silflow, C.D., Wieben, E.D., and Rosenbaum, J.L. (1980). Increased levels of mRNAs for tubulin and other flagellar proteins after amputation or shortening of *Chlamydomonas* flagella. *Cell* **20**, 469–477.

Lewin, R.A. and Burrascano, C. (1983). Another new kind of *Chlamydomonas* mutant, with impaired flagellar autotomy. *Experientia* **39**, 1397–1398.

Lewin, R.A. and Lee, K.W. (1985). Autotomy of algal flagella: electron microscope studies of *Chlamydomonas* (Chlorophyceae) and *Tetraselmis* (Prasinophyceae). *Phycologia* **24**, 311–316.

Lewin, R.A., Lee, T.-H. and Fang, L.-S. (1982). Effects of various agents on flagellar activity, flagellar autotomy and cell viability in four species of *Chlamydomonas* (Chlorophyta: Volvocales). In: *Prokaryotic and Eukaryotic Flagella* (W.B. Amos and J.G. Duckett, Eds.), *Soc. Exp. Biol. Symp. No. 35*, pp. 421–437, Cambridge University Press, London.

Li, J.B., Gerdes, J.M., Haycraft, C.J., Fan, Y., Teslovich, T.M., May-Simera, H., Li, H., Blacque, O.E., Li, L., Leitch, C.C., Lewis, R.A., Green, J.S., Parfrey, P.S., Leroux, M.R., Davidson, W.S., Beales, P.L., Guay-Woodford, L.M., Yoder, B.K., Stormo, G.D., Katsanis, N., and Dutcher, S.K. (2004). Comparative genomics identifies a flagellar and basal body proteome that includes the BBS5 human disease gene. *Cell* **117**, 541–552.

Lohret, T.A., McNally, F.J., and Quarmby, L.M. (1998). A role for katanin-mediated axonemal severing during *Chlamydomonas* deflagellation. *Mol. Biol. Cell* **9**, 421–437.

Lohret, T.A., Zhao, L., and Quarmby, L.M. (1999). Cloning of *Chlamydomonas* p60 katanin and localization to the site of outer doublet severing during deflagellation. *Cell Motil. Cytoskeleton* **43**, 221–231.

Mahjoub, M.R., Montpetit, B., Zhao, L., Finst, R.J., Goh, B., Kim, A.C., and Quarmby, L.M. (2002). The *FA2* gene of *Chlamydomonas* encodes a NIMA family kinase with roles in cell cycle progression and microtubule severing during deflagellation. *J. Cell Sci.* **115**, 1759–1768.

Mahjoub, M.R., Rasi, M.Q., and Quarmby, L.M. (2004). A NIMA-related kinase, Fa2p, localizes to a novel site in the proximal cilia of *Chlamydomonas* and mouse kidney cells. *Mol. Biol. Cell* **15**, 5172–5186.

McNally, F.J. (2000). Capturing a ring of Samurai. *Nat. Cell Biol.* **2**, E4–E7.

McNally, F.J. and Vale, R.D. (1993). Identification of katanin, an ATPase that severs and disassembles stable microtubules. *Cell* **75**, 419–429.

McNally, K.P., Bazirgan, O.A., and McNally, F.J. (2000). Two domains of p80 katanin regulate microtubule severing and spindle pole targeting by p60 katanin. *J. Cell Sci.* **113**, 1623–1633.

McNally, K.P., Buster, D., and McNally, F.J. (2002). Katanin-mediated microtubule severing can be regulated by multiple mechanisms. *Cell Motil. Cytoskeleton* **53**, 337–349.

Mohammed, B.J., Mitchell, T.J., Andrew, P.W., Hirst, R.A., and O'Callaghan, C. (1999). The effect of pneumococcal toxin, pneumolysin, on brain ependymal cilia. *Microb. Pathog.* **27**, 303–309.

Munnik, T., van Himbergen, J.A.J., ter Riet, B., Braun, F.-J., Irvine, R.F., van den Ende, H., and Musgrave, A. (1998). Detailed analysis of the turnover of poly-phosphoinositides and phosphatidic acid upon activation of phospholipases C and D in *Chlamydomonas* cells treated with non-permeabilizing concentrations of mastoparan. *Planta* **207**, 133–145.

Pan, J., Wang, Q., and Snell, W.J. (2004). An aurora kinase is essential for flagellar disassembly in *Chlamydomonas*. *Dev. Cell* **6**, 445–451.

Parker, J.D.K. and Quarmby, L.M. (2003). *Chlamydomonas fla* mutants reveal a link between deflagellation and intraflagellar transport. *BMC Cell Biol.* **4**, 11.

Paweletz, N., Mazia, D., and Finze, E.M. (1987). Fine structural studies of the bipo-larization of the mitotic apparatus in the fertilized sea urchin egg. I. The structure and behaviour of the centrosomes before fusion of the pronuclei. *Eur. J. Cell Biol.* **44**, 195–204.

Pazour, G.J., Agrin, N., Leszyk, J., and Witman, G.B. (2005). Proteomic analysis of a eukaryotic cilium. *J. Cell Biol.* **170**, 103–113.

Periz, G. and Keller, L.R. (1997). DNA elements regulating alpha1-tubulin gene induction during regeneration of eukaryotic flagella. *Mol. Cell. Biol.* **17**, 3858–3866.

Pickett-Heaps, J. and Pickett-Heaps, J. (1996). *Predatory Tactics: Survival in the Microcosmos*. NTSC videocassette, 42 minutes. Cytographics, Ascot Vale, Australia.

Piperno, G., Mead, K., and Henderson, S. (1996). Inner dynein arms but not outer dynein arms require the activity of kinesin homologue protein KHP1(FLA10) to reach the distal part of flagella in *Chlamydomonas*. *J. Cell Biol.* **133**, 371–379.

Quarmby, L.M. (1996). Ca^{2+} influx activated by low pH in *Chlamydomonas*. *J. Gen. Physiol.* **108**, 351–361.

Quarmby, L.M. (2000). Cellular Samurai: katanin and the severing of microtubules. *J. Cell Sci.* **113**, 2821–2827.

Quarmby, L.M. (2004). Cellular deflagellation. *Int. Rev. Cytol.: Surv. Cell Biol.* **233**, 47–91.

Quarmby, L.M. and Hartzell, H.C. (1994). Two distinct, calcium-mediated, signal transduction pathways can trigger deflagellation in *Chlamydomonas reinhardtii*. *J. Cell Biol.* **124**, 807–815.

Quarmby, L.M. and Lohret, T.A. (1999) Microtubule severing. *Cell Motil. Cytoskeleton* **43**, 1–9.

Quarmby, L.M. and Mahjoub, M.R. (2005). Caught Nek-ing: cilia and centrioles. *J. Cell Sci.* **118**, 5161–5169.

Quarmby, L.M. and Parker, J.D.K. (2005). Cilia and the cell cycle? *J. Cell Biol.* **169**, 707.

Quarmby, L.M., Yueh, Y.G., Cheshire, J.L., Keller, L.R., Snell, W.J., and Crain, R.C. (1992). Inositol phospholipid metabolism may trigger flagellar excision in *Chlamydomonas reinhardtii*. *J. Cell Biol.* **116**, 737–744.

Remillard, S.P. and Witman, G.B. (1982). Synthesis, transport, and utilization of specific flagellar proteins during flagellar regeneration in *Chlamydomonas*. *J. Cell Biol.* **93**, 615–631.

Rosenbaum, J.L. and Carlson, K. (1969). Cilia regeneration in protozoan flagellates. *J. Cell Biol.* **34**, 345–364.

Rosenbaum, J.L. and Child, F.M. (1967). Flagellar regeneration in protozoan flagel-lates. *J. Cell Biol.* **34**, 345–364.

Salisbury, J.L. (1995). Centrin, centrosomes, and mitotic spindle poles. *Curr. Opin. Cell Biol.* **7**, 39–45.

Salisbury, J.L., Sanders, M.A., and Harpst, L. (1987). Flagellar root contraction and nuclear movement during flagellar regeneration in *Chlamydomonas reinhardtii*. *J. Cell Biol.* **105**, 1799–1805.

Salisbury, J.L., Baron, A.T., and Sanders, M.A. (1988). The centrin-based cytoskeleton of *Chlamydomonas reinhardtii*: distribution in interphase and mitotic cells. *J. Cell Biol.* **107**, 635–641.

Sanders, M.A. and Salisbury, J.L. (1989). Centrin-mediated microtubule severing during flagellar excision in *Chlamydomonas reinhardtii*. *J. Cell Biol.* **108**, 1751–1760.

Sanders, M.A. and Salisbury, J.L. (1994). Centrin plays an essential role in microtubule severing during flagellar excision in *Chlamydomonas reinhardtii*. *J. Cell Biol.* **124**, 795–805.

Satir, B., Sale, W.S., and Satir, P. (1976). Membrane renewal after dibucaine deciliation of *Tetrahymena*. *Exp. Cell Res.* **97**, 83–91.

Stalheim, O.H. and Gallagher, J.E. (1977). Ureaplasmal epithelial lesions related to ammonia. *Infect. Immun.* **15**, 995–996.

Stephens, R.E. (1975). The basal apparatus: mass isolation from the molluscan ciliated gill epithelium and a preliminary characterization of striated rootlets. *J. Cell Biol.* **64**, 408–420.

Stolc, V., Samanta, M.P., Tongprasit, W., and Marshall, W.F. (2005). Genome-wide transcriptional analysis of flagellar regeneration in *Chlamydomonas reinhardtii* identifies orthologs of ciliary disease genes. *Proc. Natl. Acad. Sci. U.S.A.* **102**, 3703–3707.

Taillon, B., Adler, S., Suhan, J., and Jarvik, J. (1992). Mutational analysis of centrin: an EF-hand protein associated with three distinct contractile fibres in the basal body apparatus of *Chlamydomonas*. *J. Cell Biol.* **119**, 1613–1624.

Tamm, S. (1994). Ca^{2+} channels and signaling in cilia and flagella. *Trends Cell Biol.* **4**, 305–310.

Thompson, G.A., Baugh, C., and Walker, L.F. (1974). Nonlethal deciliation of *Tetrahymena* by a local anesthetic and its utility as a tool for studying cilia regeneration. *J. Cell Biol.* **61**, 253–257.

Tsien, R.Y. (1980). New calcium indicators and buffers with high selectivity against magnesium and protons: design, synthesis and properties of prototype structure. *Biochemistry* **19**, 2396–2404.

Watson, M.R. and Hopkins, J.M. (1961). Isolated cilia from *Tetrahymena pyriformis*. *Exp. Cell Res.* **28**, 280–295.

Weeks, D.P., Collis, P.S., and Gealt, M.A. (1977). Control of induction of tubulin synthesis in *Chlamydomonas reinhardi*. *Nature (London)* **268**, 667–668.

Wheeler, G.L., Joint, I., and Brownlee, C. (2008). Rapid spatiotemporal patterning of cytosolic Ca^{2+} underlies flagellar excision in *Chlamydomonas reinhardtii*. *Plant J.* **53**, 401–413.

Willoughby, R., Ecker, G., McKee, S., Riddolls, L., Vernaillen, C., Dubovi, E., Lein, D., Mahony, J.B., Chernesky, M., Nagy, E. et al. (1992). The effects of equine rhinovirus, influenza virus and herpesvirus infection on tracheal clearance rate in horses. *Can. J. Vet. Res.* **56**, 115–121.

Witman, G.B. (1986). Isolation of *Chlamydomonas* flagella and flagellar axonemes. *Methods Enzymol.* **134**, 280–290.

Witman, G.B., Carlson, K., Berliner, J., and Rosenbaum, J. (1972). *Chlamydomonas* flagella. I. Isolation and electrophoretic analysis of microtubules, matrix, membranes and mastigonemes. *J. Cell Biol.* **54**, 507–539.

Witman, G.B., Plummer, J., and Sander, G. (1978). *Chlamydomonas* flagellar mutants lacking radial spokes and central tubules. Structure, composition and function of specific axonemal components. *J. Cell Biol.* **76**, 729–747.

Wright, R.L., Salisbury, J.L., and Jarvik, J. (1985). A nucleus–basal body connector in *Chlamydomonas reinhardtii* that may function in basal body localization or segregation. *J. Cell Biol.* **101**, 1903–1912.

Wright, R.L., Adler, S.A., Spanier, J.G., and Jarvik, J.W. (1989). The nucleus–basal body connector of *C. reinhardtii*: evidence for a role in basal-body segregation and against essential roles in mitosis or in determining cell polarity. *Cell Motil. Cytoskeleton* **14**, 516–526.

Yueh, Y.G. and Crain, R.C. (1993). Deflagellation of *Chlamydomonas reinhardtii* follows a rapid transitory accumulation of inositol 1,4,5-trisphosphate and requires Ca^{2+} entry. *J. Cell Biol.* **123**, 869–875.

CHAPTER 4

Intraflagellar Transport

Douglas G. Cole

Department of Microbiology, Molecular Biology and Biochemistry,
University of Idaho, Moscow, Idaho, USA

CHAPTER CONTENTS

I. THE IDENTIFICATION OF INTRAFLAGELLAR TRANSPORT

Eukaryotic cilia and flagella typically extend away from cell bodies for a distance of a few to tens of microns. The assembly of such a structure presents the cell with a transport problem; proteins synthesized in the cell body are assembled onto cilia and flagella at the distal tip (Johnson and Rosenbaum, 1992, 1993; Dutcher, 1995). The rate of this assembly is completely dependent on the rate of delivery of multiple copies of hundreds of different protein building blocks. During flagellar growth in *Chlamydomonas*, a 5-μm long flagellum can assemble at the impressive rate of 0.4 μm/minute (Marshall et al., 2005). This maximal rate of microtubule growth requires delivery to the distal tip of at least 12,000 tubulin αβ dimers/minute or 200 αβ dimers/second in each organelle. The concomitant assembly of all other axonemal proteins contributes hundreds of additional components that must reach the distal tip every second. Although diffusion is likely to be important in the assembly process, diffusion acting alone would be inefficient and would require an unrealistic concentration gradient of hundreds of unique protein building blocks extending from the cell body to the point of assembly. Thus, the evolution of a direct transport process was needed to service the organelle.

Cilia and flagella exhibit several transport processes that are independent of dynein-based axonemal bending. Before 1992, awareness of these motilities was limited to microsphere movement, cross-linked membrane protein movement, and the gliding of cells across surfaces (reviewed in Bloodgood, 1992 and see Chapter 11). When K. Kozminski, P. Forscher, and J. Rosenbaum applied video-enhanced differential interference contrast (DIC) microscopy to better visualize these phenomena in *Chlamydomonas*, they witnessed a robust bi-directional movement of particles along the length of the flagella (see video "Chlamy_IFT_Kozminski" at http://www.elsevierdirect. com/companions/9780123708731; Kozminski et al., 1993). Since the particles were moving inside the organelle just underneath the flagellar membrane, the newly identified motility was termed intraflagellar transport or IFT (see Chapter 1, for a historical perspective).

Like axonal transport, IFT out to the distal tip is considered to be anterograde while movement back to the cell body is considered to be retrograde. In *Chlamydomonas*, anterograde IFT proceeds at ~2 μm/second while retrograde IFT proceeds at the significantly faster rate of ~3.5 μm/ second (Kozminski et al., 1993, 1995; Iomini et al., 2001; Dentler, 2005). Understanding that IFT might assist in the assembly of the organelle, it was a little surprising that both anterograde and retrograde IFT were found to be essentially unaltered regardless of whether flagella were growing, nongrowing, or undergoing resorption (Kozminski et al., 1993; Dentler, 2005). Once flagella are excised from the cell, however, IFT ceases. These early observations of continuous transport suggested that IFT might function in capacities distinct from the delivery of building supplies to the distal construction site.

Although IFT particles had appeared in select electron micrographs of *Chlamydomonas* and *Tetrahymena* (Ringo, 1967; Sattler and Staehelin, 1974; Dentler, 1980; Mesland et al., 1980; Kozminski et al., 1993), a rather elegant but technically challenging experiment was required to

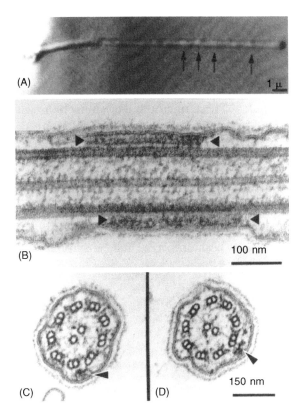

FIGURE 4.1 Chlamydomonas *IFT particles. (A) Video-enhanced DIC micrograph of a* C. moewusii *(M475) flagellum reveals IFT particles as regions of high contrast (arrows). The distal tip of the flagellum is to the right and the refractive cell body is at the edge of the field to the left. (B) EM longitudinal section of a* C. reinhardtii pf1 *flagellum showing two separate electron-dense IFT particles (between arrowheads) in close contact with both the outer doublet microtubules and the flagellar membrane. (C and D) EM transverse sections of a* pf1 *flagellum and a* fla10-1 *flagellum, respectively; the viewer is looking down the flagella from tip to base. IFT particles (arrowheads) typically show a single connection to the B-tubule of the outer doublets and multiple connections to the flagellar membrane. Reproduced from Kozminski et al. (1995) with permission of the Rockefeller University Press, New York, NY.*

confirm that the electron-dense structures shown in Figure 4.1 were, indeed, IFT particles. In order to perform both light and electron microscopy on the same cells, Kozminski et al. (1995) flat-embedded chemically fixed *Chlamydomonas* cells in resin between two Teflon® coated coverslips. The resulting thin layer of resin allowed the researchers to first capture video-enhanced DIC images that clearly showed the static flagellar membrane bulges resulting from fixed IFT particles. Subsequent thin-section electron microscopy of the same flattened resin revealed that the electron-dense lollipop-shaped particles coincided perfectly with the bulges seen by DIC microscopy, confirming their identity.

Chlamydomonas IFT particles are located between the flagellar membrane and the outer doublet microtubules and vary in length from tens to hundreds of nanometers in the direction of the long axis of the axoneme (Kozminski et al., 1993, 1995; Sloboda and Howard, 2007). Flagellar cross-sections reveal that IFT particles are approximately 50 nm across. In wild-type cells, outer doublet microtubule connections nearly always link to the B-tubule (Figure 4.1C, D); for reference, the outer and inner dynein arms decorate the A-tubules. These microtubular connections with the particles are likely to include the IFT-motor proteins, which are described below. The IFT particles also appear to have multiple connections with the flagellar membrane; the nature of these interactions is poorly understood.

The discovery of IFT in *Chlamydomonas* raised many questions. What was being moved? What were the particles made of? What molecular motors were responsible? Was IFT conserved in other ciliated organisms? Perhaps most important of all, what were the function(s) of IFT? Many answers to these questions have now been uncovered and are summarized in this chapter.

FIGURE 4.2

Chlamydomonas IFT model. Anterograde IFT particles consisting of multiple copies of complexes A and B are moved from the base of the organelle out to the distal tip by heterotrimeric kinesin-2. Retrograde transport of smaller IFT particles toward the cell body is mediated by cytoplasmic dynein 1b. The model of the dynein 1b structure is adapted from Rompolas et al. (2007). How the A and B complexes associate to form the IFT particles is unknown.

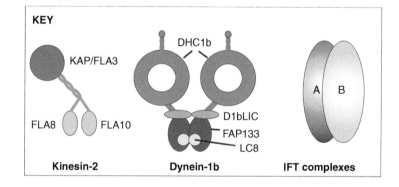

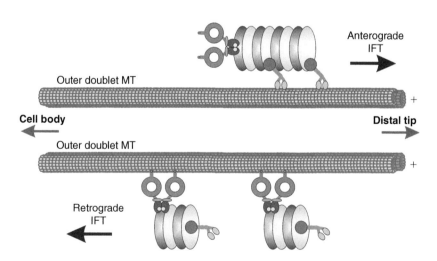

II. IFT MACHINERY

A. Motor and particle proteins in *Chlamydomonas*

Although IFT appears to be conserved in most ciliated eukaryotes, much of the IFT protein machinery has been designated as such due to pioneering research focusing on *Chlamydomonas*. As diagrammed in Figure 4.2 and tabulated in Table 4.1, kinesin-2 and cytoplasmic dynein 1b/2 serve as the respective anterograde and retrograde IFT motors. The proteinaceous IFT particles moved by these motors are encoded by at least 18 distinct genes. The IFT motor and particle proteins found in *Chlamydomonas* are described here.

B. Kinesin-2: anterograde IFT motor

The first gene in any organism to be associated with IFT was the *Chlamydomonas FLA10* gene, which encodes a kinesin-2 motor protein. The *fla10-1* mutant belongs to a collection of temperature-sensitive mutants (*fla1–fla13*) that assemble and maintain functional flagella at the permissive temperature of 20°C but are unable to maintain or generate new flagella at the restrictive temperature of 32°C (Huang et al., 1977; Adams et al., 1982). Identification and sequencing of the *FLA10* locus revealed that it encoded a kinesin-homologous protein that was initially termed KHP1 (Walther et al., 1994). KHP1, or FLA10 as it was later named, was found to group with the kinesin-II subfamily of molecular motors (Goodson et al., 1994; Dagenbach and Endow, 2004), a grouping which has more recently been designated as the kinesin-2 family (Lawrence et al., 2004; Wickstead and Gull, 2006).

The link between KHP1/FLA10 and IFT was discovered when Kozminski et al. (1995) showed that IFT disappeared in *fla10-1* flagella 90 minutes after shifting cells from 20 to 32°C. After 90 minutes at 32°C, the *fla10-1* flagella had begun shortening but still retained 80% of their permissive temperature length of 8.1 μm. At this stage, little or no IFT could be seen using DIC microscopy. Concomitantly, IFT particles were rarely identified in electron micrographs of the flagella of the same temperature-shifted *fla10-1* cells. Lastly, western blot analysis showed that the flagellar KHP1 or FLA10 protein had also disappeared from *fla10-1* flagella after 90 minutes at 32°C. While the majority of FLA10 was typically pooled near the basal bodies, immunoelectron microscopy showed that the flagellar form of the motor was found primarily between the outer doublet microtubules and flagellar membrane (Kozminski et al., 1995; Vashishtha et al., 1996). Taken together, these data indicated that the *FLA10* gene product was required for IFT and implicated direct involvement with powering anterograde IFT.

The FLA10 protein was purified from *Chlamydomonas* flagella by exploiting the kinesin-specific behavior of strong microtubule binding in the presence of AMPPNP followed by the effective release or weak binding that

Table 4.1 *Chlamydomonas* IFT motor proteins

IFT motor protein	GenBank no.	Gene size Introns	Transcript (kb) CDS[a] (kb)	Protein size Predicted mass	pI (predicted)	Deflagellation upregulation	References
Kinesin-2 subunits							
FLA10/KHP1	P46869	7.8 kb 20–21 introns	3.1 2.361	786 aa 86.7 kD	7.09	Yes	Walther et al. (1994)
FLA8	ABS50342	4.8 kb 11 introns	2.9 2.307	768 aa 86.1 kD	6.59	Yes	Miller et al. (2005), Stolc et al. (2005)
FLA3/KAP	AAW67003	8.6 kb 19 introns	3.4 2.544	847 aa 95.1 kD	5.37	Yes	Mueller et al. (2005)
Cytoplasmic dynein 1b subunits							
DHC1b	CAB56748	22.0 kb 41 introns	~14 13.005	4334 aa 481.7 kD	6.13	Yes 7.93 fold	Porter et al. (1999), Pazour et al. (1999, 2005)
D1bLIC	AAO12154	4.4 kb 10 introns	2.3 1.284	427 aa 46.6 kD	9.37	Yes 7.40 fold	Perrone et al. (2003), Hou et al. (2004)
FAP133	EDP07397	~3.5 kb 11 introns	2.5 1.677	558 aa 61.2 kD	5.34	Yes	Rompolas et al. (2007)
LC8	Q39580	0.7 kb 0 introns	0.7 0.276	91 aa 10.3 kD	6.89	Yes 4.89 fold	King and Patel-King (1995); Pazour et al. (1998, 2005)

[a]CDS, coding sequence

occurs in the presence of ATP (Vale et al., 1985; Cole et al., 1992). Like a number of other kinesin-2 motors (Cole et al., 1993; Yamazaki et al., 1995, 1996; Cole, 1999), purified FLA10 was found to be part of a heterotrimeric complex with two related, albeit distinct, motor subunits combined with a third, nonmotor subunit (Cole et al., 1998). The 87-kD FLA10 is accompanied by a related 86-kD kinesin-2 motor subunit encoded by the *FLA8* gene (Miller et al., 2005) and a 95-kD accessory subunit known as the kinesin-2-associated protein or KAP, which is encoded by the *FLA3* gene (Mueller et al., 2005). The *FLA8* gene product was initially termed FLA10H for FLA10-homologous protein in preliminary reports (Cole, 2003a). The *fla1* strain was found to be a separate allele of *FLA8* and was renamed *fla8-2* while the original *fla8* mutant was termed *fla8-1*. As with *fla10-1*, *fla8-1*, and *fla8-2*, the *fla3-1* mutant has a single base change mutation which gives rise to a single amino acid substitution.

The two kinesin-2 motor subunits have N-terminal 350-amino-acid globular kinesin motor domains that contain the microtubule-binding sites and the microtubule-activated ATPase sites. The motor domains of FLA8 and FLA10 are followed by ~325-amino-acid stalk domains that form coiled-coil secondary structure to dimerize the two motor subunits. The globular C-termini of both motor subunits serve as the docking site for the nonmotor KAP subunit and may dock directly onto specific kinesin-2 cargos. The KAP subunit contains 10 armadillo repeats which are expected to mediate transient interactions with kinesin-2 cargo (Mueller et al., 2005).

C. Cytoplasmic dynein 1b/2: retrograde IFT motor

Microtubule orientation within the axoneme dictates that retrograde IFT motility must be mediated by a minus end-directed motor, which initially narrowed down the candidate retrograde motors to the dyneins and a small number of kinesins. Biochemical analysis of the retrograde IFT motor has progressed slowly, however, due to a tendency for the retrograde motor complex to dissociate and due to a large background presence of axonemal outer and inner arm dyneins. The first substantial clue was uncovered when a *Chlamydomonas* retrograde IFT mutant, termed *fla14*, was found to result from a deletion of the dynein light chain gene that encodes LC8 (Pazour et al., 1998). With retrograde IFT events reduced nearly 100-fold, the *fla14* flagella were shorter than wild type and were packed with row upon row of IFT particles. The nearly complete loss of retrograde IFT resulted in the inability to return most IFT particles that were being delivered normally by anterograde transport. First identified in *Chlamydomonas* as a component of outer arm dynein (Piperno and Luck, 1979; Pfister et al., 1982; King and Patel-King, 1995), LC8 is a multifunctional 8-kD protein that associates with axonemal and cytoplasmic dyneins, myosin V, and additional protein complexes not associated with intracellular motility (King et al., 1996; see

Chapter 6). Thus, the loss of LC8 in the *fla14* strain could affect multiple protein complexes and processes.

A critical piece of the puzzle fell into place when two research teams independently identified *Chlamydomonas* insertional mutants in which the gene encoding a cytoplasmic dynein heavy chain known as DHC1b (homologous to DHC2 in mammals) was deleted (Pazour et al., 1999; Porter et al., 1999). Like other dynein heavy chains, DHC1b is a large (nearly 500 kD) AAA cassette protein capable of microtubule binding and microtubule-activated ATP hydrolysis. The loss of DHC1b resulted in the formation of very short, stumpy flagella of 1–2 μm. Retrograde IFT was completely absent, resulting in the flagellar loading of IFT particles and other materials which could not be returned to the cell body. Like other IFT proteins, localization of DHC1b in *Chlamydomonas* revealed a pool near the basal bodies with lesser amounts distributed throughout the flagellum (Pazour et al., 1999; Porter et al., 1999). Additional research using *Chlamydomonas* identified other subunits associated with DHC1b.

A light intermediate chain, termed D2LIC in mammals and D1bLIC in *Chlamydomonas*, was found to co-localize and biochemically co-purify with the mammalian and *Chlamydomonas* dynein 1b/2 heavy chains (Grissom et al., 2002; Perrone et al., 2003). A *Chlamydomonas d1blic* mutant was subsequently identified but the flagellar phenotype was notably weaker than that of the *dhc1b* mutants. Most *d1blic* flagella were short and unable to be used for motility; a subset, however were longer and could be used for motility, albeit at reduced swimming velocities (Hou et al., 2004). Sequence analysis of D1bLIC/D2LIC and comparison with orthologues from other ciliated organisms revealed the presence of a conserved N-terminal P-loop, a RAS signature motif, and a short C-terminal coiled-coil domain (Perrone et al., 2003). The presence of a conserved P-loop is intriguing but its function is unclear. Directed mutations affecting the P-loop did not block the ability of the D1bLIC to rescue the *Chlamydomonas d1blic* null mutant (Hou et al., 2004).

Subsequently, *Chlamydomonas* FAP133, a WD-repeat protein that shares sequence similarity to other dynein intermediate chains (Rompolas et al., 2007), was found to be an IFT-dynein subunit. The 61-kD FAP133 protein co-purifies and co-localizes with DHC1b, D1bLIC, and a subset of the flagellar LC8. Biochemical analysis of this dynein suggests that it might exist in at least two forms. One subset of the *Chlamydomonas* flagellar cytoplasmic dynein 1b sediments at ~12S, which is indicative of axonemal dyneins that have been dissociated to the point where only one heavy chain is found in each complex (Perrone et al., 2003; Hou et al., 2004, Rompolas et al., 2007). Another subset of the dynein 1b/2 complex co-sediments at ~18S, which is characteristic of dynein complexes containing two heavy chains. Formation of the smaller 12S complex is favored by increased ionic strength and mechanical shear. The IFT dynein, therefore, appears to function, at least part of the time, as a multimeric complex containing two identical heavy chains and two or more of each of the other three subunits.

D. IFT-particle proteins

1. Identification of proteins in temperature-sensitive mutants

IFT-particle proteins were first identified in and isolated from *Chlamydomonas* flagella by exploiting the *fla10-1* temperature-sensitive flagellar assembly mutant (Piperno and Mead, 1997; Cole et al., 1998). After *fla10-1* cells are shifted to the restrictive temperature of 32°C for ~2 h, the flagella begin to slowly resorb. After 90 minutes at 32°C, inactivation of the FLA10 motor subunit results in the cessation of IFT as determined by DIC microscopy (Kozminski et al., 1995). Concomitantly, the electron-dense IFT particles disappear from electron micrographs of flagellar thin sections, indicating that the majority of the IFT particles have been cleared from the *fla10-1* flagella. To identify the IFT-particle proteins, flagella were isolated from *fla10-1* maintained at either 22°C or shifted to 32°C. To compare flagellar proteins that were extracted under mild conditions, the membrane plus matrix extract was fractionated using sucrose density gradient centrifugation (Piperno and Mead, 1997; Cole et al., 1998). The resulting gradients revealed a group of co-sedimenting flagellar proteins that disappeared from flagella after *fla10-1* cells were incubated at 32°C. This same set of 16S proteins also disappeared from flagella when *fla8-1* or *fla8-2* (formerly *fla1*) cells were shifted to the restrictive temperature (Figure 4.3; Cole et al., 1998). Subsequent biochemical manipulations with the IFT-particle proteins revealed that the 16S group was easily separated into two unique complexes termed A and B (Cole et al., 1998; Piperno et al., 1998). Complex B can undergo further salt-dependent dissociation as described below.

2. IFT complex A

IFT complex A was initially thought to be composed of four distinct subunits originally termed p144, p140, p139, and p122, where the number referred to the apparent mobility of each protein (in kD) separated by sodium dodecyl sulfate-polyacrylamide gel electrophoresis (SDS-PAGE) (Cole et al., 1998). Piperno et al. (1998) identified a fifth subunit with a mobility of 43 kD and Qin and Rosenbaum (personal communication) identified a sixth subunit with apparent mobility of 121 kD. In 2000, the IFT-particle subunit nomenclature was modified such that the complex A subunits were designated IFT144, IFT140, IFT139, IFT122, IFT121, and IFT43 (Table 4.2) (Pazour et al., 2000). It is not known if each of these subunits is contained in each individual complex A. Complete immunoprecipitation of IFT139, however, does quantitatively remove all six complex A subunits from solution (Cole et al., 1998). Based on the predicted molecular masses of each gene product (Table 4.2), a complex of all six subunits would have a mass of 760 kD. Hydrodynamic measurements of complex A are consistent with a molecular mass of just 550 kD (Cole et al., 1998). Thus, it is possible that (1) each individual complex A does not contain all six subunits, (2) some subunit masses are less than that predicted by

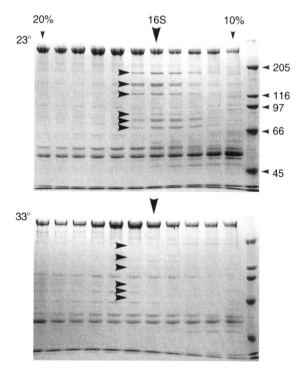

FIGURE 4.3 *Loss of 16S polypeptides in* fla8-2 *flagellar extracts of cells shifted to the restrictive temperature of 33°C. Membrane plus matrix extracts from* fla8-2 *(formerly* fla1*) mutant cells incubated at the permissive and restrictive temperatures of 23°C and 33°C were fractionated on sucrose density gradients in low ionic strength buffer. Only the portion of the gradient from 20% (left) to 10% (right) is shown on Coomassie-stained 7.5% gels. At 23°C, the ~16S IFT-particle polypeptides (arrowheads) are reduced relative to wild type (not shown) but are nearly completely depleted after incubation at 33°C (arrowheads). Similar results are observed with membrane plus matrix isolated from flagella of* fla8-1 *and* fla10-1 *cells incubated at 23°C or 33°C. Reproduced from Cole et al. (1998) with permission of the Rockefeller University Press, New York, NY.*

cDNA sequence, or (3) the hydrodynamic measurements do not accurately reflect the true mass of complex A. In any case, it seems unlikely that the biochemically isolated complex A contains multiple copies of each subunit.

3. IFT complex B

IFT complex B was initially thought to contain ~11 subunits which have since been renamed IFT172, IFT88, IFT81, IFT74/72, IFT57/55, IFT52, IFT46, IFT27, and IFT20 (Cole et al., 1998; Piperno et al., 1998; Pazour et al., 2000). Subsequent analysis has revealed three additional B subunits, IFT80, IFT25, and IFT22 (Qin and Rosenbaum, personal communication). The association of the B proteins is sensitive to increasing ionic strength. IFT172, IFT57/55, IFT80, and IFT20 dissociate from the complex with

Table 4.2 *Chlamydomonas* IFT-particle proteins

IFT-particle protein	GenBank no.	Gene size Introns	Transcript (kb) CDS[a] (kb)	Protein size Predicted mass	pI (predicted/ expt.)	Deflagellation upregulation	References
Complex A							
IFT144	ABU95019	9.1 kb 23 introns	4.7 4.068	1367 aa 150.8 kD	6.13/5.8	Yes 7.03 fold	Cole et al. (1998), Pazour et al. (2005)
IFT140	AAT95430	11.0 kb 29 introns	4.6 4.155	1384 aa 154.6 kD	5.50/6.0	Yes 3.78 fold	Cole et al. (1998), Pazour et al. (2005)
IFT139	ABU95018	15.1 kb 24 introns	5.5 4.068	1355 aa 152.0 kD	5.91/5.9	Yes 4.66 fold	Cole et al. (1998); Pazour et al. (2005)
IFT122	EDO98316	7.0 kb 13 introns	4.3 3.720	1239 aa 139.2 kD	6.71/5.9	Yes 5.92 fold	Piperno and Mead (1997), Cole et al. (1998, Pazour et al. (2005)
IFT121	ABU89876	8.5 kb 25 introns	4.3 3.675	1224 aa 136.1 kD	6.04/5.9	Yes 5.87 fold	Piperno and Mead (1997), Pazour et al. (2005)
IFT43	ABU93234	2.4 kb 4 introns	1.7 0.819	272 aa 28.8 kD	4.48/4.5	*nd*	Piperno and Mead (1997)
Complex B							
IFT172	AAT99263	9.7 kb 14 introns	5.9 5.268	1755 aa 197.6 kD	5.82/5.9	Yes	Cole et al. (1998), Pedersen et al. (2005)
IFT88	AAG37228	5.9 kb 13 introns	3.4 2.349	782 aa 86.3 kD	5.87/5.9	*nd*	Cole et al. (1998), Pazour et al. (2000)
IFT81	AAT99262	5.1 kb 12 introns	2.9 2.052	683 aa 77.1 kD	6.25/6.1	Yes 4.13 fold	Cole et al. (1998), Lucker et al. (2005), Pazour et al. (2005)
IFT80	ABQ96217	6.0 kb 12 introns	3.0 2.3	765 aa 85.7 kD	6.97/6.0	Yes 7.22 fold	Qin et al. (2001), Pazour et al. (2005)

(Continued)

Table 4.2 *Continued*

IFT-particle protein	GenBank no.	Gene size Introns	Transcript (kb) CDS[a] (kb)	Protein size Predicted mass	pI (predicted/ expt.)	Deflagellation upregulation	References
IFT74/72	AAO92260	5.1 kb 9 introns	3.1 1.926	641 aa 71.4 kD	9.02/*nd*	Yes	Cole et al. (1998), Qin et al. (2004), Lucker et al. (2005)
IFT57	ABB72789	3.3 kb 9 introns	1.9 1.410	469 aa 51.3 kD	4.99/4.9	*nd*	Cole et al. (1998)
IFT52	AAL12162	3.8 kb 4/5 introns	2.4 1.365	454 aa 50.4 kD	5.05/5.0	Yes	Cole et al. (1998), Brazelton et al. (2001), Deane et al. (2001)
IFT46	ABH06907	4.1 kb 10 introns	1.7 1.035	343 aa 37.9 kD	4.61/4.4	Yes 5.22 fold	Cole et al. (1998), Pazour et al. (2005), Hou et al. (2007)
IFT27	EDP09483	1.7 kb 4/5 introns	1.1 0.615	204 aa 22.8 kD	5.28/5.2	Yes	Cole et al. (1998), Qin et al. (2007)
IFT25		1.7 kb 3 introns	1.2 0.570	189 aa 20.4 kD	4.69/*nd*	Yes 4.56 fold	Pazour et al. (2005)
IFT22	XP_001689669	1.6 kb 4 introns	1.1 0.624	192 aa 21.3 kD	5.47/*nd*	Yes 8.71 fold	Pazour et al. (2005)
IFT20	AAM75748	1.5 kb 5 introns	0.9 0.408	135 aa 15.6 kD	4.87/4.7	*nd*	Cole et al. (1998), Pazour et al. (2002)

[a]CDS, coding sequence

increasing concentration of NaCl, leaving a biochemically intact core containing IFT88, IFT81, IFT74, IFT72, IFT52, IFT46, IFT27, IFT25, and IFT22 (Cole et al., 1998; Lucker et al., 2005). Derived from the same gene, IFT72 and IFT74 appear to be present in a 1:1 stoichiometry and are often combined into a single term, IFT74/72 (Qin et al., 2004; Lucker et al., 2005). IFT57 and IFT55 are also derived from a single gene and have been designated as either IFT57/55 or just IFT57 (Cole et al., 1998). The ratio of IFT57 to IFT55 is variable, suggesting that these two isoforms may result from a reversible post-translational modification. The remaining complex B subunits are derived from unique genes and all appear to be present at approximately one-fold stoichiometry except for IFT81, two copies of which appear to be present in each B complex (Lucker et al., 2005).

Like the IFT-motor genes and many other flagellar genes, at least 14 of the IFT-particle genes are transcriptionally upregulated following deflagellation (Table 4.2); the remaining 4 IFT-particle genes have not been similarly analyzed. Upregulation occurs in spite of the fact that only a minor fraction of the IFT proteins are typically lost with flagellar excision; most of the IFT proteins remain in a pool near the basal bodies during and immediately after excision. When regeneration of the flagellum begins, a significant loading of IFT protein into the nascent organelle is observed (Deane et al., 2001; Marshall and Rosenbaum, 2001). This sudden shift of IFT proteins into the flagellum may contribute to the upregulation.

4. Domain organization of the IFT-particle proteins

Many of the IFT-particle proteins contain repeat motifs that commonly form protein–protein binding domains (Figure 4.4). Some of these domains likely mediate interactions between specific IFT-particle subunits in order to assemble the complexes and the large particles observed in electron micrographs. Some of the domains are likely used to bind IFT cargos that are transported into or out of the flagellum. Given the relatively large number of IFT-particle subunits, the total number of unique cargo binding sites may also be large. In addition, each binding site might interact with multiple cargos, giving the IFT particles the ability to transport dozens or possibly even hundreds of different cargos. This working model of transport would require the transient docking of specific cargo proteins to specific IFT subunit proteins. Sequence analysis of the IFT-particle proteins reveals numerous domains that could very well mediate such interactions.

IFT88 and IFT139, for example, contain multiple TPR or tetratricopeptide repeats (Pazour et al., 2000; Cole, 2003a). The TPR motif is a highly degenerative ~34-amino-acid repeat that forms α-helical secondary structure. Three to four TPR α-helices fold back and forth every ~17 residues to form a binding pocket that typically associates transiently with a binding partner (Lamb et al., 1995; Das et al., 1998; Blatch and Lassle,

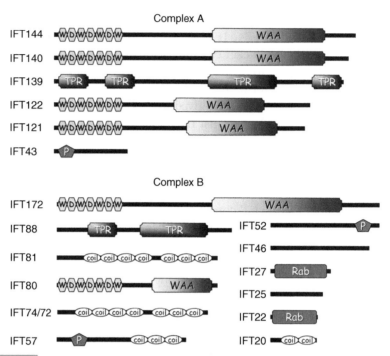

FIGURE 4.4 *IFT-particle proteins are rich in protein–protein binding motifs. Biochemical analysis of* Chlamydomonas *IFT particles have identified 6 complex A and 12 complex B proteins. Sequence analysis using the SMART algorithm reveals putative protein–protein binding motifs including WD domains (WD), tetratricopeptide repeats (TPR), and coiled-coils (coil). The WAA domains contain TPR-like repeats of unknown structure and function. IFT27 and IFT22 share significant amino acid sequence homology with the Rab-like small G proteins. IFT57, IFT52, and IFT43 contain proline-rich domains (P).*

1999; Scheufler et al., 2000). The TPR binding pocket can interact with just a few amino acids from the C-terminus, allowing TPR proteins to bind only a small portion of a target protein. Some proteins with multiple TPR domains serve as docking platforms to bring two proteins together; e.g., in human cells Hop acts to bring together Hsp70 and Hsp90 (Scheufler et al., 2000). Since IFT88 and IFT139 have multiple TPR domains, each may be capable of binding two or more cargos simultaneously.

Six of the IFT subunits contain WD repeats at their amino termini (Figure 4.4). The WD or WD40 motif is a degenerative repeat containing ~40 amino acids that usually include conserved tryptophan and aspartate residues (Smith et al., 1999; Yu et al., 2000). Although the number of repeats can vary, the most common WD domain contains seven closely spaced repeats that form a β-propeller structure first characterized in the β-subunit of the heterotrimeric G protein (Wall et al., 1995). As few as four repeats, however, can form a propeller capable of binding a second protein (Li and Roberts, 2001). Of particular note is the fact that all six IFT WD-repeat proteins also contain a second, poorly characterized, degenerate repeat sequence known as the WAA repeat because of conserved tryptophan and alanine residues. This motif is similar to, though distinct from, the canonical TPR motif (Pedersen et al., 2005). While the WAA motif is predicted to generate α-helical secondary structure, the tertiary structure is unknown. Whether or not the WAA motif forms a TPR-like binding pocket is an interesting problem that awaits structural analysis. Intriguingly, the WAA repeats are truly novel in that they appear to be specific to IFT machinery.

Four of the IFT subunits are predicted to form α-helical coiled-coil structures which typically mediate stable, long-term association between two or more proteins (Lupas et al., 1991). Homodimerization of IFT81 and heterodimerization of IFT81 and IFT74/72 are believed to be mediated by coiled-coil formation (Lucker et al., 2005). Indeed, two subunits each of IFT81 and IFT74/72 appear to form a tetrameric complex within the core of complex B. Although the mammalian orthologues of the remaining coiled-coil subunits, IFT57 and IFT20, have been shown to interact in a yeast two-hybrid assay (Baker et al., 2003), similar analysis has failed to show that *Chlamydomonas* IFT57 and IFT20 directly interact with one another (Lucker et al., 2006).

Many of the smaller IFT-particle subunits lack the obvious protein–protein binding motifs observed in the larger subunits. Amino acid sequence analysis of IFT27 and IFT22, however, reveals that these two small IFT-particle subunits are Rab-like small G proteins (Cole, 2003a; Qin et al., 2007). This suggests that these subunits could act as molecular switches that regulate specific events such as membrane trafficking (Grosshans et al., 2006).

5. Structural analysis of IFT complex B

As noted above, complex B can undergo partial salt-dependent dissociation to form a core subcomplex containing the IFT88, IFT81 × 2, IFT74/72 × 2, IFT52, IFT46, IFT27, IFT25, and IFT22 subunits. Chemical cross-linking has been combined with yeast-based two- and three-hybrid analysis to show that IFT81 forms homodimers that are capable of interacting with a pair of IFT74/72 subunits; it is not known yet whether each core contains one each of the IFT74 and IFT72 subunits (Lucker et al., 2005). Yeast-based two-hybrid and bacterial co-expression have been used to show that IFT52 and IFT46 interact (Lucker et al., 2006). Bacterial co-expression has also revealed that IFT88 can interact separately with IFT46 and IFT52 and that, when the three proteins are co-expressed, they form a ternary complex (Lucker et al., 2006). These data suggest that two subunits each of IFT81 and IFT74/72 are capable of forming a heterotetrameric subcomplex in the absence of other IFT proteins while IFT88, IFT52, and IFT46 are capable of forming a separate subcomplex also in the absence of other IFT proteins. Future analysis of *Chlamydomonas* complex B mutants should reveal if these two subcomplexes can form *in vivo*.

E. Additional IFT components in *Chlamydomonas*?

1. Have all the IFT proteins been found?

Even though the IFT machinery is best understood in *Chlamydomonas*, there may be additional IFT components that have yet to be identified. Some IFT proteins, for example, may be present at sub-stoichiometric levels and have avoided biochemical detection. Other components, like IFT172, may easily dissociate from the IFT motors or complexes and have avoided purification. Potential *Chlamydomonas* IFT candidates are described here.

2. Candidate IFT motors

Do FLA10 kinesin-2 and cytoplasmic dynein 1b act alone in driving *Chlamydomonas* anterograde and retrograde IFT, respectively? In the nematode, for example, the heterotrimeric kinesin-2 is joined by the kinesin-2-related homodimeric OSM-3, and the two motors work together to drive IFT (Signor et al., 1999; Snow et al., 2004; Cole, 2005; Ou et al., 2005a). In specific neuronal sensory cilia of *Caenorhabditis elegans*, both kinesins share anterograde duty along the doublet microtubules of the middle ciliary segments. OSM-3 alone, however, drives anterograde transport along the distal singlet microtubules. The vertebrate orthologue to OSM-3 is thought to be the homodimeric kinesin-2-related KIF17, a plus end-directed dendritic motor required for ciliary targeting of cyclic nucleotide-gated channel proteins in cultured Madin-Darby canine kidney (MDCK) cells (Setou et al., 2000; Jenkins et al., 2006). Since dominant negative constructs of KIF17 failed to affect ciliary assembly in the MDCK cells, it was hypothesized that ciliary KIF17 was not essential for general IFT but was responsible for the transport of specific cargos (Jenkins et al., 2006). Like the nematode OSM-3 though, vertebrate KIF17 may also be responsible for assembly of and transport along axonemal singlet microtubules. Morpholino knockdown of KIF17 expression in zebrafish disrupted formation of the retinal outer segments, which are highly modified cilia containing extensive singlet microtubules (Insinna et al., 2008). Evidence for the direct association of KIF17 with the traditional IFT machinery came when KIF17 was co-precipitated with IFT complex B proteins using antibody pull downs of zebrafish retinal extract (Insinna et al., 2008).

In vegetative *Chlamydomonas* cells, the distal singlet A microtubules are quite short and may not require directed transport as observed in the nematode. During mating, however, the singlet A microtubules elongate while material reminiscent of IFT particles builds up at the flagellar tip (Mesland et al., 1980; Pasquale and Goodenough, 1987). It is possible that a second kinesin motor complex is responsible for IFT along these elongated singlet microtubules. The *Chlamydomonas* nuclear genome, however, appears to lack a KIF17 or OSM-3 homologue (Richardson et al., 2006; Wickstead and Gull, 2006). Interestingly, the *Chlamydomonas* flagellar proteomic analysis, which was carried out on flagella from vegetative cells, identified just four kinesin motor subunits (Pazour et al., 2005). These included the heterotrimeric kinesin-2 proteins FLA8 and FLA10, the central-pair kinesin-9 KLP1, and FAP125 (EDP02895), an uncharacterized and ungrouped kinesin. Since FAP125 was not found in the membrane plus matrix fraction of the flagellar proteome, it is a reasonable candidate to be the second central-pair kinesin identified using pan-kinesin antibodies (Bernstein and Rosenbaum, 1994; Fox et al., 1994; Johnson et al., 1994). Kinesin motor subunits tend to migrate relatively slowly on SDS-PAGE. KLP1 is predicted to be 82.0 kD but migrates at ~96 kD (~17% larger) on SDS-PAGE (Bernstein et al., 1994). FAP125 is predicted to have a mass of 93.6 kD. Using a polyclonal antibody that recognizes diverse kinesins, Johnson et al. (1994) identified a putative second central-pair kinesin that migrated at ~110 kD by SDS-PAGE; the apparent mobility of 110 kD is 17.5% larger than the predicted mass of FAP125. This additional central-pair kinesin could be FAP125 or the gene product of one of the two additional *Chlamydomonas* kinesin-9 genes (EDP06495, EDP09383); the later two, though, were not identified in the flagellar proteome. Using the LAGSE and HYPER pan-kinesin peptide antibodies (Cole et al., 1992; Sawin et al., 1992), Fox et al. (1994), identified another putative flagellar kinesin that migrated at ~125 kD on SDS-PAGE; the identity of this protein has yet to be determined.

It is important to note, however, that at least one kinesin now known to be in the flagellum eluded earlier antibody and proteomic detection. The flagellar calmodulin-binding kinesin, KCBP (ABF50981), identified by Dymek et al. (2006), migrates at ~140 kD, making it unlikely to be one of the two central-pair kinesins described above. Based on similarity with other calmodulin-binding kinesins, KCBP might be capable of Ca^{2+}/calmodulin-sensitive minus end-directed movements within the flagellum. These could include but are not necessarily limited to retrograde IFT, gliding, and external bead movement toward the cell body (see Chapter 11). Of special note is the fact that KCBP is a flagellar kinesin that was not identified in the *Chlamydomonas* flagellar proteome (Pazour et al., 2005). This demonstrates that less abundant motors can escape detection by proteomic analyses.

Another candidate flagellar kinesin that may have eluded proteomic and other detection is the *Chlamydomonas* MCAK or kinesin-13 homologue (EDP01883), which has a predicted mass of 74.5 kD. An MCAK homologue, LmjKIN13-2, has been found at the distal tips of *Leishmania major* flagella, where the protein appears to catalyze the disassembly of the axonemal microtubules (Blaineau et al., 2007). Two additional flagellar kinesin candidates arise from a "holistic approach" to generate a kinesin phylogeny using 19 diverse eukaryotic organisms (Wickstead and Gull, 2006). This study identified four families of kinesins that are specific to ciliated organisms, kinesin-2 and kinesin-9 described above and two new families, kinesin-16 and kinesin-17. Although *Chlamydomonas* does possess kinesin-16 (EDP07970) and kinesin-17 (EDP00849) homologous proteins, neither has yet to be shown in any organism to be involved with transport in cilia or flagella.

3. Candidate IFT-particle proteins

Qilin is a candidate IFT-particle protein initially associated with cystic kidney disease in zebrafish (Sun et al., 2004). The *C. elegans* qilin mutant, *dyf-3*, phenocopies other IFT mutants with the absence of distal ciliary assembly (Ou et al., 2005b). The DYF-3::GFP fusion protein undergoes IFT-mediated transport within nematode sensory cilia. Furthermore, qilin homologues have been found in proteomic analyses of both human cilia (Ostrowski et al., 2002; Marshall, 2004) and *Chlamydomonas* flagella (Pazour et al., 2005), indicating that a ciliary function for this protein is likely to be widely conserved. Whether or not qilin is a bona fide IFT-particle subunit awaits further analysis.

Another putative IFT-particle subunit, DYF-1, was initially identified in *C. elegans*, where it is required for OSM-3-mediated anterograde IFT (Ou et al., 2005a). The *dyf-1* mutant nematode phenocopies the *osm-3* mutant where the distal portion of specific sensory cilia fail to form. Since OSM-3 still enters the sensory cilia of the *dyf-1* mutant but fails to move in a productive manner, DYF-1 is believed to regulate OSM-3 activity. A DYF-1 homologous protein (FAP259, NCBI accession number EDP03884) was identified in the *Chlamydomonas* flagellar proteome though the low number of unique peptides (2) suggests that it might not be present at a stoichiometric ratio relative to other IFT-particle proteins of a similar size (Pazour et al., 2005). Since no OSM-3 kinesin gene has been found in the *Chlamydomonas* genome, it appears that DYF-1 functions in the *Chlamydomonas* flagellum in an OSM-3-independent manner.

Another candidate IFT subunit, DYF-13, is conserved in ciliated organisms and is required for assembly of sensory cilia in *C. elegans* (Starich et al., 1995; Blacque et al., 2005). As observed with qilin, the *dyf-13* mutant phenocopies other IFT mutants and the GFP-tagged protein moves along sensory cilia as if it was associated with the IFT machinery. DYF-13 genes are conserved in ciliated organisms including *Chlamydomonas* (EDO99269) though the protein was not identified in the *Chlamydomonas* flagellar proteome. Absence from the proteomic analysis suggests that flagellar levels of DYF-13 are low and, if present, likely to be sub-stoichiometric to the established A and B proteins.

BBS7 and BBS8 are proteins associated with Bardet–Biedl syndrome (reviewed in Blacque and Leroux, 2005; also see Chapters 2 and 15). Loss of BBS7 or BBS8 in *C. elegans* results in a decoupling of complexes A and B IFT (Ou et al., 2005a). In these mutants, complex A is moved by the heterotrimeric kinesin-2 at slow velocities and only in the middle segment.

Complex B is moved by OSM-3 at faster velocities in both the distal and middle segments. In order to test if the complexes were being physically held together by BBS7/8, *bbs7* and *bbs8* mutants were examined in the IFT-motor mutant backgrounds where either the kinesin-2 or the OSM-3 was no longer present or functional (Pan et al., 2006). When kinesin-2 was absent, both complexes A and B moved along together at the faster rate of OSM-3. When OSM-3 was absent, both complexes moved together slowly at the kinesin-2 rate. Thus, BBS7/8 are not essential for physically holding the complexes together but they do serve to stabilize their association. *BBS7* and *BBS8* homologous genes are present in the *Chlamydomonas* nuclear genome but the proteins (EDP09445 and EDO97459, respectively) were not identified in the flagellar proteome, suggesting that algal flagellar levels are either low or nonexistent. The mammalian BBS7 and BBS8 have been shown to be associated with a complex known as the BBSome that also contains stoichiometric levels of BBS1, BBS2, BBS4, BBS5, and BBS9 (Nachury et al., 2007). GFP-tagged BBS4 was found to move along mammalian primary cilia at IFT velocities, suggesting that the BBSome may represent a rather large IFT cargo. It will be interesting to see if the BBSome is conserved in other ciliated organisms.

III. VISUALIZATION OF IFT

A. Subcellular localization of IFT machinery

1. Punctate flagellar distribution

Early biochemical fractionations revealed that the *Chlamydomonas* IFT motor and particle proteins were present in both cell bodies and excised flagella (Piperno and Mead, 1997; Cole et al., 1998; Pazour et al., 1999; Porter et al., 1999). As shown with anti-IFT52 in Figure 4.5A,

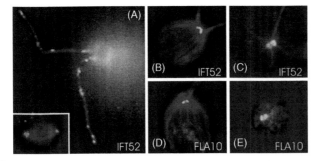

FIGURE 4.5 *IFT proteins are distributed throughout flagella and pooled near the basal bodies. Double immunofluorescence micrographs of IFT proteins (green) and tubulin (red); regions of overlap are yellow. (A) IFT52 staining shows punctate distribution within flagella. (B–E) Most of the IFT52 and FLA10 protein appears to be concentrated in a pool near the basal bodies. (A, inset) During mitosis, IFT52 stays close to the basal bodies as they become the centrioles located at the poles of the mitotic spindle. Reproduced from Deane et al. (2001) with permission from Blackwell Publishing, Oxford, UK.*

immunofluorescence of fixed wild-type cells have typically shown a punctate distribution of the IFT motors and the particles throughout the flagellum (Kozminski et al., 1995; Cole et al., 1998; Deane et al., 2001; Marshall and Rosenbaum, 2001). Immunoelectron microscopy has shown that the flagellar IFT machinery is consistently found between the outer doublet microtubules and the outer membrane, consistent with the current model of IFT where kinesin-2 and dynein 1b are responsible for moving the particles to and from the distal end, respectively (Kozminski et al., 1995; Deane et al., 2001).

2. An IFT pool near the basal bodies

When flagella are removed from *Chlamydomonas*, only a minor fraction (10–20%) of the IFT-particle proteins are typically removed from the total cellular pool. The majority of the IFT machinery is found in a poorly defined pool at the base of the flagellum near the basal bodies (Figure 4.5B–E). The basal body region pooling was first observed with the FLA10 (KHP1) motor subunit of kinesin-2 (Vashishtha et al., 1996) and later observed with the DHC1b motor subunit of dynein 1b (Pazour et al., 1999) and numerous IFT-particle proteins (Cole et al., 1998; Deane et al., 2001; Hou et al., 2007; Qin et al., 2007). Although staining of the motors and the complexes A and B partially overlap in this region, it is notable that the cell body distributions of the three are not completely coincident. This phenomenon is illustrated in Figure 4.6 where staining of the two complex B subunits IFT46 and IFT172 completely overlap while the complex A IFT139 staining only partially overlaps with that of IFT46 and presumably the rest of complex B.

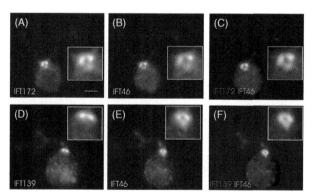

FIGURE 4.6 *Cellular distribution of IFT complex A differs from that of complex B. Wild-type cells were double-labeled either with antibodies to IFT172 (A) and IFT46 (B) or with antibodies to IFT139 (D) and IFT46 (E). The merged images are shown in C and F; the insets show enlargements of the basal body regions. IFT172 usually colocalizes with IFT46. IFT139, however, only partially colocalizes with IFT46 in the anterior part of the basal body region. Reproduced from Hou et al. (2007) with permission of the Rockefeller University Press, New York, NY.*

The basal body region in which the IFT proteins pool may act as a staging area for the IFT machinery and their cargos prior to entry into the flagellum. Immunogold labeling of IFT52 clearly decorated the transition fibers that emanate from the basal bodies and extend to the plasma membrane (Figure 4.7; Deane et al., 2001). It has been hypothesized that the transition fibers may function as part of a flagellar pore where decisions are made as to whether specific proteins will enter or exit the organelle.

3. Spindle pole localization during mitosis and cell division

Just prior to mitosis and cell division, *Chlamydomonas* flagella resorb and the basal bodies migrate to become the centrioles for the mitotic spindles. Much of the IFT machinery appears to co-migrate to the spindle poles such that they never stray far from the basal bodies/centrioles. This was first observed with the FLA10 kinesin-2 IFT motor (Vashishtha et al., 1996), and then later observed with IFT-particle proteins such as IFT52, which is shown in the inset in Figure 4.5A (Deane et al., 2001). It is possible that some of the IFT proteins play functional roles during mitosis and cell division; null mutations in the *IFT88*, *IFT52*, and *FLA10* genes, however, do not slow down mitosis or cell division (Pazour et al., 2000; Deane et al., 2001; Matsuura et al., 2002). It is possible that much of the IFT machinery stays with each basal body/centriole in order to generate a fairly even division of the existing IFT proteins to be inherited by the two daughter cells. This strategy would also force the IFT machinery to be in the right location when the centrioles return to the cell surface to once again function as flagellar basal bodies.

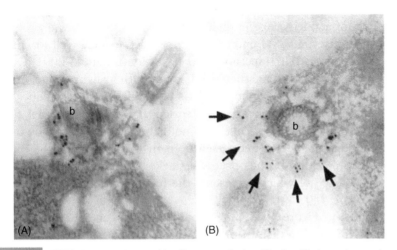

FIGURE 4.7   *IFT52 decorates the transition fibers near the basal bodies. Electron micrographs of silver-enhanced immunogold labeling of IFT52 reveals specific decoration of the transition fibers, particularly where the fibers interact with the membrane. Reproduced from Deane et al. (2001) with permission from Blackwell Publishing, Oxford, UK.*

B. Movements

1. Video-enhanced DIC vs. fluorescence microscopy

IFT was unknown to biology prior to 1991, in part, because video-enhanced DIC microscopy was not developed until 1981 (Allen et al., 1981; Inoue, 1981) and, in part, because the application of video-enhanced DIC microscopy to visualize IFT in other model organisms has proven either difficult or impossible. As an alternative, fluorescently tagged proteins have been used to monitor movement. This was first achieved in *C. elegans* using GFP fused to the nematode KAP and to the IFT52 orthologue known as OSM-6 (Collet et al., 1998; Orozco et al., 1999). Both GFP-tagged proteins were observed to move within chemosensory cilia in the anterograde direction at 0.65 μm/second, confirming that, like their *Chlamydomonas* counterparts, these two nematode proteins were involved with IFT. Since 1999, fluorescent tags have been used to track assorted IFT proteins in a variety of model organisms. Included in this list is *Chlamydomonas*, where the movement of KAP-GFP (kinesin-2) and IFT27-GFP (complex B) have been monitored (Mueller et al., 2005; Qin et al., 2007) (see videos "KAP-GFP" and "IFT27-GFP" at http://www.elsevierdirect.com/companions/9780123708731).

2. Kymography

An important advance in analyzing IFT was the development of kymography to carefully track the movement of many, if not all, IFT particles along the entire length of the organelle. Kymography was first applied to IFT when Piperno and coworkers (Piperno et al., 1998; Iomini et al., 2001) analyzed a new set of *Chlamydomonas* temperature-sensitive flagellar assembly mutants, *fla15–fla18*. To capture the movement of IFT particles in the field of a microscope, a digital camera records frames on the order of tens of milliseconds; 33 and 67 ms/frame have worked well for analysis of *Chlamydomonas* IFT (Piperno et al., 1998; Iomini et al., 2001; Dentler, 2005). Successive frames are then overlaid but with a slight and regular displacement perpendicular to the long axis of the flagellum. To reduce background noise, pixels are averaged and scaled. It is also useful that flagella be paralyzed so that the organelle does not bend or move from side-to-side; motile flagella can be restricted, however, in 0.75% low-melting agarose (Mueller et al., 2005). In the offset and stacked images of flagella, diagonal lines appear, each of which represents the movement of an individual IFT particle (Figure 4.8).

As shown in Figure 4.8A, anterograde particles generate diagonal lines that move toward the distal tip (up and to the right, black arrows) while retrograde particles move in the opposite direction (down and to the right, white arrows). The angle of each track reflects the velocity of that particular particle. Note that most of the anterograde tracks are parallel and, since these particles move slower than the retrograde particles, the anterograde

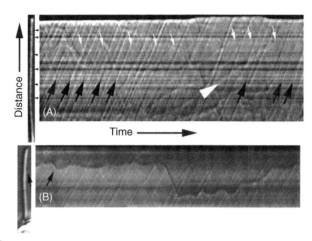

FIGURE 4.8 *Kymographic analysis of IFT-particle movement in* Chlamydomonas. *(A) The left panel shows a single image of a flagellum with the distal tip at the top; IFT-particle bulges are indicated by arrowheads. The panel on the right is a kymogram generated from 600 images taken over 40 seconds. The tracks of the larger anterograde IFT particles (base to tip, black arrows) and the smaller retrograde IFT particles (tip to base, white arrows) are clearly evident. Tracks of very large, slowly and discontinuous moving particles may be extraflagellar particles (white arrowhead). (B) The discontinuous and reversible movement of a large extraflagellar particle (arrows) is superimposed on anterograde and retrograde particle tracks. Reproduced with modification from Dentler (2005) with permission of the Rockefeller University Press, New York, NY.*

tracks are less steep than the retrograde tracks. The width of each track is a measure of the length of the "membrane bulge" observed by the DIC microscopy and, therefore, is believed to directly correlate to the size of that particular IFT particle. Most anterograde particles move continuously to the distal tip, but a few reverse direction before reaching the end. Less frequent than the standard IFT movement is the occasional large track corresponding to large particles, some of which are extraflagellar, that move slowly and often bi-directionally along the long axis of the organelle (Figure 4.8A). Some extraflagellar particles, however, appear to move along established IFT tracks, suggesting that these particular objects are being moved, at least temporarily, by the IFT machinery (Figure 4.8B).

3. The five phases of IFT

Pulse labeling experiments with ^{35}S established that the IFT-particle proteins turn over more rapidly than most other flagellar matrix proteins and more rapidly than salt-extracted axonemal proteins such as inner and outer dynein arms (Piperno and Mead, 1997). These experiments help illustrate a model of IFT where particles are continually turning over in the cell cytoplasm and a significant fraction of particles entering the flagellum contain newly synthesized proteins. Furthermore, the process of IFT can be partitioned into at least five different phases (Iomini et al., 2001) consisting of anterograde particle assembly, anterograde transport, distal tip reorganization, retrograde transport, and finally retrograde particle disassembly or reorganization.

The first phase of IFT occurs at the base of the organelle where anterograde particles assemble and flagellar-bound cargos such as axonemal building blocks are loaded onto the particles. The precise location(s) of the anterograde assembly and cargo loading is unknown but may occur at or near the transition fibers where IFT-particle proteins appear to be docked (Figure 4.7). Indeed, it has been hypothesized that the transition fibers help delineate a flagellar pore which functions to filter out nonflagellar proteins (Deane et al., 2001). Anterograde particle assembly and/or loading of cargos could also occur just inside the flagellum at the transition zone.

The second phase of IFT involves kinesin-2-mediated anterograde transport. During this movement, dynein 1b will be carried by IFT but the retrograde motor is likely to be temporarily inactivated by an unidentified regulator such as a kinase, phosphatase, etc. During this phase IFT cargos such as radial spoke complexes and dynein arms are moved out to the distal tip.

The third phase of IFT involves the release of IFT cargos and the reorganization of IFT particles at the distal tip (Sloboda, 2005; Pedersen et al., 2006). Reorganization of the particles must occur because multiple kymographic studies have shown that wild-type cells exhibit nearly twice as many retrograde events or particles than anterograde (Iomini et al., 2001; Dentler, 2005). With more retrograde than anterograde events, the IFT particles must undergo some sort of downsizing, which could be as simple as anterograde particles dissociating into two smaller, albeit separate, retrograde-bound particles. The particles arriving at the tip might also undergo a more complex rearrangement where individual IFT complexes dissociate and re-associate to form retrograde particles. It is also imagined, but not well documented, that axonemal building blocks are unloaded at the tip while axonemal turnover products are loaded onto the particles. It is possible that some IFT cargos are so weakly associated with the IFT particle that they continually associate and dissociate along the length of the organelle. This could be particularly advantageous for proteins that can enter or exit their site of function from somewhere along the middle of the organelle.

The fourth phase consists of the dynein 1b-mediated retrograde transport of IFT particles and any cargos they may carry. The plus end-directed kinesin-2, for example, requires directed retrograde transport in order to avoid distal-tip accumulation. Axonemal turnover products such as radial spoke complexes are also removed from the flagellum by retrograde transport (Qin et al., 2004).

What happens to the IFT particles during the fifth phase of movement when they reach the base of the organelle? Outside of the organelle, the regular train-like appearance of large IFT particles has not been documented. Furthermore, pulse labeling reveals that at least some of the anterograde particles entering the flagellum contain newly synthesized protein (Piperno and Mead, 1997). Thus, it appears that some sort of disassembly or reorganization of IFT particles must occur once the particles reach the base of the organelle. It should be noted that although kymography allows careful analysis of movement along the organelle, it does not reveal much information at the tip or base. Presently, little is known about how IFT particles are assembled, reorganized and ultimately, disassembled.

It is tempting to think that IFT might vary depending on the net assembly or disassembly of the organelle. Indeed, localization of kinesin-2 and IFT-particle proteins reveals that the short stubs of newly regenerating flagella have more concentrated pools of these proteins than do full-length flagella (Marshall and Rosenbaum, 2001; Deane et al., 2001). However, quantitation of immunofluorescent anti-IFT52 labeling revealed that the total flagellar concentration of IFT-particle proteins was independent of organelle length (Marshall and Rosenbaum, 2001). Moreover, kymographic analysis revealed that the frequency of particles entering and exiting the flagellum does not vary as a function of assembly, steady state, and disassembly of the organelle (Dentler, 2005). Thus, the frequency of IFT events is essentially constant. The higher concentration of IFT particles in regenerating flagella may reflect slower turnaround at the flagellar tip.

IV. FUNCTIONAL ANALYSIS OF *CHLAMYDOMONAS* IFT

A. IFT mutants

1. Necessity for IFT

Much of what is known about IFT function has come from genetic and cell biological analysis of IFT mutants using model organisms such as *C. elegans* and *Chlamydomonas*. Many mutations have resulted in ciliary and flagellar assembly defects (reviewed in Rosenbaum and Witman, 2002; Cole, 2003b; Scholey, 2003; Snell et al., 2004). Thus, an essential role for IFT in the assembly of the organelle appears to be strongly conserved in most ciliated eukaryotic organisms. In select organisms and cell types, however, where the axoneme is assembled within the cell body, the necessity of IFT has been lost. In *Drosophila*, for example, IFT is required for assembly of sensory cilia but is not required for formation of the extremely long sperm flagellum (Han et al., 2003).

2. Kinesin-2 mutants

As detailed above and shown in Table 4.3, temperature-sensitive mutations in the genes encoding the anterograde kinesin motor subunits FLA10 and

Table 4.3 *Chlamydomonas* IFT Mutants

IFT subunit	Mutant strain	Mutation	Flagellar phenotype	References
FLA10	*fla10-1*	N329K	ts assembly; anterograde IFT lost at 32°C	Huang et al. (1977), Adams et al. (1982), Walther et al. (1994)
	fla10-7	–	ts assembly; anterograde IFT lost at 32°C	Huang et al. (1977), Adams et al. (1982)
	fla10-14	E24K, L196F	ts assembly; anterograde IFT lost at 32°C	Adams et al. (1982), Lux and Dutcher (1991)
	fla10-15	–	ts assembly; anterograde IFT lost at 32°C	Adams et al. (1982), Lux and Dutcher (1991)
	fla10-2 (2F10)	Null	Bald	Matsuura et al. (2002)
FLA8	*fla8-1*	F55S	ts assembly; anterograde IFT lost at 32°C	Huang et al. (1977), Adams et al. (1982), Miller et al. (2005)
	fla8-2 (fla1)	E21K	ts assembly; anterograde IFT lost at 32°C	Huang et al. (1977), Adams et al. (1982), Miller et al. (2005)
KAP	*fla3-1*	F753L	ts assembly; anterograde IFT lost at 32°C	Huang et al. (1977), Adams et al. (1982), Mueller et al. (2005)
DHC1b	*dhc1b-1*	Null	Stumpy; no retrograde IFT	Pazour et al. (1999)
	stf1-1	Null	Stumpy; no retrograde IFT	Porter et al. (1999)
	stf1-2	Null	Stumpy; no retrograde IFT	Porter et al. (1999)
D1bLIC	*d1blic*	Null	Variable length; retrograde IFT disrupted	Hou et al. (2004)
LC8	*fla14*	Null	Short; retrograde IFT disrupted	Pazour et al. (1998)
IFT172	*fla11-1*	L1615P	ts assembly; EB1 mislocalized at 32°C	Adams et al. (1982), Pedersen et al. (2005)
IFT88	*ift88-1*	Null	Bald	Pazour et al. (2000)
IFT52	*bld1-1*	Null	Bald	Brazelton et al. (2001), Deane et al. (2001)
	bld1-2 (Y18)	Null	Bald	Brazelton et al. (2001)
IFT46	*ift46-1*	Null	Stumpy	Hou et al. (2007)
	sup_{ift46}1	3′ expression	Variable length; missing outer dynein arms	Hou et al. (2007)

FLA8 and the nonmotor accessory subunit, FLA3 (KAP), allow the cell to assemble full-length flagella at the permissive temperature of 22°C. After incubation of *fla8* and *fla10* mutants at the restrictive temperature of 32°C, IFT ceases and the flagella slowly resorb to form quite short appendages (Adams et al., 1982; Kozminski et al., 1995). Interestingly, relatively moderate levels of external calcium (~300 μM) are sufficient to cause *fla10* mutants at the restrictive temperature to lose flagella by deflagellation instead of resorption (Parker and Quarmby, 2003). A null mutation of the *FLA10* gene results in complete loss of ability to make even short stumps, indicating that *Chlamydomonas* flagellar assembly is absolutely dependent on the heterotrimeric kinesin-2 (Matsuura et al., 2002). This is similar to the mouse where loss of either kinesin motor subunit, KIF3A or KIF3B, results in the complete absence of cilia, a condition associated with *situs inversus* and embryonic lethality (Nonaka et al., 1998; Marszalek et al., 1999). Algal and mammalian IFT contrasts with that of *C. elegans*, however, where the homodimeric OSM-3 kinesin and heterotrimeric kinesin-2 are partially redundant with respect to driving anterograde movement (Snow et al., 2004; Ou et al., 2005a).

Even though the *fla8-1*, *fla8-2*, *fla10-1*, and *fla10-14* temperature-sensitive mutants exhibit no cytokinesis defects, they do have a small but significant increase in the rate of chromosome loss (Miller et al., 2005). Because the null *fla10-2* mutant strain does not demonstrate a similar increase in chromosome loss, it was hypothesized that the temperature-sensitive mutations in either *fla8* or *fla10* result in a gain-of-function activity. Since the *Chlamydomonas* kinesin-2 is associated with the mitotic spindle near the centrioles (Vashishtha et al., 1996; Deane et al., 2001), it is tempting to think that the mutant kinesin poisons the process of chromosomal segregation.

How the point mutations affect the temperature-sensitive kinesin-2 motor domains was explored through suppression analysis (Miller et al., 2005). Interestingly, both *fla8-2* and *fla10-14* contain mutations that result in the substitution of the same conserved glutamate with a lysine residue; a small insertion in the FLA10 sequence offsets the amino acid numbering by three amino acids (Table 4.3). In contrast, the *fla8-1* mutation results in the substitution of a conserved phenylalanine at position 55 with a serine residue. UV-induced mutagenesis was used to generate intragenic revertants that kept their flagella at the restrictive temperature. With all three mutant strains, true revertants were identified where the amino acid sequence returned to the wild-type state. A second mutation resulting in a leucine to phenylalanine substitution at position 196 within the *fla10-14* motor domain was determined to not affect the biological activity of the motor. Mapping the *fla8-1*, *fla8-2*, and *fla10-14* mutations onto the KIF1A crystal structure revealed that these substituted amino acids were quite close to the P-loop of the kinesin active site (Miller et al., 2005). Importantly, all of the amino acid substitutions that yielded intragenic suppressions also mapped close to this same area near the P-loop. Since this region of the kinesin motor domain is poorly understood, it should be interesting to determine how the original mutations and pseudorevertant mutations affect the microtubule-activated enzymatic activities.

Although the nonmotor KAP subunit is thought to assist in mediating cargo attachment to kinesin-2, its actual role is not well understood. Some clues to its function come from analysis of the *fla3-1* temperature-sensitive KAP mutant (Mueller et al., 2005). In the *fla3-1* mutant, for example, kinesin-2 fails to localize in a concentrated pool near the basal bodies. This contrasts with the *fla10-1* mutant where, even though total cellular kinesin-2 is reduced, the motor still localizes to the basal body region (Cole et al., 1998). The KAP subunit, therefore, is needed for efficient targeting of kinesin-2 to this region. Also in contrast to *fla10-1*, the shift to the restrictive temperature does not cause *fla3-1* cells to significantly reduce cellular levels of kinesin-2 (Mueller et al., 2005). Thus, KAP seems to be more important for targeting the motor than for stabilization. As in *fla10-1*, anterograde IFT particles move in *fla3-1* flagella at wild-type velocities at the permissive temperature of 21°C (Iomini et al., 2001; Mueller et al., 2005). Unlike *fla10-1*, however, the *fla3-1* particles appear larger and pause more frequently, which is consistent with the idea that the mutant KAP affects docking between the IFT particles and the kinesin-2 motor.

3. Dynein 1b mutants

As detailed above and shown in Table 4.3, mutations in dynein 1b subunit genes result in disruptions of retrograde IFT, resulting in short or stubby (severely shortened) flagella. Because anterograde IFT continues to operate, the shortened flagella of dynein 1b mutants fill with IFT particles and other debris. There are differences, however, in the phenotypes when specific dynein subunits are absent. Loss of the heavy chain, DHC1b, results in a severe assembly phenotype where very short flagella stubs are filled with IFT particles (Pazour et al., 1999; Porter et al., 1999). Loss of the light intermediate chain, D1bLIC, results in a less severe phenotype where cells display variable length flagella (Hou et al., 2004). Both anterograde and retrograde IFT operate in the *d1blic* mutant, but an apparent defect in cargo binding by dynein 1b results in IFT-particle accumulation at the distal ends of stumpy flagella and randomly along the length of longer flagella (Hou et al., 2004). Assembly of the *d1blic* axoneme appears to be fairly normal since structural features such as dynein arms and radial spokes are present. Loss of the LC8 subunit in the *fla14* strain results in the assembly of short flagella that contain large accumulations of IFT particles (Pazour et al., 1998). Unlike the *d1blic* mutant, *fla14* axonemes are missing a significant fraction of dynein arms and radial spokes, probably because LC8 also is a component of these structures (see Chapters 6 and 7). The above results indicate that defects in the retrograde motor do not immediately disrupt anterograde IFT.

4. IFT complex mutants

The identification and characterization of *Chlamydomonas* complex B mutants has progressed more quickly than that of complex A mutants. It may be that disruptions in complex A genes specifically affect retrograde IFT and, thus, result in less severe assembly defects as observed with *C. elegans* complex A mutants (Perkins et al., 1986; Qin et al., 2001; Bell et al., 2006; Blacque et al., 2006; Efimenko et al., 2006). Little detail is known, however, about the specific functions of the two complexes. Nevertheless, analysis of specific IFT-particle mutants using *Chlamydomonas* and other model organisms has revealed that some of these proteins have distinct roles.

Chlamydomonas mutant strains with disruptions in the *IFT88* (Pazour et al., 2000) and *IFT52* (Brazelton et al., 2001; Deane et al., 2001) genes are completely bald while *IFT46* mutant cells (Hou et al., 2007) have short, stumpy flagella. A partial suppressor mutation of *ift46* ($sup_{ift46}1$) allows assembly of flagella in the absence of full-length IFT46, but the axonemes specifically lack the outer dynein arms, indicating that IFT46 is involved in outer arm transport (Hou et al., 2007). A yeast-based two-hybrid screen found that mammalian IFT46 interacts directly with the mammalian homologue of ODA16 (AAZ77789) (Ahmed et al., 2006), a *Chlamydomonas* protein that, like known IFT-particle proteins, is distributed along the flagella with a concentrated pool at the basal bodies. Interaction of bacterially expressed proteins verified that the *Chlamydomonas* IFT46 and ODA16 interact directly (Ahmed et al., 2006). Since the *ODA16* gene is essential for the assembly of outer arm dyneins (Ahmed and Mitchell, 2005), these two studies suggest that IFT46 interacts with ODA16 in order to achieve efficient axonemal assembly of outer arm dyneins.

The temperature-sensitive flagellar assembly mutant *fla11-1* contains a point mutation in the *IFT172* gene, resulting in the substitution of a proline for a conserved leucine at residue 1615 (Pedersen et al., 2005). Like many other *fla* mutants, *fla11-1* cells are able to assemble flagella at the permissive temperature of 22°C but are unable to assemble or maintain flagella at restrictive temperature of 32°C (Adams et al., 1982). Kymographic analysis performed at 22°C revealed that, unlike IFT phenotypes observed with the anterograde mutants *fla10-1* and *fla8-1*, retrograde IFT in *fla11-1* is reduced both in velocity and frequency. In addition, complex A but not complex B proteins build up near the distal ends of the *fla11-1* flagella after shifting to 32°C (Pedersen et al., 2003). This indicates that IFT172 may play a role in the reorganization of IFT particles that occurs at the distal tip of the organelle. Further involvement of IFT172 with tip dynamics was revealed when EB1 (AAO62368) localization at the distal end was shown to be lost from *fla11-1* flagella at the restrictive temperature (Pedersen et al., 2003, 2005). EB1 is typically found at the plus ends of dynamic microtubules where it is thought to modulate the activity of specific microtubule-binding proteins (reviewed in Vaughan, 2005). First shown to be present at the axonemal tips of *Chlamydomonas* flagella (Pedersen et al., 2003), EB1 has since been observed to similarly decorate the distal ends of fibroblast primary cilia (Schrøder et al., 2007).

IFT172 was shown to interact biochemically with EB1 but only in the absence of other IFT-particle proteins (Pedersen et al., 2005). Immunoprecipitation of EB1 from flagellar extracts also brought down a fraction of IFT172 but did not pull down other complex A or B subunits. Conversely, immunoprecipitation of complex B using anti-IFT74/72 brought down all of complex B including a fraction of IFT172 but did not pull down EB1. Immunoprecipitation of IFT172 brought down EB1 and complex B but not complex A. Bacterially expressed EB1 fused to glutathione S-transferase was used with glutathione resin to also pull down IFT172 but not other complex A or B subunits from flagellar extracts (Pedersen et al., 2005). Interestingly, a single point mutation within EB1 where cysteine at position 228 was substituted with tyrosine blocked the interaction of EB1 with IFT172. Taken together, these results imply that IFT172 can not interact simultaneously with both EB1 and IFT complex B.

Exactly why EB1 localization at the distal tip is dependent on IFT172 is not clear. EB1 does not disappear from flagellar tips when anterograde IFT is shut down in temperature-shifted *fla8-1* or *fla10-1* strains. It is also not clear whether or not the loss of EB1 is directly responsible for the flagellar assembly phenotype. In support of EB1 being ciliogenic, however, is the finding that reduction of EB1 in fibroblasts significantly reduces the number of cilia assembled by the cell (Schrøder et al., 2007).

B. Transport of axonemal building blocks and turnover products

Since IFT is required for ciliary and flagellar assembly and since that assembly occurs at the distal tip, it has been long thought that IFT physically transports axonemal building blocks from the cell body out to the site of assembly. Some axonemal proteins such as dynein arms and radial spokes form multimeric complexes in the cell body before entering the organelle (Fok et al., 1994; Fowkes and Mitchell, 1998; Qin et al., 2004). A requirement of IFT for assembly of dynein onto the axoneme was first demonstrated by Piperno and coworkers with elegant dikaryon rescue experiments where *fla10-1* cells were mated with either *fla10-1 ida4* or *fla10-1 oda2* cells shortly after shifting to the restrictive temperatures (Piperno et al., 1996). After successful mating and cell fusion, the resulting dikaryon cells had four flagella, two that were missing either outer (*oda2*) or inner (*ida4*) dynein arms and two that were wild type with respect to dynein arms. Anterograde IFT, however, was blocked at 32°C in all four flagella because both cells carried the *fla10-1* mutation. Under these circumstances, the inner arm dynein was unable to assemble onto the *ida4* flagella at 32°C, demonstrating a requirement for IFT. In contrast, the outer arm dynein was able to assemble onto the *oda2* flagella at 32°C, indicating that FLA10 was not required for efficient assembly of outer arm dynein onto the axoneme. These results contrast with the analysis of the suppressed *ift46* mutant, $sup_{ift46}1$, where a specific defect in assembly of the outer dynein arms was observed (see above; Hou et al., 2007). In addition, IFT46 appears to interact directly with the outer arm assembly protein ODA16 (Ahmed et al., 2006). It is intriguing to hypothesize that IFT complex B participates in a FLA10-independent event that results in axonemal assembly of outer arms.

A strong case for IFT-mediated radial spoke transport has been made. A subset of the radial spoke complex proteins assemble into a 12S cytoplasmic or cell body complex prior to entry into the flagellum (Qin et al., 2004). The 12S cytoplasmic complex contrasts with a significantly larger 20S radial spoke complex that is constantly being removed from the axoneme in some sort of turnover. These two forms provide easily discernable pre-axonemal (12S) and post-axonemal (20S) fractions of the radial spoke. In *bld1* and *bld2* cells that never assemble axonemes, for example, only the 12S form of the radial spokes can be found in the cell bodies. This contrasts with wild-type cells where both 12S and 20S radial spokes are present in the cell body. To test for the involvement of IFT, Qin et al. (2004) shifted the *fla10-1* mutant to 32°C to selectively ablate anterograde IFT. Under these conditions, the flagellar levels of 12S radial spokes was significantly reduced, demonstrating that IFT was required to transport the 12S radial spokes into the organelle. In related experiments, the cell bodies of *fla14* and *dhc1b* mutants were found to contain only 12S radial spoke complexes. This is particularly significant because *fla14* cells are able to assemble half-length flagella with axonemes that undergo turnover, producing 20S radial spoke complexes within the mutant flagella. None of these 20S complexes, however, are able to make it into the cell body. Thus, retrograde IFT is required to transport the 20S radial spoke complexes out of the flagellum and into the cell body. In summary, anterograde IFT is required to transport 12S radial spoke complexes into the flagellum while retrograde IFT is required to transport 20S radial spoke complexes out of the flagellum.

C. Mediation of flagellar signaling

Cilia and flagella function as sensory organelles in many tissues (reviewed in Pazour and Witman, 2003; Pan et al., 2005; Marshall and Nonaka, 2006; Scholey and Anderson, 2006; Singla and Reiter, 2006; Yoder, 2006).

Specific receptors including the somatostatin receptor (Handel et al., 1999), progesterone receptor (Teilmann et al., 2006), platelet-derived growth-factor receptor α (Schneider et al., 2005), and the sonic hedgehog receptor, Smoothened (Corbit et al., 2005), have all been localized to primary and/or motile cilia. In some cases, the loss of cilia significantly inhibits the cellular response to ligand stimulation (Huangfu et al., 2003; Haycraft et al., 2005; Schneider et al., 2005; Huangfu and Anderson, 2006). In the case of Smoothened, IFT has been shown to be directly involved with the ciliary targeting and placement of the receptor and the downstream targets Gli1 and Gli2 (Haycraft et al., 2005; Liu et al., 2005; May et al., 2005). However, to date *Chlamydomonas* serves as the only model organism where the effect of IFT ablation can be studied in full-length cilia or flagella.

The existence of temperature-sensitive *Chlamydomonas* flagellar assembly mutants such as *fla10-1* and *fla8-1* affords the ability to allow cells to assemble full-length flagella under permissive conditions prior to a temperature shift that inactivates the anterograde IFT motor. For a brief time after IFT disappears, the flagella remain intact. In all other model organisms, complete disruption of anterograde IFT has blocked assembly of the organelle. This phenomenon was exploited to analyze the role of IFT in signal transduction during the mating of *Chlamydomonas* (Wang et al., 2006).

When mating-type *plus* and *minus* gametes come in contact, agglutinin binding (Adair et al., 1983) brings the flagella of opposite mating types together, initiating a signaling pathway that leads to cell fusion (reviewed in Pan and Snell, 2000; see Chapter 12). Flagellar adhesion leads to morphological changes including cell wall release (Goodenough and Weiss, 1975), flagellar tip elongation (Mesland et al., 1980), and erection of an actin-based microvillus fusion organelle (Goodenough and Weiss, 1975). One of the earliest biochemical responses to adhesion is the flagellar activation of a protein tyrosine kinase (PTK) that activates a flagellar-specific cGMP-dependent protein kinase (PKG) (Wang and Snell, 2003; Wang et al., 2006). This leads to activation of a flagellar adenylyl cyclase and elevated levels of flagellar cAMP that stimulate gamete activation (Pasquale and Goodenough, 1987; Pan and Snell, 2000).

Shifting *fla10-1* gametes to the restrictive temperature for 40 minutes was sufficient to ablate IFT but did not significantly affect flagellar length or flagella adhesion (Piperno et al., 1996; Pan and Snell, 2002; Wang and Snell, 2003). However, activation of PTK, elevation of flagellar cAMP, and cell fusion failed to occur. These studies indicate that the downstream portion of the signaling pathway following flagellar adhesion during mating is FLA10-dependent. Furthermore, IFT-particle proteins and the PKG isolated from adhering flagella partially co-sediment in sucrose density gradients, suggesting that the two may become associated upon adhesion (Wang et al., 2006). The role of IFT in *Chlamydomonas* mating is described in more detail in Chapter 12.

D. Association with the cell cycle

The *Chlamydomonas IFT27* gene encodes a Rab-like small G protein (Qin et al., 2007). IFT27 co-localizes and co-fractionates with IFT complex B and, unlike many Rab proteins, does not partition into membrane fractions during extraction and isolation. GFP-tagged IFT27 moves in *Chlamydomonas* flagella at the normal anterograde and retrograde velocities of 1.9 and 3.3 μm/second, respectively (Qin et al., 2007). As expected for an IFT protein, RNA interference (RNAi) of IFT27 resulted in a ciliogenic phenotype. When protein levels were reduced 80%, 40% of the cells had no flagella while another 20% had flagella of 5 μm or less (Qin et al., 2007). As a positive control, RNAi knockdowns of IFT52 showed similar partial assembly defects even though the null mutant of IFT52 (*bld1*) is completely aflagellate (Brazelton et al., 2001; Deane et al., 2001). Western blot analysis of whole cells showed that complexes A and B protein levels were reduced concomitantly with IFT27, whereas knockdowns of IFT52 did not reduce either complex A proteins or IFT27. These results suggest that IFT27 may regulate the expression of the other IFT-particle proteins.

Knockdown of IFT27 in *Chlamydomonas* also resulted in a dramatic defect in cell division (Qin et al., 2007). This was an unexpected result because none of the previously characterized IFT-particle mutants (Table 4.3) had shown defects in cell division. Multiple transformations using two distinct RNAi constructs resulted in many transformed colonies that remained small, turned yellow, and then died, indicating that IFT27 is essential for cell growth. Of the colonies that survived but maintained low levels of IFT27, cell growth was significantly slower than for wild-type cells. When cells were under selective pressure to express IFT27 RNAi, the cells failed to increase density over 2 weeks in liquid media. IFT27 RNAi cells grown without the selective pressure gradually recovered both expression of IFT27 and the ability to divide. In addition to slow growth, IFT27 knockdown generated significant numbers of individual cells displaying classical cell-division defects such as multinucleation, multiple sets of flagella, and large irregular clumping or chaotic division. Although IFT27 is the first IFT-particle protein to be associated with the cell cycle, several other ciliary proteins have also been linked to cell cycle control including the *Chlamydomonas* flagellar NIMA-related kinases, CNK2 and FA2 (Mahjoub et al., 2002, 2004; Bradley and Quarmby, 2005, and see Chapter 3) Associated with flagellar shortening and detachment, CNK2 and FA2 affect cell size and the G2/M transition of the cell cycle, respectively. Indeed, evidence is growing that, in ciliated cells that that must undergo cell division, the cilia appear to regulate progression of the cell cycle (reviewed in Quarmby and Parker, 2005; Pan and Snell, 2007).

V. SUMMARY

First identified in *Chlamydomonas*, IFT is now known to be an ancient process that has been remarkably well conserved in many ciliated eukaryotes (Rosenbaum et al., 1999; Rosenbaum and Witman, 2002; Scholey, 2003; Ou et al., 2007). For example, putative orthologues for every algal IFT protein have been identified in human and mouse. Some model organisms such as *C. elegans* and *Drosophila* appear to be missing specific IFT proteins but most of the machinery appears to be present. In all ciliated organisms IFT appears to be responsible for moving many cargos including soluble and membrane-associated proteins. In addition to ciliogenesis, IFT functions in signaling pathways and as a regulator of the cell cycle. Future experiments should address the molecular basis of IFT function with respect to these various activities. Especially important will be an understanding of how IFT is regulated.

REFERENCES

Adair, W.S., Hwang, C., and Goodenough, U.W. (1983). Identification and visualization of the sexual agglutinin from the mating-type plus flagellar membrane of *Chlamydomonas*. *Cell* **33**, 183–193.

Adams, G.M.W., Huang, B., and Luck, D.J.L. (1982). Temperature-sensitive, assembly defective flagella mutants of *Chlamydomonas reinhardtii*. *Genetics* **100**, 579–586.

Ahmed, N.T. and Mitchell, D.R. (2005). ODA16p, a *Chlamydomonas* flagellar protein needed for dynein assembly. *Mol. Biol. Cell* **16**, 5004–5012.

Ahmed, N.T., Lucker, B.F., Cole, D.G., and Mitchell, D.R. (2006). The *Chlamydomonas* ODA16 locus, needed for outer dynein arm assembly, encodes an intraflagellar transport adaptor. *Mol. Biol. Cell* **17**. abstract no. 1612 (CD-ROM).

Allen, R.D., Allen, N.S., and Travis, J.L. (1981). Video-enhanced contrast, differential interference contrast (AVEC-DIC) microscopy: a new method capable of analyzing microtubule-related motility in the reticulopodial network of *Allogromia laticollaris*. *Cell Motil.* **1**, 291–302.

Baker, S.A., Freeman, K., Luby-Phelps, K., Pazour, G.J., and Besharse, J.C. (2003). IFT20 links kinesin II with a mammalian intraflagellar transport complex that is conserved in motile flagella and sensory cilia. *J. Biol. Chem.* **278**, 34211–34218.

Bell, L.R., Stone, S., Yochem, J., Shaw, J.E., and Herman, R.K. (2006). The molecular identities of the *Caenorhabditis elegans* intraflagellar transport genes *dyf-6*, *daf-10* and *osm-1*. *Genetics* **173**, 1275–1286.

Bernstein, M. and Rosenbaum, J.L. (1994). Kinesin-like proteins in the flagella of *Chlamydomonas*. *Trends Cell Biol.* **4**, 236–240.

Bernstein, M., Beech, P.L., Katz, S.G., and Rosenbaum, J.L. (1994). A new kinesin-like protein (Klp1) localized to a single microtubule of the *Chlamydomonas* flagellum. *J. Cell Biol.* **125**, 1313–1326.

Blacque, O.E. and Leroux, M.R. (2005). Bardet-Biedl syndrome: an emerging pathomechanism of intracellular transport. *Cell. Mol. Life Sci.* **63**, 2145–2161.

Blacque, O.E., Perens, E.A., Boroevich, K.A., Inglis, P.N., Li, C., Warner, A., Khattra, J., Holt, R.A., Ou, G., Mah, A.K., McKay, S.J., Huang, P., Swoboda, P., Jones, S.J., Marra, M.A., Baillie, D.L., Moerman, D.G., Shaham, S., and Leroux, M.R. (2005). Functional genomics of the cilium, a sensory organelle. *Curr. Biol.* **15**, 935–941.

Blacque, O.E., Li, C., Inglis, P.N., Esmail, M.A., Ou, G., Mah, A.K., Baillie, D.L., Scholey, J.M., and Leroux, M.R. (2006). The WD repeat-containing protein IFTA-1 is required for retrograde intraflagellar transport. *Mol. Biol. Cell* **17**, 5053–5062.

Blaineau, C., Tessier, M., Dubessay, P., Tasse, L., Crobu, L., Pages, M., and Bastien, P. (2007). A novel microtubule-depolymerizing kinesin involved in length control of a eukaryotic flagellum. *Curr. Biol.* **17**, 778–782.

Blatch, G.L. and Lassle, M. (1999). The tetratricopeptide repeat: a structural motif mediating protein–protein interactions. *BioEssays* **21**, 932–939.

Bloodgood, R.A. (1992). Directed movements of ciliary and flagellar membrane components: a review. *Biol. Cell* **76**, 291–301.

Bradley, B.A. and Quarmby, L.M. (2005). A NIMA-related kinase, Cnk2p, regulates both flagellar length and cell size in *Chlamydomonas*. *J. Cell Sci.* **118**, 3317–3326.

Brazelton, W.J., Amundsen, C.D., Silflow, C.D., and Lefebvre, P.A. (2001). The *bld1* mutation identifies the *Chlamydomonas osm-6* homolog as a gene required for flagellar assembly. *Curr. Biol.* **11**, 1591–1594.

Cole, D.G. (1999). Kinesin-II, the heteromeric kinesin. *Cell Mol. Life Sci.* **56**, 217–226.

Cole, D.G. (2003a). The intraflagellar transport machinery of *Chlamydomonas reinhardtii*. *Traffic* **4**, 435–442.

Cole, D.G. (2003b). Intraflagellar transport in the unicellular green alga, *Chlamydomonas reinhardtii*. *Protist* **154**, 181–191.

Cole, D.G. (2005). Intraflagellar transport: keeping the motors coordinated. *Curr. Biol.* **15**, R798–R801.

Cole, D.G., Cande, W.Z., Baskin, R.J., Skoufias, D.A., Hogan, C.J., and Scholey, J.M. (1992). Isolation of a sea urchin egg kinesin-related protein using peptide antibodies. *J. Cell Sci.* **101**, 291–301.

Cole, D.G., Chinn, S.W., Wedaman, K.P., Hall, K., Vuong, T., and Scholey, J.M. (1993). Novel heterotrimeric kinesin-related protein purified from sea urchin eggs. *Nature* **366**, 268–270.

Cole, D.G., Diener, D.R., Himelblau, A.L., Beech, P.L., Fuster, J.C., and Rosenbaum, J.L. (1998). *Chlamydomonas* kinesin-II-dependent intraflagellar transport (IFT): IFT particles contain proteins required for ciliary assembly in *Caenorhabditis elegans* sensory neurons. *J. Cell Biol.* **141**, 993–1008.

Collet, J., Spike, C.A., Lundquist, E.A., Shaw, J.E., and Herman, R.K. (1998). Analysis of *osm-6*, a gene that affects sensory cilium structure and sensory neuron function in *Caenorhabditis elegans*. *Genetics* **148**, 187–200.

Corbit, K.C., Aanstad, P., Singla, V., Norman, A.R., Stainier, D.Y., and Reiter, J.F. (2005). Vertebrate smoothened functions at the primary cilium. *Nature* **437**, 1018–1021.

Dagenbach, E.M. and Endow, S.A. (2004). A new kinesin tree. *J. Cell Sci.* **117**, 3–7.

Das, A.K., Cohen, P.W., and Barford, D. (1998). The structure of the tetratricopeptide repeats of protein phosphatase 5: implications for TPR-mediated protein–protein interactions. *EMBO J.* **17**, 1192–1199.

Deane, J.A., Cole, D.G., Seeley, E.S., Diener, D.R., and Rosenbaum, J.L. (2001). Localization of intraflagellar transport protein IFT52 identifies basal body transitional fibers as the docking site for IFT particles. *Curr. Biol.* **11**, 1586–1590.

Dentler, W.L. (1980). Structures linking the tips of ciliary and flagellar microtubules to the membrane. *J. Cell Sci.* **42**, 207–220.

Dentler, W. (2005). Intraflagellar transport (IFT) during assembly and disassembly of *Chlamydomonas* flagella. *J. Cell Biol.* **170**, 649–659.

Dutcher, S.K. (1995). Flagellar assembly in two hundred and fifty easy-to-follow steps. *Trends Genet.* **11**, 398–404.

Dymek, E.E., Goduti, D., Kramer, T., and Smith, E.F. (2006). A kinesin-like calmodulin-binding protein in *Chlamydomonas*: evidence for a role in cell division and flagellar functions. *J. Cell Sci.* **119**, 3107–3116.

Efimenko, E., Blacque, O.E., Ou, G., Haycraft, C.J., Yoder, B.K., Scholey, J.M., Leroux, M.R., and Swoboda, P. (2006). *Caenorhabditis elegans* DYF-2, an orthologue of human WDR19, is a component of the intraflagellar transport machinery in sensory cilia. *Mol. Biol. Cell* **17**, 4801–4811.

Fok, A.K., Wang, H., Katayama, A., Aihara, M.S., and Allen, R.D. (1994). 22S axonemal dynein is preassembled and functional prior to being transported to and attached on the axonemes. *Cell Motil. Cytoskeleton* **29**, 215–224.

Fowkes, M.E. and Mitchell, D.R. (1998). The role of preassembled cytoplasmic complexes in assembly of flagellar dynein subunits. *Mol. Biol. Cell* **9**, 2337–2347.

Fox, L.A., Sawin, K.E., and Sale, W.S. (1994). Kinesin-related proteins in eukaryotic flagella. *J. Cell Sci.* **107**, 1545–1550.

Goodenough, U.W. and Weiss, R.L. (1975). Gametic differentiation in *Chlamydomonas reinhardtii.* III. Cell wall lysis and microfilament-associated mating structure activation in wild-type and mutant strains. *J. Cell Biol.* **67**, 623–637.

Goodson, H.V., Kang, S.J., and Endow, S.A. (1994). Molecular phylogeny of the kinesin family of microtubule motor proteins. *J. Cell Sci.* **107**, 1875–1884.

Grissom, P.M., Vaisberg, E.A., and McIntosh, J.R. (2002). Identification of a novel light intermediate chain (D2LIC) for mammalian cytoplasmic dynein 2. *Mol. Biol. Cell* **13**, 817–829.

Grosshans, B.L., Ortiz, D., and Novick, P. (2006). Rabs and their effectors: achieving specificity in membrane traffic. *Proc. Natl. Acad. Sci. U.S.A.* **103**, 11821–11827.

Han, Y.G., Kwok, B.H., and Kernan, M.J. (2003). Intraflagellar transport is required in *Drosophila* to differentiate sensory cilia but not sperm. *Curr. Biol.* **13**, 1679–1686.

Handel, M., Schulz, S., Stanarius, A., Schreff, M., Erdtmann-Vourliotis, M., Schmidt, H., Wolf, G., and Hollt, V. (1999). Selective targeting of somatostatin receptor 3 to neuronal cilia. *Neuroscience* **89**, 909–926.

Haycraft, C.J., Banizs, B., Aydin-Son, Y., Zhang, Q., Michaud, E.J., and Yoder, B.K. (2005). Gli2 and Gli3 localize to cilia and require the intraflagellar transport protein polaris for processing and function. *PLoS Genet.* **1**, e53.

Hou, Y., Pazour, G.J., and Witman, G.B. (2004). A dynein light intermediate chain, D1bLIC, is required for retrograde intraflagellar transport. *Mol. Biol. Cell* **15**, 4382–4394.

Hou, Y., Qin, H., Follit, J.A., Pazour, G.J., Rosenbaum, J.L., and Witman, G.B. (2007). Functional analysis of an individual IFT protein: IFT46 is required for transport of outer dynein arms into flagella. *J. Cell Biol.* **176**, 653–665.

Huang, B., Rifkin, M.R., and Luck, D.J.L. (1977). Temperature-sensitive mutations affecting flagellar assembly and function in *Chlamydomonas reinhardtii*. *J. Cell Biol.* **72**, 67–85.

Huangfu, D. and Anderson, K.V. (2006). Signaling from Smo to Ci/Gli: conservation and divergence of Hedgehog pathways from *Drosophila* to vertebrates. *Development* **133**, 3–14.

Huangfu, D., Liu, A., Rakeman, A.S., Murcia, N.S., Niswander, L., and Anderson, K.V. (2003). Hedgehog signalling in the mouse requires intraflagellar transport proteins. *Nature* **426**, 83–87.

Inoue, S. (1981). Video image processing greatly enhances contrast, quality, and speed in polarization-based microscopy. *J. Cell Biol.* **89**, 346–356.

Insinna, C., Pathak, N., Perkins, B., Drummond, I., and Besharse, J.C. (2008). The homodimeric kinesin, Kif17, is essential for vertebrate photoreceptor sensory outer segment development. *Dev. Biol.* **316**, 160–170.

Iomini, C., Babaev-Khaimov, V., Sassaroli, M., and Piperno, G. (2001). Protein particles in *Chlamydomonas* flagella undergo a transport cycle consisting of four phases. *J. Cell Biol.* **153**, 13–24.

Jenkins, P.M., Hurd, T.W., Zhang, L., McEwen, D.P., Brown, R.L., Margolis, B., Verhey, K.J., and Martens, J.R. (2006). Ciliary targeting of olfactory CNG channels requires the CNGB1b subunit and the kinesin-2 motor protein, KIF17. *Curr. Biol.* **16**, 1211–1216.

Johnson, K.A. and Rosenbaum, J.L. (1992). Polarity of flagellar assembly in *Chlamydomonas*. *J. Cell Biol.* **119**, 1605–1611.

Johnson, K.A. and Rosenbaum, J.L. (1993). Flagellar regeneration in *Chlamydomonas*: a model system for studying organelle assembly. *Trends Cell Biol.* **3**, 156–161.

Johnson, K.A., Haas, M.A., and Rosenbaum, J.L. (1994). Localization of a kinesin-related protein to the central pair apparatus of the *Chlamydomonas reinhardtii* flagellum. *J. Cell Sci.* **107**, 1551–1556.

King, S.M. and Patel-King, R.S. (1995). The $M(r) = 8,000$ and 11,000 outer arm dynein light chains from *Chlamydomonas* flagella have cytoplasmic homologues. *J. Biol. Chem.* **270**, 11445–11452.

King, S.M., Barbarese, E., Dillman, J.F., III, Patel-King, R.S., Carson, J.H., and Pfister, K.K. (1996). Brain cytoplasmic and flagellar outer arm dyneins share a highly conserved Mr 8,000 light chain. *J. Biol. Chem.* **271**, 19358–19366.

Kozminski, K.G., Johnson, K.A., Forscher, P., and Rosenbaum, J.L. (1993). A motility in the eukaryotic flagellum unrelated to flagellar beating. *Proc. Natl. Acad. Sci. U. S. A.* **90**, 5519–5523.

Kozminski, K.G., Beech, P.L., and Rosenbaum, J.L. (1995). The *Chlamydomonas* kinesin-like protein FLA10 is involved in motility associated with the flagellar membrane. *J. Cell Biol.* **131**, 1517–1527.

Lamb, J.R., Tugendreich, S., and Hieter, P. (1995). Tetratricopeptide repeat interactions: to TPR or not to TPR? *Trends Biochem. Sci.* **20**, 257–259.

Lawrence, C.J., Dawe, R.K., Christie, K.R., Cleveland, D.W., Dawson, S.C., Endow, S.A., Goldstein, L.S., Goodson, H.V., Hirokawa, N., Howard, J., Malmberg, R.L., McIntosh, J.R., Miki, H., Mitchison, T.J., Okada, Y., Reddy, A.S., Saxton, W.M., Schliwa, M., Scholey, J.M., Vale, R.D., Walczak, C.E., and Wordeman, L. (2004). A standardized kinesin nomenclature. *J. Cell Biol.* **167**, 19–22.

Li, D. and Roberts, R. (2001). WD-repeat proteins: structure characteristics, biological function, and their involvement in human diseases. *Cell Mol. Life Sci.* **58**, 2085–2097.

Liu, A., Wang, B., and Niwander, L.A. (2005). Mouse intraflagellar transport proteins regulate both the activator and repressor functions of Gli transcription factors. *Development* **132**, 3103–3111.

Lucker, B.F., Behal, R.H., Qin, H., Siron, L.C., Taggart, W.D., Rosenbaum, J.L., and Cole, D.G. (2005). Characterization of the intraflagellar transport complex B core: direct interaction of the IFT81 and IFT74/72 subunits. *J. Biol. Chem.* **280**, 27688–27696.

Lucker, B.F., Miller, M.S., Qin, H., Blackmarr, P., Ferrell, M., Rosenbaum, J.L., and Cole, D.G. (2006). Architectural analysis of the *Chlamydomonas* intraflagellar transport complex B. *Mol. Biol. Cell* **17**. abstract no. 1611 (CD-ROM)

Lupas, A., Van Dyke, M., and Stock, J. (1991). Predicting coiled coils from protein sequences. *Science* **252**, 1162–1164.

Lux, F.G. and Dutcher, S.K. (1991). Genetic interactions at the *FLA10* locus: suppressors and synthetic phenotypes that affect the cell cycle and flagellar function in *Chlamydomonas reinhardtii*. *Genetics* **128**, 549–561.

Mahjoub, M.R., Montpetit, B., Zhao, L., Finst, R.J., Goh, B., Kim, A.C., and Quarmby, L.M. (2002). The *FA2* gene of *Chlamydomonas* encodes a NIMA family kinase with roles in cell cycle progression and microtubule severing during deflagellation. *J. Cell Sci.* **115**, 1759–1768.

Mahjoub, M.R., Qasim Rasi, M., and Quarmby, L.M. (2004). A NIMA-related kinase, Fa2p, localizes to a novel site in the proximal cilia of *Chlamydomonas* and mouse kidney cells. *Mol. Biol. Cell* **15**, 5172–5186.

Marshall, W.F. (2004). Human cilia proteome contains homolog of zebrafish polycystic kidney disease gene qilin. *Curr. Biol.* **14**, R913–R914.

Marshall, W.F. and Nonaka, S. (2006). Cilia: tuning in to the cell's antenna. *Curr. Biol.* **16**, R604–R614.

Marshall, W.F. and Rosenbaum, J.L. (2001). Intraflagellar transport balances continuous turnover of outer doublet microtubules: implications for flagellar length control. *J. Cell Biol.* **155**, 405–414.

Marshall, W.F., Qin, H., Rodrigo Brenni, M., and Rosenbaum, J.L. (2005). Flagellar length control system: testing a simple model based on intraflagellar transport and turnover. *Mol. Biol. Cell* **16**, 270–278.

Marszalek, J.R., Ruiz-Lozano, P., Roberts, E., Chien, K.R., and Goldstein, L.S. (1999). Situs inversus and embryonic ciliary morphogenesis defects in mouse mutants lacking the KIF3A subunit of kinesin-II. *Proc. Natl. Acad. Sci. U. S. A.* **96**, 5043–5048.

Matsuura, K., Lefebvre, P.A., Kamiya, R., and Hirono, M. (2002). Kinesin-II is not essential for mitosis and cell growth in *Chlamydomonas*. *Cell Motil. Cytoskeleton* **52**, 195–201.

May, S.R., Ashique, A.M., Karlen, M., Wang, B., Shen, Y., Zarbalis, K., Reiter, J., Ericson, J., and Peterson, A.S. (2005). Loss of the retrograde motor for IFT disrupts localization of Smo to cilia and prevents the expression of both activator and repressor functions of Gli. *Dev. Biol.* **287**, 378–389.

Mesland, D.A., Hoffman, J.L., Caligor, E., and Goodenough, U.W. (1980). Flagellar tip activation stimulated by membrane adhesions in *Chlamydomonas* gametes. *J. Cell Biol.* **84**, 599–617.

Miller, M.S., Esparza, J.M., Lippa, A.M., Lux, F.G., III, Cole, D.G., and Dutcher, S.K. (2005). Mutant kinesin-2 motor subunits increase chromosome loss. *Mol. Biol. Cell* **16**, 3810–3820.

Mueller, J., Perrone, C.A., Bower, R., Cole, D.G., and Porter, M.E. (2005). The FLA3 KAP subunit is required for localization of kinesin-2 to the site of flagellar assembly and processive anterograde intraflagellar transport. *Mol. Biol. Cell* **16**, 1341–1354.

Nachury, M.V., Loktev, A.V., Zhang, Q., Westlake, C.J., Peranen, J., Merdes, A., Slusarski, D.C., Scheller, R.H., Bazan, J.F., Sheffield, V.C., and Jackson, P.K. (2007). A core complex of BBS proteins cooperates with the GTPase Rab8 to promote ciliary membrane biogenesis. *Cell* **129**, 1201–1213.

Nonaka, S., Tanaka, Y., Okada, Y., Takeda, S., Harada, A., Kanai, Y., Kido, M., and Hirokawa, N. (1998). Randomization of left-right asymmetry due to loss of nodal cilia generating leftward flow of extraembryonic fluid in mice lacking KIF3B motor protein. *Cell* **95**, 829–837.

Orozco, J.T., Wedaman, K.P., Signor, D., Brown, H., Rose, L., and Scholey, J.M. (1999). Movement of motor and cargo along cilia. *Nature* **398**, 674.

Ostrowski, L.E., Blackburn, K., Radde, K.M., Moyer, M.B., Schlatzer, D.M., Moseley, A., and Boucher, R.C. (2002). A proteomic analysis of human cilia: identification of novel components. *Mol. Cell Proteomics* **1**, 451–465.

Ou, G., Blacque, O.E., Snow, J.J., Leroux, M.R., and Scholey, J.M. (2005a). Functional coordination of intraflagellar transport motors. *Nature* **436**, 583–587.

Ou, G., Qin, H., Rosenbaum, J.L., and Scholey, J.M. (2005b). The PKD protein qilin undergoes intraflagellar transport. *Curr. Biol.* **15**, R410–R411.

Ou, G., Koga, M., Blacque, O.E., Murayama, T., Ohshima, Y., Schafer, J.C., Li, C., Yoder, B.K., Leroux, M.R., and Scholey, J.M. (2007). Sensory ciliogenesis in *Caenorhabditis elegans:* Assignment of IFT components into distinct modules based on transport and phenotypic profiles. *Mol. Biol. Cell* **18**, 1554–1569.

Pan, J. and Snell, W.J. (2000). Signal transduction during fertilization in the unicellular green alga, *Chlamydomonas*. *Curr. Opin. Microbiol.* **3**, 596–602.

Pan, J. and Snell, W.J. (2002). Kinesin-II is required for flagellar sensory transduction during fertilization in *Chlamydomonas*. *Mol. Biol. Cell* **13**, 1417–1426.

Pan, J. and Snell, W. (2007). The primary cilium: keeper of the key to cell division. *Cell* **129**, 1255–1257.

Pan, J., Wang, Q., and Snell, W.J. (2005). Cilium-generated signaling and cilia-related disorders. *Lab. Invest.* **85**, 452–463.

Pan, X., Ou, G., Civelekoglu-Scholey, G., Blacque, O.E., Endres, N.F., Tao, L., Mogilner, A., Leroux, M.R., Vale, R.D., and Scholey, J.M. (2006). Mechanism of transport of IFT particles in *C. elegans* cilia by the concerted action of kinesin-II and OSM-3 motors. *J. Cell Biol.* **174**, 1035–1045.

Parker, J.D. and Quarmby, L.M. (2003). *Chlamydomonas fla* mutants reveal a link between deflagellation and intraflagellar transport. *B.M.C. Cell Biol.* **4**, 11.

Pasquale, S.M. and Goodenough, U.W. (1987). Cyclic AMP functions as a primary sexual signal in gametes of *Chlamydomonas reinhardtii*. *J. Cell Biol.* **105**, 2279–2292.

Pazour, G.J. and Rosenbaum, J.L. (2002). Intraflagellar transport and cilia-dependent diseases. *Trends Cell Biol.* **12**, 551–555.

Pazour, G.J. and Witman, G.B. (2003). The vertebrate primary cilium is a sensory organelle. *Curr. Opin. Cell Biol.* **15**, 105–110.

Pazour, G.J., Wilkerson, C.G., and Witman, G.B. (1998). A dynein light chain is essential for the retrograde particle movement of intraflagellar transport (IFT). *J. Cell Biol.* **141**, 979–992.

Pazour, G.J., Dickert, B.L., and Witman, G.B. (1999). The DHC1b (DHC2) isoform of cytoplasmic dynein is required for flagellar assembly. *J. Cell Biol.* **144**, 473–481.

Pazour, G.J., Dickert, B.L., Vucica, Y., Seeley, E.S., Rosenbaum, J.L., Witman, G.B., and Cole, D.G. (2000). *Chlamydomonas IFT88* and its mouse homologue, polycystic kidney disease gene *tg737*, are required for assembly of cilia and flagella. *J. Cell Biol.* **151**, 709–718.

Pazour, G.J., Baker, S.A., Deane, J.A., Cole, D.G., Dickert, B.L., Rosenbaum, J.L., Witman, G.B., and Besharse, J.C. (2002). The intraflagellar transport protein, IFT88, is essential for vertebrate photoreceptor assembly and maintenance. *J. Cell Biol.* **157**, 103–113.

Pazour, G.J., Agrin, N., Leszyk, J., and Witman, G.B. (2005). Proteomic analysis of a eukaryotic cilium. *J. Cell Biol.* **170**, 103–113.

Pedersen, L.B., Geimer, S., Sloboda, R.D., and Rosenbaum, J.L. (2003). The microtubule plus end-tracking protein EB1 is localized to the flagellar tip and basal bodies in *Chlamydomonas reinhardtii*. *Curr. Biol.* **13**, 1969–1974.

Pedersen, L.B., Miller, M.S., Geimer, S., Leitch, J.M., Rosenbaum, J.L., and Cole, D.G. (2005). *Chlamydomonas* IFT172 is encoded by FLA11, interacts with CrEB1, and regulates IFT at the flagellar tip. *Curr. Biol.* **15**, 262–266.

Pedersen, L.B., Geimer, S., and Rosenbaum, J.L. (2006). Dissecting the molecular mechanisms of intraflagellar transport in *Chlamydomonas*. *Curr. Biol.* **16**, 450–459.

Perkins, L.A., Hedgecock, E.M., Thomson, J.N., and Culotti, J.G. (1986). Mutant sensory cilia in the nematode *Caenorhabditis elegans*. *Dev. Biol.* **117**, 456–487.

Perrone, C.A., Tritschler, D., Taulman, P., Bower, R., Yoder, B.K., and Porter, M.E. (2003). A novel dynein light intermediate chain colocalizes with the retrograde motor for intraflagellar transport at sites of axoneme assembly in *Chlamydomonas* and mammalian cells. *Mol. Biol. Cell* **14**, 2041–2056.

Pfister, K.K., Fay, R.B., and Witman, G.B. (1982). Purification and polypeptide composition of dynein ATPases from *Chlamydomonas* flagella. *Cell Motil.* **2**, 525–547.

Piperno, G. and Luck, D.J. (1979). Axonemal adenosine triphosphatases from flagella of *Chlamydomonas reinhardtii*. Purification of two dyneins. *J. Biol. Chem.* **254**, 3084–3090.

Piperno, G. and Mead, K. (1997). Transport of a novel complex in the cytoplasmic matrix of *Chlamydomonas* flagella. *Proc. Natl. Acad. Sci. U. S. A.* **94**, 4457–4462.

Piperno, G., Mead, K., and Henderson, S. (1996). Inner dynein arms but not outer dynein arms require the activity of kinesin homologue protein KHP1 (FLA10) to reach the distal part of flagella in *Chlamydomonas*. *J. Cell Biol.* **133**, 371–379.

Piperno, G., Siuda, E., Henderson, S., Segil, M., Vaananen, H., and Sassaroli, M. (1998). Distinct mutants of retrograde intraflagellar transport (IFT) share similar morphological and molecular defects. *J. Cell Biol.* **143**, 1591–1601.

Porter, M.E., Bower, R., Knott, J.A., Byrd, P., and Dentler, W. (1999). Cytoplasmic dynein heavy chain 1b is required for flagellar assembly in *Chlamydomonas*. *Mol. Biol. Cell* **10**, 693–712.

Qin, H., Rosenbaum, J.L., and Barr, M.M. (2001). An autosomal recessive polycystic kidney disease gene homolog is involved in intraflagellar transport in *C. elegans* ciliated sensory neurons. *Curr. Biol.* **11**, 457–461.

Qin, H., Diener, D.R., Geimer, S., Cole, D.G., and Rosenbaum, J.L. (2004). Intraflagellar transport (IFT) cargo: IFT transports flagellar precursors to the tip and turnover products to the cell body. *J. Cell Biol.* **164**, 255–266.

Qin, H., Wang, Z., Diener, D., and Rosenbaum, J. (2007). Intraflagellar transport protein 27 is a small G protein involved in cell-cycle control. *Curr. Biol.* **17**, 193–202.

Quarmby, L.M. and Parker, J.D. (2005). Cilia and the cell cycle? *J. Cell Biol.* **169**, 707–710.

Richardson, D.N., Simmons, M.P., and Reddy, A.S. (2006). Comprehensive comparative analysis of kinesins in photosynthetic eukaryotes. *B.M.C. Genomics* **7**, 18.

Ringo, D.L. (1967). Flagellar motion and fine structure of the flagellar apparatus in *Chlamydomonas*. *J. Cell Biol.* **33**, 543–571.

Rompolas, P., Pedersen, L.B., Patel-King, R.S., and King, S.M. (2007). *Chlamydomonas* FAP133 is a dynein intermediate chain associated with the retrograde intraflagellar transport motor. *J. Cell Sci.* **120**, 3653–3665.

Rosenbaum, J.L. and Witman, G.B. (2002). Intraflagellar transport. *Nat. Rev. Mol. Cell Biol.* **3**, 813–825.

Rosenbaum, J.L., Cole, D.G., and Diener, D.R. (1999). Intraflagellar transport: the eyes have it. *J. Cell Biol.* **144**, 385–388.

Sattler, C.A. and Staehelin, L.A. (1974). Ciliary membrane differentiations in *Tetrahymena pyriformis*. *Tetrahymena* has four types of cilia. *J. Cell Biol.* **62**, 473–490.

Sawin, K.E., Mitchison, T.J., and Wordeman, L.G. (1992). Evidence for kinesin-related proteins in the mitotic apparatus using peptide antibodies. *J. Cell Sci.* **101**, 303–313.

Scheufler, C., Brinker, A., Bourenkov, G., Pegoraro, S., Moroder, L., Bartunik, H., Hartl, F.U., and Moarefi, I. (2000). Structure of TPR domain-peptide complexes: critical elements in the assembly of the Hsp70-Hsp90 multichaperone machine. *Cell* **101**, 199–210.

Schneider, L., Clement, C.A., Teilmann, S.C., Pazour, G.J., Hoffmann, E.K., Satir, P., and Christensen, S.T. (2005). PDGFRαα signaling is regulated through the primary cilium in fibroblasts. *Curr. Biol.* **15**, 1861–1866.

Scholey, J.M. (2003). Intraflagellar transport. *Annu. Rev. Cell Dev. Biol.* **19**, 423–443.

Scholey, J.M. and Anderson, K.V. (2006). Intraflagellar transport and cilium-based signaling. *Cell* **125**, 439–442.

Schrøder, J.M., Schneider, L., Christensen, S.T., and Pedersen, L.B. (2007). EB1 is required for primary cilia assembly in fibroblasts. *Curr. Biol.* **17**, 1134–1139.

Setou, M., Nakagawa, T., Seog, D.H., and Hirokawa, N. (2000). Kinesin superfamily motor protein KIF17 and mLin-10 in NMDA receptor-containing vesicle transport. *Science* **288**, 1796–1802.

Signor, D., Wedaman, K.P., Rose, L.S., and Scholey, J.M. (1999). Two heteromeric kinesin complexes in chemosensory neurons and sensory cilia of *Caenorhabditis elegans*. *Mol. Biol. Cell* **10**, 345–360.

Singla, V. and Reiter, J.F. (2006). The primary cilium as the cell's antenna: signaling at a sensory organelle. *Science* **313**, 629–633.

Sloboda, R.D. (2005). Intraflagellar transport and the flagellar tip complex. *J. Cell Biochem.* **94**, 266–272.

Sloboda, R.D., and Howard, L. (2007). Localization of EB1, IFT polypeptides, and kinesin-2 in *Chlamydomonas* flagellar axonemes via immunogold scanning electron microscopy. *Cell Motil. Cytoskeleton* **64**, 446–460.

Smith, T.F., Gaitatzes, C., Saxena, K., and Neer, E.J. (1999). The WD repeat: a common architecture for diverse functions. *Trends Biochem. Sci.* **24**, 181–185.

Snell, W.J., Pan, J., and Wang, Q. (2004). Cilia and flagella revealed: from flagellar assembly in *Chlamydomonas* to human obesity disorders. *Cell* **117**, 693–697.

Snow, J.J., Ou, G., Gunnarson, A.L., Walker, M.R., Zhou, H.M., Brust-Mascher, I., and Scholey, J.M. (2004). Two anterograde intraflagellar transport motors cooperate to build sensory cilia on *C. elegans* neurons. *Nat. Cell Biol.* **6**, 1109–1113.

Starich, T.A., Herman, R.K., Kari, C.K., Yeh, W.H., Schackwitz, W.S., Schuyler, M.W., Collet, J., Thomas, J.H., and Riddle, D.L. (1995). Mutations affecting the chemosensory neurons of *Caenorhabditis elegans*. *Genetics* **139**, 171–188.

Stolc, V., Samanta, M.P., Tongprasit, W., and Marshall, W.F. (2005). Genome-wide transcriptional analysis of flagellar regeneration in *Chlamydomonas reinhardtii* identifies orthologs of ciliary disease. *Proc. Natl. Acad. Sci. U. S. A.* **102**, 3703–3707.

Sun, Z., Amsterdam, A., Pazour, G.J., Cole, D.G., Miller, M.S., and Hopkins, N. (2004). A genetic screen in zebrafish identifies cilia genes as a principal cause of cystic kidney. *Development* **131**, 4085–4093.

Teilmann, S.C., Clement, C.A., Thorup, J., Byskov, A.G., and Christensen, S.T. (2006). Expression and localization of the progesterone receptor in mouse and human reproductive organs. *J. Endocrinol.* **191**, 525–535.

Vale, R.D., Reese, T.S., and Sheetz, M.P. (1985). Identification of a novel force-generating protein, kinesin, involved in microtubule-based motility. *Cell* **42**, 39–50.

Vashishtha, M., Walther, Z., and Hall, J.L. (1996). The kinesin-homologous protein encoded by the *Chlamydomonas FLA10* gene is associated with basal bodies and centrioles. *J. Cell Sci.* **109**, 541–549.

Vaughan, K.T. (2005). TIP maker and TIP marker; EB1 as a master controller of microtubule plus ends. *J. Cell Biol.* **171**, 197–200.

Wall, M.A., Coleman, D.E., Lee, E., Iniguez-Lluhi, J.A., Posner, B.A., Gilman, A.G., and Sprang, S.R. (1995). The structure of the G protein heterotrimer G_i alpha 1 beta 1 gamma 2. *Cell* **83**, 1047–1058.

Walther, Z., Vashishtha, M., and Hall, J.L. (1994). The *Chlamydomonas FLA10* gene encodes a novel kinesin-homologous protein. *J. Cell Biol.* **126**, 175–188.

Wang, Q. and Snell, W.J. (2003). Flagellar adhesion between mating type plus and mating type minus gametes activates a flagellar protein-tyrosine kinase during fertilization in *Chlamydomonas*. *J. Biol. Chem.* **278**, 32936–32942.

Wang, Q., Pan, J., and Snell, W.J. (2006). Intraflagellar transport particles participate directly in cilium-generated signaling in *Chlamydomonas*. *Cell* **125**, 549–562.

Wickstead, B. and Gull, K. (2006). A "holistic" kinesin phylogeny reveals new kinesin families and predicts protein functions. *Mol. Biol. Cell* **17**, 1734–1743.

Yamazaki, H., Nakata, T., Okada, Y., and Hirokawa, N. (1995). KIF3A/B: a heterodimeric kinesin superfamily protein that works as a microtubule plus end-directed motor for membrane organelle transport. *J. Cell Biol.* **130**, 1387–1399.

Yamazaki, H., Nakata, T., Okada, Y., and Hirokawa, N. (1996). Cloning and characterization of KAP3: a novel kinesin superfamily-associated protein of KIF3A/3B. *Proc. Natl. Acad. Sci. U.S.A.* **93**, 8443–8448.

Yoder, B.K. (2006). More than just the postal service: novel roles for IFT proteins in signal transduction. *Dev. Cell* **10**, 541–542.

Yu, L., Gaitatzes, C., Neer, E., and Smith, T.F. (2000). Thirty-plus functional families from a single motif. *Protein Sci.* **9**, 2470–2476.

CHAPTER 5

Flagellar Length Control

Paul A. Lefebvre
Department of Plant Biology, University of Minnesota,
St. Paul, Minnesota, USA

CHAPTER CONTENTS

I. INTRODUCTION

Chlamydomonas cells need efficient flagellar motility to perform the three key tasks facing a photosynthetic microalga: finding optimal light conditions, avoiding toxic environments, and mating. Efficient flagellar motility requires

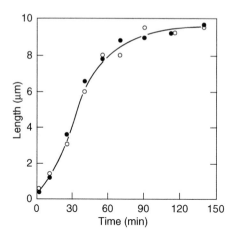

FIGURE 5.1

Flagellar elongation after deflagellation. Wild-type cells (21gr+) were deflagellated by mechanical shear, and the length of flagella at different times was determined using phase contrast microscopy. Filled and open circles represent independent experiments.

that flagella be just the right length for optimal swimming. As a result, *Chlamydomonas* and other flagellated and ciliated organisms have evolved a specific and sophisticated mechanism for regulating flagellar length.

Chlamydomonas cells rapidly regenerate their flagella after amputation, regrowing full-length flagella within 45–60 minutes (Rosenbaum et al., 1969) (Figure 5.1). This regrowth is accompanied by a rapid and specific induction of synthesis of flagellar proteins (reviewed in Lefebvre and Rosenbaum, 1986; see Chapter 3). This induction reflects increased transcription as well as increased transcript stability and perhaps enhanced efficiency of translation of flagellar protein mRNAs (Keller et al., 1984; Baker et al., 1986). Although flagellar proteins are synthesized in the cell body, early "growth zone" experiments demonstrated that the subunits are added at the distal tips of the regenerating flagella (Rosenbaum et al., 1969; Witman, 1975).

Is there is a specific mechanism to control flagellar length, separate from the regulation of the amount of flagellar proteins produced by the cell? Early experiments on flagellar regrowth demonstrated that flagellar assembly is not simply a reflection of the amount of flagellar protein available in the cell for growth. When flagella were amputated in the absence of protein synthesis, which was blocked by treating the cells with cycloheximide, flagella re-grew to up to two-thirds normal length (Rosenbaum et al., 1969). This critical experiment showed that cells with full-length flagella (ca. 13 μm) maintained a pool of assembly-competent flagellar proteins sufficient to assemble substantially longer flagella than were actually formed. Clearly some mechanism determined the partitioning of flagellar proteins between the assembled and unassembled state.

Cells also can be shown to monitor the assembled state of their flagella. When one of the two flagella is amputated (the so-called "long-zero" experiment), the other immediately begins to resorb into the cell (Rosenbaum et al., 1969) (Figure 5.2). The missing flagellum begins to

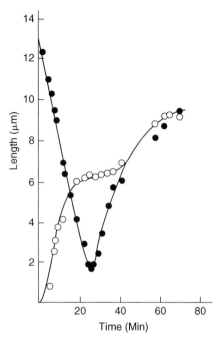

FIGURE 5.2 *"Long-zero" resorption and regrowth of flagella after amputation of one flagellum. Cells of the paralyzed flagella mutant pf16 were deflagellated by mechanical shear at low intensity, such that a percentage of the cells lost only one of the two flagella. The length of each flagellum on a single cell at different times was recorded from time-lapse photomicrographs taken with a phase microscope. Open circles: the regrowth of the amputated flagellum; filled circles: the resorption and then regrowth of the remaining (unamputated) flagellum.* **Source:** *From Rosenbaum et al. (1969).*

regrow at the same time, leading to a state in which one flagellum is elongating and the other is resorbing on a single cell. The rate of regeneration of the growing flagellum is somehow coupled to the length of the resorbing flagellum. In some cases the shortening flagellum resorbed so rapidly that it became shorter than the growing flagellum (Figure 5.2). Remarkably, in such cells, the regenerating flagellum paused in its growth until the resorbing flagellum stopped shortening and began to regrow. When the two flagella reached the same length, the two flagella then regenerated with rapid kinetics to their original length (Rosenbaum et al., 1969).

II. GENETIC EVIDENCE FOR A FLAGELLAR LENGTH CONTROL MECHANISM

The strongest evidence that cells actively control flagellar length comes from experiments utilizing mutants with long flagella. Two of the first flagellar mutants described in *Chlamydomonas*, *lf1* and *lf2*, were identified in a screen of mutants with defective swimming (McVittie, 1972). These

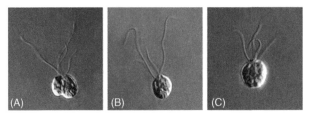

FIGURE 5.3   *Dikaryon rescue of* lf2 *by wild-type cells.* lf2 *gametes were mated with wild-type gametes. Panel A: cell fusion has begun. Panel B: 15 minutes after mating, dikaryons with two long flagella and two normal length flagella are present. Panel C: 40 minutes after mating, all four flagella are of wild-type length.*

mutants assembled flagella up to three times normal length. Most importantly, these mutants allowed a clear demonstration that flagellar length is tightly regulated. After fusion of mating-type *plus* and *minus* gametes, temporary dikaryons are formed that have two nuclei and four flagella in a common cytoplasm (Figure 5.3). When either *lf1* or *lf2* mutants were crossed to wild-type cells, the extra-long flagella were resorbed to wild-type length within minutes of fusion of the two cells, but in the process the flagella from the wild-type cell did not lengthen (Starling and Randall, 1971). Somehow the dikaryon cell sensed the excessive length of the flagella from the *lf* mutant cell, and rapidly shortened the long flagella to the appropriate length. The fact that the flagella from the wild-type cell did not lengthen indicates that the cells did not average their flagellar lengths, but rather they restored the long flagella to a predetermined length. This report was the first to describe "dikaryon rescue" of mutant phenotypes in *Chlamydomonas*, a technique that has proven to be very useful for quickly determining the dominance or recessiveness of mutants, and for identifying proteins affected by mutations disrupting flagellar function (Luck et al., 1977). Similar experiments with mutants with short flagella (*ca.* one-half length) have been described, with the *shf* mutant phenotype also being recessive in temporary dikaryons (Jarvik et al., 1984; Kuchka and Jarvik, 1987).

Four genes, *LF1*, *LF2*, *LF3* and *LF4*, have been identified in which mutation can lead to a long-flagella phenotype (McVittie, 1972; Barsel et al., 1988; Asleson and Lefebvre, 1998). Multiple mutant alleles have been isolated for each, suggesting that there are not a large number of other genes that can be mutated to produce a long-flagella phenotype. Each of the genes has been cloned, and the gene product characterized. They fall into two classes based on their protein interactions and cellular localization. *LF4* encodes a flagellar protein that also accumulates in the cell body, whereas *LF1*, *LF2* and *LF3* encode interacting proteins found predominantly if not exclusively in the cell body. The *LF2* and *LF4* genes encode a CDK-like kinase and a MAP kinase respectively (Berman et al., 2003; Tam et al., 2007).

III. LF4: AN UNUSUAL MAP KINASE

Most *lf* mutants isolated to date affect the *LF4* gene (Asleson and Lefebvre, 1998). Null mutants produced by insertional mutagenesis and deletion of the *LF4* locus have long flagella (Berman et al., 2003). *LF4* encodes an unusual MAP kinase (AAO86688) of previously unknown function. Two mammalian MAP kinases of unknown function, known as MAK and MOK (Miyata et al., 1999), have sequence similarity to the *LF4* gene product and to each other. MAP kinases with sequence similarity to LF4 have also been reported to be involved in flagellar length control in *Leishmania* (Bengs et al., 2005; Erdman et al., 2006), and in the control of the length of sensory cilia in *C. elegans* (Burghoorn et al., 2007).

Because the *lf4* mutants completely lack expression of the LF4 MAP kinase, the protein is presumed to function by shortening flagella. Oddly, however, the predominant form of the protein in flagella is enzymatically inactive, whereas the protein in the cell body is enzymatically active (Berman et al., 2003). MAP kinases require the phosphorylation of both a tyrosine and a serine or threonine residue in the activation lip (Canagarajah et al., 1997). The LF4 enzyme in wild-type flagella is not phosphorylated on either amino acid, and it has been shown by in-gel kinase assays to be inactive. The form in the cell body is doubly phosphorylated and enzymatically active. The inactivation of the protein in the flagella is likely to be maintained by the action of one or more phosphatase enzymes in the flagella, as mixing experiments with flagellar extracts and active (phosphorylated) LF4 show that an enzymatic activity in flagella dephosphosphorylates and inactivates LF4 (Wilson and Lefebvre, unpublished observations). These observations suggest that the LF4 MAP kinase enzyme may be transiently, and perhaps locally, activated in flagella that have assembled beyond some preset length. The active enzyme would then induce shortening until the flagella shortened below the target length.

IV. LF1, LF2, LF3: PARTNERS IN THE LENGTH REGULATORY COMPLEX

The *LF1*, *LF2* and *LF3* genes encode proteins (AAP83163, ABK34487 and AAO62545, respectively) that interact in the regulation of flagellar length. All three are found predominantly if not exclusively in the cell body, in punctate structures called the Length Regulatory Complex (LRC) (Tam et al., 2007) (Figure 5.4). *LF2* encodes a protein kinase with sequence similarity to members of the cyclin-dependent kinase (CDK) family, although it is not directly regulated by cyclins because it lacks the PSTAIRE amino-acid motif required for cyclin binding. The *LF2* gene product is needed both for flagellar assembly and for flagellar length control, as the null mutant

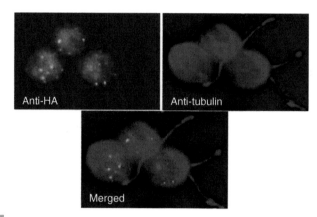

FIGURE 5.4 *The LF3 protein localizes in the cytoplasm as punctate spots. A construct encoding an epitope- (HA-) tagged LF3 was transformed into lf3 mutant cells to rescue the mutant phenotype. Green: HA epitope (localizes LF3). Red: anti-alpha tubulin antibody staining.*

phenotype of *lf2* mutants is not long flagella, but short and unequal length flagella or "*ulf*". An insertional mutant allele of *lf2* as well as two different insertional alleles of *lf3* show similar *ulf* phenotypes. The single *lf1* mutant allele is a nonsense mutation in the amino terminal 20% of the gene, and is therefore likely to be a null allele (Nguyen et al., 2005).

Other members of the CDK family require accessory proteins other than cyclin for activation, but all require some accessory protein for activity. LF1, LF2 and LF3 clearly interact, suggesting that LF1, LF3 or both act as accessory proteins for the LF2 kinase. First, all three proteins co-sediment in the same fractions of sucrose gradients of proteins from the cell body. None of the proteins is found in significant amounts in the flagella. Second, all three proteins localize to punctate structures throughout the cell, as observed using immunofluorescence microscopy of epitope-tagged gene constructs. Third, using yeast two-hybrid assays, the *LF1*, *LF2* and *LF3* gene products interact in all pairwise combinations, but none of these interact with the gene product of *LF4* (Tam et al., 2007). Finally, any double mutant combination of *lf1*, *lf2* and *lf3* mutant alleles produces a flagella-less phenotype (Barsel et al., 1988), highlighting a possible functional interaction between the gene products.

Analysis of a series of mutant alleles of *lf2* suggests that LF2 protein kinase activity is needed for both flagellar assembly and length control (Tam et al., 2007). As discussed above, insertional mutants that totally abolish *LF2* gene expression assemble short, unequal length flagella. Several other mutant alleles were sequenced, and their phenotypic consequences ranged from a partial lengthening, with no flagellar regeneration defects, to a more severe defect in which flagella grew very long (more than twice normal length) and regenerated very poorly after deflagellation. This allelic series of phenotypes of increasing severity was interpreted to reflect different levels of protein kinase activity for the products of the different *lf2* alleles based on experiments using *in vitro* mutagenesis. In these experiments, two different changes (lysine to arginine at amino acid 41 and glycine to valine at amino acid 21) were introduced that were predicted to reduce protein kinase activity without altering the overall structure of the protein. When transformed into *lf2* null mutants, each of these rescued the *ulf* phenotype and produced a long-flagella phenotype. When the double mutant incorporating both changes was produced and introduced into *lf2* null mutant cells, however, no phenotypic effects were observed (Tam et al., 2007). The conclusion reached from these experiments and from the sequencing of alleles with increasingly severe mutant phenotypes is that when the LF2 kinase has reduced activity, cells can assemble flagella, but they cannot control their length, whereas the complete abolition of the kinase activity produces the null mutant phenotype of unequal length, short flagella. Low levels of LF2 activity may be needed for assembling flagella, with higher levels needed to control their length.

As observed for *lf2* alleles, *lf3* mutant alleles show a gradation in severity from the null phenotype (short, unequal length flagella) to presumptive weak alleles which have uniformly long flagella and rapid flagellar regeneration after amputation. As noted above, the only *lf1* mutant allele is presumed to be a null mutant. Whether the effects of the *lf1* and *lf3* mutants are manifested through reduced LF2 kinase activity is not yet known, because *in vitro* enzyme assays for the LF2 kinase have not yet been developed.

V. WHAT OTHER GENES ARE INVOLVED IN FLAGELLAR LENGTH CONTROL?

A. *lf* mutant suppressor mutations

Although there may only be four genes that can produce a long-flagella phenotype upon mutagenesis, there are clearly many other genes involved in flagellar length control. For example, starting with double *lf* mutant flagella-less strains, it was possible to isolate extragenic suppressor mutations that restored wild-type flagellar length (Asleson and Lefebvre, 1998). The mutations caused no phenotypic consequences by themselves, as the suppressor-containing strains were wild-type in flagellar length and regeneration kinetics. Although the affected genes have not been isolated, it is unlikely that they work as informational suppressors, because the mutations they suppress include transition mutations and various splice junction defects. The mutations may act as bypass suppressors of flagellar length defects.

B. Short-flagella mutants

Mutations in three different genes, *SHF1*, *SHF2* and *SHF3*, cause cells to assemble only short flagella, approximately one-half length (Kuchka and Jarvik, 1987). These mutants have not been extensively characterized, and the affected genes are not known. Based on the relative map locations of the *lf* and *shf* genes, no locus has both long- and short-flagella mutant alleles. Not included among the mutants designated as "*shf*" are paralyzed flagella mutants, such as the dynein outer arm mutant *pf22* that assembles flagella between one-half and two-thirds length (Huang et al., 1979). Interestingly, the short-flagella phenotype of *pf22* is epistatic to the long-flagella phenotype of *lf3*, suggesting that the *pf22* mutation inhibits flagellar assembly, rather than alters flagellar length control (Barsel et al., 1988).

C. Glycogen synthase kinase 3

Glycogen synthase kinase 3 (GSK3; AAT40314) is a flagellar protein whose enzymatic activity is inhibited by lithium (Wilson and Lefebvre, 2004). *Chlamydomonas* flagella elongate up to 40% beyond normal length when treated with lithium (Nakamura et al., 1987; Periz et al., 2007; Tuxhorn et al., 1998). The extent of flagellar elongation is comparable to the amount of unassembled flagellar proteins maintained in the cytoplasm of untreated cells. In fact, analysis of flagellar protein pools remaining in the cell body after lithium treatment indicated that proteins from the pool were used to assemble flagella beyond the normal length. Possibly GSK3 activity is involved in partitioning flagellar proteins between the assembled and unassembled state. It is difficult to dissect the specific flagellar function of GSK3, however, because there is only a single GSK3 gene in *Chlamydomonas*, and even slight reductions in GSK3 expression using RNAi were lethal (Wilson and Lefebvre, 2004). Given that there are many flagella-less mutants that are viable, it seems likely that the lethal effects of GSK3 depletion are not due to effects on flagellar assembly but are an indication of an essential role for the kinase in cellular function.

D. CNK2

Two members of the never-in-mitosis (NIMA) family of protein kinases have been shown to play a role in flagellar homeostasis. In *fa2* mutants, defects in a NIMA kinase block flagellar microtubule excision (see Chapter 3) and also delay the G2/M transition during the cell cycle (Mahjoub et al., 2002). The second NIMA kinase, CNK2 (AAQ64683), plays a role in regulating flagellar length (Bradley and Quarmby, 2005). CNK2 is an axonemal protein in *Chlamydomonas*. Overexpression of epitope-tagged wild-type CNK2 produces cells with short flagella (approximately 50% of normal length), whereas partial inhibition of CNK2 levels using RNAi causes flagella to elongate by up to 25% beyond wild-type length. Although this elongation is less than that observed with any of the *lf* mutants, it does indicate a role for CNK2 in the control of flagellar length. This role may be related to defective control of cell size, as overexpression of CNK2 leads to a reduction in cell size, and conversely, reduction of CNK2 levels leads to enlargement of cells (Bradley and Quarmby, 2005).

VI. MODELS FOR FLAGELLAR LENGTH CONTROL

A. Early models

As soon as the early flagellar regeneration and long-zero experiments established that flagellar assembly in *Chlamydomonas* was tightly regulated, models were developed to attempt to explain the correlation between flagellar regeneration rates and final flagellar length. One of the first of these explained the deceleratory kinetics of flagellar regeneration by careful analysis of potential rates of diffusion of flagellar protein precursors (Levy, 1974). It was proposed that there is a constant pool of flagellar protein precursors (or of a small molecule necessary for flagellar assembly) at the base of the flagellum, that precursors are supplied to the growing tip by diffusion, and that the concentration of precursors at the tip decreases as the flagellum elongates. Modeling based on these assumptions resulted in a decreasing rate of elongation with increasing flagellar length. It was further proposed that precursors are lost by degradation or leakage through the membrane at the tip of the flagellum, which at some flagellar length would result in the precursor concentration at the tip dropping to zero and growth ceasing, thus explaining why a flagellum reaches a maximum length. With the discovery that flagellar proteins are actively transported by regulated intraflagellar transport both into and out of the flagella (see below), there is little to support a diffusion-based model.

A clever model that could readily explain deceleratory kinetics of flagellar regeneration, as well as the slow lengthening of flagella that occurs throughout the cell cycle, postulated that two elements running the length of the flagella, but with opposite polarity, were slowly crosslinked during flagellar growth by lateral structures that matured over time and constrained further growth (Child, 1978). No candidate structures for any of these predicted structures were identified, however, and when it became clear that all of the microtubules had the same polarity, this model lost much of its attractiveness.

B. Balance point model

Intraflagellar transport, or IFT, is the regulated transport of flagellar components in both the anterograde (away from the cell) and retrograde (toward the cell) directions (see Chapter 4). With the discovery of this process, a model invoking a balance between assembly and disassembly to explain regulation of flagellar length was proposed (Marshall and Rosenbaum, 2001). Marshall and Rosenbaum used the incorporation of epitope-tagged subunits into flagella after mating to demonstrate that flagella turn over rapidly at their distal ends, and that this turnover can be blocked by colchicine, indicating that microtubule assembly and disassembly are actively occurring even when flagella have reached steady-state length. The balance between assembly and disassembly was directly demonstrated by blocking anterograde IFT (and therefore assembly at the tip) using the mutant *fla10*, which has a temperature-sensitive mutation in one of the subunits of the anterograde motor kinesin-2 (Walther et al., 1994). At the non-permissive temperature, incorporation of subunits at the tip of the flagella ceased, but disassembly continued, leading to flagellar shortening. *In vivo* labeling experiments in cells with full-length flagella also demonstrated that proteins in intact flagella turn over rapidly (Song and Dentler, 2001).

The model postulates that the rate of flagellar assembly is length dependent, as demonstrated by the deceleratory kinetics of assembly after deflagellation. The rate of flagellar disassembly is postulated to be length independent, as demonstrated by the linear kinetics of flagellar resorption when *fla10* mutants are shifted to the non-permissive temperature. The length of flagella at which the rate of assembly is equal to the rate of disassembly would then be the steady-state flagellar length. As predicted by the model, *fla10* cells grown at a precisely controlled intermediate temperature, part way between the permissive and restrictive temperatures, have flagella of approximately one-half length, consistent with a decrease in the assembly rate with no change in the disassembly rate. Other observations consistent with the model are that the rate of flagellar growth is a function of the length of the flagella and not time after deflagellation and that flagella have the same amount of IFT proteins at all times after deflagellation. Marshall et al. (2005) have also found support for the balance point model in the observation that cells with variable numbers of flagella (*vfl* mutants) have different lengths of flagella. In *vfl* mutants the length of flagella was found to be inversely proportional to the number of flagella on the cell, presumably because a limited pool of essential IFT proteins was partitioned among more flagella, thereby decreasing the rate of assembly. In contrast, Kuchka and Jarvik (1982) reported, using *vfl2* mutants, that the length of flagella was independent of flagellar number. One possible source for this discrepancy is the use of non-dividing gametes by Marshall et al., whereas Kuchka and Jarvik studied dividing vegetative cells.

As yet the identities of the proteins controlling flagellar assembly and disassembly have not been established. For assembly, some property or properties of the IFT machinery will very likely be under regulation as part of length control. For example, null mutants in *lf3* accumulate large clumps of IFT proteins in bulbous protrusions of the membranes on the *ulf* flagella, perhaps indicating that the *lf3* lesion causes a defect in retrograde IFT (Tam et al., 2003). Identification of the substrate molecules for the LF2 and LF4 protein kinases, along with the substrates for GSK3 and CNK2 will perhaps point toward the proteins that directly control flagellar assembly and disassembly in the regulation of flagellar length.

VII. QUESTIONS ABOUT FLAGELLAR LENGTH CONTROL

A. Is there a sensor for flagellar length?

The behavior of long-flagella mutants during dikaryon formation strongly suggests that when the cytoplasm of an *lf* mutant cell fuses with the cytoplasm of a wild-type cell, some signal triggers the long flagella to recognize their excess length and shorten to a predetermined (i.e. wild-type) length (Figure 5.3). Thus, cells appear to have a sensor that monitors the length of the flagella, and mating restores the appropriate length in response to a signal from the sensor. The balance point equilibrium model, however, provides an explanation for the appearance of a "correct" length based solely on the rates of assembly and disassembly of the flagella. In this model, *lf* mutants would have long flagella because their rates of flagellar assembly were too rapid, and/or their rates of flagellar disassembly were too slow. In a dikaryon experiment, proteins from the wild-type cell would have to quickly modify one or both rates in the long flagella to induce flagellar resorption to the equilibrium length. For at least *lf2* and *lf4* mutants, a kinase activity must be required for this modification of assembly and/or disassembly rates. The theoretical differences in assembly/disassembly rates in wild-type versus *lf* mutant flagella are not reflected in visible differences in velocities of IFT particles during anterograde or retrograde movement (Dentler, 2005). It is possible that alterations in IFT-mediated flagellar assembly and disassembly do not involve visible differences in the particles or their rates of movement, but instead involve differences in the loading and unloading of key regulatory proteins. Support for this possibility comes from the observation that during chemically-induced flagellar shortening, the loading of radial spoke proteins onto IFT particles was reduced, even though the particles continued to enter the flagella and undergo anterograde motion (Pan and Snell, 2005).

If there is a sensor for measuring flagellar length, then IFT may be involved in the measurement. For example, if a key regulatory protein is a cargo for IFT, and if it is in some way modified at a constant rate per unit time in transit along the flagellum, then the protein may acquire excessive levels of modification during transit along a long flagellum. The fact that at least four kinases (LF2, LF4, GSK3 and CNK2) are involved in flagellar length control would be consistent with one logical candidate, phosphorylation, for a length sensor. It should also be noted, as suggested by Marshall et al. (2005), that the balance point equilibrium model and a length sensor are not mutually exclusive, but could act coordinately in a complex system of sensing and regulating the extent of flagellar assembly.

B. What role do the regulatory systems controlling flagellar shortening play in controlling flagellar length?

Flagellar shortening and excision are regulated by signal transduction through a protein kinase of the aurora kinase family called CALK (AAF97501), for *Chlamydomonas* aurora-like kinase (Pan and Snell, 2000). Signals that induce flagellar excision by the cell, such as lowering the pH below 4.5, or induce flagellar shortening, such as treatment with 20 mM sodium pyrophosphate (NaPPi), activate CALK by phosphorylation (Pan et al., 2004). The signal to activate CALK is somehow related to the assembled state of the flagella, as indicated by the fact that two flagella-less mutants show constitutive CALK activation. The activation of CALK is a very early step in the process of flagellar resorption, as shown by the fact that when the temperature-sensitive IFT mutant *fla10* was shifted to the non-permissive temperature, phosphorylation of CALK preceded any flagellar resorption.

The link between CALK-mediated flagellar resorption and flagellar length control, however, is not clear. None of the *lf* mutants have lesions in the CALK gene, and it does not map near any of the three *shf* mutant loci. As discussed earlier, because null mutants deleting the MAP kinase encoded by the *LF4* gene have extra-long flagella, LF4 MAP kinase activity is likely to be involved in flagellar shortening. In *lf4* null mutants, however, NaPPi induces CALK phosphorylation and flagellar shortening with kinetics indistinguishable from those observed using wild-type cells, indicating that CALK-induced flagellar shortening does not require the *LF4* gene product.

VIII. CONCLUSION AND FUTURE DIRECTIONS

Although seven genes have been shown to contain genetic lesions that result in flagella of abnormal length (four *lf* genes and three *shf* genes), there are clearly many more gene products that play a role in flagellar length control. Because there are multiple alleles for three of the four *lf* gene loci, it is unlikely that more than a few other genes can be mutated to produce a long-flagella phenotype. Even for *lf2* and *lf3*, the null phenotype is not long flagella, but rather stumpy, unequal length flagella.

Biochemical approaches to flagellar length control are likely to be the most important avenue for discovering additional proteins involved in the process. For example, the protein target(s) for the LF2 and LF4 protein kinases are expected to play a role in controlling flagellar length, as is the putative flagellar phosphatase that maintains the LF4 MAP kinase in an inactive state. Detailed comparison of protein differences between the flagella from *lf* mutant cells and wild-type cells may identify proteins that function in length control.

REFERENCES

Asleson, C.M. and Lefebvre, P.A. (1998). Genetic analysis of flagellar length control in *Chlamydomonas reinhardtii*: a new long-flagella locus and extragenic suppressor mutations. *Genetics* **148**, 693–702.

Baker, E.J., Keller, L.R., Schloss, J.A., and Rosenbaum, J.L. (1986). Protein synthesis is required for rapid degradation of tubulin mRNA and other deflagellation-induced RNAs in *Chlamydomonas reinhardi*. *Mol. Cell Biol.* **6**, 54–61.

Barsel, S.E., Wexler, D.E., and Lefebvre, P.A. (1988). Genetic analysis of long-flagella mutants of *Chlamydomonas reinhardtii*. *Genetics* **118**, 637–648.

Bengs, F., Scholz, A., Kuln, D., and Wiese, M. (2005). LmxMPK9, a mitogen-activated protein kinase homologue affects flagellar length in *Leishmania mexicana*. *Mol. Microbiol.* **55**, 1606–1615.

Berman, S.A., Wilson, N.F., Haas, N.A., and Lefebvre, P.A. (2003). A novel MAP kinase regulates flagellar length in *Chlamydomonas*. *Curr. Biol.* **13**, 1145–1149.

Bradley, B.A. and Quarmby, L.M. (2005). A NIMA-related kinase, Cnk2p, regulates both flagellar length and cell size in *Chlamydomonas*. *J. Cell Sci.* **118**, 3317–3326.

Burghoorn, J., Dekkers, M.P., Rademakers, S., de Jong, T., Willemsen, R., and Jansen, G. (2007). Mutation of the MAP kinase *DYF-5* affects docking and undocking of kinesin-2 motors and reduces their speed in the cilia of *Caenorhabditis elegans*. *Proc. Natl. Acad. Sci. U.S.A.* **104**, 7157–7162.

Canagarajah, B.J., Khokhlatchev, A., Cobb, M.H., and Goldsmith, E.J. (1997). Activation mechanism of the MAP kinase ERK2 by dual phosphorylation. *Cell* **90**, 859–869.

Child, F.M. (1978). The elongation of cilia and flagella: a model involving antagonistic growth zones. In: *Cell Reproduction: Dan Mazia Dedicatory Volume* (E.R. Dirkson, D. Prescott and C. Fox, Eds.). *ICN-UCLA Symp. Mol. Cell Biol.* **12**, 337–350.

Dentler, W. (2005). Intraflagellar transport (IFT) during assembly and disassembly of *Chlamydomonas* flagella. *J. Cell Biol.* **170**, 649–659.

Erdman, M., Scholz, A., Melzer, I.M., Schmetz, C., and Wiese, M. (2006). Interacting protein kinases involved in the regulation of flagellar length. *Mol. Biol. Cell* **17**, 2035–2045.

Huang, B., Piperno, G., and Luck, D.J.L. (1979). Paralyzed flagella mutants of *Chlamydomonas reinhardtii* defective for axonemal doublet microtubule arms. *J. Biol. Chem.* **254**, 3091–3099.

Jarvik, J.W., Reinhardt, F.D., Kuchka, M.R., and Adler, S.A. (1984). Altered flagellar size control in *shf-1* short flagellar mutants of *Chlamydomonas reinhardtii*. *J. Protozool.* **31**, 199–204.

Keller, L.R., Schloss, J.A., Silflow, C.D., and Rosenbaum, J.L. (1984). Transcription of alpha- and beta-tubulin genes *in vitro* in isolated *Chlamydomonas reinhardi* nuclei. *J. Cell Biol.* **98**, 1138–1143.

Kuchka, M.R. and Jarvik, J.W. (1982). Analysis of flagellar size control using a mutant of *Chlamydomonas reinhardtii* with a variable number of flagella. *J. Cell Biol.* **92**, 170–175.

Kuchka, M.R. and Jarvik, J.W. (1987). Short-flagella mutants of *Chlamydomonas reinhardtii*. *Genetics* **115**, 685–691.

Lefebvre, P.A. and Rosenbaum, J.L. (1986). Regulation of the synthesis and assembly of ciliary and flagellar proteins during regeneration. *Annu. Rev. Cell Biol.* **2**, 517–546.

Levy, E.M. (1974). Flagellar elongation: an example of controlled growth. *J. Theor. Biol.* **43**, 133–149.

Luck, D., Piperno, G., Ramanis, Z., and Huang, B. (1977). Flagellar mutants of *Chlamydomonas*: studies of radial spoke-defective strains by dikaryon and revertant analysis. *Proc. Natl. Acad. Sci. U.S.A.* **74**, 3456–3460.

Mahjoub, M.R., Montpetit, B., Zhao, L., Finst, R.J., Goh, B., Kim, A.C., and Quarmby, L.M. (2002). The *FA2* gene of *Chlamydomonas* encodes a NIMA family kinase with roles in cell cycle progression and microtubule severing during deflagellation. *J. Cell Sci.* **115**, 1759–1768.

Marshall, W.F. and Rosenbaum, J.L. (2001). Intraflagellar transport balances continuous turnover of outer doublet microtubules: implications for flagellar length control. *J. Cell Biol.* **155**, 405–414.

Marshall, W.F., Qin, H., Rodrigo Brenni, M., and Rosenbaum, J.L. (2005). Flagellar length control system: testing a simple model based on intraflagellar transport and turnover. *Mol. Biol. Cell.* **16**, 270–278.

McVittie, A.C. (1972). Flagellum mutants of *Chlamydomonas reinhardtii*. *J. Gen. Microbiol.* **71**, 525–540.

Miyata, Y., Akashi, M., and Nishida, E. (1999). Molecular cloning and characterization of a novel member of the MAP kinase superfamily. *Genes Cell.* **4**, 299–309.

Nakamura, S., Takino, H., and Kojima, M.K. (1987). Effect of lithium on flagellar length in *Chlamydomonas reinhardtii*. *Cell Struct. Funct.* **12**, 369–374.

Nguyen, R.L., Tam, L.-W., and Lefebvre, P.A. (2005). The *LF1* gene of *Chlamydomonas reinhardtii* encodes a novel protein required for flagellar length control. *Genetics* **169**, 1415–1424.

Pan, J. and Snell, W.J. (2000). Regulated targeting of a protein kinase into an intact flagellum. An aurora/Ipl1p-like protein kinase translocates from the cell body into the flagella during gamete activation in *Chlamydomonas*. *J. Biol. Chem.* **275**, 24106–24114.

Pan, J. and Snell, W.J. (2005). *Chlamydomonas* shortens its flagella by activating axonemal disassembly, stimulating IFT particle trafficking, and blocking anterograde cargo loading. *Dev. Cell* **9**, 431–438.

Pan, J., Wang, Q., and Snell, W.J. (2004). An aurora kinase is essential for flagellar disassembly in *Chlamydomonas*. *Dev. Cell* **6**, 445–451.

Periz, G., Dharia, D., Miller, S.H., and Keller, L.R. (2007). Flagellar elongation and gene expression in *Chlamydomonas reinhardtii*. *Eukaryot. Cell* **6**, 1411–1420.

Rosenbaum, J.L., Moulder, J.E., and Ringo, D.L. (1969). Flagellar elongation and shortening in *Chlamydomonas*. The use of cycloheximide and colchicine to study the synthesis and assembly of flagellar proteins. *J. Cell Biol.* **41**, 600–619.

Song, L. and Dentler, W.L. (2001). Flagellar protein dynamics in *Chlamydomonas*. *J. Biol. Chem.* **276**, 29754–29763.

Starling, D. and Randall, J. (1971). The flagella of temporary dikaryons of *Chlamydomonas reinhardii*. *Genet. Res.* **18**, 107–113.

Tam, L.-W., Dentler, W.L., and Lefebvre, P.A. (2003). Defective flagellar assembly and length regulation in *LF3* null mutants in *Chlamydomonas*. *J. Cell Biol.* **163**, 597–607.

Tam, L.-W., Wilson, N.F., and Lefebvre, P.A. (2007). A CDK-related kinase regulates the length and assembly of flagella in *Chlamydomonas*. *J Cell Biol.* **176**, 819–829.

Tuxhorn, J., Daise, T., and Dentler, W.L. (1998). Regulation of flagellar length in *Chlamydomonas*. *Cell Motil. Cytoskeleton.* **40**, 133–146.

Walther, Z., Vashishtha, M., and Hall, J.L. (1994). The *Chlamydomonas FLA10* gene encodes a novel kinesin-homologous protein. *J Cell Biol.* **126**, 175–188.

Wilson, N.F. and Lefebvre, P.A. (2004). Regulation of flagellar assembly by glycogen synthase kinase 3 in *Chlamydomonas reinhardtii*. *Eukaryot. Cell.* **3**, 1307–1319.

Witman, G.B. (1975). The site of *in vivo* assembly of flagellar microtubules. *Ann. N.Y. Acad. Sci.* **253**, 178–191.

CHAPTER 6

Axonemal Dyneins: Assembly, Structure, and Force Generation

Stephen M. King[1] and Ritsu Kamiya[2]

[1]Department of Molecular, Microbial and Structural Biology,
University of Connecticut Health Center, Farmington, Connecticut, USA
[2]Department of Biological Sciences, Graduate School of Science,
University of Tokyo, Hongo, Bunkyo-ku, Tokyo, Japan

CHAPTER CONTENTS

I. INTRODUCTION

Axonemal dyneins from *Chlamydomonas* flagella have been studied for over 30 years, and the powerful combination of genetic analysis, biochemical tractability, and physiological measurements available with this model organism have resulted in their being some of the best characterized members of this class of minus-end directed motor. These enzymes comprise the arms associated with the axonemal doublet microtubules (Figure 6.1A–C) and are composed of multiple components organized around the heavy chain (HC) motor units. The complexity of the axonemal dynein system is clearly illustrated by the observation that of the 16 distinct dynein HCs now known in *Chlamydomonas*, all but two have been identified as integral components of the axoneme. The deep knowledge base now available concerning these enzymes and the tools that exist for their analysis makes the *Chlamydomonas* system an excellent paradigm for understanding both ciliary/flagellar motors and the closely related enzymes present in cytoplasm. Here, we concentrate on describing the assembly, structure, and function of *Chlamydomonas* axonemal inner and outer arm dyneins, but have also included studies on cytoplasmic and axonemal dyneins from other organisms where they help elucidate conserved structural and/or functional features of these motors.

II. ARRANGEMENT OF DYNEINS IN THE AXONEME

A. Components of the 96-nm axonemal repeat

The 9 + 2 structure of the axoneme has a basic longitudinal repeat unit of 96 nm (Figure 6.1D). Within this unit, an outer doublet microtubule bears four outer arm dyneins, one each of several different kinds of inner arm dyneins, one dynein regulatory complex (DRC, see Chapter 9), two radial spokes, and, most likely, one inter-doublet link (Nicastro et al., 2005). What defines this repeat length in the outer doublet is not known, although it is conceivable that some protein(s) on the doublet function as a "ruler."

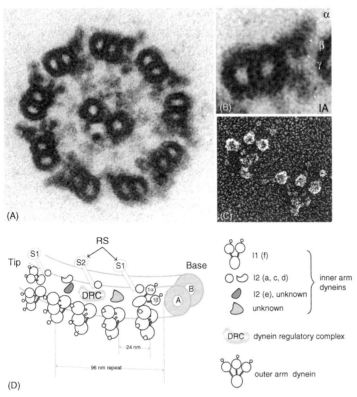

FIGURE 6.1 *The ultrastructure of axonemal dyneins. (A) Axonemal cross-section revealing the inner and outer dynein arms associated with the outer doublet microtubules. No outer arm dyneins are bound to doublet no. 1 (at bottom), and instead, this doublet is connected to the adjacent one by a distinct linker structure of unknown composition (Hoops and Witman, 1983); it remains uncertain whether this doublet contains a full complement of inner arms. This micrograph was kindly provided by Dr. Ken-ichi Wakabayashi. Reprinted from Sakato and King (2004), with permission from Elsevier. (B) Enlarged view of dynein arms associated with one doublet microtubule. The inner arms are indicated by "IA." The regions of the outer arm occupied by the α, β, and γ HCs are also indicated; these were determined by analysis of mutants lacking either the entire α HC or the β or γ HC motor domains (Liu et al., 2008; Sakakibara et al., 1991, 1993). Reprinted from Sakato and King (2004), with permission from Elsevier. (C) Quick-freeze deep-etch micrograph of two intact outer arm dynein particles adsorbed to a mica surface (Goodenough and Heuser, 1984). Each dynein consists of a complex basal region and three globular heads, ~14 nm in diameter, derived from the AAA+ domains. The 10-nm coiled-coil stalks terminating in the microtubule-binding domain protrude from the AAA+ rings. This micrograph was kindly provided by Dr. John Heuser. (D) Diagram illustrating the arrangement of the four outer arms, the inner arms, DRC, and radial spokes within the 96-nm axonemal repeat. Reprinted from Sakato and King (2004), with permission from Elsevier.*

In the axonemes of sea urchin sperm and other organisms, tektin has been proposed to play this role in axonemal assembly (Norrander et al., 1996). Tektin is a protein with a high probability of forming coiled coils, similar to the subunits of intermediate filaments. However, in *Chlamydomonas*, tektin (BAC77347) is apparently involved in the docking of an inner arm dynein and is not essential for the formation of the doublet microtubules (Yanagisawa and Kamiya, 2004). Thus, it is unlikely that tektin plays a fundamental role in determining the repeat length.

Outer arm dyneins are arranged linearly with an interval of 24 nm, a length corresponding to three tubulin dimers (Figure 6.1D). This spacing is apparently defined by the outer dynein arm-docking complex (ODA-DC), a protein complex located at the base of each outer arm dynein (Takada and Kamiya, 1994; Koutoulis et al., 1997; Takada et al., 2002; Casey et al., 2003a), and by molecular interactions between adjacent outer arm dyneins (Nicastro et al., 2006; Ishikawa et al., 2007). This idea is supported by the observation that the ODA-DC assembled from recombinant proteins can bind to cytoplasmic microtubules with a spacing of 24 nm (K. Wakabayashi, unpublished observations). What determines the outer arm docking site around the circumference of the outer doublet is unknown.

Each species of inner arm dynein is placed at a distinct locus in the 96-nm repeat unit. The localization of a particular dynein species has been studied by comparing electron microscopy (EM) images of wild-type and mutant axonemes lacking particular dyneins (Mastronarde et al., 1992). In this way, the locations of dynein species I1/f and species c (see below) have been determined among the electron densities in the 96-nm repeat unit. In the case of species I1/f, locations of the head domains of the I1/f 1α and I1/f 1β HCs, and that of the stem domains have been assessed by analyzing axonemes that retain parts of this dynein (Perrone et al., 2000). Images of the *ida4* axoneme that lacks species a, c, and d suggested that, relative to the other dyneins, these are located nearest to the axonemal center, with almost even distances between them (Mastronarde et al., 1992). Species c has been localized at the base of the S2 spoke (Yagi et al., 2005). Analysis of mutant axonemes deficient in the DRC also located dynein species e near the S2 radial spoke (Gardner et al., 1994). The localization of subspecies b and g are unknown because no mutants lacking them are available at present. The mechanism of inner arm localization is unclear. However, by analogy with the ODA-DC, some structure on the outer doublet microtubule may provide multiple weak binding sites for the N-terminal portion of each HC and the intermediate chain/light chain (IC/LC) complex (see below).

B. Distinct dyneins in the proximal and distal regions of the flagellum

SDS-PAGE analyses of short flagella, such as those in the process of re-growing following amputation or of short-flagella mutants, have suggested that the composition of inner arm dyneins in the proximal and distal regions of the axoneme differ somewhat from that in the main intermediate region (Piperno and Ramanis, 1991). Specifically, Piperno and Ramanis proposed that the inner arms in the middle portion of the axoneme consist of three kinds of dynein (termed I1, I2, and I3), whereas the proximal portion contains I3′ in place of I3 and the distal portion contains I2′ in place of I2. Since later studies discovered more diverse HC species and more complex organization of inner arm dyneins (Kagami and Kamiya, 1992; Mastronarde et al., 1992), their observations need to be re-interpreted. If the distribution of the inner arm dyneins is as proposed, and if we consider the relationship between I1–I3 and dynein species a–g (Table 6.1), it follows that the proximal portion of the axoneme contains inner arm dyneins of species I1(f), I2(c, e), and I3′(b); the central portion has I1(f), I2(c, e), and I3(g); while I1(f), I2′(a, d), and I3(g) are present in the distal portion. Since image analysis of *ida4* (missing dynein species a, c, and d; see Table 6.3) and *ida9* (missing species c) axonemes indicates that dynein species a, c, and d are present in the same 96-nm unit (Yagi et al., 2005), the proposed arrangement cannot be entirely correct. However, it is likely that some dyneins are located only in specialized regions of the axoneme; in fact, a minor species of monomeric inner arm dynein (see below) has been localized to a proximal region by immunofluorescence microscopy (T. Yagi, unpublished observation). Biased HC localization also has been reported for the outer dynein arm in human cilia (Fliegauf et al., 2005). A hypothetical dynein specifically located at the base of the axoneme is predicted to play an important role in bend initiation, as only the basal portion of the axoneme is capable of initiating bending with a sufficiently large curvature and bend angle (Brokaw, 1994).

Table 6.1   Subunit composition of major inner arm dyneins[a]

Species	a	b	c	d	e	f	g
Alternative name	I2′	I3′	I2A	I2′	I2B	I1	I3
HC	DHC6	DHC5	DHC9	DHC2	DHC8	DHC1 (1α) DHC10 (1β)	DHC7
IC						IC140 IC138 IC97 ODA7	
LC	Actin p28	Actin Centrin	Actin p28	p44 Actin p38 p28	Actin Centrin	Tctex1 Tctex2b LC7a LC7b LC8	Actin Centrin

[a] *Three other inner arm dynein species, containing DHC3, DHC4, and DHC11 have been found (T. Yagi, unpublished). These dyneins are present in the axoneme in much smaller amounts than the seven species detailed here and their composition is unclear.*

III. DYNEIN GENETICS

A. Mutant isolation

The first dynein-deficient mutants were isolated by Huang et al. (1979), who screened cells for paralyzed-flagella phenotypes. Three mutants, *pf13*, *pf22*, and *pf23*, enabled these researchers to identify the proteins specifically contained in the inner and outer arm dyneins. Later, it was shown that many dynein-deficient mutants have slow-swimming, rather than paralyzed-flagella, phenotypes. Both slow-swimming and paralyzed mutants were isolated by screening cells that tend to grow at the bottom of test-tube cultures after mutagenesis with chemicals or UV, or following insertion of a plasmid encoding a selectable marker.

B. Mutant classes

With the exception of several *pf* strains (*pf13*, *pf22*, and *pf23*) and *fla14* (which encodes LC8, a subunit of several dyneins and many other protein complexes), all mutants lacking the outer and inner arm dyneins are termed *oda* and *ida*, respectively. Currently, 16 genetically independent *oda* mutants and 9 *ida* mutants have been reported (Tables 6.2 and 6.3). Most of the *oda* mutants lack the entire outer arm dynein and display the same slow-swimming phenotype due to flagellar beating at greatly reduced beat frequency, as detailed below in section III.D. The mutant *oda11*, a null mutant of the α HC gene, retains an outer arm complex that lacks only that HC and an associated LC (Sakakibara et al., 1991). The motility of this mutant is intermediate between that of wild type and other *oda* mutants. Several *oda* mutants deficient in LCs have reduced numbers of outer arms rather than a complete absence. The paralyzed-flagella mutants *pf13* and *pf22* were originally reported as strains lacking the outer arm dynein (Huang et al., 1979), but more recent studies reveal that they lack both outer arm and inner arm subunits (Piperno et al., 1992) (Kagami and Kamiya, unpublished observations). Hence all mutants that lack only outer arm dynein equally display slow-swimming phenotypes.

The nine *ida* strains lack various combinations of inner arm dyneins. All display slow-swimming phenotypes due to some change in their waveform. The mutant *pf23* is paralyzed and has short flagella because it lacks the dimeric inner arm dynein (I1/f) in addition to a few monomeric dyneins (see below).

Null mutation of the β or γ HC of outer arm dynein results in the absence of the entire outer arm, and likewise, null mutations of the I1/f 1α or I1/f 1β HCs result in the absence of the entire inner arm I1/f dynein. However, mutants that produce C-terminal truncated forms of these HCs have also been isolated and shown to assemble partial dynein arms. For example, *oda4-s7* has a mutation that results in production of a 160-kD N-terminal peptide

| Table 6.2 | *Chlamydomonas* outer arm mutants |

Strain	Missing structure/components	Protein encoded by wild-type gene	References[a]
oda1	Outer arm, ODA-DC	DC2	[1], [2], [16]
oda2/pf28	Outer arm	γ HC	[1], [3], [4]
oda3	Outer arm, ODA-DC	DC1	[1], [2], [5]
oda4	Outer arm	β HC	[1], [6], [7]
oda5	Outer arm	ODA5	[1], [17]
oda6	Outer arm	IC2	[1], [8]
oda7	Outer arm	ODA7	[1], [21]
oda8	Outer arm		[1]
oda9	Outer arm	IC1	[1], [9]
oda10	Outer arm		[1]
oda11	α HC + LC5 + LIS1	α HC	[10], [11]
oda12	Outer arm	LC2	[12]
oda13		LC6	[12], [13]
oda14	Outer arm (partial)	DC3	[5], [12], [14]
oda15	Outer arm (partial)	LC7a	[18]
oda16	Outer arm (partial)	ODA16	[19]
pf13	Outer arm, inner arm dynein (*c*)		[15]
pf22	Outer arm, inner arm dyneins		[15]
fla14	Only short flagella stubs assemble – defects in outer arms, inner arms, radial spokes, B-tubule projections, and retrograde IFT	LC8	[13], [20]

[a][1] Kamiya (1988); [2] Takada and Kamiya (1994); [3] Mitchell and Rosenbaum (1985); [4] Wilkerson et al. (1994); [5] Koutoulis et al. (1997); [6] Luck and Piperno (1989); [7] Mitchell and Brown (1994); [8] Mitchell and Kang (1991); [9] Wilkerson et al. (1995); [10] Sakakibara et al. (1991); [11] Mitchell and Brown (1997); [12] Pazour et al. (1999); [13] King and Patel-King (1995b); [14] Casey et al. (2003a); [15] Huang et al. (1979); [16] Takada et al. (2002); [17] Wirschell et al. (2004); [18] DiBella et al. (2004a); [19] Ahmed and Mitchell (2005); [20] Pazour et al. (1998); [21] Freshour et al. (2007).

of the β HC (Sakakibara et al., 1993). This mutant retains about 70% of the outer dynein arms with a normal composition of ICs and LCs, and displays reduced motility comparable to that of mutants lacking the entire outer arm structure, indicating the importance of the β HC in outer arm function. The identification of this and other mutants clearly demonstrates that the N-terminal portion of the HC is important for docking of axonemal dyneins onto the doublet microtubules.

Table 6.3	*Chlamydomonas* inner arm mutants		
Strain	**Missing components[b]**	**Mutated gene**	**References[a]**
ida1/pf9	f	1α (DHC1)	[1]–[6]
ida2	f	1β (DHC10)	[1], [7]
ida3	f		[1]
ida4	a, c, d	p28	[1], [2], [8]
ida5	a, c, d, e	Actin	[9], [10]
ida6	e		[9]
ida7	f	IC140	[11]
ida8/bop2	152 kD		[12], [13[c]]
ida9	c	DHC9	[14]
pf23	a, c, d, f		[15]
pf3	e, regulatory proteins		[16], [17]
bop5	(f)	IC138	[18[c]]

[a][1] Kamiya et al. (1991); [2] Kagami and Kamiya (1992); [3] Piperno et al. (1990); [4] Myster et al. (1997); [5] Myster et al. (1999); [6] Luck and Piperno (1989); [7] Perrone et al. (2000); [8] LeDizet and Piperno (1995a); [9] Kato et al. (1993); [10] Kato-Minoura et al. (1997); [11] Perrone et al. (1998); [12] Le et al. (1999); [13] King et al. (1994); [14] Yagi et al. (2005); [15] Huang et al. (1979); [16] Piperno et al. (1994); [17] Gardner et al. (1994); [18] Hendrickson et al. (2004).

[b]Dyneins in parentheses are not completely absent.

[c]See Chapter 9 for further discussion of suppressor mutations.

The $sup_{pf}1$ (*sup-pf-1*) mutations in the β HC of outer arm dynein result in short deletions within the microtubule-binding stalk region and have been shown to restore motility in mutants lacking the central pair or radial spokes (Huang et al., 1982; Porter et al., 1994). The $sup_{pf}2$ (*sup-pf-2*) mutations also suppress flagellar paralysis and occur within the outer arm γ HC (Huang et al., 1982; Rupp et al., 1996). Although in this case the precise alterations remain uncertain, these strains exhibit an allele-specific reduction in the number of assembled outer arms (up to 45–50%) with an apparent bias toward loss on specific doublet microtubules. This and other mutations which serve to ameliorate the paralyzed-flagella phenotypes of central pair/radial spoke-deficient mutants have been termed "suppressors." Functions of suppressor mutations are detailed in Chapter 9, and the structural basis for how a mutation in outer arm dynein might suppress paralyzed-flagella mutations is discussed briefly in section IV.D.1.c.

C. Genetic analysis

The genetic relationship between different dynein-deficient mutants can be examined by standard procedures of genetic analysis, i.e., mating with appropriate mutants and analyzing the resulting tetrads. In many cases, mating

Table 6.4	Motility and flagellar length of dynein mutants				
	Swimming velocity[a,e] (μm/second)	Beat frequency[b,e] (Hz)	Distance moved per beat cycle[c] (μm)	Flagella length (μm)	Missing inner arm species
wt	136.0 ± 12.0	51 ± 13	2.67	11.1 ± 1.9[d]	
oda1	46.7 ± 5.9	23 ± 3	2.07	11.4 ± 2.0[d]	
ida1	77.8 ± 9.2	45 ± 6	1.72	10.9 ± 1.6[d]	f
ida4	79.8 ± 7.2	50 ± 12	1.60	9.4 ± 1.6[d]	a, c, d
ida5	75.0 ± 10.9	42 ± 16	1.79	12.4 ± 1.2[d]	a, c, d, e
ida9	110.0 ± 8.0	52 ± 11	2.11	11.8 ± 1.8	c

[a]*Average and standard deviation were calculated from the data for more than 50 cells. Temperature: 20°C.*

[b]*Median frequency and standard deviation estimated from the power spectrum from a population of cells. The beat frequency data display greater deviations than the swimming velocity data. This is possibly due to the fact that the velocity was measured using cells that swam smoothly in straight paths, while beat frequency was measured using a total population of cells. Temperature: 20°C.*

[c]*Average swimming velocity divided by average beat frequency.*

[d]*From Kato-Minoura et al. (1997).*

[e]*Swimming velocity and beat frequency in cells swimming forward. Cells displaying circling movements with only one flagellum beating were not used for measurements. From Yagi et al. (2005).*

between two different mutants results in the formation of quadriflagellate temporary dikaryons that display higher motility than the original strains due to cytoplasmic complementation. This phenomenon is called temporary dikaryon rescue, and can conveniently be used for determining the relationship between a new isolate and known mutants. However, in some cases, such as the mating between *oda1* and *oda3*, or between *oda5* and *oda10*, temporary dikaryons do not undergo clear rescue (Kamiya, 1988). This is most likely because the protein products missing in the two mutants form a complex and either product becomes unstable in the cytoplasm if not associated with the other (Takada et al., 2002; Wirschell et al., 2004).

D. Motility phenotypes of dynein mutants

Both *oda* and *ida* mutants display lowered swimming velocities compared to wild-type cells (Table 6.4). However, the reason for the decreased velocity differs; the lowered velocity in *oda* strains is due to the reduction in beat frequency, while that in *ida* mutants is due to a reduction in the angle of the principal bend in the flagellar waveform (Brokaw and Kamiya, 1987; Kamiya et al., 1991; Yagi et al., 2005). Thus, outer arms appear to be important for the flagella to beat at high frequency, while the inner arm dyneins are required for flagella to beat with the proper waveform. Because of the low flagellar beat frequency, *oda* mutants display jerky swimming; in contrast, due to a higher beat frequency, *ida* mutants swim smoothly.

Although mutants lacking either outer arm dynein or some of the inner arm dyneins are motile, double mutants lacking the outer arm and inner arm species I1/f, the outer arm and several monomeric inner arm dyneins, or dimeric I1/f dynein and several monomeric dyneins, are nonmotile (Kamiya et al., 1991; Kamiya, 2002). Mutants lacking outer arm dynein and I1/f usually have short flagella, although several strains with additional mutations that suppress this phenotype also exist. These observations suggest a functional inter-dependence between different types of dyneins.

Cells demembranated with nonionic detergent (cell models) can be reactivated to beat in the presence of ATP. The movement of cell models is similar to that of live cells; in wild-type cell models, the maximal beat frequency is typically 40–50 Hz, i.e., 70–80% of that of live cells, while in *oda* cell models, the beat frequency can be 25–30 Hz, i.e., 90–100% that of live *oda* cells (at 25°C). Isolated and demembranated flagella (axonemes) also display beating *in vitro* in appropriate buffer solutions containing ATP (Bessen et al., 1980).

In vitro axonemal movements display marked changes depending on the Ca^{2+} concentration. Most notably, when Ca^{2+} is raised from below 10^{-6} to 10^{-4} M, axonemes change waveform from an asymmetric, ciliary-type pattern to a symmetric, flagellar-type beat (Hyams and Borisy, 1978; Bessen et al., 1980). This waveform switching is used when the cell displays the photophobic response, a transient backward movement upon sudden stimulation by strong light (see Chapter 13). Mutant axonemes lacking outer arm dynein have difficulty in this waveform switching, indicating that outer arm dynein at least partially contributes to this mechanism (Kamiya and Okamoto, 1985). Mutants lacking inner arm dyneins undergo waveform conversion almost normally (R. Kamiya, unpublished observations).

Another Ca^{2+}-dependent change in axonemal activity is the alteration in the power balance between the *cis*-flagellum (the one nearest to the eyespot) and the *trans*-flagellum (the one farthest from the eyespot). At Ca^{2+} concentrations lower than 10^{-7} M, the *cis*-flagellum is dominant, and at 10^{-7}–10^{-6} M Ca^{2+}, the *trans*-flagellum becomes dominant (Kamiya and Witman, 1984). This change in flagellar dominance has been suggested to be important for the phototactic behavior of the cell (Horst and Witman, 1993; Okita et al., 2005).

The primary function of dyneins in the flagellar beat mechanism is to cause sliding between outer doublets (Satir, 1968; Summers and Gibbons, 1971; Sale and Satir, 1977). This activity can be examined under the microscope by briefly treating fragmented axonemes with an appropriate protease to break the inter-doublet links and then measuring the rate of doublet microtubule extrusion following ATP addition (Summers and Gibbons, 1971; Witman et al., 1978; Okagaki and Kamiya, 1986). Such assays have shown that the maximal velocity of microtubule sliding in wild-type axonemes is ~18 μm/second, while that of *oda1* axonemes is ~4 μm/second. Interestingly, the sliding velocities in mutant axonemes lacking several inner arm dyneins are close to the wild-type velocity. These results indicate that outer arm dynein causes intrinsically faster sliding than inner arm dyneins (Kurimoto and Kamiya, 1991).

Flagellar beating produces the force that propels the cell. Mutants lacking dyneins swim slower than wild-type cells because their flagella produce less propulsive force. The swimming velocity of a cell also varies with the viscosity of the medium, in a manner that differs from one dynein-deficient mutant to another (Minoura and Kamiya, 1995; Yagi et al., 2005). The propulsive force at various viscosities can be determined from the product of the swimming velocity and the viscosity. Such estimates show that force production in wild-type cells increases at slightly increased viscosities, but that a similar increase is not evident in the *oda1* mutant (Figure 6.2). The maximal force production in wild-type cells is about three times as large as that in *oda1* cells at the same viscosity. At higher viscosities, while the force production in wild-type cells decreases sharply, that in the *oda1* mutant persists fairly constantly. Mutants lacking monomeric dyneins, such as *ida4* and *ida9*, are very sensitive to viscous load. These observations indicate that force production by outer arm dynein is regulated such that it increases when the cell needs greater force at slightly increased viscosity. Furthermore, these studies revealed that outer arm dynein has the potential to produce twice as much force as inner arm dyneins and that the monomeric inner arm dynein(s) are important for production of force in viscous media.

As discussed in detail in Chapters 7 and 8, the radial spokes and central pair complex regulate axonemal beating by controlling the generation of force by dynein. This idea first came from the finding that the "suppressor" mutations, which can restore motility in mutants lacking the central pair or radial spokes, are mutations in dynein itself or in the DRC, a protein

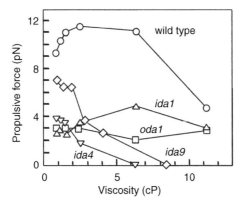

FIGURE 6.2 *The effect of viscosity on dynein-generated propulsive force. The propulsive forces (pN) produced by wild-type and mutant cells swimming in media of different viscosities is illustrated. Propulsive force was estimated from the product of the average swimming velocity, viscosity of the medium, and the radius of the cell body. Medium viscosity was changed with Ficoll. The data are from Minoura and Kamiya (1995) and Yagi et al. (2005).*

complex that is apparently involved in dynein regulation (Huang et al., 1982; Porter et al., 1992). The discovery of suppressors implies that the axoneme can beat without the central pair or radial spokes if dynein motor function is activated. Some of the suppressor mutations may well be those that cause dynein activation. For example, the $sup_{pf}1$ mutation in the outer arm β HC may have constitutively activated outer arm dynein either due to a direct activating effect on that HC or by allowing an inhibitory system controlling another HC to be circumvented.

Another line of evidence that activated dyneins can bypass the central pair/radial spoke defects has been obtained from reactivation experiments. As described in section VII.A.1, the ability of dynein to translocate microtubules *in vitro* is greatly enhanced in the presence of a low concentration of ATP, or in the presence of ADP in addition to ATP. When mutant axonemes lacking the central pair or radial spokes are placed under such conditions, they display beating (Omoto et al., 1996). Outer arm dynein is necessary for this movement to occur, suggesting that it is the activated outer arm that functions to suppress paralysis.

IV. COMPOSITION AND PHYSICAL PROPERTIES OF PURIFIED DYNEINS

A. Two groups of dyneins: monomeric and oligomeric motors

From a structural and compositional perspective, dyneins may be divided into two general groups that are defined by the number of HC motor units and by very distinct sets of associated components (reviewed in Sakato and King, 2004). Several species of inner arm dyneins consist of single HC motors associated with one molecule of actin and either the Ca^{2+}-binding protein centrin or the p28 LC. Outer arm dynein, inner arm dynein I1/f and cytoplasmic dyneins contain two or three HCs, WD-repeat ICs, and members of at least three distinct families of LC (LC8, Tctex1, and LC7/roadblock); the ICs and members of the three LC families associate to form an IC/LC complex. Some dyneins contain additional components beyond this core composition such as the supplementary IC in inner arm I1/f, the light intermediate chains (LICs) of cytoplasmic dyneins, and the regulatory LCs associated with the outer arm HCs. The polypeptides comprising isolated axonemal dyneins from *Chlamydomonas* and their properties are listed in Tables 6.1 and 6.5[1] and a model and protein–protein interaction map for the outer dynein arm is shown in Figure 6.3.

---

[1]The nomenclature for dynein components used here is specific for *Chlamydomonas* and great care must be taken when identifying the orthologous components of dyneins from other species. For example, the *Chlamydomonas* outer arm dynein γ HC is most closely related to the α HC of the sea urchin outer arm, and the highly conserved protein termed LC8 in *Chlamydomonas* is known as LC6 in sea urchin. Similarly, sea urchin LC1 is a member of the Tctex1/Tctex2 family, whereas in *Chlamydomonas* LC1 designates a leucine-rich repeat protein.

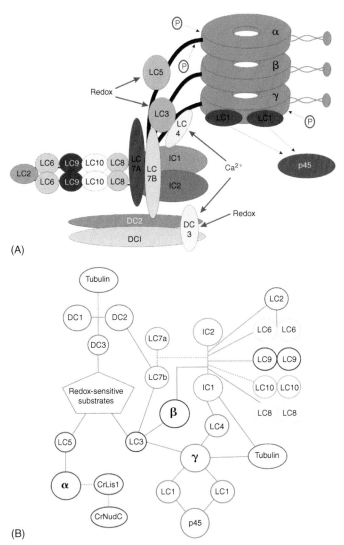

FIGURE 6.3 *Model and interaction map for outer arm dynein. (A) Model for the arrangement of proteins within the outer dynein arm. Cryoelectron tomographic reconstructions indicate that the AAA[+] rings for the three HCs overlap each other (Nicastro et al., 2006; Ishikawa et al., 2007; Oda et al., 2007). Furthermore, the ICs and their associated LCs are known to be located at the base of the particle (King and Witman, 1990), although further details of how these various proteins are related within the* in situ *arms remain unclear. Interaction with the docking complex involves the LC7b protein and may also be mediated by other components (DiBella et al., 2004a). Multiple regulatory inputs impinge on various components of the motor including redox signals through LC3, LC5, and DC3 (Patel-King et al., 1996; Casey et al., 2003b; Wakabayashi and King, 2006), Ca[2+] binding by LC4 and DC3 (King and Patel-King, 1995a; Casey et al., 2003b), and phosphorylation of both the α and γ HCs (Piperno and Luck, 1981; King and Witman, 1994; Wagner et al., 2006). (B) Map for the* Chlamydomonas *outer dynein arm revealing the protein–protein interactions that occur within this complex enzyme. Associations for which there is direct experimental evidence in* Chlamydomonas *are indicated by solid lines, whereas those interactions that are presumed based on analysis of other dyneins are indicated by broken lines.*

Table 6.5 Proteins present in purified axonemal dyneins

Dynein	Protein	Mass (kD)	pI	Stoichiometry	Accession number	Properties
Outer arm	α HC	504	5.29	1	Q39610	ATPase/microtubule motor, contains Kelch + immunoglobulin domains at N-terminus, phosphorylated at multiple sites
	β HC	520	5.40	1	Q39565	ATPase/microtubule motor
	γ HC	513	6.56	1	Q39575	ATPase/microtubule motor, contains IQ motifs in N-terminal region, phosphorylated on AAA3
	IC1	76	6.58	1	Q39578	WD-repeat protein, binds α-tubulin, associates with multiple LCs
	IC2	63	5.40	1	P27766	WD-repeat + coiled-coil protein, associates with multiple LCs
	LC1	22	5.54	2	AAD41040	Leucine-rich repeat, binds γ HC motor domain
	LC2	16	6.01	1	AAB58383	Tctex2, interacts with ICs and LC6
	LC3	17	7.79	1	Q39592	Thioredoxin, associates with β and γ HCs
	LC4	18	4.22	1	Q39584	Ca^{2+} binding, γ HC-associated
	LC5	14	8.34	1	Q39591	Thioredoxin, α HC-associated
	LC6	14	6.80	2	Q39479	LC8 homologue, dimer interacts with ICs and LC2
	LC7a	12	7.85	1	AAD45881	Roadblock homologue, interacts with ICs
	LC7b	11	5.81	1	EDP03034	Roadblock homologue, interacts with DC2 and LC3
	LC8	10	7.51	4	Q39580	Highly conserved, dimer interacts with ICs, also present in inner arm I1/f and radial spokes
	LC9	13	4.37	2	AAZ95589	Tctex1 homologue, dimeric, interacts with IC1 and IC2
	LC10	12	5.28	2	EDP00562	LC8 homologue, dimeric
	DC1	83	5.79	1	AAC49732	Docking complex, coiled-coil protein
	DC2	62	5.74	1	AAK72125	Docking complex, coiled-coil protein
	DC3	21	5.51	1	AAP49435	Docking complex, binds Ca^{2+} in a redox-sensitive manner
Inner arm I1/f	Iα	523	5.21	1	Q9SMH3	ATPase/microtubule motor
	Iβ	510	6.23	1	Q9MBF8	ATPase/microtubule motor
	IC140	110	4.70	1	AAD45352	WD-repeat protein
	IC138	111	5.80	1	AAU93505	WD-repeat protein involved in phosphorylation-based regulation
	IC97	~90	n.d.	1	EDP03678	Related to vertebrate Las1; interacts with IC140, IC138, and tubulin
	Tctex1	12	5.17	2	AAC18035	Dimeric, LC9 homologue
	Tctex2b	14	5.81	n.d.	DAA05278	LC2 homologue
	LC7a	12	7.85	n.d.	AAD45881	Shared with outer arm dynein
	LC7b	11	5.81	n.d.	EDP03034	Shared with outer arm dynein, interacts with IC138
	LC8	10	7.51	>2?	Q39580	Shared with outer arm dynein
	ODA7	47	4.47	1?	ABI63572	Purifies with the outer arm in the absence of I1/f; putative outer-inner arm linker

Monomeric inner arms species	DHC6	n.d.	n.d.	1	AAC49519[a]	ATPase/microtubule motor, monomeric species a
	DHC5	n.d.	n.d.	1	AAC49518[a]	ATPase/microtubule motor, monomeric species b
	DHC9	465	5.76	1	BAE19786	ATPase/microtubule motor, monomeric species c
	DHC2	n.d.	n.d.	1	AAC49515[a]	ATPase/microtubule motor, monomeric species d
	DHC8	n.d.	n.d.	1	AAC49512	ATPase/microtubule motor, monomeric species e
	DHC7	n.d.	n.d.	1	AAC49520[a]	ATPase/microtubule motor, monomeric species g
	DHC4	n.d.	n.d.	1	AAC49517[a]	ATPase/microtubule motor, minor monomeric species of unknown composition
	DHC11	n.d.	n.d.	1	CAB39162[a]	ATPase/microtubule motor, minor monomeric species of unknown composition
	DHC3	n.d.	n.d.	1	AAC49516[a]	ATPase/microtubule motor, minor monomeric species of unknown composition
	Actin	42	5.24	1	P53498	Essential for assembly of species a, c, d, e and one minor dynein
	NAP	41	5.66	1?	AAC49834	Novel actin-related protein that can functionally replace actin for species b and g assembly
	p28	29	7.13	2	Q39604	Essential for assembly of species a, c, and d; dimeric and binds N-terminal region of HC
	Centrin	19	4.52	1	P05434	Ca^{2+}-binding protein
	p38	41	5.29	1?	BAG07147	Associates with species d only
	p44	44	5.10	1?	BAF98914	Associates with species d only

[a]Entries with these accession numbers contain only partial HC sequences.

B. Biochemical purification and fractionation

Axonemal dyneins were initially extracted from demembranated flagella by treatment with 0.6 M NaCl or KCl in the presence of 1 mM ATP (Piperno and Luck, 1979; Pfister et al., 1982; King et al., 1986) (Figure 6.4A). Subsequent studies revealed that high concentrations (10 mM) of ATP are also sufficient to remove these structures from the axoneme (Goodenough and Heuser, 1984). Following extraction, crude dynein samples were then fractionated by sucrose density gradient centrifugation and hydroxyapatite or ion exchange chromatography to yield highly purified enzymes (Figure 6.4B–D). In sucrose gradients, the inner arms are obtained as two distinct peaks sedimenting at ~17 S (inner arm I1/f) and ~10 S (monomeric inner arms termed I2/3 by Piperno and Luck, 1981). When extracted from *oda1* axonemes, this latter group can be separated by ion exchange chromatography into at least six distinct particles (designated a, b, c, d, e, and g) that contain single HCs, actin, and either centrin or p28 (Kagami et al., 1990); a typical elution pattern and HC composition is shown in Figure 6.5. In addition, dynein species d contains two other proteins of 44 and 38 kD (see below). Because no mutants are available that lack dynein species b and g, the subunit composition of these enzymes has not been firmly established and they also may have subunits in addition to actin and centrin. Some minor inner-arm species have been found to resist high-salt extraction (T. Yagi and R. Kamiya, unpublished observations).

The initial isolations of outer arm dynein were performed in the absence of Mg^{2+} and using ultracentrifuge rotors with large tube size that subjects the particles to high hydrostatic pressure. This resulted in dissociation of this dynein into three subparticles: the 18 S $\alpha\beta$ HC complex that also contains the IC/LC assembly, the 12 S γ HC particle, and a 7 S trimeric ODA-DC (Piperno and Luck, 1979; Pfister et al., 1982; Takada and Kamiya, 1994). Electron backscattering measurements made by scanning transmission EM suggested that the 18 and 12 S particles have masses of 1220 ± 117 and 460 ± 51 kD, respectively (Witman et al., 1983). The upper limits of these mass values (especially for the 18 S particle) are close to those calculated based on our current understanding of composition and sequence (1351 and 577 kD for the $\alpha\beta$ and γ subunits, respectively).

The $\alpha\beta$ HC complex may be further dissociated by low ionic strength dialysis to yield two subparticles consisting of the α HC and LC5, and the β HC, LC3, and the IC/LC complex. These two particles have distinct sedimentation coefficients (13.1 S and 15.8 S, respectively) and may be purified using additional sucrose density gradient steps following restoration of the ionic strength to physiological levels (Pfister and Witman, 1984). It was subsequently determined that intact 23 S outer arm dynein containing all three HCs and the ODA-DC could be obtained by addition of Mg^{2+} to the buffer and use of an ultracentrifuge rotor with smaller tube size to reduce the imposed hydrostatic pressure (Takada et al., 1992; Nakamura

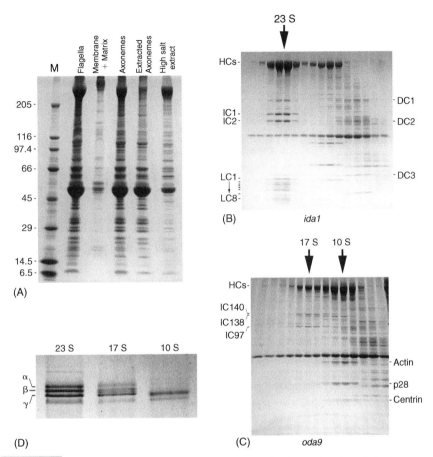

FIGURE 6.4 *Extraction and purification of axonemal dyneins. (A) Electrophoretic analysis of fractionated* Chlamydomonas *flagella in a 5–15% acrylamide gradient. Flagella were first treated with 1% IGEPAL CA-630 to solubilize the membrane and flagellar matrix components. Subsequently, the resulting axonemes were treated with 0.6 M NaCl to extract the outer and inner dynein arms in addition to other axonemal structures. From Patel-King et al. (2004). Copyright of the American Society for Cell Biology. (B, C) High salt extracts of axonemes from mutants* ida1 *(B; lacking inner arm I1/f) and* oda9 *(C; lacking the outer arm) were fractionated in 5–20% sucrose density gradients in the presence of Mg²⁺. Intact outer arm dynein sediments at ~23S (* ida1 *gradient) whereas inner arm I1/f and the monomeric inner arms migrate at ~17 and 10 S, respectively (* oda9 *gradient). The IC, LC, and ODA-DC components of these complexes are indicated. From DiBella et al. (2004a). Copyright of the American Society for Cell Biology. (D) Dynein HCs from the 23, 17, and 10 S regions of a sucrose gradient loaded with a wild-type extract were resolved in a 4% acrylamide, 4 M urea gel. The outer arm HCs are well resolved and are indicated at left. The band below the* γ *HC consists of proteolytic fragments of the* α *and* β *HCs that derive from cleavage between AAA5 and AAA6; these products were previously referred to as bands 10 and 11 (Pfister et al., 1982). The 17 and 10 S fractions contain distinct HCs of the I1/f and monomeric dyneins.*

et al., 1997). The requirement for Mg^{2+} remains unexplained, although it potentially may reflect binding of this cation by the DC3 component of the ODA-DC. In addition, pretreatment of the axonemal sample with 0.6 M K acetate prior to 0.6 M NaCl removes many salt-extractable components

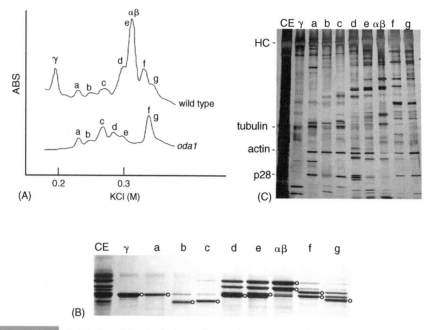

FIGURE 6.5 *Purification of dyneins by ion exchange chromatography. (A) Elution patterns of wild-type and* oda1 *axonemal high-salt extracts separated by ion exchange chromatography on a Mono-Q column. The peaks corresponding to each dynein species are indicated. The αβ and γ subparticles derive from the dissociation of outer arm dynein; the peaks labeled a–g indicate the inner arm dynein species. (B, C) SDS-PAGE analysis of the peak fractions from wild-type extracts. Panel B shows only the high molecular weight dynein HC region. Bands with circular marks are those of the HCs constituting the dynein species indicated in (A). Panel C reveals the lower molecular weight species associated with these HCs; CE indicates the crude high-salt extract.*

that otherwise contaminate the dynein preparation (Nakamura et al., 1997).

Importantly, from both biochemical and genetic studies, it is now clear that other flagellar components associate directly with dyneins *in vivo*, but become dissociated during extraction and/or purification; these are discussed below in section V.

C. Structural organization of isolated dynein particles and their relationship to *in situ* arms

Many studies have used various EM techniques, including thin sections, negative stain, quick-freeze/deep-etch, scanning transmission EM, and cryo-EM tomographic reconstructions, to define the structure of purified *Chlamydomonas* dyneins and/or determine how the various domains are arranged *in situ* within the confines of the axonemal superstructure. In longitudinal thin sections, the outer arms are evident as tilted overlapping structures in a configuration that is also seen by negative stain (Witman and Minervini, 1982). Further details of this arrangement were revealed using the quick-freeze/deep-etch method (Goodenough and Heuser, 1982). In transverse sections (Figure 6.1A, B), the outer arm appears as a club-shaped structure with a tapered base at the A-tubule. Analysis of mutants lacking either the α HC or the β HC motor domain (Sakakibara et al., 1991, 1993) revealed that the region near the B-tubule can be subdivided into three sections each derived from a dynein motor domain, with that from the α HC being outermost and with the γ HC motor unit nearest the inner arms (Figure 6.1B).

In negative stain, isolated outer arm αβ subparticles contain two globular domains interconnected by stems that join at a common base (Witman et al., 1983); in some images of *Tetrahymena* outer arm dyneins, an accumulation of stain was visible in the center of each domain suggesting a depression or pit (Marchese-Ragona and Johnson, 1988). Reconstruction and image averaging of negatively stained *Dictyostelium* dynein HCs to a resolution of 25 Å revealed that the globular head consists of a planar ring of seven subdomains (Samsó et al., 1998; Samsó and Koonce, 2004). This general organization was also confirmed by quick-freeze/deep-etch EM of intact outer arms, which identified additional substructures within the particle including subdomains near the base that presumably derive from the ICs and/or LCs (Goodenough and Heuser, 1984). Immuno-EM of purified αβ HC dimers bound to monoclonal antibodies raised against an outer arm IC (King et al., 1985; King and Witman, 1990) and analysis of fractionated sea urchin outer arm dyneins lacking the IC/LC complex (Sale et al., 1985) both revealed that the ICs are indeed located toward the base of the isolated particle. The work of Goodenough and Heuser (1984) also identified the coiled-coil stalks terminating in globular microtubule-binding domains that protrude from the globular head of each dynein HC (Figure 6.1C); this feature was confirmed in negative stain images of the monomeric inner arm dynein species c (Burgess et al., 2003). The latter study suggested that the stalks change flexibility in response to different nucleotide-bound states (see section VII.B).

Electron tomography of metal-replicated outer arms in sperm axonemes of the cecidomid dipteran insect *Monoarthropalpus flavus* revealed that the globular heads are stacked against each other and are each a slightly different distance from the B-tubule (Lupetti et al., 2005). Tomographic analysis of frozen hydrated *Chlamydomonas* axonemes has determined that all three outer arm HC motor units are stacked against each other and that the AAA$^+$ rings are arranged parallel to the axonemal long axis and have equivalent access to the surface lattice of the doublet microtubule (Nicastro et al., 2005, 2006; Ishikawa et al., 2007 and see Chapter 9); this arrangement is also observed for outer arm dyneins that have been rebound via both structural and ATP-sensitive domains to taxol-stabilized microtubules (Oda et al., 2007).

These cryo-EM studies have also confirmed the existence of linkers between the outer arms that had been suggested by previous work (Goodenough and Heuser, 1984; Burgess et al., 1991; Lupetti et al., 2005). Currently, there is some disagreement on how the densities joined by the linkages between outer arms should be assigned to particular dynein components. In the tomographic reconstructions by Nicastro et al. (2006), these densities appear to span between the base of the α and β HC motor units (apparently near the junction of AAA1 and the N-terminal stem) of one outer arm and the γ HC AAA$^+$ ring of the adjacent outer arm (see Chapter 9). In contrast, Oda et al. (2007) suggest that this linkage traverses between the β HC motor unit of one dynein and the basal assembly (presumably formed from the IC/LC complex) of its neighbor. A third model suggested by Ishikawa et al. (2007) is that the linker represents the γ HC N-terminal stem. In this case, two potential arrangements are proposed in which (1) all HC motor units in one arm derive from the same dynein; or (2) the γ HC head of one dynein associates with the adjacent arm. As this outer arm linker is more complex in *Chlamydomonas* than in sea urchin (Nicastro et al., 2006), an additional possibility is that it derives, at least in part, from the IC/LC complex and that the *Chlamydomonas*-specific elaborations involve components (such as LC6) that are not present in sea urchin.

Nicastro et al. (2006) also identified linkages between the outer arm and inner arm I1/f and between the outer arm and the DRC (these interactions and their significance are discussed more fully in Chapter 9). As there are four outer arms within the 96-nm axonemal repeat, this implies that they cannot all be equivalent, and it is likely that functional and possibly compositional differences are present, although these remain to be defined.

D. The outer dynein arm

1. Heavy chains

a. Overview

A phylogenetic tree illustrating the relationship of the 16 different dynein HCs found in the *Chlamydomonas* genome is shown in Figure 6.6A. The *Chlamydomonas* outer dynein arm contains three HCs (α, β, and γ) which are the products of distinct genes and as described below have very different assembly, enzymatic, and motor properties. Unlike kinesins and myosins, which are distantly related structurally to the GTPase subunit of heterotrimeric G proteins (Kull et al., 1998), dynein HCs are members of the ancient AAA$^+$ (ATPases associated with cellular activities) family of ATPases that includes the bacterial Clp proteases, mammalian *N*-ethyl-maleimide-sensitive vesicle fusion protein (NSF), the microtubule-severing protein katanin, and the RuvB DNA pump that is required for the resolution of Holliday junctions (Patel and Latterich, 1998; Neuwald et al., 1999; Ogura and Wilkinson, 2001). Dynein HCs have a mass of ~520 kD and

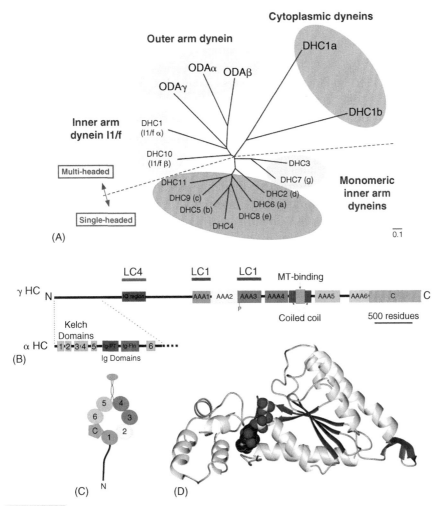

FIGURE 6.6 *Phylogeny and organization of dynein heavy chains. (A) Phylogenetic tree illustrating the relationship between all the* Chlamydomonas *dynein HCs; the inner arm dyneins to which HCs have been assigned are indicated in parentheses (Yagi and Kamiya, unpublished). Sequences of the HC motor domains were aligned with CLUSTALW and the tree constructed using TREEVIEW. This analysis revealed a major division between HCs that exist as monomeric motors compared to those that function as oligomers. The DHC1b HC is the retrograde motor for IFT (see Chapter 4), whereas the HC labeled DHC1a may represent the* Chlamydomonas *version of the major cytoplasmic dynein isoform. However, the sequence of this latter protein is quite distinct from cytoplasmic dyneins of other organisms and as yet there is no experimental evidence to indicate that this is indeed a* bona fide *cytoplasmic motor. (B) Diagram of the γ HC indicating the location of the various domains discussed in the text and the regions involved in binding the LC1 and LC4 light chains. Also shown is the domain structure of the N-terminal region of the α HC which is completely distinct from that of other dynein HCs. (C) A model illustrating the general arrangement of the various subdomains is shown. Following the N-terminal stem, the motor domain consists of seven subdomains (six AAA+ domains and the C-terminal unit) arranged in a toroidal structure. The coiled-coil segment and microtubule-binding domain protrude from between AAA4 and AAA5. The color-coding of the various domains is as for panel B. (D) Structural model of the AAA1 domain of the outer arm γ HC (King, unpublished) was calculated using SWISS-MODEL (Peitsch, 1996) based on the known structures of several AAA+ proteins. This conserved domain consists of an α/β region that binds nucleotide and an α helical subdomain. ATP is shown as a space-filling model (gray, carbon; blue, nitrogen; red, oxygen; yellow, phosphorous).*

contain ~4500 residues. They possess a complex domain organization that in general consists of an N-terminal region of ~1800 residues followed by a motor unit containing a heptameric ring (Samsó et al., 1998; Samsó and Koonce, 2004) of six AAA$^+$ domains and a C-terminal unit of uncertain function (King, 2000); the microtubule-binding domain at the tip of a coiled-coil stalk protrudes from between the fourth and fifth AAA$^+$ domains (Figure 6.6B, C). An alternative model based on EM localization of green and blue fluorescent protein tags inserted within the motor unit proposes a hexameric ring of AAA$^+$ domains with the C-terminal segment spanning from the AAA1/AAA6 junction toward the MT-binding stalk (Roberts et al., 2007).

b. The N-terminal region

This region of the dynein HC forms part of the stem of the complex and is required for assembly into the particle. The functional roles played by this domain remain somewhat uncertain, but are known to include association with multiple protein partners, and the transduction of regulatory signals; as this region in dynein c appears to curve around and interact with one side of the AAA$^+$ ring (Burgess et al., 2003) it may also play an essential role in the motor mechanism (discussed further below in section VII.B).

In cytoplasmic dynein, the N-terminal domain has been proposed to dimerize (Mazumdar et al., 1996) through a ~300-residue segment that mediates HC–HC interactions (Habura et al., 1999); this same region is also apparently involved in HC–IC and HC–LIC associations (Habura et al., 1999; Tynan et al., 2000). Currently, there are no data from *Chlamydomonas* to support the direct interaction of the outer arm HCs with each other, and the proposed cytoplasmic dynein HC dimerization region is not well conserved in the axonemal enzymes. Interestingly, the cytoplasmic dynein HC constructs were expressed in cell systems that would also have contained considerable quantities of the LC8 protein and potentially other dynein components, raising the possibility that the interactions observed were not direct. Using increasing levels of detergents to progressively dissociate the outer arm, Mitchell and Rosenbaum (1986) revealed a direct interaction between the β HC and IC2. However, it remains unclear whether the α and γ HCs also interact directly with one or more of these IC components or whether their association is mediated through the LCs that also form part of this complex.

Outer arm dynein HCs are distinct in that they are directly associated with a series of LCs (Piperno and Luck, 1979; Pfister et al., 1982) that appear to play a role in the regulation of motor function (King, 2002a). Both α and β HCs bind thioredoxin-related proteins (LC5 and LC3, respectively) within their N-terminal region (Patel-King et al., 1996; Harrison et al., 2002). The γ HC N-terminal domain also interacts directly with LC3 and with the Ca^{2+}-binding protein LC4 (Pfister et al., 1982; Sakato et al., 2007). This latter interaction involves a region encompassing residues

875–1173 that contains two IQ motifs that are known to bind calmodulin-like proteins (Rhoads and Friedberg, 1997). Furthermore, two copies of the leucine-rich repeat (LRR) protein LC1 associate with the AAA$^+$ domains of the γ HC motor unit (Benashski et al., 1999). These LC associations and their significance are described in more detail below in sections IV.D.3.b–d. The γ HC contains a structural (ATP-insensitive) microtubule-binding site that is most likely located within the N-terminal region (Sakakibara and Nakayama, 1998).

As noted previously by Mitchell and Brown (1994), the N-terminal ~650 residues of the α HC are distinct from the analogous regions of the β and γ HCs and are separated from the remainder of the molecule by a repeated degenerate EPAA motif. Indeed, more recent domain analysis (S.M. King, unpublished) has revealed the presence of six kelch domains (named for the *Drosophila* F actin-binding protein; see Prag and Adams, 2003, for review) and two immunoglobulin-like folds (one of the Ig/filamin class and one of the Ig/plexin/transcription factor class) (Figure 6.6B); a dynein HC containing N-terminal kelch domains is also present in *Plasmodium* suggesting that this is not a *Chlamydomonas*-specific modification. Kelch domains form four β strands as do the WD-repeats (see section IV.D.2) and, based on sequence analysis (Prag and Adams, 2003) and the structure of a fungal galactose oxidase (pdb accession 1GOF; Ito et al., 1991), are usually organized as a six-bladed β-propeller: note that the fungal protein is actually unusual in that it contains seven repeats. The Ig domains consist of two antiparallel β sheets arranged as a β sandwich. Both Ig folds are located between the fifth and sixth kelch domains, suggesting that they protrude to the same side of the β propeller. The interactions mediated by this region are uncertain at present.

c. AAA$^+$ domains and the motor unit

The dynein motor unit consists of six AAA$^+$ domains, a coiled-coil region containing the microtubule-binding site, and a C-terminal unit of unknown function. High resolution structural information is available for a variety of AAA$^+$ proteins—e.g., DNA polymerase III δ′ clamploader (Guenther et al., 1997), NSF (Yu et al., 1998), and the Hs1VU ATP-dependent protease (Wang et al., 2001). In general each AAA$^+$ domain forms two subdomains: an α/β type structure and a helical region. A model for AAA1 from the γ HC is shown in Figure 6.6D. The former subdomain usually acts as a nucleotide-binding module and contains the GX$_4$GKT/S motif (P-loop or Walker A box) that forms ligands with the phosphates of ATP, and the acidic Walker B box that coordinates Mg^{2+}. The helical region provides ligands that detect the presence of the terminal γ-phosphate and also contains the sensor segment that undergoes a conformational change following nucleotide hydrolysis.

Table 6.6	P-loop motifs within heavy chains of oligomeric dynein motors			
Heavy chain	P1	P2	P3	P4
α	GPAGTGKT	GPTGTGRT	GGAGVGKT	GVGGSGKQ
β	GPAGTGKT	GAAGCGKT	GNTGTGKS	GVGGSGKQ
γ	GPAGTGKT	GPSGSGKS	GGPGTAKT	GVGGSGKQ
1α	GPAGTGKT	GQTGGGKT	GESGTAKS	GVGGSGKQ
1β	GPAGTGKT	GRTGSGKS	GNVGVGKT	GVGGSGRK
	* * * * * * * *	* * *	* * *	* * * * * *
Human cytoplasmic dynein 1 HC (DYNC1H1)	GPAGTGKT	GPSGSGKS	GPPGSGKT	GVSGAGKT
	* * * * * * * *	* * *	* * *	* * * *

Within dynein, the first AAA$^+$ domain is most highly conserved and indeed the P-loop motif is invariant across all dyneins (Table 6.6). Even among the *Chlamydomonas* outer arm HCs, the sequences of these motifs in the second and third domains are poorly conserved. In contrast, the fourth P-loop is almost completely conserved in the axonemal dyneins (except for the I1/f 1β HC) but not in their cytoplasmic counterparts; for the axonemal enzymes this motif ends in glutamine rather than the canonical serine/threonine. Furthermore, analysis of NSF (Lenzen et al., 1998; Yu et al., 1998) revealed that a lysine residue from one AAA$^+$ domain can also interact with and stabilize nucleotide bound to the adjacent subunit, thus allowing for ATP-dependent structural alterations to be detected and propagated around the ring. Both cytoplasmic and axonemal dyneins contain a basic residue at this position in AAA5 which may therefore potentially interact with and stabilize nucleotide bound to AAA4 (King, 2000).

Within the dynein HC, sequence analysis predicts that the last two AAA$^+$ domains are unlikely to bind nucleotide as they lack intact P-loops (Walker A box motifs). However, based on equilibrium dialysis measurements of sea urchin outer arm dynein *in vitro* (Mocz and Gibbons, 1996), the first four modules may potentially all bind nucleotide, although it remains uncertain whether this actually occurs *in vivo*, as fluorescence anisotropy measurements of methylanthraniloyl-ATP binding identified only two sites with physiologically relevant association constants (Mocz et al., 1998). Furthermore, photoinsertion sites for azido nucleotide analogues were mapped to two regions within the α and β HCs (King et al., 1989): one near the first AAA$^+$ domain and a second in one of the more C-terminal modules (probably AAA4).

Located between the fourth and fifth AAA$^+$ domains are two regions that have a high probability of forming an antiparallel coiled coil ~10 nm in length. Between these two segments is a small region of ~150 residues that forms a globular unit at the tip of the coiled-coil stalk and has been assigned as the microtubule-binding site (Koonce and Tikhonenko, 2000). Somewhat surprisingly, this region is not particularly well conserved, although all dyneins appear competent to bind microtubules from diverse sources. For example, this region of the γ HC contains a segment of ~40 residues with a very high coiled-coil prediction that is not present in other dynein HCs.

In vitro studies of sea urchin dynein have revealed that there is one register of the antiparallel coiled coil that allows for high affinity interaction with microtubules (Gibbons et al., 2005). This observation led to the suggestion that ATP-driven alterations in the coiled-coil register by half a heptad repeat might be used to regulate microtubule-binding affinity. Thermodynamic modeling suggests that a combination of sliding, bending, and twisting of the coiled coil can promulgate a significant allosteric change in free energy along the length of this structure (Hawkins and McLeish, 2006), and furthermore could lead directly to an alteration in its inherent flexibility as predicted from electron microscopic analysis of dyneins in rigor vs. ATP/vanadate (Burgess et al., 2003). Potentially, twisting of the coiled coil might provide the underlying basis for the torque generated by dyneins that is evidenced by microtubule rotation or bending during translocation (Vale and Toyoshima, 1988; Kagami and Kamiya, 1992; Kikushima and Kamiya, 2008). Intriguingly, two mutations (*sup$_{pf}$1-1* and *sup$_{pf}$1-2*) that suppress flagellar paralysis due to lack of the central pair and radial spokes (Huang et al., 1982) result in small deletions within the first coiled-coil region of the β HC (Porter et al., 1994) and potentially function by disrupting the regulation of microtubule binding by this motor. This suggestion is further supported by the observation that these strains have a reduced ability to undergo the flagellar quiescence that occurs during the transition from swimming to surface gliding motility (Mitchell et al., 2004 and see Chapter 11).

The C-terminal ~40-kD region of the HC is predicted to fold as a globular unit and has been proposed (King, 2000) to contribute the seventh subdomain within the heptameric ring observed by EM (Samsó et al., 1998). This arrangement would place the C-terminal region in close proximity of AAA1 (Figure 6.6C). The C-terminal domain together with AAA6 may be readily cleaved from the HCs by proteolysis although it does not dissociate unless denatured (King and Witman, 1988a; Inaba et al., 1991); more recently, this region has been suggested to act as an autoregulatory domain (Hook et al., 2005). Intriguingly, this region is apparently absent from the cytoplasmic dynein HC of *Saccharomyces cerevisiae* (Eshel et al., 1993), suggesting that in budding yeast this enzyme may be degenerate and lack functional or regulatory attributes present in other dyneins.

Although the AAA$^+$ rings of isolated dynein particles do not appear to interact when examined by EM, there is a clear association of the heads in glutaraldehyde-fixed samples (Goodenough and Heuser, 1984) and a continuity of density in the arms observed in thin sections (Figure 6.1). Furthermore, cryo-electron tomographic studies indicate that these rings overlap each other *in situ* and lie in the longitudinal plane of the flagellum (Lupetti et al., 2005; Nicastro et al., 2005; Ishikawa et al., 2007; Oda et al., 2007). Thus, it is quite possible that the AAA$^+$ units of different HCs interact directly, which may have important mechanistic consequences for the coordination of HC motor activity.

d. Enzymatic and functional properties of outer arm heavy chains

When assayed individually, the α, β, and γ HCs have ATPase activities of 0.6, 7.3, and 1.1 μmol phosphate released/minute/mg, respectively (Pfister et al., 1982; Pfister and Witman, 1984). Reassociation of the α and β HCs downregulates the activity of the latter to yield a combined activity of 2.3 μmol phosphate released/minute/mg that is very close to that of the unfractionated $\alpha\beta$ HC dimer (2.0 μmol phosphate released/minute/mg) (Pfister et al., 1982; Pfister and Witman, 1984). Furthermore, rebinding of the γ HC to the $\alpha\beta$ dimer leads to the further suppression of activity (Nakamura et al., 1997). These individual enzymes also exhibit distinct pH optima: the $\alpha\beta$ HC dimer has a sharp optimum at pH 5.5 and a broader peak of activity from pH 7.5 to 9.0, whereas the γ HC shows a single optimum at pH 10, suggesting that the ionization of a basic residue is involved in ATPase activation (Gatti et al., 1991). Together, the available data provide strong support for the hypothesis that the enzymatic properties of these individual motor units are coordinated within the outer arm.

Both GTP and ITP are poor substrates for *Chlamydomonas* dyneins (Watanabe and Flavin, 1976; Piperno and Luck, 1979), which suggests that the moiety attached to the C-6 position of the purine ring is important for nucleotide recognition by dynein. The photoactive nucleotide probes 2-N_3ATP and 8-N_3ATP have been used to identify the ATP-binding subunits of the outer arm (Pfister et al., 1984, 1985; King et al., 1989). Kinetic measurements revealed that the 8-substituted nucleotide was a poor substrate for dynein (Pfister et al., 1984, 1985), whereas the 2-azido compound and also a 2-chloro derivative were hydrolyzed at rates close to that of ATP (King et al., 1989; Omoto and Nakamaye, 1989). The most likely explanation for these observations is that dynein exhibits a strong preference for nucleotide in the *anti* conformation (with the base away from the ribose sugar) compared to the *syn* form (with the base stacked above the sugar ring). This may also explain the enhanced ATPase activity observed for dyneins in the presence of Mn^{2+} compared to Mg^{2+} (King and Witman, 1989), as the former cation acts to stabilize the *anti* conformation (Sloan and Mildvan, 1976).

Treatment of axonemes with sulfhydryl oxidizing agents such as dithionitrobenzoic acid leads to the activation of ATPase activity at low ($\sim 10^{-5}$ M) oxidant concentrations, but complete inhibition at higher levels ($> 10^{-3}$ M). Activation does not occur in mutant axonemes lacking outer arms, and fractionation of purified dynein revealed that this effect was due to the specific enhancement of the γ HC ATPase (Harrison et al., 2002) (see section IV.D.3.b). Similar effects of sulfhydryl modification using *N*-ethylmaleimide and *p*-mercuribenzoate were observed on the ATPase activity of sea urchin and *Tetrahymena* dyneins, suggesting a conserved role for the modified cysteine residues in outer arm function (Ogawa and Mohri, 1972; Shimizu and Kimura, 1974; Gibbons and Fronk, 1979).

The outer dynein arm must respond to the rise in intraflagellar Ca^{2+} that occurs during the photophobic response and leads to waveform conversion (Bessen et al., 1980; Kamiya and Okamoto, 1985; Mitchell and Rosenbaum, 1985). However, ATP-dependent microtubule binding by intact purified outer arm dynein is apparently unaffected by increasing Ca^{2+} (Sakato and King, 2003). If tubulin is rebound to the structural microtubule-binding site, this dynein does not bind microtubules. However, when the α HC is removed genetically (by the *oda11* mutation), the resulting $\beta\gamma$ HC particle exhibits a Ca^{2+}-sensitive increase in ATP-dependent microtubule binding. This response is abrogated by the subsequent removal of the β HC motor domain in an *oda4-s7 oda11* double mutant (Sakato and King, 2003). Intriguingly, lack of the α HC also results in the loss of the frequency imbalance observed between the *cis* and *trans* flagella (Kamiya and Hasegawa, 1987; Rüffer and Nultsch, 1987; Sakakibara et al., 1991). Thus it is possible that the α HC acts as a conduit for a negative signal that controls the response of other HCs to Ca^{2+} (Sakato and King, 2003) and/or that leads to distinct motor activities in the two flagella.

A defining enzymatic property of all dyneins is their exquisite sensitivity to low (submicromolar) concentrations of metavanadate, which acts as a γ-phosphate analogue in an ADP-V_i complex within the active site (Gibbons et al., 1978). Usefully, when such dynein HC-ADP-V_i complexes are irradiated with 365-nm ultraviolet light, the HC backbone is specifically cleaved at a single site located within the first P-loop (termed the V1 site) (Lee-Eiford et al., 1986; King and Witman, 1987). This results in the complete inhibition of ATPase activity. The active sites of the α and γ HCs are distinct from that of the β HC in that V1 cleavage will occur in the presence of vanadate alone (King and Witman, 1987, 1988b). Analysis of vanadate-mediated photocleavage intermediates of myosin where the half-life of the complex is several days (rather than a few minutes for dynein) revealed that photocleavage is a two-step process that involves vanadate-promoted photo-oxidation of an active-site residue (Cremo et al., 1991).

The HCs may be cleaved at additional sites (termed V2) by irradiation in the presence of Mn^{2+} (which acts to suppress V1 cleavage) and higher concentrations (0.1–1.0 mM) of vanadate (King and Witman, 1987; Tang and Gibbons, 1987). At these levels, metavanadate polymerizes in solution and the chromophore for V2 cleavage is likely a cyclic tri- or tetrameric species (Chasteen, 1983; King and Witman, 1989). In *Chlamydomonas*, the number of V2 cleavage sites ranges from one (for the β HC) to three (for the α HC) (King and Witman, 1987, 1988b). These sites all occur C-terminal of the V1 site and are presumed to represent cleavage within one or more of the other AAA$^+$ domains. Based on mass estimates of the fragments generated, it is likely that V2 cleavage occurs within the highly conserved fourth

AAA$^+$ domain of all HCs and that the additional one or two sites seen in the γ and α HCs, respectively, occur within AAA2 and/or AAA3; however, this suggestion has yet to be confirmed experimentally.

In addition to vanadate, dynein ATPases may be inhibited by a variety of other compounds that act as nonhydrolyzable nucleotide analogues such as adenyl-5'-yl imidodiphosphate (AMP-PNP), adenylyl(β,γ-methylene) diphosphonate (AMP-PCP), and adenosine 5'-O-(3-thiotriphosphate) (ATPγS) (Penningroth et al., 1982); all these compounds act as general competitive inhibitors of ATP-utilizing enzymes. Furthermore, two additional compounds have been identified that have more specific effects on dyneins. Erythro-9-{3-(2-hydroxynonyl)}adenine (EHNA), a structural analogue of adenosine but not ATP, acts as a noncompetitive inhibitor ($K_i = 7.8 \times 10^{-5}$ M) suggesting that it does not bind the ATP hydrolytic site in AAA1, but rather one of the other AAA$^+$ domains that exert allosteric control (Bouchard et al., 1981). A similar conclusion may be drawn concerning the action of purealin, a brominated bioactive secondary metabolite isolated from the Okinawan sea sponge *Psammaplysilla purea* (Nakamura et al., 1985). This compound appears to be an uncompetitive dynein inhibitor; i.e., it binds only to the enzyme–substrate complex not the free enzyme (Fang et al., 1997). However, purealin does not block vanadate-mediated photolysis at the V1 site, suggesting that its binding site is located elsewhere than AAA1.

e. Heavy chain phosphorylation

Both long-term- and pulse-labeling of *Chlamydomonas* with ^{32}P-phosphate revealed that the α HC is the only component of the 18 S $\alpha\beta$ HC particle that is phosphorylated *in vivo* (Piperno and Luck, 1981; King and Witman, 1994). This HC is modified on at least six residues (mostly serine), and examination of mapped proteolytic fragments revealed that these sites are located within two distinct regions. The first is a ~20-kD section immediately N-terminal of the first AAA$^+$ domain, while the second is a ~90-kD region that encompasses the sixth AAA$^+$ domain and the C-terminal unit (King and Witman, 1994). The rapid incorporation of label into the α HC suggests that these sites are undergoing relatively rapid cycles of acylation/deacylation *in vivo*. Furthermore, label is readily incorporated into the α HC when isolated axonemes are incubated with [γ-^{32}P] ATP, suggesting that the kinases, and perhaps also the phosphatases, involved are intimately associated with the outer arm within the axonemal superstructure (King and Witman, 1994). In addition, an analysis of the *Chlamydomonas* phosphoproteome using mass spectrometry has revealed that the outer arm γ HC is phosphorylated on Ser2456 that is located within the third AAA$^+$ domain (Wagner et al., 2006). There is clear evidence for control of inner arm dynein function by phosphorylation of the IC138 intermediate chain (see section IV.E.3.c and Chapter 9), and phosphorylation of the cytoplasmic dynein HC within rat optic nerve has been correlated with activation of motor function (Dillman and Pfister, 1994). Together, these observations suggest that α and γ HC phosphorylation may play a role in the control of outer arm dynein activity although there is as yet no direct evidence in support of this idea in *Chlamydomonas*.

2. Intermediate chains

a. Overview

All oligomeric dynein motors include at least two ICs that contain seven WD repeats, which fold to form a β propeller structure within the C-terminal region (Figure 6.7). Each WD repeat provides three β strands to one blade of the propeller and one strand to the adjacent blade (Figure 6.7B, C). This interlocking design is similar to that formed by the kelch domains and yields a highly stable toroidal structure (Figure 6.7B) that provides for multiple protein–protein interaction surfaces. The N-terminal domains and extensions C-terminal of the WD repeats are remarkably diverse in sequence, which presumably reflects the distinct roles played by each of these components. These ICs interact with LCs of the LC8, Tctex1/Tctex2, and LC7/Robl classes to form an IC/LC subcomplex and are essential for assembly and stability of the dynein particle. Within cytoplasmic dynein, the related IC74 proteins (DYNC1I1 and DYNC1I2) have been implicated in cargo attachment and in binding dynactin (see Pfister et al., 2006 for review), and it is likely that the outer arm IC components are important for dynein attachment within the axonemal superstructure.

Two WD-repeat proteins (IC1 and IC2) are present within the purified *Chlamydomonas* outer dynein arm (Piperno and Luck, 1979; Pfister et al., 1982; Mitchell and Kang, 1991; Wilkerson et al., 1995), and both fractionation (Mitchell and Rosenbaum, 1986) and crosslinking (King et al., 1991) studies suggest that they interact directly. Interestingly, *Chlamydomonas* does not contain an orthologue of the additional IC consisting of a thioredoxin module followed by three nucleoside diphosphate kinase catalytic units and a C-terminal domain that is found in the outer arm dyneins of sea urchin and vertebrate sperm flagella (Ogawa et al., 1996; Padma et al., 2001; Sadek et al., 2001); note however that thioredoxin-related LCs are present in the *Chlamydomonas* outer arm (see section IV.D.3.b) and nucleoside diphosphate kinases are associated with the radial spokes and outer doublet microtubules (Patel-King et al., 2004 and see Chapter 7).

b. Intermediate chain 1

The *Chlamydomonas* IC1 protein is encoded at the *ODA9* locus and is essential for assembly of the outer arm (Wilkerson et al., 1995). This protein has close orthologues in other species; e.g., it shares ~45% identity with the IC2 proteins from the sea urchin *Anthocidaris crassispina* and the ascidian *Ciona intestinalis*. The C-terminal portion of the molecule consists

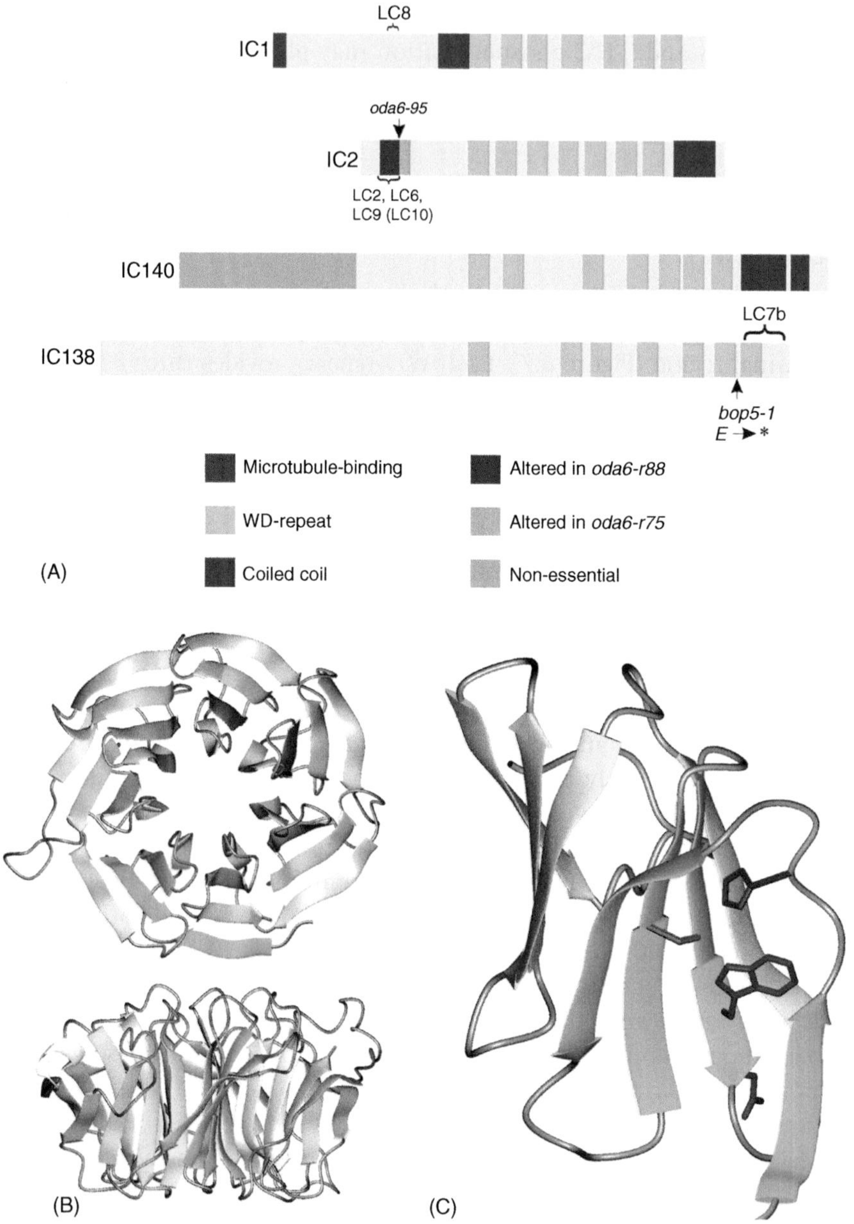

FIGURE 6.7 *Organization of WD-repeat intermediate chains. (A) Diagram illustrating the domain structure of the WD-repeat proteins associated with the outer arm (IC1 and IC2) and inner arm I1/f (IC140 and IC138). The color code for each region is indicated as are the LC- and microtubule-binding regions and several mutations discussed in the text. (B) Two views of a β propeller structure formed from seven WD repeats (from the G_β subunit of heterotrimeric G proteins; accession 1GP2; Sondek et al., 1996); two repeats are colored in yellow and green. The toroidal structure formed can participate in multiple protein–protein interactions. (C) Two WD repeats (in green and yellow) are shown. Each repeat forms the outer strand of one blade of the β propeller and the three innermost strands of the adjacent blade. On one repeat, the tryptophan and aspartic acid residue side chains are indicated in red and orange, as are the serine (pink) and histidine (blue) side chains that form a hydrogen bond network with tryptophan to stabilize the structure.*

of the WD-repeat β propeller, and analysis of truncated proteins suggest that this region interacts directly with IC2 (King et al., 1995). The N-terminal region of IC1 has a variety of intriguing features. At the extreme N-terminus there is a highly basic region (Wilkerson et al., 1995) that is separated from the rest of the molecule by a segment rich in glutamine and proline residues that is thought to form a flexible linker. Zero-length crosslinking revealed that IC1 interacts directly with α-tubulin *in situ* (King et al., 1991), and further mapping studies revealed that this crosslink involved the IC1 N-terminal region; deletion of residues 5–29 resulted in a severe decrease in axoneme-binding activity (King et al., 1995). Interestingly, a second segment involved in microtubule binding was also identified immediately N-terminal of the first WD repeat. This implies that IC1 helps mediate ATP-insensitive attachment of the outer arm to the doublet microtubules, although it is clearly insufficient without the ODA-DC.

Crosslinking also demonstrated that IC1 interacts directly with LC9, which is a member of the Tctex1/Tctex2 family (DiBella et al., 2005). In cytoplasmic dynein, Tctex1 has been proposed to bind the IC via a (K/R)(K/R)xx(K/R) motif (Mok et al., 2001); within IC1 this motif ($_{18}$KKTRK$_{22}$) is present near the N-terminus. It is likely that this protein also interacts with several other LCs of the IC/LC complex, and fractionation studies suggest a direct association with LC2 (Tctex2) (Mitchell and Rosenbaum, 1986). Moreover, IC1 may be involved in Ca^{2+} regulation of dynein function as it associates with the calmodulin homologue LC4 only in the presence of Ca^{2+} (Sakato et al., 2007 and see section IV.D.3.c). IC1 is also thought to interact with the highly conserved LC8 protein. Again, in the cytoplasmic dynein IC the equivalent interaction has been proposed to occur through a (K/R)xTQT motif (Fan et al., 2001 and see section IV.D.3.d). Although this precise motif does not occur in *Chlamydomonas* IC1, it is present in *A. crassispina* IC2 and alignment of the two IC sequences suggests that $_{184}$RETFT$_{188}$ is the equivalent segment in the *Chlamydomonas* IC1 protein. Furthermore, a 60-residue region of cytoplasmic dynein IC74 (residues 151–211 of the rat 2C isoform) has been demonstrated to allow for dimerization in the absence of LC binding (Lo et al., 2006). However, this region cannot be readily identified within the *Chlamydomonas* ICs, raising the possibility that IC associations are mediated through one or more of the LCs.

An additional intriguing feature of IC1 is that it becomes heavily photolabeled by both 8-azido and 2-azido nucleotide analogues in a UV- and ATP-dependent manner, whereas the related IC2 protein does not (Pfister et al., 1985; King et al., 1989). Surprisingly, with 2-azido ATP, incorporation of label occurred in the presence of Mg^{2+} but not with Mn^{2+}, even though Mn^{2+} enhances the ATPase activity of the αβ HC complex, probably by stabilizing the *anti* conformation of the nucleotide. Sequence analysis has failed to identify a nucleotide-binding site within IC1 and so it remains unclear why this protein becomes labeled.

c. Intermediate chain 2

IC2 is encoded at the *ODA6* locus and as for IC1 is absolutely essential for assembly of the outer arm (Kamiya, 1988; Mitchell and Kang, 1991). This protein is ~50% identical to *A. crassispina* IC3 and *C. intestinalis* IC1, and defines a second subclass of outer arm ICs (King et al., 1985; Ogawa et al., 1995). Immuno-EM using monoclonal antibodies against two different epitopes revealed that IC2 is located at the base of the isolated αβ subunit (King and Witman, 1990). This IC consists of an N-terminal region involved in LC binding (see below), a seven WD-repeat β propeller, and a C-terminal segment containing a region of ~56 residues that is predicted to form a coiled coil. Although the target of this coiled-coil region is currently unknown, it may interact with either the coiled-coil proteins DC1 and DC2 of the ODA-DC or with ODA5, which is also essential for outer arm assembly (see section V.B).

The original *oda6-95* mutation (Kamiya, 1988) introduces a frame-shift within the N-terminal region that leads to out-of-frame translation of 14 residues and a premature stop (Mitchell and Kang, 1993). Isolation of two distinct classes of intragenic pseudorevertants that introduce additional frame-shifts to restore the reading frame has revealed intriguing data concerning the function of this region (Mitchell and Kang, 1993). One class of revertant (exemplified by *oda6-r75*) has a second frame-shift C-terminal of the original which results in a full-length protein containing only 10 residues that differ from the wild-type protein. This version of IC2 is competent to allow for assembly of the outer arm and also for the restoration of beat frequency to essentially wild-type levels. A second revertant (*oda6-r88*) has a frame-shift N-terminal of the original *oda6* mutation and leads to inclusion of 23 residues that are distinct from the wild-type protein. Although this IC2 variant is competent for assembly of the outer arm, its inclusion does not result in restoration of beat frequency to wild-type levels (Mitchell and Kang, 1993). Subsequent analysis revealed that *oda6-r88* dynein (but not that from *oda6-r75*) completely lacks LC2, LC6, and LC9 (DiBella et al., 2005), and is also grossly deficient for LC10 (Rompolas et al., 2005), implying that these LCs associate with this small N-terminal region of the IC or with a neighboring section whose structure is altered in the mutant protein. These data further imply that the IC/LC complex exerts a significant effect on HC motor function.

3. Light chains

a. Overview

Based on their locations, the outer arm contains two general groups of LCs: those associated directly with the HC motors (LC1, LC3, LC4, and LC5) and those that form part of the IC/LC complex at the base of the structure (LC2, LC6, LC7A, LC7B, LC8, LC9, and LC10). These components appear to function in both dynein assembly within the flagellum and in the direct regulation of motor activity.

b. Thioredoxins (LC3 and LC5) and redox-based regulation

The outer dynein arm contains two proteins (LC3 and LC5) that are members of the thioredoxin family of sulfhydryl oxidoreductases (Patel-King et al., 1996). The presence of thioredoxins within outer arm dynein is not exclusive to *Chlamydomonas* and similar modules have been identified in metazoan dyneins. Both proteins have perfect copies of the WCGPCK thioredoxin active-site motif that contains a redox-sensitive vicinal dithiol. Proteins with vicinal dithiols exhibit a high affinity for trivalent metals such as arsenic as they form a dithioarsine ring structure that is stable in the absence of reducing agents (Kalef and Gitler, 1994). Following salt extraction, the outer arm, but not the inner arms that lack thioredoxins, could be purified on a phenylarsine oxide column and eluted with dithiothreitol (Patel-King et al., 1996). Furthermore, molecular modeling and analysis of *in vivo* redox state by thiol derivatization suggest that LC3 has a second redox-active vicinal dithiol (Patel-King et al., 1996; Harrison et al., 2002; Wakabayashi and King, 2006).

Both thioredoxin-like proteins are present within the 18 S αβ HC subparticle (Piperno and Luck, 1979; Pfister et al., 1982), and biochemical fractionation and genetic studies revealed that LC5 is associated with the α HC N-terminal domain (Pfister and Witman, 1984; Sakakibara et al., 1991; Harrison et al., 2002). The LC3 protein was originally found to cofractionate with the isolated β HC (Pfister and Witman, 1984); however, it is now clear that *in situ* this LC interacts with both the β and γ HCs (Harrison et al., 2002; Sakato et al., 2007), again via their N-terminal domains (Sakakibara et al., 1993; Sakato et al., 2007). In addition, crosslinking experiments have revealed a direct association between the LC3 thioredoxin and the LC7b component of the IC/LC complex that in turn interacts with the ODA-DC (DiBella et al., 2004a).

The identification of redox-active proteins in the outer arm and the demonstration that γ HC ATPase can be activated by sulfhydryl oxidation raised the question of whether this dynein is regulated *in vivo* by the modulation of intraflagellar redox poise. Alterations in flagellar beat frequency and duration of the photophobic response following changes in light intensity (photokinesis) requires the outer dynein arm (Pazour et al., 1995; Moss and Morgan, 1999; Casey et al., 2003b; Wakabayashi and King, 2006) and also NADPH generated by photosystem I (Wakabayashi and King, 2006). This response can be mimicked by placing cells under oxidative or reductive stress. The beat frequency of reactivated wild-type cell models containing outer arms was reduced when performed in the presence of a high ratio of oxidized:reduced glutathione; mutants lacking the outer arm show little response to alterations in redox state. Furthermore, when the α HC and LC5 were removed (using the *oda11* mutant), cell models reactivated with a beat frequency lower than wild type under reducing conditions but with a frequency greater than wild type when the medium was made more oxidizing, indicating an alteration in the axonemal response to changes in redox poise (Wakabayashi and King, 2006).

Determination of the *in vivo* redox state of LC3 and LC5 following light–dark transitions demonstrated that LC5 is essentially always reduced but that ~10% of flagellar LC3 exists in the oxidized state (Wakabayashi and King, 2006). Furthermore, use of a two-dimensional electrophoretic procedure that identifies proteins interacting through mixed disulfides revealed that *in vivo* both LC3 and LC5 have multiple binding partners and that these change in response to alterations in light intensity (a similar result was obtained for the redox-sensitive DC3 protein of the ODA-DC; see section IV.D.4.c). This suggests that these proteins are continually undergoing transient disulfide bond formation within the flagellum. An analysis of *oda11* mutant flagella identified a similar number of proteins interacting with LC3, suggesting that these thioredoxins likely do not interact with each other as part of an intra-dynein redox cascade (Wakabayashi and King, 2006).

c. Light chain 4 and Ca^{2+} regulation

A single copy of the LC4 protein associates with the γ HC (Pfister et al., 1982). This protein is a member of the calmodulin family and exhibits 42% sequence identity with *Chlamydomonas* calmodulin (King and Patel-King, 1995a). Although structure predictions suggest the presence of four helix-loop-helix motifs, only the two near the N-terminus conform to the EF-hand consensus for Ca^{2+}-binding loops. Direct measurement indicates that LC4 can bind one mole Ca^{2+}/mole with a K_{Ca} of ~2.7×10^{-5} M. This high-affinity metal-binding site is specific for Ca^{2+} and is unaffected by the addition of millimolar levels of Mg^{2+}. LC4 may be crosslinked to the N-terminal region of the γ HC and associates with a region containing two IQ motifs within this domain. Indeed, LC4 exhibits both Ca^{2+}-dependent and Ca^{2+}-independent interactions, which presumably allow for both permanent attachment to the HC and Ca^{2+}-dependent change in conformation (Sakato et al., 2007). In sucrose gradients, the isolated γ subunit exhibits a small shift in S value in the presence of Ca^{2+}, suggesting that it undergoes a change in conformation. Furthermore, crosslinking indicates that, within the isolated intact dynein particle, LC4 interacts with IC1 only in the presence of Ca^{2+}; interestingly, the yield of crosslinked product is reduced by the addition of ATP and vanadate, suggesting a further subtle structural change (Sakato et al., 2007). This change in conformation has been confirmed by negative stain EM of isolated γ HC subparticles, which revealed that the N-terminal HC stem is able to adopt multiple conformations in the presence of Ca^{2+} as it exhibits a large angular dispersion about an inflection point ~2/3 the way along its length; in contrast, this region appears much more rigid in the absence of ligand (Sakato et al., 2007).

The photoshock response is initiated by an increase in intraflagellar Ca^{2+} from pCa 6 to pCa 4 (Bessen et al., 1980; Hyams and Borisy, 1978) and is either missing or aberrant in mutants lacking outer arms (Kamiya and Okamoto, 1985; Mitchell and Rosenbaum, 1985). The location of LC4 within the complex, its Ca^{2+}-binding affinity, the observation that it undergoes a Ca^{2+}-dependent alteration in conformation, and the Ca^{2+}-dependent activation of ATP-sensitive microtubule binding by the $\beta\gamma$ HC particle makes this LC a strong candidate for an outer arm Ca^{2+} sensor. As discussed below (section IV.D.4.c), the second Ca^{2+}-binding component of the outer arm (DC3) does not appear to play a role in waveform conversion.

d. Motor domain-associated light chain 1

There are two copies of the 22-kD LC1 protein associated with the γ HC. Using a combination of crosslinking and vanadate-mediated photolysis, the interaction sites were mapped to the first and third AAA^+ domains of the γ HC (Benashski et al., 1999; Wu et al., 2000), both of which are thought to bind nucleotide (Figure 6.6B). 1-Ethyl-3-(3-dimethylaminopropyl)carbodiimide (EDC) crosslinking of isolated axonemes revealed that LC1 also interacts with an as yet unidentified integral axonemal component of ~45 kD (p45) as part of a ternary complex with the γ HC. Furthermore, analysis of an *oda6-r88 oda12-1* double mutant demonstrated that LC1 can assemble in the complete absence of its target HC (DiBella et al., 2005). This suggests that LC1 mediates the association of the γ HC motor unit with another axonemal substructure. Given the cryo-EM structural data (Nicastro et al., 2006; Ishikawa et al., 2007; Oda et al., 2007), it is possible that p45 is located on the A-tubule to which the outer arm is attached rather than the B-tubule of the adjacent doublet.

LC1 is a member of the SDS22 subclass of the LRR family (Kajava, 1998; Benashski et al., 1999). Each repeat (of ~22 residues) provides two β strands and an α helix (Figure 6.8A), and is constructed around an 11-residue core having the consensus sequence LxxLxLxxNxL; all these conserved residues (including asparagine) are buried within the interior of the protein. NMR structural studies revealed that LC1 consists of an N-terminal helix, six $\beta\beta\alpha$ motifs derived from the LRRs (the first and last repeats lack the second β strand) that form a central barrel consisting of two β sheets and an α helical face, and a C-terminal helical domain that protrudes from the main protein axis (accession numbers 1DS9 and 1M9L) (Figure 6.8B) (Wu et al., 2000, 2003). As the interaction between LC1 and the γ HC is stable in 0.6 M salt, it is most likely hydrophobic in nature (Pfister et al., 1982). In contrast, the interaction between LC1 and p45 is not preserved following high-salt extraction and therefore involves an ionic interface (Benashski et al., 1999; Wu et al., 2000). Examination of the LC1 molecular surface identified a single hydrophobic patch centered on the larger β sheet face and a charged region on the opposite side of the central barrel (Figure 6.8C). Backbone dynamics

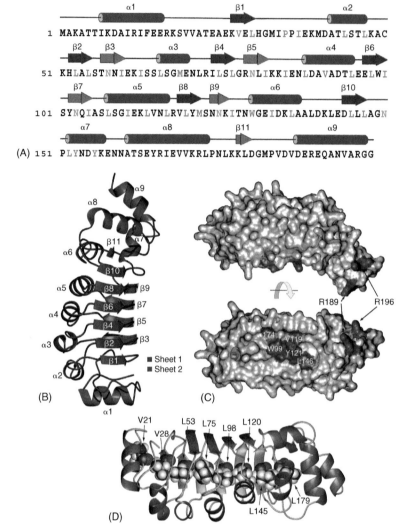

FIGURE 6.8 *Structural features of the motor domain-associated light chain. (A) Primary and secondary structure of LC1. In general, each leucine-rich repeat contributes two β strands (together these form two distinct β sheet faces) and an α helical segment; the first and last repeats form only a single strand and helix because of degeneracy in repeat 1 and the altered orientation of the repeat 6 helix (α7) with respect to the α2–α6 face. Conserved hydrophobic and asparagine residues within each repeat are indicated in orange and green, respectively. The LRR segment is capped by a single helix α1 at the N-terminus, while at the C-terminus an additional helical domain (α8 and α9) is present. Adapted from Wu and King (2003). (B) Ribbon structure of LC1. The LRRs form a central barrel comprising two β sheet surfaces and an α helical face. The C-terminal helical domain protrudes from the main protein axis. Reprinted with permission from Wu et al. (2003). Copyright of the American Chemical Society. (C) Two views of the LC1 molecular surface related by a 90° rotation about the long axis are shown. A single hydrophobic patch (colored green) is located on the larger β sheet (sheet 1) and is centered on Trp99. This region is thought to bind the γ HC, which would place the C-terminal region within the motor domain. Intriguingly, helix α9 contains two arginine residues (orange-red) that could make ionic contacts within the HC. Reprinted from Sakato and King (2004), with permission from Elsevier. (D) Ribbon diagram of LC1 indicating the hydrophobic (leucine and valine) residues within the large β sheet face that exhibit significant backbone dynamics as determined by NMR spectroscopy. Each of these residues represents one of the conserved components of the LRR motif and together they form a flexible leucine spine that may play a significant role in high-affinity binding of LC1 to its target HC. Reprinted from Wu and King (2003).*

analysis of LC1 revealed that although the molecule is mostly rigid, a single leucine residue within each strand of the large β sheet exhibits enhanced flexibility, forming a mobile leucine spine (Figure 6.8D) (Wu et al., 2003). Similarly, ionic residues on the opposite surface also show increased backbone motion. Together, these regions provide two slightly deformable binding surfaces that may allow for high-affinity interactions between LC1 and its targets (Wu and King, 2003).

If the hydrophobic region of LC1 indeed mediates association with the γ HC, then the C-terminal helical region must protrude into the AAA$^+$ domains. The terminal helix (α9) contains two arginine residues (Arg189, Arg196) that may therefore mediate ionic interactions with the HC or possibly with nucleotide, perhaps acting as "arginine fingers" similar to those found in the GAPs that activate Ras and Rho GTPase activity (Rittinger et al., 1997; Scheffzek et al., 1997). Arg189, but not Arg196, is conserved in the LC1 homologues of *Plasmodium* and *Giardia*, whereas Arg196 is present in the trypanosome and *Leishmania* proteins; Arg189 is altered to lysine in these latter organisms. In contrast, the LC1 homologues in vertebrates, insects, and sea urchins have a distinctive, more acidic, C-terminal sequence; although this region contains one lysine, that residue does not align precisely with either Arg189 or Arg196. Intriguingly, two residues (Met182 and Asp185) located within the loop between helices α8 and α9 also show enhanced backbone flexibility and may serve to control the orientation of the terminal helix (Wu et al., 2003; Wu and King, 2003).

The prediction that Arg189 and/or Arg196 are functionally important is supported by the observation that γ subunit ATPase is specifically activated at pH 10, suggesting that ionization of a basic residue is necessary (Gatti et al., 1991). Furthermore, expression of LC1 bearing the R189A mutation in a wild-type background results in a subtle decrease in beat frequency but a dramatic alteration of swimming behavior; cells constantly change direction in an apparently random fashion (R.S. Patel-King and S.M. King, unpublished observations). An additional indication of the importance of LC1 in dynein function comes from the observation that reduction of LC1 expression by RNA interference in *Trypanosoma brucei* results in defects in outer arm dynein assembly (Baron et al., 2007). This leads to the alteration of flagellar beat pattern and reversal of swimming direction.

e. The light chain 8 family (LC6, LC8, LC10)

Two members of this family (LC6 and LC8) were initially identified in *Chlamydomonas* outer arm dynein by biochemical means (Piperno and Luck, 1979; Pfister et al., 1982) and were subsequently shown to be related at the sequence level (King and Patel-King, 1995b) (Figure 6.9A). Stoichiometry measurements suggest that there are at least four copies of LC8 and two of LC6 per outer arm particle. Furthermore, sequence analysis revealed that LC8 has been extraordinarily highly conserved (King and Patel-King, 1995b),

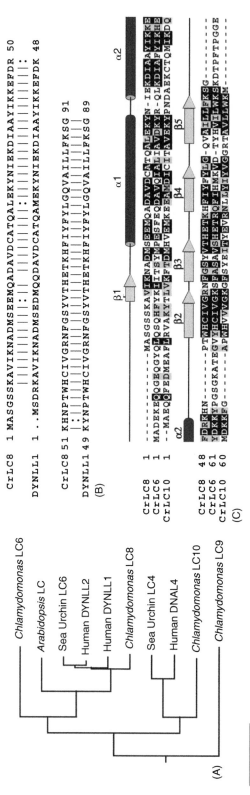

FIGURE 6.9 *Phylogeny and sequence alignment of the LC8 family of light chains. (A) Neighbor-joining phylogenetic tree illustrating the relationship between the LC8 family proteins. The tree was constructed using Chlamydomonas LC9 (a Tctex1) as the out-group; this protein is structurally related to LC8, but shows little primary sequence similarity. (B) Alignment of Chlamydomonas LC8 and the human orthologue DYNLL1. These two proteins are ~90% identical and differ mainly at the extreme N-terminus. (C) CLUSTALW alignment of the three Chlamydomonas members of the LC8 family; the secondary structure of LC8 is indicated above the alignment. The regions most highly conserved correspond to the helical and β strand segments while most differences occur either in the N-terminal segment or in the loop regions between structural elements.*

showing almost 90% identity with the human orthologues DYNLL1 (Figure 6.9B) and DYNLL2 (Pfister et al., 2005), which differ by only six residues and are also found in cytoplasmic dynein (King et al., 1996a). More distantly related proteins are present in higher plants (King and Patel-King, 1995b), which do not contain dynein (Lawrence et al., 2001),[2] and it has now become clear that LC8 functions in diverse multimeric complexes including myosin V (Espindola et al., 2000), neuronal nitric oxide synthase (Jaffrey and Snyder, 1996), and many others (see Pfister et al., 2006 for review). Indeed, total loss-of-function defects in LC8 lead to embryonic lethality in *Drosophila* through the induction of apoptosis (Dick et al., 1996a), and to inhibition of growth in *Aspergillus* due to the failure of nuclear movement (Beckwith et al., 1998). Unlike metazoans, lack of LC8 is not lethal in *Chlamydomonas*, and indeed the LC8 null mutant *fla14* grows at wild-type rates (Pazour et al., 1998); likewise, mutants of the budding yeast orthologue are also viable (Dick et al., 1996b). LC8 has also been identified within inner arm I1/f (Harrison et al., 1998) and the radial spokes (Yang et al., 2001). In addition, the *fla14* mutant assembles only very short flagellar stubs that accumulate intraflagellar transport (IFT) proteins, suggesting that LC8 also plays a role in retrograde IFT (Pazour et al., 1998); it has been found that LC8 associates with FAP133, which is a dynein IC associated with the retrograde transport motor (Rompolas et al., 2007).

In contrast, in *Chlamydomonas*, LC6 is present exclusively within the outer arm and to date no obvious LC6 orthologue has been identified in other species. The LC6-null mutant *oda13* exhibits only a minor swimming defect (King and Patel-King, 1995b; Pazour and Witman, 2000; DiBella et al., 2005).

Examination of the *Chlamydomonas* genome has revealed a third member (LC10) of this LC class that is 33% identical to LC8 (Figure 6.9A, C); a close homologue of LC10 (termed DNAL4) exhibiting 50% sequence identity is present in mammals. Within the *Chlamydomonas* flagellum, LC10 is exclusively associated with the outer dynein arm and presumably forms part of the IC/LC complex although crosslinking experiments failed to identify its binding partners (Rompolas et al., 2005). The LC10 gene is located close to LC2 and is also completely deleted in the LC2 null mutant *oda12-1*, although not in *oda12-2* that only lacks the 3′ end of the LC2 gene (Pazour et al., 1999; Rompolas et al., 2005). A Toc-2 transposon (Day, 1995) is located at the 3′ end of the LC10 gene but does not appear to be transcribed (Rompolas et al., 2005).

---

[2] The *Arabidopsis* genome does not contain dynein sequences (Lawrence et al., 2001). However, multiple partial dynein HC sequences (King, 2002b) were present in the original release of the *Oryza sativa* L. ssp. *indica* (rice) genome (Yu et al., 2002). Subsequently, these were deemed to derive from contamination of the genomic DNA sample used and the sequences have been removed from the databases.

In mammals, the LC8 proteins DYNLL1 and DYNLL2 both contain 89 residues whereas the *Chlamydomonas* protein has two additional residues near the N-terminus. This protein exists as a symmetric dimer (Benashski et al., 1997) and both NMR and X-ray structures are available with several different bound ligands (Liang et al., 1999; Fan et al., 2001) (Figure 6.10, upper panel). Each monomer folds to form an N-terminal β strand followed by two helices and four C-terminal β strands. All the strands interact to form an antiparallel β sheet that mediates dimerization. Importantly, this is a strand-switched interface, where the β3 strand from one monomer hydrogen bonds to strands from the other monomer. LC8 undergoes a pH-induced dimer–monomer transition *in vitro* following ionization of His55 although, given the acidic conditions under which this occurs, it is uncertain whether this is physiologically relevant (Liang et al., 1999; Nyarko et al., 2005). Two symmetric hydrophobic grooves are formed at the dimer interface and structural studies have revealed that the IC of cytoplasmic dynein as well as other proteins interact via this region, forming a sixth antiparallel strand. For mammalian cytoplasmic dynein, LC8 binds the consensus (K/R)xTQT (Fan et al., 2001) within the N-terminal region of the IC while a completely different motif is used to interact with neuronal nitric oxide synthase (Fan et al., 1998; Liang et al., 1999). A slightly altered version of the first motif is present in IC1 (see above); however, these motifs are not readily recognizable in *Chlamydomonas* IC2, and it remains unclear precisely how LC8 interacts with this component. Although all LC8 peptide ligands investigated to date bind within the inter-monomer grooves, the α helical surfaces of the LC8 dimer have also been extraordinarily highly conserved and it seems likely that they too are involved in specific interactions with target proteins.

f. The Tctex1/Tctex2 family (LC2 and LC9)

Tctex1 and Tctex2 were originally identified within the murine *t* complex, which is a region of 30–40 Mb on chromosome 17 (Lader et al., 1989; Huw et al., 1995). Variant forms (termed *t* haplotypes) contain large inversions that suppress recombination with wild type and exhibit the remarkable property that heterozygous (+/*t*) males transmit the *t* haplotype form of the chromosome to >95% of their progeny (Silver, 1993). Homozygosity for a *t* haplotype leads to embryonic lethality due to the presence of recessive lethal factors, while hemizygotes (containing two *t* alleles with complementing lethal factors) are male-sterile. This transmission ratio distortion or meiotic drive is due to the action of a responder on a series of distorter proteins (Lyon, 1984). The responder has been identified as a sperm motility kinase (the *t* mutant form has defective kinase function) (Herrmann et al., 1999) and the strongest distorter is now known to be a dynein HC (Fossella et al., 2000; Samant et al., 2002). Tctex1 and Tctex2 remain as candidates for two of the other distorter loci. Tctex1 was first found as a LC (DYNLT1) of cytoplasmic dynein (King et al., 1996b) and a homologue subsequently

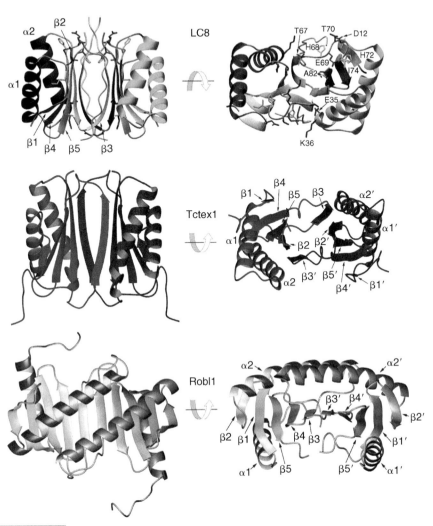

FIGURE 6.10 *Structural features of the intermediate chain-associated light chains present in oligomeric axonemal dynein motors. Ribbon diagrams of the dimeric human LC8 (DYNLL1), Chlamydomonas Tctex1, and murine Roadblock 1 (DYNLRB1) LCs are shown; the PDB accession numbers are 1CMI, 1XDX, and 1Y4O, respectively. The right-hand structures are related to those on the left by a 90° rotation about the indicated axis. LC8 is shown with bound peptide (light blue) derived from neuronal nitric oxide synthase (Liang et al., 1999). The secondary structural elements for each LC are indicated. For LC8, residues that show decreased backbone dynamics following peptide binding (Fan et al., 2002) are also illustrated. Note that LC8 and Tctex1 both dimerize across a strand-switched β sheet interface and are closely related at the structural level, while members of the LC7/Roadblock class have a completely different organization. The LC8 structures are reprinted from Wu and King (2003), and those for Chlamydomonas Tctex1 from Wu et al. (2005) with permission from Elsevier.*

was identified within *Chlamydomonas* inner arm I1/f (Harrison et al., 1998); a second Tctex1 version was found in outer arm dyneins from sea urchin (Kagami et al., 1998) and from *Chlamydomonas* (where it is termed LC9) (DiBella et al., 2005). Similarly, a Tctex2 homologue was identified in the *Chlamydomonas* outer arm (as LC2) (Patel-King et al., 1997; Pazour et al., 1999) and a closely related protein (Tctex2b) is also present in inner arm I1/f (DiBella et al., 2004b). These observations provided the first clues that defects in axonemal dynein-based motility might underlie the transmission ratio distortion of *t* haplotypes.

Within the outer arm, LC2 and LC9 share 35 and 46% sequence identity with mammalian Tctex2 and Tctex1, respectively. Phylogenetic analysis (Figure 6.11A) revealed that these proteins are related even though LC2 is monomeric (at least *in vitro*) whereas LC9 and Tctex1 form homodimers (DiBella et al., 2001). Structural studies of mammalian Tctex1 (Williams et al., 2005) and the orthologue from *Chlamydomonas* inner arm I1/f (accession number 1XDX) (Wu et al., 2005) (Figure 6.10, center panel) confirmed that this protein is dimeric and revealed a close structural similarity with the LC8 dimer, although there is little obvious sequence conservation between these LC classes. Each monomer consists of two helices and five β strands that form a strand-switched interface. The major difference between Tctex1 and LC8 is that the former is elongated by ~10Å along the axis defined by the inter-monomer groove.

Previous studies in cytoplasmic dynein suggest that Tctex1 associates with the dynein ICs via a (K/R)(K/R)xx(K/R) motif (Mok et al., 2001) and perfect copies of this motif exist within both *Chlamydomonas* IC1 and IC2. Chemical shift mapping identified a series of Tctex1 residues involved in binding this motif (Mok et al., 2001), all of which are located near one end of the inter-monomer grooves (Wu et al., 2005). In mammals, Tctex1 has been implicated in binding specific cargoes such as rhodopsin (Tai et al., 1999), although there is no evidence that it plays a similar role in the axonemal enzymes. No strain defective for LC9 currently exists; however, an LC2 null mutant does not assemble functional outer arms (Pazour et al., 1999; DiBella et al., 2005).

Crosslinking experiments revealed that LC2 interacts with LC6 and is part of the IC/LC complex (DiBella et al., 2001). More recently, examination of dyneins containing a mutant form of IC2 (encoded by the *oda6* pseudo-revertant *oda6-r88*; Mitchell and Kang, 1993), which has 23 residues within the N-terminal region that are different from wild type, revealed that several LCs including LC2, LC6, and LC9 are all completely missing (DiBella et al., 2005) and that LC10 is reduced significantly (Rompolas et al., 2005). Importantly, although the outer arm can assemble in the *oda6-r88* mutant (Mitchell and Kang, 1993), these motors appear dysfunctional as this strain exhibits a flagellar beat frequency much lower than that of wild-type cells (Mitchell and Kang, 1993; DiBella et al., 2005). This suggests that LCs located in the basal region of the enzyme influence motor function.

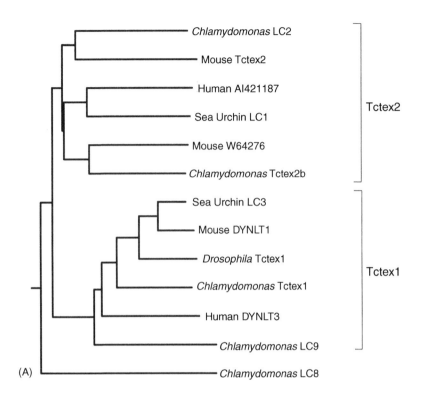

```
DYNLT1   1 MEDFQ.ASEETAFVVDEVSSIVKEAIESAIGGNAYQHSKVNQWTTNVLEQ 49
             ||   | || ||| |:||.|:||.|:. :   |  .||.|||.. ||
CrTctex1 1 MEGVDPAVEEAAFVADDVSNIIKESIDAVLQNQQYSEAKVSQWTSSCLEH 50

DYNLT1   50 TLSQLTKLGRPFKYIVTCVIMQKNGAGLHSASSCFWDSSTDGSCTVRWEN 99
              : .|| | :||||:|||:||||||||||.|.||.|||.|||| ||||||
CrTctex1 51 CIKRLTALNKPFKYVVTCIIMQKNGAGLHTAASCWWDSTTDGSRTVRWEN 100

DYNLT1   100 KTMYCIVSTFGLSI 113
               |.|||| . |||.|
CrTctex1 101 KSMYCICTVFGLAI 114
```

FIGURE 6.11 *Phylogenetic analysis of the Tctex1/Tctex2 light chain family. (A) Sequences of* Chlamydomonas, *sea urchin, and mammalian Tctex1 and Tctex2-related proteins were aligned using CLUSTALW. The neighbor-joining tree calculated based on this alignment reveals that Tctex1 and Tctex2 define two distinct subgroups within this LC family. The structurally related protein LC8 was used as the out-group. (B) Sequence alignment of murine Tctex1 (DYNLT1) and the related protein from* Chlamydomonas *inner arm I1/f. The sequence identity between these proteins is highest in the C-terminal region, which forms the β sheet involved in formation of the strand-switched dimer interface.*

g. The LC7/Roadblock family (LC7a and LC7b)

Studies in *Drosophila* and *Chlamydomonas* identified a third class of LC that is shared by both cytoplasmic and axonemal dyneins (Bowman et al., 1999). Sequence analysis has suggested that these proteins belong to an ancient lineage that extends into the bacterial kingdom; e.g., in *Myxococcus xanthus* the LC7/roadblock-related protein MglB has been implicated in NTPase regulation as its target MglA is related to the Ras GTPase (Spormann and Kaiser, 1999; Koonin and Aravind, 2000). The *Drosophila* mutant *roadblock* (*robl*) exhibits multiple phenotypes including defects in axonal transport and mitosis. A homozygous *robl*-null mutation results in pupal lethality whereas an EMS-induced allele (*robl^z*), which has a 193-bp deletion affecting a splice site, produces a truncated poisonous product and leads to death at the larval stage. *Drosophila* contains six members of this family, including the bithoraxoid protein that has been implicated in abdominal para-segment development, whereas *Chlamydomonas* and mammals contain only two (Bowman et al., 1999; DiBella et al., 2004a); note however that the mammalian hepatitis B X-interacting protein is apparently structurally related to this LC class and interacts with Robl1 and Robl2 *in vitro* (Nikulina and King, 2002).

Similar to the LC8 and Tctex1 LC classes, LC7/Robl proteins are also dimeric and exist as either homo- or heterodimers *in situ* (Nikulina et al., 2004). Each monomer consists of an N-terminal α1 helix, two β strands, a second helix (α2), and three additional strands at the C-terminus. The NMR solution structure of a member of this dynein LC class (dimeric murine Robl1) was originally solved by Song et al. (2005) (accession 1Y4O) (Figure 6.10, lower panel). The β strands form a five-strand antiparallel sheet, with the α2 helix arranged orthogonally on one face with the α1 helix almost parallel to the β strands on the opposite face. The two monomers associate through the β3 strands to form an extended β sheet; there are also extensive contacts along the length of the α2 helices. The opposite face contains a large cleft formed between the curved β sheet and α1 helices.

The two *Chlamydomonas* members of this LC class (LC7a and LC7b) exhibit 47% sequence identity (Bowman et al., 1999; DiBella et al., 2004a). Both are dramatically reduced in mutants lacking the outer arms, but are only completely absent from flagella of double mutants that are also missing inner arm I1/f. LC7a and LC7b copurify with the outer arm in sucrose gradients of extracts from the *ida1* mutant; conversely, these proteins copurify with inner arm I1/f when extracted from *oda* mutants (DiBella et al., 2004a). These proteins are dimeric; therefore, based on their apparent stoichiometry (Table 6.5), they presumably form heterodimers within axonemal dyneins, although this remains to be confirmed.

Crosslinking studies indicate that LC7b interacts directly with the LC3 thioredoxin associated with the β and γ HCs and also with the DC2 component of the ODA-DC (DiBella et al., 2004a). Mapping studies in cytoplasmic dynein revealed that Robl binds to a region of the IC immediately N-terminal of the WD-repeat region (Susalka et al., 2002). However, analysis of the *Chlamydomonas* mutant *bop5-1* (Hendrickson et al., 2004), expressing a truncated form of IC138 from inner arm I1/f (and see below), indicated interaction with the last repeat or a segment C-terminal of it. Due to the interlocking nature of the WD-repeat β propeller (Figure 6.7B), these regions would actually be closely aligned in the folded structure and so may both contribute to binding this LC class. NMR mapping studies revealed that the cytoplasmic dynein IC region binds to the hydrophobic cleft delineated by the $\alpha 1$ helices of Robl1 (Song et al., 2005).

The LC7a null mutant *oda15* exhibits the slow, jerky swimming phenotype characteristic of strains completely lacking outer dynein arms (Pazour and Witman, 2000). However, this strain still assembles considerable quantities of dynein proteins onto the axoneme (DiBella et al., 2004a). Interestingly, these are arranged in runs with large gaps between them, and many microtubule segments are completely devoid of dyneins, indicating the cooperative nature of dynein arm assembly. The LC7a-minus dyneins are unstable following extraction and dissociate into the $\alpha\beta$ HC subunit, the γ HC complex, and the ODA-DC in sucrose gradients. Furthermore, LC7b becomes completely detached from these other subparticles, indicating that LC7a is required for its stable association with the remainder of the dynein arm.

4. The outer arm docking complex

a. Overview

Although the $\alpha\beta$ and γ outer arm dynein subunits exhibit motor activity and can interact with doublet microtubules via the γ HC and IC1 (King et al., 1991; Sakakibara and Nakayama, 1998), these components are insufficient to allow for rebinding of the outer arm onto the flagellar axoneme. Biochemical studies revealed that a dissociated 7S factor, the ODA-DC, restored rebinding activity (Takada and Kamiya, 1994). In the presence of Mg^{2+} and following ultracentrifugation at low hydrostatic pressure, the ODA-DC remains associated with the other subunits and indeed would appear to play an essential role in their association (Takada et al., 1992). The ODA-DC is a trimeric structure consisting of two coiled-coil proteins (DC1 and DC2) and the Ca^{2+}-binding calmodulin homologue DC3. It presumably interacts directly with the flagellar outer doublet microtubules, although it remains unclear what specifies the site at which this complex binds. Furthermore, this structure would appear to play a role in the control of outer arm function, as the characteristic frequency imbalance between the *cis* and *trans* flagella is lost when wild-type dyneins are rebound to mutant axonemes lacking the ODA-DC (Takada and Kamiya, 1997). This suggests that one or more ODA-DC component(s) are modified differently within the two flagella (e.g., DC2 is phosphorylated at multiple sites).

b. Coiled-coil proteins (DC1 and DC2)

The major structural components of the ODA-DC, DC1 and DC2, are encoded at the *ODA3* and *ODA1* loci, respectively (Koutoulis et al., 1997; Takada et al., 2002). These proteins have calculated masses of 83 and 62 kD, although DC1 migrates anomalously with $M_r = 105,000$ in SDS gels (Takada and Kamiya, 1994). Both DC1 and DC2 are absolutely essential for outer arm assembly and are predicted to contain large segments of coiled coil within the N-terminal 400–500 residues. These proteins associate directly in the cytoplasm. Immuno-EM revealed that, when assembled in the flagellum, DC1 repeats every 24 nm (Wakabayashi et al., 2001). In the *oda6* mutant that lacks outer arms, immunofluorescence labeling with an anti-DC1 antibody decreased toward the flagellar tip, suggesting that there is a reciprocal effect of the outer arm on ODA-DC stability. In agreement with this, immunoblots indicate that the amount of DC2 is reduced in isolated axonemes of the *oda9* mutant, which also lacks outer arms, compared to wild-type axonemes (Wirschell et al., 2004). The DC2 protein is phosphorylated on at least two sites, giving rise to a minimum of three different isoforms that may exhibit distinct regulatory properties (Luck and Piperno, 1989).

c. The Ca^{2+}-binding component (DC3)

The third ODA-DC component, DC3, is related to calmodulin and contains four EF hands (Casey et al., 2003a). Two of these EF hands fit the consensus for Ca^{2+}-binding loops and one also contains two cysteine residues. Direct measurement of Ca^{2+} binding *in vitro* revealed that DC3 binds one Ca^{2+} with a K_d of 1×10^{-5} M; this binding was redox-sensitive as it occurred only under reducing conditions and was abrogated upon sulfhydryl oxidation (Casey et al., 2003b). As has been observed for EF hands within some other proteins, DC3 can bind Mg^{2+} as well as Ca^{2+}, although with much lower affinity. Given that flagella presumably contain millimolar levels of Mg^{2+}, DC3 may well have this metal bound under low Ca^{2+} conditions, but then exchange it for Ca^{2+} following the influx that signals flagellar reversal. It is also possible that Mg^{2+} binding by DC3 is responsible for the requirement of this cation in keeping the ODA-DC and $\alpha\beta$ and γ subunits associated *in vitro*. As described above for LC3 and LC5, DC3 also participates in mixed disulfide formation with other flagellar components in a manner that is sensitive to alterations in redox state (Wakabayashi and King, 2006).

The intron-less DC3 gene is located at the *ODA14* locus and a null mutant can only partially assemble outer arms (Casey et al., 2003a). However, both DC1 and DC2 assemble normally in this mutant, indicating that DC3 is not essential for their integration within the axonemal superstructure. The photoshock response is altered in the *oda14* mutant as cells stop in response to a bright flash of light but do not move backwards (Casey et al., 2003a). However, rescue of the *oda14* mutant with an altered version of DC3 that is unable to bind Ca^{2+} restored both outer arm assembly and the photoshock response (Casey et al., 2003b). This implies that DC3 does not act as an outer arm Ca^{2+} sensor for waveform conversion.

E. Inner arm I1/f

1. An inner arm dynein organized around two heavy chains

The I1/f inner arm dynein contains two distinct HCs, three ICs, and five different LCs, only two of which are specific to this particular motor. Following high salt extraction of axonemes, all these components comigrate in sucrose density gradients at ~17–18 S (Figure 6.4C, D). Furthermore, as described in more detail in Chapter 9, this dynein occupies a key position in the regulatory pathways controlling flagellar beating and the phototactic response.

2. Heavy chains

a. Overview

The I1/f 1α HC is encoded by the DHC1 gene at the *IDA1* locus and I1/f 1β by DHC10 at the *IDA2* locus (Porter et al., 1996; Perrone et al., 2000) (Tables 6.1 and 6.3). Phylogenetic analysis indicates that these HCs are similar to each other and more closely related to the outer arm HCs than to those of the monomeric inner-arm dyneins (Figure 6.6A). Both HCs are essential for assembly of the I1/f complex, and the *ida1* and *ida2* mutants missing the DHC1 and DHC10 genes, respectively, totally lack the I1/f dynein. However, transformation of a DHC10-null mutant with DHC10 constructs encoding <20% of the HC can partially rescue the motility defects by reassembly of an I1/f complex containing an N-terminal 1β HC fragment and a full-length 1α HC (Perrone et al., 2000); the transformants also retain the ICs and LCs. These data suggest that the N-terminal portion of the 1β HC is necessary and sufficient for assembly and docking of the I1/f complex onto the outer doublet microtubules. Together with observations on the outer arm HCs and information on the structural organization of dynein HCs, these findings have led to the general idea that the N-terminal stem region of the HCs functions in the assembly of dynein motors and their attachment to appropriate cargoes.

b. Enzymatic and functional properties

The inner dynein arms have an inherently low basal ATPase activity (Kagami and Kamiya, 1990). Motor activity of isolated I1/f dynein has been examined using *in vitro* motility assay experiments (this is described in detail in section VII). Examination of mutant versions of I1/f that lack either the 1α or 1β HCs of this dynein revealed that both proteins are capable of translocating microtubules *in vitro* (S. Toba, H. Sakakibara, and K. Oiwa, personal communication) and contribute to motility *in vivo* (Myster et al., 1999; Perrone et al., 2000). The velocity of gliding is very slow compared with that displayed by other dyneins: about 1.5 μm/second for the 1α1β heterodimer and 1α HC, and 3 μm/second for the 1β HC. How this slow *in vitro* gliding velocity can be reconciled with the intrinsic rate of microtubule sliding in the axoneme, which can be as high as 20 μm/second, is an interesting unresolved question.

3. Intermediate chains

a. Overview

Inner arm I1/f contains three distinct ICs. Two of these (IC140 and IC138) are WD-repeat proteins related to the IC1 and IC2 proteins within the outer arm, while the third (IC97) does not contain obvious structural or functional motifs and its presence appears to be an I1/f-specific adaptation.

b. Intermediate chain 140

IC140 is one of two WD-repeat proteins within inner arm I1/f and contains seven of these motifs within the C-terminal region (Figure 6.7A). The C-terminal 53 kD of IC140, which encompasses part of the WD-repeat region and a terminal coiled-coil segment, is capable of specifically rebinding to mutant axonemes that lack this dynein (Yang and Sale, 1998). Crosslinking indicates that it associates directly with IC97 (Yang and Sale, 1998; Wirschell et al., 2005). IC140 is encoded at the *IDA7* locus and defects lead to the complete failure of inner arm I1/f assembly (Perrone et al., 1998). The *ida7* defect could be completely rescued by transformation with both the full-length IC140 gene and, surprisingly, with a shorter version that lacked the first four exons and consequently expressed a truncated protein lacking the N-terminal 283 residues (Perrone et al., 1998) (Figure 6.7A). Further analysis suggests that this form of IC140 would lack a putative Tctex1 interaction site but would retain a potential binding site for LC8 (S.M. King, unpublished).

c. Intermediate chain 138

This 111-kD phosphoprotein is encoded at the *BOP5* locus (Hendrickson et al., 2004) and plays a central role in the regulation of flagellar motor activity through a pathway that includes the radial spokes and central pair complex in addition to the products of the *MIA1* and *MIA2* loci (King and Dutcher, 1997 and see Chapter 9). IC138 contains seven WD repeats within the C-terminal half of the protein (Figure 6.7A) and crosslinking studies indicate that it interacts directly with both IC97 (Wirschell et al., 2005) and LC7b (DiBella et al., 2004a). The *bop5-1* point mutation results in cells that swim slower than wild-type cells although they exhibit normal flagellar beat frequency. This mutation converts a codon encoding a glutamic acid residue located between the sixth and seventh WD repeats to a premature stop. Consequently, the *bop5-1* form of IC138 is unable to form the complete β propeller structure; even so, this altered protein can assemble within the axoneme although the LC7b component fails to associate (Hendrickson et al., 2004). Furthermore, analysis of an insertional allele at this locus that removes ~90% of the IC138 coding region revealed that the I1/f dynein can still assemble although it lacks both IC138 and IC97 (Bower et al., 2005) and presumably also LC7b. This suggests that IC138 is required for IC97 association but, surprisingly, that neither protein is necessary for incorporation of the rest of the structure.

d. Intermediate chain 97

The third IC within inner arm I1/f has a mass of 90–100 kD and has been termed IC97 or IC110 (Smith and Sale, 1991; Kagami and Kamiya, 1992;

Porter et al., 1992). This protein does not contain obvious structural motifs (Wirschell et al., 2005), although intriguingly, it does appear to share significant similarity with a vertebrate axonemal protein (Las1) (Wirschell et al., 2005) that has been implicated in lung tumor formation in mice (Zhang et al., 2003). Zero-length crosslinking revealed that IC97 interacts directly with both α- and β-tubulin (Wirschell et al., 2005), which suggests that this component may play a role similar to that of IC1 within the outer arm (King et al., 1991). However unlike IC1, IC97 is not essential for assembly of the I1/f dynein as the *bop5* mutant expressing a truncated form of IC138 also lacks IC97 yet incorporates most other components of this dynein into the axoneme (Wirschell et al., 2005).

4. Light chains

a. Overview

This inner arm contains five distinct LCs, two of which are specific to this particular motor. The other three LCs (LC7a, LC7b, and LC8) are also present in the outer arm and have been discussed above (see sections IV.D.3.e and IV.D.3.g).

b. I1/f-specific Tctex1

This inner arm Tctex1 is more closely related to the canonical murine protein DYNLT1 than is LC9 from the outer arm, exhibiting 60% (vs. 48%) sequence identity (Harrison et al., 1998) (Figure 6.11A, B). The C-terminal region involved in formation of the β sheet interface (see section IV.D.3.f) is most highly conserved whereas the α helical faces are less so (Wu et al., 2005); this pattern is similar to that observed for the structurally-related LC8 family proteins. There are currently no functional data available concerning the role of this Tctex1 component. However, its initial identification within the I1/f dynein provided essential evidence that transmission ratio distortion of the *t* haplotypes was due to effects on axonemal rather than cytoplasmic dyneins. This inner arm protein is the closest *Chlamydomonas* homologue of the Tctex1 protein present in *Drosophila* (Figure 6.11A). In that organism, homozygous defects do not lead to lethality as is the case for LC8, but rather to male sterility, suggesting that this protein class indeed plays an important role in development of sperm and/or in axonemal dynein function (Caggese et al., 2001; Li et al., 2004).

c. I1/f-Specific Tctex2b

The Tctex2b protein copurifies with inner arm I1/f and is dramatically reduced, but not completely absent, in axonemes from strains (*ida1* and *pf28pf30*) that lack this dynein (DiBella et al., 2004b). The *TCTEX2b* gene is located within ~2 kb of *PF16*, which encodes a component of the central pair apparatus, and a null *pf16* insertional allele is also completely missing the Tctex2b LC. Although inner arm I1/f is assembled in this strain and also in a strain rescued for PF16 but not Tctex2b, the dynein is unstable following high salt extraction and can be fractionated into several distinct particles in sucrose density gradients. For example, Tctex1, the other I1/f-specific LC, is completely dissociated and sediments at the top of the gradient. Thus, Tctex2b appears to be involved in stabilizing this dynein complex via salt-insensitive interactions. The lack of this LC also has a subtle effect on dynein motor function as evidenced by a small but significant decrease in the rate of microtubule sliding disintegration of axonemes from strains that lack only this protein (DiBella et al., 2004b).

F. Monomeric inner arms

1. Heavy chains

a. Phylogeny and relative abundance

Analysis of the *Chlamydomonas* genome has revealed the presence of nine HC genes in addition to those encoding components of the outer arm, inner arm I1/f, the cytoplasmic dynein (CAB56748) that powers retrograde IFT, and a second putative cytoplasmic dynein (EDP05194) (Porter et al., 1996; Perrone et al., 2000) (T. Yagi and R. Kamiya, unpublished and see Figure 6.6A). Biochemical analyses of the high-salt extract from axonemes identified six HCs belonging to these monomeric dynein species (Kagami and Kamiya, 1992) (Figure 6.5). A subsequent re-examination of the fractionated dyneins resulted in the further identification of three minor species of HCs (DHC3, DHC4, DHC11), each of which probably constitutes a novel monomeric dynein (T. Yagi and R. Kamiya, unpublished observations). The HC of each species from the fractionated samples has been determined by mass spectrometry and correlated to a particular HC gene (Tables 6.1 and 6.5) (T. Yagi and R. Kamiya, unpublished). The phylogenetic relationship between these HCs (Figure 6.6A) indicates that the monomeric HCs form a single class that can be further subdivided into two distinct groupings. The two DHCs that comprise the smaller group (delineated in yellow in Figure 6.6A) have some peculiar features: DHC3 is unusually large due to several inserted sequences in the motor domain, and DHC7 is one of the two major DHCs (the other one is DHC5) present in all previously isolated dynein-deficient mutants. Thus, these DHCs may perform some special function in the axoneme.

The novel dynein species present in the axoneme in amounts smaller than that of the major monomeric forms tend to remain in the axoneme after salt-extraction. Immunofluorescence observations indicate that at least one of these minor dyneins displays biased localization toward the flagellar base (T. Yagi, unpublished). As a general class, the monomeric HCs are heavily phosphorylated *in vivo* (Piperno and Luck, 1981), although it remains uncertain whether all currently known species are modified in this way.

b. Enzymatic and functional properties

As described in section VII, all of the major monomeric dyneins have been found capable of force generation in an assay wherein brain microtubules and ATP are introduced into a glass chamber coated with these dyneins. All appear to translocate microtubules with the plus-end leading, as is expected for minus-end directed motor proteins. Velocity varies with the species of dynein and with various factors including dynein density and ATP concentration. At 0.1 mM ATP, gliding velocities of 2–4 μm/second have been recorded for dyneins a, b, e, and g, and 5–7 μm/second for dyneins c and d (Kagami and Kamiya, 1992). In subsequent studies, the translocation velocity was found to be greatly increased when ADP was present in addition to ATP (Yagi, 2000; Kikushima et al., 2004). Therefore, some of the earlier results may have been underestimates. In addition, all these monomeric dyneins (except species b) also impart torque during translocation and cause microtubules to rotate at rates of up to ~10 Hz, although the rotational pitch differs between the different dyneins.

2. Actin and NAP

All monomeric dyneins have actin as a subunit and this protein is essential for the assembly and/or targeting of these motors (Piperno and Luck, 1979; Kato-Minoura et al., 1997). The presence of actin in inner arm dynein has been confirmed in other organisms including the Antarctic rock cod *Notothenia coriiceps* (King et al., 1997) and *Tetrahymena* (Muto et al., 1994), and appears to be a general feature in a wide variety of organisms. Chemical crosslinking suggests that actin is present in monomeric form, which is consistent with the observation that phalloidin, which specifically binds F-actin, does not stain flagella. Dynein-bound actin is indirectly associated with the N-terminal part of the HC through interaction with p28 and centrin (Yanagisawa and Kamiya, 2001). *Chlamydomonas* has two actin genes, one coding for conventional actin and the other coding for an unconventional actin called NAP (novel actin-related protein) which has 64% amino-acid sequence identity with conventional actin (Lee et al., 1997; Kato-Minoura et al., 1998). A mutant, *ida5*, that is defective for the conventional actin gene lacks dyneins a, c, d, and e (Kato-Minoura et al., 1997); also, it has greatly reduced amounts of minor dynein species (T. Yagi, unpublished). Intriguingly, *ida5* axonemes retain dynein species b and g because NAP is abundantly expressed in this mutant and can functionally substitute for the conventional actin subunit in these motors. However, NAP apparently cannot replace actin in other dyneins, resulting in their absence and implying a distinct structural difference between these two groups of monomeric inner arms.

In wild-type cells, NAP is expressed only in negligibly small amounts; however, it is expressed as much as conventional actin immediately following deflagellation (Hirono et al., 2003), suggesting that NAP plays some role in the flagellar assembly process.

3. Centrin

Centrin (previously also termed caltractin by Lee et al., 1991) is a calmodulin-like Ca^{2+}-binding protein present in essentially all eukaryotic cells. As for calmodulin, centrin contains four helix-loop-helix motifs that all conform to the EF-hand consensus for Ca^{2+} binding (Taillon et al., 1992); two of these motifs exhibit high affinity for Ca^{2+} ($K_{Ca} = 1.2 \times 10^{-6}$M) and two have much lower affinity ($K_{Ca} = 1.6 \times 10^{-4}$M) (Weber et al., 1994). In *Chlamydomonas*, centrin is present in the distal and proximal fibers connecting the two basal bodies (Salisbury et al., 1988), the bridge between the basal body and the nucleus (Wright et al., 1989), in the proximal transitional region of the axoneme (Sanders and Salisbury, 1989) (see Chapters 2 and 3), and in species b, e, and g of inner arm dynein (Piperno et al., 1990, 1992; Kagami and Kamiya, 1992).

Chemical crosslinking experiments suggest that centrin is associated with the N-terminal portion of the HC and also with actin (Yanagisawa and Kamiya, 2001). However, its association with actin has not been confirmed by immunoprecipitation assays (H. Yanagisawa and R. Kamiya, unpublished observations). In dynein, centrin appears to exist in a monomeric state. A mutant, *vfl1* (for <u>v</u>ariable <u>f</u>lagellar number), has a mutation in the centrin gene. This strain exhibits defects in arrangement of the basal bodies, and undergoes cell division into unequally-sized daughter cells with variable numbers of flagella (see Chapter 2). The inner arm dyneins of this mutant appear unaffected (H. Yanagisawa, unpublished observations). No mutants lacking this gene have been isolated, and it is conceivable that centrin null mutations are lethal.

4. The p28 light chain

The p28 LC is found in all monomeric dyneins except those containing centrin (Table 6.1) (Piperno et al., 1992; LeDizet and Piperno, 1995a, b; Yanagisawa and Kamiya, 2001). It is present as a dimer and associates with the N-terminal region of the HC and also with actin (Yanagisawa and Kamiya, 2001). Recombinant p28 produced in *Escherichia coli* dimerizes *in vitro* and binds both monomeric and polymeric actin (H. Yanagisawa, unpublished results). Furthermore, when mixed with F-actin, p28 induces bundling of actin filaments (H. Yanagisawa, unpublished results). Homologues of p28 have been found in other eukaryotes with motile cilia and flagella, including sea urchin (Gingras et al., 1996; Kastury et al., 1997) and humans (Kastury et al., 1997). Intriguingly, a protein sharing 40% sequence identity with *Chlamydomonas* p28 is present in *Caenorhabditis elegans*, which only has nonmotile sensory cilia, suggesting that p28 has an additional function unrelated to inner arm dyneins.

In *Chlamydomonas*, p28 is encoded at the *IDA4* locus, and several mutant alleles affecting intron/exon splice sites have been identified that lead to defects in inner arm assembly and changes in flagellar waveform (LeDizet and Piperno, 1995a). As axonemes of the *ida4* mutants lack dyneins a, c, and d, p28 must be important for the stability of these dyneins or for their attachment to the outer doublets (Kagami and Kamiya, 1992; LeDizet and Piperno, 1995a).

5. Proteins specifically associated with inner arm species d

Comparison of wild-type and mutant axonemes that lack specific dyneins revealed that monomeric dynein species d contains 38- and 44-kD proteins (designated p38 and p44, respectively) not present in the other enzymes. These components are greatly diminished in, but not completely absent from, the axonemes of *ida4* and *ida5*, mutants that lack species d and other monomeric dyneins. Cloning of the cDNA for p38 (Yamamoto et al., 2006) revealed that homologues of this dynein subunit are present in many organisms with motile cilia and flagella, but not in organisms with only nonmotile cilia, such as the nematode *C. elegans*. Experiments using a p38-specific antibody indicate that this protein is associated with dynein species d when extracted from wild-type axonemes with high salt.

The cDNA of p44 has been cloned (Yamamoto et al., 2008); this protein also is conserved among ciliated organisms. Immunolocalization indicates that p44 is associated with dynein species d, and is present in *ida4* and *ida5* axonemes in reduced amounts. The presence of small amounts of p38 and p44 in the axonemes lacking dynein d suggests that these proteins function as part of the structure that anchors dynein d to the outer doublets.

V. OTHER COMPONENTS THAT AFFECT DYNEIN ASSEMBLY AND/OR FUNCTION

A. Additional dynein-interacting proteins

Both genetic and biochemical studies have revealed that several additional components have profound effects on the assembly and potentially also on regulation of the outer and/or inner dynein arms, although these proteins do not copurify with these enzymes following their removal from the axoneme. Those components which have been identified and characterized are discussed below, while several others are known only from mutant studies (including *oda8*, *oda10*, *pf13*, and *pf22*) and remain to be identified.

B. ODA5 and flagellar adenylate kinase

The products of the *ODA5*, *ODA8*, and *ODA10* genes are of significant interest as they are required for outer arm assembly but do not encode known dynein proteins. Furthermore, these mutants do not complement each other in temporary dikaryons, suggesting that they may encode components of the same complex whose expression is coordinated (Kamiya, 1988). The *ODA5* gene was identified by cloning a tagged insertional mutant allele (Wirschell et al., 2004). The first ~400 residues of the 66-kD ODA5 protein (AAS10183) are predicted to form five discrete coiled coils, the 142–595 region is alanine-rich and the more C-terminal 434–608 segment has a high glycine content. This protein is extracted from the axoneme by 0.6 M salt and sediments at ~5 S in sucrose density gradients. ODA5 is missing in *oda10* axonemes, suggesting that the *ODA10* gene product is necessary for its incorporation into the axoneme; in contrast, ODA5 assembles at wild-type levels in *oda8* and in mutants that lack the ODA-DC and/or outer arm (Wirschell et al., 2004). Intriguingly, the *oda5* strain incorporates significantly reduced levels of a flagellar adenylate kinase that contains three catalytic modules that perform the reaction $2ADP \rightleftharpoons ATP + AMP$. This suggests that ODA5 or an associated protein may participate in anchoring adenylate kinase near the outer arms, which would allow for efficient use of both anhydride bonds in ATP. It was also observed that a WD-repeat protein (FAP52; EDP05376) is reduced in *oda5* flagella, but that protein has not been characterized further (Wirschell et al., 2004).

Characterization of outer arm dynein from *Ciona* sperm has revealed several novel ICs that have significant similarity to DC2 (Hozumi et al., 2006); similar additional ICs were observed in purified trout sperm dynein (Gatti et al., 1989). Phylogenetic analysis revealed that *Chlamydomonas* ODA5 groups with these additional DC2-like ICs. This suggests that the basic requirement for this protein class has been conserved, and that in *Chlamydomonas*, but not *Ciona*, dissociation simply occurs as a consequence of high salt extraction.

C. The ODA7 protein associates with both inner and outer arms

The *ODA7* locus was cloned following a genomic walk and found to encode a 47-kD polypeptide (ABI63572) that contains six LRR repeats near the N-terminus (Freshour et al., 2007). Although this region exhibits similarity to the LRR barrel domain of LC1, the C-terminal segments are not related. Clear ODA7 homologues have been identified in vertebrates, insects, and protists, indicating that this protein has a conserved role in dynein assembly. ODA7 incorporates into the flagellum independent of the outer arm, ODA-DC, and ODA5/adenylate kinase complex and is mostly solubilized following detergent treatment; the protein remaining in the axoneme is extracted by 0.6 M NaCl. Intriguingly, ODA7 levels are reduced only in mutants that lack both the outer arm and inner arm I1/f. When isolated from wild-type or outer arm-less axonemes, ODA7 copurifies with inner arm I1/f. In contrast, when I1/f assembly is impaired, ODA7 sediments either as two distinct peaks (one of ~15 S and a second near the top of the gradient) or copurifies with the outer arm depending upon the hydrostatic pressure to which the sample is subjected. Together these data have led to the proposal that ODA7 may derive from the linker observed between the inner and outer arms by cryo-electron tomography (Freshour et al., 2007). However, as this linker occurs only once in every 96-nm axonemal repeat, it is unclear why all outer arms in the repeat fail to incorporate rather than just the arm immediately adjacent to I1/f. One possibility is that ODA7 also functions to promote or regulate outer arm assembly in the cytoplasm prior to transport into the flagellum (Fowkes and Mitchell, 1998; Freshour et al., 2007) (see section VI).

D. The flagellar matrix protein ODA16

Two slow-swimming *oda16* mutants have been isolated and both show a severe reduction in the number of outer arms (Ahmed and Mitchell, 2005). As was observed for *oda15* (DiBella et al., 2004a), arms assemble cooperatively in long runs when present. The assembled arms in *oda16* are functional, as an *oda16 oda2* double mutant exhibits a further reduction in beat frequency compared to either single mutant. Conversely, this implies that ODA16 may also function in assembly of an inner arm (Ahmed and Mitchell, 2005). ODA16 (AAZ77789) contains eight WD repeats but no other obvious motifs, and homologues are present only in other organisms that assemble motile cilia/flagella. Dikaryon rescue experiments indicate that this protein is not a component of the dynein arm, the ODA-DC or the ODA5 complex, and wild-type levels of ODA16 are present in flagella lacking outer arms. ODA16 is completely solubilized following detergent extraction of flagella and also when matrix components are released following freeze/thaw cycles which is characteristic of the IFT machinery. This raised the possibility that ODA16 may be an adaptor necessary for attachment of outer arm dynein to IFT particles. In support of this, the IFT-particle protein IFT46 is required for outer arm dynein transport in *Chlamydomonas* (Hou et al., 2007), and the mammalian orthologues of ODA16 and IFT46 have been shown to interact directly in a yeast two-hybrid assay (Ahmed et al., 2006).

E. The lissencephaly protein orthologue LIS1 and associated NUDC

Defects in the mammalian lissencephaly protein Lis1 lead to a "smooth brain" phenotype and have very severe neurological consequences (Wynshaw-Boris and Gambello, 2001). Lis1 acts to regulate cytoplasmic dynein by interacting directly with the HC motor unit and also the N-terminal stem (Sasaki et al., 2000; Tai et al., 2002). In *Aspergillus*, the Lis1 orthologue NudF functions in the dynein-driven nuclear migration pathway (Xiang et al., 1995). Immunofluorescence studies have revealed that Lis1 is present in motile cilia of murine tracheal and oviductal epithelial cells but is absent from immotile primary cilia (Pedersen et al., 2007). An apparent orthologue has been identified in *Chlamydomonas* (Pedersen et al., 2007). *Chlamydomonas* LIS1 (ABG33844) is mainly associated with the axoneme following detergent extraction and is missing in mutants lacking outer arms. LIS1 is also absent from *oda11* flagella, which suggests that it interacts with the α HC or possibly the LC5 thioredoxin and requires their presence for axonemal assembly. *Chlamydomonas* LIS1 contains seven WD repeats as does mammalian Lis1, although it lacks the LisH domain that is apparently involved in Lis1 dimerization and which is required for cytoplasmic dynein activation. However, because the outer arm is a heterotrimer of HCs rather than being a HC homodimer as is cytoplasmic dynein, dimerization may be unnecessary for the axonemal function of LIS1.

Mammalian Lis1 also binds two other proteins (NudC and NudE) (Morris et al., 1998) that were initially identified by their effects on fungal nuclear movement (Morris, 2000). NudC is required for the stability of cytoplasmic dynein (Zhou et al., 2006) and NudE plays a role in dynein targeting to microtubules (Hoffmann et al., 2001). Analysis of the *Chlamydomonas* genome has failed to identify an orthologue of NudE; however, *Chlamydomonas* does contain a NudC-like protein (EDO98115) that shares 41% identity with the mammalian form. Expression of *Chlamydomonas* NUDC is upregulated upon deflagellation (Stolc et al., 2005), suggesting that it is a flagellar protein. Consistent with this, mammalian NudC is highly expressed in ciliated epithelial cells (Gocke et al., 2000). *In vitro* biochemistry has demonstrated that *Chlamydomonas* LIS1 binds directly to both recombinant rat NudC and also to the native protein from brain extracts (Pedersen et al., 2007). Thus it is likely that LIS1 interacts with NUDC in *Chlamydomonas*.

VI. PREASSEMBLY OF DYNEINS IN THE CYTOPLASM

Complex multimeric systems such as dynein arms and radial spokes (see Chapter 7) undergo preassembly within the cytoplasm prior to transport and incorporation into the axoneme. In *Chlamydomonas*, immunoprecipitation of dynein subunits from the cytoplasm of various *oda* mutants revealed that the HCs and ICs exist as a preassembled structure (Fowkes and Mitchell, 1998); the ODA-DC and ODA5 complex are thought to assemble independently prior to their association to form the complete structure. Interestingly, the α HC is missing from the cytoplasm of *oda7*, suggesting that the ODA7 is required for the stability of this HC. Furthermore, *oda7* cytoplasm allows for association of the β and γ HCs but not for incorporation of IC1/2 (Fowkes and Mitchell, 1998). This suggests either that these HCs can interact directly or that other components such as LCs mediate the association. Similarly, preassembled axonemal dyneins have also been isolated from the cytoplasm of *Paramecium* (Fok et al., 1994), suggesting that this is a general feature of these axonemal enzymes. In *Chlamydomonas*, LC4 has been shown to bind to the γ HC in cytoplasm, indicating that LCs also are preassembled (Sakato et al., 2007). One possible exception to this is the LC1 protein associated with the γ HC motor domain, as analysis of an *oda6-r88 oda12-1* double mutant revealed that this protein may be incorporated into the axoneme in the absence of its target HC (DiBella et al., 2005).

VII. MOLECULAR MECHANISM OF FORCE GENERATION

A. Biophysical properties

1. In vitro *motility*

Fractionated outer and inner arm dyneins can be assayed for their motile activity *in vitro* (Paschal et al., 1987; Vale and Toyoshima, 1988). Typically, a given dynein sample is added to an observation chamber made of a glass slide and a cover slip followed by perfusion with a buffer, so that the glass surfaces are coated with dynein. Microtubule translocation is induced by adding taxol-stabilized microtubules together with ATP to the chamber, and observed using dark-field or differential interference contrast optics or, when the microtubules are fluorescently labeled, a fluorescence microscope.

For outer arm dyneins, the maximal velocity of microtubule translocation has been reported to be ~5 μm/second for the αβ HC dimer, 4 μm/second for the γ HC subunit, but only 1 μm/second for the β HC (Sakakibara and Nakayama, 1998); the motility properties of intact 23 S outer arm dynein, containing all three HCs, have not been reported. Furthermore, β HC-driven motility required that the particle be preincubated with tubulin, presumably to block ATP-insensitive microtubule binding, and also needed the addition of methylcellulose to reduce Brownian motion, which implies a relatively low duty ratio (for an explanation of duty ratio, see below). All these velocities are lower than that of microtubule sliding in disintegrating axonemes, which is ~15–20 μm/second. Therefore, the velocity measured using this kind of assay does not quantitatively reflect the motor activity within the axoneme. The outer arm dyneins in the axoneme are attached to the doublet microtubules in a densely-packed linear array and are aligned in the same direction, whereas dyneins on the glass surface are randomly positioned. Such a difference in motor orientation may underlie the observed discrepancy in velocity.

Chlamydomonas inner arm dyneins have been more extensively studied using *in vitro* motility assays than has the outer arm dynein. In a study that used fractionated inner arm dynein species a–g, all but species I1/f were found to translocate microtubules with maximal velocities of 2–7 μm/second at 0.1 mM ATP (Kagami and Kamiya, 1992). Species I1/f displays much slower translocation of about 0.1 µm/second (Smith and Sale, 1991); more recently, a higher velocity of 1.2–1.6 µm/second has been observed with this dynein (Kotani et al., 2007). Interestingly, species a, c, d, e, and g were observed to generate microtubule rotation around the long axis during translocation, implying that the motor applies torque to the microtubule (Kagami and Kamiya, 1992). Subsequent analyses also showed that species d and g cause clockwise microtubule circling in the plane of gliding, suggesting that these dyneins can bend microtubules (Kikushima and Kamiya, 2008).

The velocity of microtubule translocation driven by axonemal dyneins depends on the concentration of both ATP and ADP. In most cases, it is greater in the presence of ADP and ATP than in ATP alone. For inner arm dynein species a and e, microtubule translocation does not occur at 0.1 mM ATP if the hydrolysis product ADP is completely removed from the medium with an ATP-regenerating system (Yagi, 2000). When such an inactive sample is perfused with a solution containing both ATP and ADP, microtubules start gliding, and the translocation velocity slowly increases over a few minutes to a plateau level (Kikushima et al., 2004; Kikushima and Kamiya, 2008). These observations suggest that the activity of dynein is regulated by slow binding/release of ADP and ATP. Presumably, the three nucleotide-binding sites in AAA domains 2–4, rather than the unique catalytic site in AAA1, are involved in this regulation.

2. Single-molecule experiments

The motile properties of inner arm dynein species c, a monomeric dynein HC, have been studied at the single-molecule level using optical tweezers (Sakakibara et al., 1999). Two features are important.

First, this dynein has been found to have a duty ratio as low as 0.16. Duty ratio is the fraction of the total time of a complete ATP-hydrolysis cycle during which a motor protein is engaged in force production. Thus, a motor protein with a low duty ratio is normally expected to spend most of the time dissociated from the microtubule. This type of motor is suited to a system in which multiple motors are engaged in moving a common object, such as an actin filament in muscle or a microtubule in the axoneme. In such a system, the movements produced by individual motors sum up and result in high translocation velocity, because each motor produces movement within a short period of time that is independent of the activity of other motors. Thus, motor proteins with low duty ratios are well-adapted for high-speed movements within the context of a defined superstructure (Howard, 1997).

The second important finding, reported by Sakakibara et al. (1999), is that, quite unexpectedly if we consider the above generalization, species c displays translocation for a long distance without detaching from the microtubule. Such long-distance movement, involving many cycles of ATP hydrolysis without releasing from the microtubule, is termed processive. For myosins and kinesins, processivity usually has been observed with dimeric motor proteins, in which two heads regularly alternate binding and walk on a single filament for a long distance (for review see Higuchi and Endow, 2002). As an exception, a monomeric kinesin, KIF1A, has been found to display processive features. The likely reason is that this kinesin continues to weakly associate with the microtubule through ionic interactions; KIF1A has a positively charged domain, which can interact with the negatively charged C-terminal region of tubulin to overcome thermally driven dissociation (Okada and Hirokawa, 2000). The observed processivity of inner arm dynein species c could also be based on such an interaction. This dynein has been found to be very important for force production under viscous load (Yagi et al., 2005) and its processive nature may well be related to the requirement for strong force production under high-load conditions.

B. Structural changes during the mechanochemical cycle

Evidence that dynein HCs undergo a change in conformation during the mechanochemical cycle has come from several sources. For example, proteolytic studies of sea urchin outer arm dynein revealed that the digestion pattern in the presence of ATP or ADP plus vanadate was quite distinct from that obtained in the presence of ATP alone or nonhydrolyzable nucleotide analogues (Inaba and Mohri, 1989). Mapping studies indicate that the proteolyzed sites occur over a large region of the HC (Inaba et al., 1991), suggesting that there is major change in conformation that occurs following ATP hydrolysis to ADP·P_i.

Indeed, negative stain EM studies of isolated dynein c (Burgess et al., 2003) have indicated that for this dynein the N-terminal stem domain is curled around one face of the AAA^+ ring and that the orientation of these two regions undergoes a rotation in response to nucleotide; i.e., there is motion about the junction between AAA1 and the N-terminal region that results in the latter domain altering its interactions with the AAA^+ ring. In addition, these studies revealed that the microtubule-binding stalk undergoes a change in orientation with respect to the AAA^+ ring in ADP·vanadate vs. the apo form; there also appears to be an alteration in the flexibility of this region. In contrast, cryo-EM studies of the outer arm *in situ* indicate that the N-terminal domains of these HCs are arranged perpendicular to the plane of the AAA^+ ring (Nicastro et al., 2006; Ishikawa et al., 2007; Oda et al., 2007), which would suggest an alternate power stroke. When the tomographic reconstructions for outer arms obtained in the absence of nucleotide or following ATP·vanadate addition were compared (Oda et al., 2007), the AAA^+ ring of the γ HC did not appear to change orientation with respect to the microtubule long axis, whereas the α HC density could not be located in the relaxed state and so presumably became more mobile. In contrast, the β HC motor unit shifted 3.7Å toward the B-tubule and adopted a prominent 44° tilt. This transition reduced the distance of the β HC motor unit to the B-tubule from 14 nm to 10 nm, and could allow for a ~6-nm movement by the tip of the ~15-nm microtubule-binding stalk.

A number of possibilities may account for the apparent discrepancy in these studies. It could simply result from artifactual alterations that occur during isolation and/or negative staining of dynein c, or the inability to resolve ring rotation or the HC N-terminal domains and/or microtubule-binding stalks in the tomograms. Alternatively, this disparity may reflect either an inherent difference between monomeric inner arm and oligomeric outer arm dynein HCs, or represent a specific modification of dynein c perhaps related to its ability to function under high viscous load.

The general idea that the AAA$^+$ ring rotates with respect to the N-terminal stem has received direct experimental support from fluorescence resonance energy transfer analysis of altered forms of *Dictyostelium* cytoplasmic dynein HC containing blue- and green fluorescent protein moieties (BFP and GFP). These studies indicate that a GFP attached to the N-terminal region near AAA1 is close to BFP incorporated in the C-terminal domain in the apo or ADP-bound states, but that GFP moves close to BFP associated with AAA2 following ATP binding (Kon et al., 2005). These measurements were made under zero load and it will clearly be of interest to determine whether this changes under high-load conditions. A study examining the orientation of the N-terminal domain with bound Fab antibody fragment with respect to the microtubule-binding stalk did not define a preferred orientation, suggesting that full rotation of the AAA$^+$ ring about the linkage to the N-terminal region can occur (Meng et al., 2006). Whether this represents uncoupling of the two domains during preparation for EM or a fundamental property of this complex remains to be defined.

It has been reported that cytoplasmic dynein step size decreases from as much as 32 nm to 8 nm in response to load imposed by an optical trap, and that this occurs with an increase in the stall force from <0.4 pN to 1.1 pN (Mallik et al., 2004). These observations raised the concept of a complex relationship between power output and the distance moved per ATP hydrolyzed that could involve significant changes in conformation of the AAA$^+$ ring. However, in these experiments, the measured velocity of movement was rather low (~50 nm/second) when compared to both *in vivo* measurements of dynein-driven motility and the rates of microtubule gliding observed using *in vitro* assays (>1 µm/second) (Paschal et al., 1987). A later report using cytoplasmic dynein that had been further purified by microtubule affinity followed by ATP release suggested that the maximal force output per motor is 7–8 pN with a step size of 8 nm that is independent of the imposed load (Toba et al., 2006). This force output and step size are consistent with those measured for the 22 S outer arm axonemal dynein from *Tetrahymena* cilia (Hirakawa et al., 2000). This study by Toba et al. (2006)

provides support for a hand-over-hand mechanism of dynein action whereby the binding sites for the two heads on the microtubule are 8 nm apart and the individual motor domains thus move 16 nm per stroke.

VIII. CONCLUSIONS AND PROSPECTS

Chlamydomonas has proven to be an exceptional model organism for studying flagellar dyneins, and based on genetics and biochemistry we now have a reasonably clear understanding of the composition of various dynein particles and of their approximate location, although there is clearly more to be learned in this regard. In addition, progress has been made in defining the assembly process, and some of the key regulatory molecules are now apparent. Although cryo-EM has brought us a higher resolution image of dynein arrangement within the axoneme than previously available, one obvious challenge is to obtain atomic level structural information for these complexes and to use that in combination with genetic, biochemical, and biophysical studies to understand motor and regulatory processes *in situ*. Furthermore, the powerful genetic screens available in *Chlamydomonas* continue to make significant contributions, and we can anticipate that these approaches will provide additional exciting insights into mechanism in the future.

ACKNOWLEDGMENTS

We thank John Heuser and Ken-ichi Wakabayashi for electron micrographs, Miho Sakato for electrophoretic analysis of dynein heavy chains, and Kazuhiro Oiwa, Ramila S. Patel-King, Panteleimon Rompolas, Hitoshi Sakakibara, Miho Sakato, Shiori Toba, Ken-ichi Wakabayashi, Toshiki Yagi, and Haru-aki Yanagisawa for sharing unpublished data. Our laboratories are supported by the Japanese Ministry of Education, Culture, Sports, Science and Engineering (to RK) and by grant GM51293 from the National Institutes of Health (to SMK). SMK is an investigator of the Patrick and Catherine Weldon Donaghue Medical Research Foundation.

REFERENCES

Ahmed, N.T. and Mitchell, D.R. (2005). ODA16p, a *Chlamydomonas* flagellar protein needed for dynein assembly. *Mol. Biol. Cell* **16**, 5004–5012.

Ahmed, N.T., Lucker, B., Cole, D.G., and Mitchell, D.R. (2006). The *Chlamydomonas ODA16* locus, needed for outer arm dynein assembly, encodes an intraflagellar transport adaptor. *Mol. Biol. Cell* **17**. Abstract #1612. (CD-ROM)

Baron, D.M., Kabututu, Z.P., and Hill, K.L. (2007). Stuck in reverse: loss of LC1 in *Trypanosoma brucei* disrupts outer dynein arms and leads to reverse flagellar beat and backward movement. *J. Cell Sci.* **120**, 1513–1520.

Beckwith, S.M., Roghi, C.H., Liu, B., and Morris, N.R. (1998). The "8-kD" cyto-plasmic dynein light chain is required for nuclear migration and for dynein heavy chain localization in *Aspergillus nidulans*. *J. Cell Biol.* **143**, 1239–1247.

Benashski, S.E., Harrison, A., Patel-King, R.S., and King, S.M. (1997). Dimerization of the highly conserved light chain shared by dynein and myosin V. *J. Biol. Chem.* **272**, 20929–20935.

Benashski, S.E., Patel-King, R.S., and King, S.M. (1999). Light chain 1 from the *Chlamydomonas* outer dynein arm is a leucine-rich repeat protein associated with the motor domain of the γ heavy chain. *Biochemistry* **38**, 7253–7264.

Bessen, M., Fay, R.B., and Witman, G.B. (1980). Calcium control of waveform in isolated flagellar axonemes of *Chlamydomonas*. *J. Cell Biol.* **86**, 446–455.

Bouchard, P., Penningroth, S.M., Cheung, A., Gagnon, C., and Bardin, C.W. (1981). Erythro-9-[3-(2-hydroxynonyl)]adenine is an inhibitor of sperm motility that blocks dynein ATPase and protein carboxylmethylase activities. *Proc. Natl. Acad. Sci. U.S.A.* **78**, 1033–1036.

Bower, R., Perrone, C.A., Mueller, J., O'Toole, E.T., Wirschell, M., Fox, L., Sale, W.S., and Porter, M.E. (2005). Localization of a regulatory intermediate chain (IC) complex at the base of the I1 dynein. *Mol. Biol. Cell* **16**. Abstract #967. (CD-ROM)

Bowman, A.B., Patel-King, R.S., Benashski, S.E., McCaffery, J.M., Goldstein, L.S., and King, S.M. (1999). *Drosophila* roadblock and *Chlamydomonas* LC7: a con-served family of dynein-associated proteins involved in axonal transport, flagel-lar motility, and mitosis. *J. Cell Biol.* **146**, 165–180.

Brokaw, C.J. (1994). Control of flagellar bending: a new agenda based on dynein diversity. *Cell Motil. Cytoskeleton* **28**, 199–204.

Brokaw, C.J. and Kamiya, R. (1987). Bending patterns of *Chlamydomonas* flagella: IV. Mutants with defects in inner and outer dynein arms indicate differences in dynein arm function. *Cell Motil. Cytoskeleton* **8**, 68–75.

Burgess, S.A., Dover, S.D., and Woolley, D.M. (1991). Architecture of the outer arm dynein ATPase in an avian sperm flagellum, with further evidence for the B-link. *J. Cell Sci.* **98**, 17–26.

Burgess, S.A., Walker, M.L., Sakakibara, H., Knight, P.J., and Oiwa, K. (2003). Dynein structure and power stroke. *Nature* **421**, 715–718.

Caggese, C., Moschetti, R., Ragone, G., Barsanti, P., and Caizzi, R. (2001). *dtctex-1*, the *Drosophila melanogaster* homolog of a putative murine *t*-complex distorter encoding a dynein light chain, is required for production of functional sperm. *Mol. Genet. Genomics* **265**, 436–444.

Casey, D.M., Inaba, K., Pazour, G.J., Takada, S., Wakabayashi, K., Wilkerson, C.G., Kamiya, R., and Witman, G.B. (2003a). DC3, the 21-kD subunit of the outer dynein arm-docking complex (ODA-DC), is a novel EF-hand protein impor-tant for assembly of both the outer arm and the ODA-DC. *Mol. Biol. Cell* **14**, 3650–3663.

Casey, D.M., Yagi, T., Kamiya, R., and Witman, G.B. (2003b). DC3, the smallest subunit of the *Chlamydomonas* flagellar outer dynein arm-docking complex, is a redox-sensitive calcium-binding protein. *J. Biol. Chem.* **278**, 42652–42659.

Chasteen, N. (1983). The biochemistry of vanadium. *Struct. Bond.* **53**, 105–138.

Cremo, C.R., Grammer, J.C., and Yount, R.G. (1991). Vanadate-mediated photo-cleavage of myosin. *Methods Enzymol.* **196**, 442–449.

Day, A. (1995). A transposon-like sequence with short terminal inverted repeats in the nuclear genome of *Chlamydomonas reinhardtii*. *Plant Mol. Biol.* **28**, 437–442.

DiBella, L.M., Benashski, S.E., Tedford, H.W., Harrison, A., Patel-King, R.S., and King, S.M. (2001). The Tctex1/Tctex2 class of dynein light chains. Dimerization, differential expression, and interaction with the LC8 protein family. *J. Biol. Chem.* **276**, 14366–14373.

DiBella, L.M., Sakato, M., Patel-King, R.S., Pazour, G.J., and King, S.M. (2004a). The LC7 light chains of *Chlamydomonas* flagellar dyneins interact with components required for both motor assembly and regulation. *Mol. Biol. Cell* **15**, 4633–4646.

DiBella, L.M., Smith, E.F., Patel-King, R.S., Wakabayashi, K., and King, S.M. (2004b). A novel Tctex2-related light chain is required for stability of inner dynein arm I1 and motor function in the *Chlamydomonas* flagellum. *J. Biol. Chem.* **279**, 21666–21676.

DiBella, L.M., Gorbatyuk, O., Sakato, M., Wakabayashi, K., Patel-King, R.S., Pazour, G.J., Witman, G.B., and King, S.M. (2005). Differential light chain assembly influences outer arm dynein motor function. *Mol. Biol. Cell* **16**, 5661–5674.

Dick, T., Ray, K., Salz, H.K., and Chia, W. (1996a). Cytoplasmic dynein (*ddlc1*) mutations cause morphogenetic defects and apoptotic cell death in *Drosophila melanogaster*. *Mol. Cell Biol.* **16**, 1966–1977.

Dick, T., Surana, U., and Chia, W. (1996b). Molecular and genetic characterization of SLC1, a putative *Saccharomyces cerevisiae* homolog of the metazoan cytoplasmic dynein light chain 1. *Mol. Gen. Genet.* **251**, 38–43.

Dillman, J.F. and Pfister, K.K. (1994). Differential phosphorylation *in vivo* of cytoplasmic dynein associated with anterogradely moving organelles. *J. Cell Biol.* **127**, 1671–1681.

Eshel, D., Urrestarazu, L.A., Vissers, S., Jauniaux, J.C., van Vliet-Reedijk, J.C., Planta, R.J., and Gibbons, I.R. (1993). Cytoplasmic dynein is required for normal nuclear segregation in yeast. *Proc. Natl. Acad. Sci. U.S.A.* **90**, 11172–11176.

Espindola, F.S., Suter, D.M., Partata, L.B., Cao, T., Wolenski, J.S., Cheney, R.E., King, S.M., and Mooseker, M.S. (2000). The light chain composition of chicken brain myosin-Va: calmodulin, myosin-II essential light chains, and 8-kDa dynein light chain/PIN. *Cell Motil. Cytoskeleton* **47**, 269–281.

Fan, J.-S., Zhang, Q., Li, M., Tochio, H., Yamazaki, T., Shimizu, M., and Zhang, M. (1998). Protein inhibitor of neuronal nitric oxide synthase, PIN, binds to a 17-amino acid residue fragment of the enzyme. *J. Biol. Chem.* **273**, 33472–33481.

Fan, J.-S., Zhang, Q., Tochio, H., Li, M., and Zhang, M. (2001). Structural basis of diverse sequence-dependent target recognition by the 8 kDa dynein light chain. *J. Mol. Biol.* **306**, 97–108.

Fan, J.S., Zhang, Q., Tochio, H., and Zhang, M. (2002). Backbone dynamics of the 8 kDa dynein light chain dimer reveals molecular basis of the protein's functional diversity. *J. Biomol. NMR* **23**, 103–114.

Fang, Y.I., Yokota, E., Mabuchi, I., Nakamura, H., and Ohizumi, Y. (1997). Purealin blocks the sliding movement of sea urchin flagellar axonemes by selective inhibition of half the ATPase activity of axonemal dyneins. *Biochemistry* **36**, 15561–15567.

Fliegauf, M., Olbrich, H., Horvath, J., Wildhaber, J.H., Zariwala, M.A., Kennedy, M., Knowles, M.R., and Omran, H. (2005). Mislocalization of DNAH5 and DNAH9 in respiratory cells from patients with primary ciliary dyskinesia. *Am. J. Respir. Crit. Care Med.* **171**, 1343–1349.

Fok, A.K., Wang, H., Katayama, A., Aihara, M.S., and Allen, R.D. (1994). 22S axonemal dynein is preassembled and functional prior to being transported to and attached on the axonemes. *Cell Motil. Cytoskeleton* **29**, 215–224.

Fossella, J., Samant, S.A., Silver, L.M., King, S.M., Vaughan, K.T., Olds-Clarke, P., Johnson, K.A., Mikami, A., Vallee, R.B., and Pilder, S.H. (2000). An axonemal dynein at the *Hybrid Sterility 6* locus: implications for *t* haplotype-specific male sterility and the evolution of species barriers. *Mamm. Genome* **11**, 8–15.

Fowkes, M.E. and Mitchell, D.R. (1998). The role of preassembled cytoplasmic complexes in assembly of flagellar dynein subunits. *Mol. Biol. Cell* **9**, 2337–2347.

Freshour, J., Yokoyama, R., and Mitchell, D.R. (2007). *Chlamydomonas* flagellar outer row dynein assembly protein Oda7 interacts with both outer row and I1 inner row dyneins. *J. Biol. Chem.* **282**, 5404–5412.

Gardner, L.C., O'Toole, E., Perrone, C.A., Giddings, T., and Porter, M.E. (1994). Components of a "dynein regulatory complex" are located at the junction between the radial spokes and the dynein arms in *Chlamydomonas* flagella. *J. Cell Biol.* **127**, 1311–1325.

Gatti, J.L., King, S.M., Moss, A.G., and Witman, G.B. (1989). Outer arm dynein from trout spermatozoa. Purification, polypeptide composition, and enzymatic properties. *J. Biol. Chem.* **264**, 11450–11457.

Gatti, J.L., King, S.M., and Witman, G.B. (1991). The ATPases of *Chlamydomonas* outer arm dynein differ in their pH and cationic requirements. In: *Comparative Spermatology 20 Years After* (B. Baccetti, Ed.), pp. 373–375. Raven Press, New York.

Gibbons, I.R. and Fronk, E. (1979). A latent adenosine triphosphatase form of dynein 1 from sea urchin sperm flagella. *J. Biol. Chem.* **254**, 187–196.

Gibbons, I.R., Cosson, M.P., Evans, J.A., Gibbons, B.H., Houck, B., Martinson, K.H., Sale, W.S., and Tang, W.J. (1978). Potent inhibition of dynein adenosinetriphosphatase and of the motility of cilia and sperm flagella by vanadate. *Proc. Natl. Acad. Sci. U.S.A.* **75**, 2220–2224.

Gibbons, I.R., Garbarino, J.E., Tan, C.E., Reck-Peterson, S.L., Vale, R.D., and Carter, A.P. (2005). The affinity of the dynein microtubule-binding domain is modulated by the conformation of its coiled-coil stalk. *J. Biol. Chem.* **280**, 23960–23965.

Gingras, D., White, D., Garin, J., Multigner, L., Job, D., Cosson, J., Huitorel, P., Zingg, H., Dumas, F., and Gagnon, C. (1996). Purification, cloning, and sequence analysis of a M_r = 30,000 protein from sea urchin axonemes that is important for sperm motility. Relationship of the protein to a dynein light chain. *J. Biol. Chem.* **271**, 12807–12813.

Gocke, C., Osmani, S., and Miller, B. (2000). The human homologue of *Aspergillus* nuclear migration gene *nudC* is preferentially expressed in dividing cells and ciliated epithelia. *Histochem. Cell Biol.* **114**, 293–301.

Goodenough, U.W. and Heuser, J.E. (1982). Substructure of the outer dynein arm. *J. Cell Biol.* **95**, 798–815.

Goodenough, U. and Heuser, J. (1984). Structural comparison of purified dynein proteins with *in situ* dynein arms. *J. Mol. Biol.* **180**, 1083–1118.

Guenther, B., Onrust, R., Sali, A., O'Donnell, M., and Kuriyan, J. (1997). Crystal structure of the δ' subunit of the clamploader complex of *E. coli* DNA polymerase III. *Cell* **91**, 335–345.

Habura, A., Tikhonenko, I., Chisholm, R.L., and Koonce, M.P. (1999). Interaction mapping of a dynein heavy chain. Identification of dimerization and intermediate-chain binding domains. *J. Biol. Chem.* **274**, 15447–15453.

Harrison, A., Olds-Clarke, P., and King, S.M. (1998). Identification of the *t* complex-encoded cytoplasmic dynein light chain Tctex1 in inner arm I1 supports the involvement of flagellar dyneins in meiotic drive. *J. Cell Biol.* **140**, 1137–1147.

Harrison, A., Sakato, M., Tedford, H.W., Benashski, S.E., Patel-King, R.S., and King, S.M. (2002). Redox-based control of the γ heavy chain ATPase from *Chlamydomonas* outer arm dynein. *Cell Motil. Cytoskeleton* **52**, 131–143.

Hawkins, R. and McLeish, T. (2006). Dynamic allostery of protein α helical coiled coils. *J. R. Soc. Interface* **22**, 125–138.

Hendrickson, T.W., Perrone, C.A., Griffin, P., Wuichet, K., Mueller, J., Yang, P., Porter, M.E., and Sale, W.S. (2004). IC138 is a WD-repeat dynein intermediate chain required for light chain assembly and regulation of flagellar bending. *Mol. Biol. Cell* **15**, 5431–5442.

Herrmann, B.G., Koschorz, B., Wertz, K., McLaughlin, J., and Kispert, A. (1999). A protein kinase encoded by the *t* complex responder gene causes non-Mendelian inheritance. *Nature* **402**, 141–146.

Higuchi, H. and Endow, S. (2002). Directionality and processivity of molecular motors. *Curr. Opin. Cell Biol.* **14**, 50–57.

Hirakawa, E., Higuchi, H., and Toyoshima, Y.Y. (2000). Processive movement of single 22S dynein molecules occurs only at low ATP concentrations. *Proc. Natl. Acad. Sci. U.S.A.* **97**, 2533–2537.

Hirono, M., Uryu, S., Ohara, A., Kato-Minoura, T., and Kamiya, R. (2003). Expression of conventional and unconventional actins in *Chlamydomonas reinhardtii* upon deflagellation and sexual adhesion. *Eukaryot. Cell* **2**, 486–493.

Hoffmann, B., Zuo, W., Liu, A., and Morris, N.R. (2001). The lis1-related protein NudF of *Aspergillus nidulans* and its interaction partner NudE bind directly to specific subunits of dynein and dynactin and to α- and γ-tubulin. *J. Biol. Chem.* **276**, 38877–38884.

Hook, P., Mikami, A., Shafer, B., Chait, B.T., Rosenfeld, S.S., and Vallee, R.B. (2005). Long range allosteric control of cytoplasmic dynein ATPase activity by the stalk and C-terminal domains. *J. Biol. Chem.* **280**, 33045–33054.

Hoops, H.J. and Witman, G.B. (1983). Outer doublet heterogeneity reveals structural polarity related to beat direction in *Chlamydomonas* flagella. *J. Cell Biol.* **97**, 902–908.

Horst, C.J. and Witman, G.B. (1993). *ptx1*, a nonphototactic mutant of *Chlamydomonas*, lacks control of flagellar dominance. *J. Cell Biol.* **120**, 733–741.

Hou, Y., Qin, H., Follit, J.A., Pazour, G.J., Rosenbaum, J.L., and Witman, G.B. (2007). Functional analysis of an individual IFT protein: IFT46 is required for transport of outer dynein arms into flagella. *J. Cell Biol.* **176**, 653–665.

Howard, J. (1997). Molecular motors: structural adaptations to cellular functions. *Nature* **389**, 561–567.

Hozumi, A., Satouh, Y., Makino, Y., Toda, T., Ide, H., Ogawa, K., King, S.M., and Inaba, K. (2006). Molecular characterization of *Ciona* sperm outer arm dynein reveals multiple components related to outer arm docking complex protein 2. *Cell Motil. Cytoskeleton* **63**, 591–603.

Huang, B., Piperno, G., and Luck, D.J.L. (1979). Paralyzed flagellar mutants of *Chlamydomonas reinhardtii* defective for axonemal doublet microtubule arms. *J. Biol. Chem.* **254**, 3091–3099.

Huang, B., Ramanis, Z., and Luck, D.J. (1982). Suppressor mutations in *Chlamydomonas* reveal a regulatory mechanism for flagellar function. *Cell* **28**, 115–124.

Huw, L.Y., Goldsborough, A.S., Willison, K., and Artzt, K. (1995). Tctex2: a sperm tail surface protein mapping to the *t*-complex. *Dev. Biol.* **170**, 183–194.

Hyams, J.S. and Borisy, G.G. (1978). Isolated flagellar apparatus of *Chlamydomonas*: characterization of forward swimming and alteration of waveform and reversal of motion by calcium ions *in vitro*. *J. Cell Sci.* **33**, 235–253.

Inaba, K. and Mohri, H. (1989). Dynamic conformational changes of 21 S dynein ATPase coupled with ATP hydrolysis revealed by proteolytic digestion. *J. Biol. Chem.* **264**, 8384–8388.

Inaba, K., Ogawa, K., and Mohri, H. (1991). Mapping of ATP-dependent trypsin-sensitive sites on the beta chain of outer-arm dynein from sea urchin sperm flagella. *J. Biochem. (Tokyo)* **110**, 795–801.

Ishikawa, T., Sakakibara, H., and Oiwa, K. (2007). The architecture of outer dynein arms *in situ*. *J. Mol. Biol.* **368**, 1249–1258.

Ito, N., Phillips, S., Stevens, C., Ogel, Z., McPherson, M., Keen, J., Yadav, K., and Knowles, P. (1991). Novel thioether bond revealed by a 1.7 Å crystal structure of galactose oxidase. *Nature* **350**, 87–90.

Jaffrey, S.R. and Snyder, S.H. (1996). PIN: an associated protein inhibitor of neuronal nitric oxide synthase. *Science* **274**, 774–777.

Kagami, O. and Kamiya, R. (1990). Strikingly low ATPase activities in flagellar axonemes of a *Chlamydomonas* mutant missing outer dynein arms. *Eur. J. Biochem.* **189**, 441–446.

Kagami, O. and Kamiya, R. (1992). Translocation and rotation of microtubules caused by multiple species of *Chlamydomonas* inner-arm dynein. *J. Cell Sci.* **103**, 653–664.

Kagami, O., Takada, S., and Kamiya, R. (1990). Microtubule translocation caused by three subspecies of inner-arm dynein from *Chlamydomonas* flagella. *FEBS Lett.* **264**, 179–182.

Kagami, O., Gotoh, M., Makino, Y., Mohri, H., Kamiya, R., and Ogawa, K. (1998). A dynein light chain of sea urchin sperm flagella is a homolog of mouse Tctex 1, which is encoded by a gene of the *t* complex sterility locus. *Gene* **211**, 383–386.

Kajava, A.V. (1998). Structural diversity of leucine-rich repeat proteins. *J. Mol. Biol.* **277**, 519–527.

Kalef, E. and Gitler, C. (1994). Purification of vicinal dithiol-containing proteins by arsenical-based affinity chromatography. *Methods Enzymol.* **233**, 395–403.

Kamiya, R. (1988). Mutations at twelve independent loci result in absence of outer dynein arms in *Chlamydomonas reinhardtii*. *J. Cell Biol.* **107**, 2253–2258.

Kamiya, R. (2002). Functional diversity of axonemal dyneins as studied in *Chlamydomonas* mutants. *Int. Rev. Cytol.* **219**, 115–155.

Kamiya, R. and Hasegawa, E. (1987). Intrinsic difference in beat frequency between the two flagella of *Chlamydomonas reinhardtii*. *Expt. Cell Res.* **173**, 299–304.

Kamiya, R. and Okamoto, M. (1985). A mutant of *Chlamydomonas reinhardtii* that lacks the flagellar outer dynein arm but can swim. *J. Cell Sci.* **74**, 181–191.

Kamiya, R. and Witman, G.B. (1984). Submicromolar levels of calcium control the balance of beating between the two flagella in demembranated models of *Chlamydomonas*. *J. Cell Biol.* **98**, 97–107.

Kamiya, R., Kurimoto, E., and Muto, E. (1991). Two types of *Chlamydomonas* flagellar mutants missing different components of inner-arm dynein. *J. Cell Biol.* **112**, 441–447.

Kastury, K., Taylor, W.E., Gutierrez, M., Ramirez, L., Coucke, P.J., Van Hauwe, P., Van Camp, G., and Bhasin, S. (1997). Chromosomal mapping of two members of the human dynein gene family to chromosome regions 7p15 and 11q13 near the deafness loci DFNA 5 and DFNA 11. *Genomics* **44**, 362–364.

Kato, T., Kagami, O., Yagi, T., and Kamiya, R. (1993). Isolation of two species of *Chlamydomonas reinhardtii* flagellar mutants, *ida5* and *ida6*, that lack a newly identified heavy chain of the inner dynein arm. *Cell Struct. Funct.* **18**, 371–377.

Kato-Minoura, T., Hirono, M., and Kamiya, R. (1997). *Chlamydomonas* inner-arm dynein mutant, *ida5*, has a mutation in an actin-encoding gene. *J. Cell Biol.* **137**, 649–656.

Kato-Minoura, T., Uryu, S., Hirono, M., and Kamiya, R. (1998). Highly divergent actin expressed in a *Chlamydomonas* mutant lacking the conventional actin gene. *Biochem. Biophys. Res. Commun.* **251**, 71–76.

Kikushima, K., and Kamiya, R. (2008). Clockwise translocation of microtubules by flagellar inner-arm dyneins *in vitro*. *Biophys. J.* **94**, 4014–4019.

Kikushima, K., Yagi, T., and Kamiya, R. (2004). Slow ADP-dependent acceleration of microtubule translocation produced by an axonemal dynein. *FEBS Lett.* **563**, 119–122.

King, S.J. and Dutcher, S.K. (1997). Phosphoregulation of an inner dynein arm complex in *Chlamydomonas reinhardtii* is altered in phototactic mutant strains. *J. Cell Biol.* **136**, 177–191.

King, S.J., Inwood, W.B., O'Toole, E.T., Power, J., and Dutcher, S.K. (1994). The *bop2-1* mutation reveals radial asymmetry in the inner dynein arm region of *Chlamydomonas reinhardtii*. *J. Cell Biol.* **126**, 1255–1266.

King, S.M. (2000). AAA domains and organization of the dynein motor unit. *J. Cell Sci.* **113**, 2521–2526.

King, S.M. (2002a). Dynein motors: structure, mechanochemistry and regulation. In: *Molecular Motors* (M. Schliwa, Ed.), pp. 45–78. Wiley-VCH Verlag GmbH, Weinheim.

King, S.M. (2002b). Dyneins motor on in plants. *Traffic* **3**, 930–931.

King, S.M. and Patel-King, R.S. (1995a). Identification of a $Ca^{(2+)}$-binding light chain within *Chlamydomonas* outer arm dynein. *J. Cell Sci.* **108**, 3757–3764.

King, S.M. and Patel-King, R.S. (1995b). The $M_{(r)} = 8,000$ and 11,000 outer arm dynein light chains from *Chlamydomonas* flagella have cytoplasmic homologues. *J. Biol. Chem.* **270**, 11445–11452.

King, S.M. and Witman, G.B. (1987). Structure of the α and β heavy chains of the outer arm dynein from *Chlamydomonas* flagella. Masses of chains and sites of ultraviolet-induced vanadate-dependent cleavage. *J. Biol. Chem.* **262**, 17596–17604.

King, S.M. and Witman, G.B. (1988a). Structure of the α and β heavy chains of the outer arm dynein from *Chlamydomonas* flagella. Location of epitopes and protease-sensitive sites. *J. Biol. Chem.* **263**, 9244–9255.

King, S.M. and Witman, G.B. (1988b). Structure of the γ heavy chain of the outer arm dynein from *Chlamydomonas* flagella. *J. Cell Biol.* **107**, 1799–1808.

King, S.M., and Witman, G.B. (1989). Molecular structure of *Chlamydomonas* outer arm dynein. In: *Cell Movement: The Dynein ATPases*, Vol. 1 (F.D. Warner, P. Satir, and I.R. Gibbons, Eds.), pp. 61–75. Alan R. Liss, Inc., New York.

King, S.M. and Witman, G.B. (1990). Localization of an intermediate chain of outer arm dynein by immunoelectron microscopy. *J. Biol. Chem.* **265**, 19807–19811.

King, S.M. and Witman, G.B. (1994). Multiple sites of phosphorylation within the α heavy chain of *Chlamydomonas* outer arm dynein. *J. Biol. Chem.* **269**, 5452–5457.

King, S.M., Otter, T., and Witman, G.B. (1985). Characterization of monoclonal antibodies against *Chlamydomonas* flagellar dyneins by high-resolution protein blotting. *Proc. Natl. Acad. Sci. U.S.A.* **82**, 4717–4721.

King, S.M., Otter, T., and Witman, G.B. (1986). Purification and characterization of *Chlamydomonas* flagellar dyneins. *Methods Enzymol.* **134**, 291–306.

King, S.M., Haley, B.E., and Witman, G.B. (1989). Structure of the α and β heavy chains of the outer arm dynein from *Chlamydomonas* flagella. Nucleotide binding sites. *J. Biol. Chem.* **264**, 10210–10218.

King, S.M., Wilkerson, C.G., and Witman, G.B. (1991). The M_r 78,000 intermediate chain of *Chlamydomonas* outer arm dynein interacts with α-tubulin *in situ*. *J. Biol. Chem.* **266**, 8401–8407.

King, S.M., Patel-King, R.S., Wilkerson, C.G., and Witman, G.B. (1995). The 78,000-$M_{(r)}$ intermediate chain of *Chlamydomonas* outer arm dynein is a microtubule-binding protein. *J. Cell Biol.* **131**, 399–409.

King, S.M., Barbarese, E., Dillman, J.F., Patel-King, R.S., Carson, J.H., and Pfister, K.K. (1996a). Brain cytoplasmic and flagellar outer arm dyneins share a highly conserved M_r 8,000 light chain. *J. Biol. Chem.* **271**, 19358–19366.

King, S.M., Dillman, J.F., Benashski, S.E., Lye, R.J., Patel-King, R.S., and Pfister, K.K. (1996b). The mouse *t*-complex-encoded protein Tctex-1 is a light chain of brain cytoplasmic dynein. *J. Biol. Chem.* **271**, 32281–32287.

King, S.M., Marchese-Ragona, S.P., Parker, S.K., and Detrich, H.W. (1997). Inner and outer arm axonemal dyneins from the Antarctic rockcod *Notothenia coriiceps*. *Biochemistry* **36**, 1306–1314.

Kon, T., Mogami, T., Ohkura, R., Nishiura, M., and Sutoh, K. (2005). ATP hydrolysis cycle-dependent tail motions in cytoplasmic dynein. *Nat. Struct. Mol. Biol.* **12**, 513–519.

Koonce, M.P. and Tikhonenko, I. (2000). Functional elements within the dynein microtubule-binding domain. *Mol. Biol. Cell* **11**, 523–529.

Koonin, E.V. and Aravind, L. (2000). Dynein light chains of the Roadblock/LC7 group belong to an ancient protein superfamily implicated in NTPase regulation. *Curr. Biol.* **10**, R774–776.

Kotani, N., Sakakibara, H., Burgess, S.A., Kojima, H., and Oiwa, K. (2007). Mechanical properties of inner-arm dynein-f (dynein I1) studied with *in vitro* motility assays. *Biophys. J.* **93**, 886–894.

Koutoulis, A., Pazour, G.J., Wilkerson, C.G., Inaba, K., Sheng, H., Takada, S., and Witman, G.B. (1997). The *Chlamydomonas reinhardtii ODA3* gene encodes a protein of the outer dynein arm docking complex. *J. Cell Biol.* **137**, 1069–1080.

Kull, F.J., Vale, R.D., and Fletterick, R.J. (1998). The case for a common ancestor: kinesin and myosin motor proteins and G proteins. *J. Muscle Res. Cell Motil.* **19**, 877–886.

Kurimoto, E. and Kamiya, R. (1991). Microtubule sliding in flagellar axonemes of *Chlamydomonas* mutants missing inner- or outer-arm dynein: velocity measurements on new types of mutants by an improved method. *Cell Motil. Cytoskeleton* **19**, 275–281.

Lader, E., Ha, H.S., O'Neill, M., Artzt, K., and Bennett, D. (1989). *tctex-1*: a candidate gene family for a mouse *t* complex sterility locus. *Cell* **58**, 969–979.

Lawrence, C.J., Morris, N.R., Meagher, R.B., and Dawe, R.K. (2001). Dyneins have run their course in plant lineage. *Traffic* **2**, 362–363.

Le, T., Perrone, C., de Cathelineau, A., O'Toole, E., and Porter, M. (1999). Recovery of a new inner arm dynein gene (*IDA8*) in *Chlamydomonas*. *Mol. Biol. Cell* **10**, 368a. (Abstract)

LeDizet, M. and Piperno, G. (1995a). *ida4-1*, *ida4-2*, and *ida4-3* are intron splicing mutations affecting the locus encoding p28, a light chain of *Chlamydomonas* axonemal inner dynein arms. *Mol. Biol. Cell* **6**, 713–723.

LeDizet, M. and Piperno, G. (1995b). The light chain p28 associates with a subset of inner dynein arm heavy chains in *Chlamydomonas* axonemes. *Mol. Biol. Cell* **6**, 697–711.

Lee, V.D., Finstad, S.L., and Huang, B. (1997). Cloning and characterization of a gene encoding an actin-related protein in *Chlamydomonas*. *Gene* **197**, 153–159.

Lee, V.D., Stapleton, M., and Huang, B. (1991). Genomic structure of *Chlamydomonas* caltractin: evidence for intron insertion suggests a probable genealogy for the EF-hand superfamily of proteins. *J. Mol. Biol.* **221**, 175–191.

Lee-Eiford, A., Ow, R.A., and Gibbons, I.R. (1986). Specific cleavage of dynein heavy chains by ultraviolet irradiation in the presence of ATP and vanadate. *J. Biol. Chem.* **261**, 2337–2342.

Lenzen, C.U., Steinmann, D., Whiteheart, S.W., and Weiss, W.I. (1998). Crystal structure of the hexamerization domain of *N*-ethylmaleimide-sensitive fusion protein. *Cell* **94**, 525–536.

Li, M.-G., Serr, M., Newman, E.A., and Hays, T.S. (2004). The *Drosophila* Tctex-1 light chain is dispensable for essential cytoplasmic dynein functions but is required during spermatid differentiation. *Mol. Biol. Cell* **15**, 3005–3014.

Liang, J., Jaffrey, S.R., Guo, W., Snyder, S.H., and Clardy, J. (1999). Structure of the PIN/LC8 dimer with a bound peptide. *Nat. Struct. Biol.* **6**, 735–740.

Liu, Z., Takazaki, H., Nakazawa, Y., Sakato, M., Yagi, T., Yasunaga, T., King, S.M., and Kamiya, R. (2008). Partially functional outer-arm dynein in a novel *Chlamydomonas* mutant expressing a truncated γ heavy chain. *Eukaryotic Cell* **7**, 1136–1145.

Lo, K.W.-H., Kan, H.-M., and Pfister, K.K. (2006). Identification of a novel region of the cytoplasmic dynein intermediate chain important for dimerization in the absence of the light chains. *J. Biol. Chem.* **281**, 9552–9559.

Luck, D.J.L., and Piperno, G. (1989). Dynein arm mutants of *Chlamydomonas*. In: *Cell Movement: The Dynein ATPases*, Vol. 1 (F.D. Warner, P. Satir, and I.R. Gibbons, Eds.), pp. 49–60. Alan R. Liss, Inc., New York.

Lupetti, P., Lanzavecchia, S., Mercati, D., Cantele, F., Dallai, R., and Mencarelli, C. (2005). Three-dimensional reconstruction of axonemal outer dynein arms *in situ* by electron tomography. *Cell Motil. Cytoskeleton* **62**, 69–83.

Lyon, M.F. (1984). Transmission ratio distortion in mouse *t* haplotypes is due to multiple distorter genes acting on a responder locus. *Cell* **37**, 621–628.

Mallik, R., Carter, B.C., Lex, S.A., King, S.J., and Gross, S.P. (2004). Cytoplasmic dynein functions as a gear in response to load. *Nature* **427**, 649–652.

Marchese-Ragona, S.P. and Johnson, K.A. (1988). STEM analysis of the dynein ATPase. *Electron Microsc. Rev.* **1**, 141–153.

Mastronarde, D.N., O'Toole, E.T., McDonald, K.L., McIntosh, J.R., and Porter, M.E. (1992). Arrangement of inner dynein arms in wild-type and mutant flagella of *Chlamydomonas*. *J. Cell Biol.* **118**, 1145–1162.

Mazumdar, M., Mikami, A., Gee, M.A., and Vallee, R.B. (1996). *In vitro* motility from recombinant dynein heavy chain. *Proc. Natl. Acad. Sci. U.S.A.* **93**, 6552–6556.

Meng, X., Samsó, M., and Koonce, M.P. (2006). A flexible linkage between the dynein motor and its cargo. *J. Mol. Biol.* **357**, 701–706.

Minoura, I. and Kamiya, R. (1995). Strikingly different propulsive forces generated by different dynein-deficient mutants in viscous media. *Cell Motil. Cytoskeleton* **31**, 130–139.

Mitchell, B.F., Grulich, L.E., and Mader, M.M. (2004). Flagellar quiescence in *Chlamydomonas*: characterization and defective quiescence in cells carrying *sup-pf-1* and *sup-pf-2* outer dynein arm mutations. *Cell Motil. Cytoskeleton* **57**, 186–196.

Mitchell, D.R. and Brown, K.S. (1994). Sequence analysis of the *Chlamydomonas* α and β dynein heavy chain genes. *J. Cell Sci.* **107**, 635–644.

Mitchell, D.R. and Brown, K.S. (1997). Sequence analysis of the *Chlamydomonas reinhardtii* flagellar α dynein gene. *Cell Motil. Cytoskeleton* **37**, 120–126.

Mitchell, D.R. and Kang, Y. (1991). Identification of *oda6* as a *Chlamydomonas* dynein mutant by rescue with the wild-type gene. *J. Cell Biol.* **113**, 835–842.

Mitchell, D.R. and Kang, Y. (1993). Reversion analysis of dynein intermediate chain function. *J. Cell Sci.* **105**, 1069–1078.

Mitchell, D.R. and Rosenbaum, J.L. (1985). A motile *Chlamydomonas* flagellar mutant that lacks outer dynein arms. *J. Cell Biol.* **100**, 1228–1234.

Mitchell, D.R. and Rosenbaum, J.L. (1986). Protein–protein interactions in the 18S ATPase of *Chlamydomonas* outer dynein arms. *Cell Motil. Cytoskeleton* **6**, 510–520.

Mocz, G. and Gibbons, I.R. (1996). Phase partition analysis of nucleotide binding to axonemal dynein. *Biochemistry* **35**, 9204–9211.

Mocz, G., Helms, M.K., Jameson, D.M., and Gibbons, I.R. (1998). Probing the nucleotide binding sites of axonemal dynein with the fluorescent nucleotide analogue 2'(3')-O-(-N-methylanthraniloyl)-adenosine 5'-triphosphate. *Biochemistry* **37**, 9862–9869.

Mok, Y.-K., Lo, K.W.-H., and Zhang, M. (2001). Structure of Tctex-1 and its interaction with cytoplasmic dynein intermediate chain. *J. Biol. Chem.* **276**, 14067–14074.

Morris, N.R. (2000). A rough guide to a smooth brain. *Nat. Cell Biol.* **2**, E201–202.

Morris, S.M., Albrecht, U., Reiner, O., Eichele, G., and Yu-Lee, L.Y. (1998). The lissencephaly gene product Lis1, a protein involved in nuclear migration, interacts with a nuclear movement protein NudC. *Curr. Biol.* **8**, 603–606.

Moss, A.G. and Morgan, D.D. (1999). Rapid analysis of *Chlamydomonas reinhardtii* flagellar beat activity with a LSCM. *Microsc. Anal.* **34**, 7–9.

Muto, E., Edamatsu, M., Hirono, M., and Kamiya, R. (1994). Immunological detection of actin in the 14S ciliary dynein of Tetrahymena. *FEBS Lett.* **343**, 173–177.

Myster, S.H., Knott, J.A., O'Toole, E., and Porter, M.E. (1997). The *Chlamydomonas Dhc1* gene encodes a dynein heavy chain subunit required for assembly of the I1 inner arm complex. *Mol. Biol. Cell* **8**, 607–620.

Myster, S.H., Knott, J.A., Wysocki, K.M., O'Toole, E., and Porter, M.E. (1999). Domains in the 1α dynein heavy chain required for inner arm assembly and flagellar motility in *Chlamydomonas*. *J. Cell Biol.* **146**, 801–818.

Nakamura, H., Wu, H., Kobayashi, J.i., Nakamura, Y., Ohizumi, Y., and Hirata, Y. (1985). Purealin, a novel enzyme activator from the Okinawan marine sponge. *Tetrahedron Lett.* **26**, 4517–4520.

Nakamura, K., Wilkerson, C.G., and Witman, G.B. (1997). Functional interaction between *Chlamydomonas* outer arm dynein subunits: the γ subunit suppresses the ATPase activity of the αβ dimer. *Cell Motil. Cytoskeleton* **37**, 338–345.

Neuwald, A.F., Aravind, L., Spouge, J.L., and Koonin, E.V. (1999). AAA$^+$: a class of chaperone-like ATPases associated with the assembly, operation, and disassembly of protein complexes. *Genome Res.* **9**, 27–43.

Nicastro, D., McIntosh, J.R., and Baumeister, W. (2005). 3D structure of eukaryotic flagella in a quiescent state revealed by cryo-electron tomography. *Proc. Natl. Acad. Sci. U.S.A.* **102**, 15889–15894.

Nicastro, D., Schwartz, C., Pierson, J., Gaudette, R., Porter, M.E., and McIntosh, J.R. (2006). The molecular architecture of axonemes revealed by cryoelectron tomography. *Science* **313**, 944–948.

Nikulina, K. and King, S.M. (2002). The roadblock light chains of cytoplasmic dynein associate with *Hepatitis B* X-interacting protein. *Mol. Biol. Cell* **13**, 182a.

Nikulina, K., Patel-King, R.S., Takebe, S., Pfister, K.K., and King, S.M. (2004). The roadblock light chains are ubiquitous components of cytoplasmic dynein that form homo- and heterodimers. *Cell Motil. Cytoskeleton* **57**, 233–245.

Norrander, J.M., Perrone, C.A., Amos, L.A., and Linck, R.W. (1996). Structural comparison of tektins and evidence for their determination of complex spacings in flagellar microtubules. *J. Mol. Biol.* **257**, 385–397.

Nyarko, A., Cochrun, L., Norwood, S., Pursifull, N., Voth, A., and Barbar, E. (2005). Ionization of His 55 at the dimer interface of dynein light-chain LC8 is coupled to dimer dissociation. *Biochemistry* **44**, 14248–14255.

Oda, T., Hirokawa, N., and Kikkawa, M. (2007). Three-dimensional structures of the flagellar dynein-microtubule complex by cryoelectron microscopy. *J. Cell Biol.* **177**, 243–252.

Ogawa, K. and Mohri, H. (1972). Studies on the flagellar ATPase from sea urchin spermatozoa. I. Purification and some properties of the enzyme. *Biochim. Biophys. Acta* **256**, 142–155.

Ogawa, K., Kamiya, R., Wilkerson, C.G., and Witman, G.B. (1995). Interspecies conservation of outer arm dynein intermediate chain sequences defines two intermediate chain subclasses. *Mol. Biol. Cell* **6**, 685–696.

Ogawa, K., Takai, H., Ogiwara, A., Yokota, E., Shimizu, T., Inaba, K., and Mohri, H. (1996). Is outer arm dynein intermediate chain 1 multifunctional?. *Mol. Biol. Cell* **7**, 1895–1907.

Ogura, T. and Wilkinson, A.J. (2001). AAA$^+$ superfamily of ATPases: common structure-diverse function. *Genes Cells* **6**, 575–597.

Okada, Y. and Hirokawa, N. (2000). Mechanism of the single-headed processivity: diffusional anchoring between the K-loop of kinesin and the C terminus of tubulin. *Proc. Natl. Acad. Sci. U.S.A.* **97**, 640–645.

Okagaki, T. and Kamiya, R. (1986). Microtubule sliding in mutant *Chlamydomonas* axonemes devoid of outer or inner dynein arms. *J. Cell Biol.* **103**, 1895–1902.

Okita, N., Isogai, N., Hirono, M., Kamiya, R., and Yoshimura, K. (2005). Phototactic activity in *Chlamydomonas* "non-phototactic" mutants deficient in Ca^{2+}-dependent control of flagellar dominance or in inner-arm dynein. *J. Cell Sci.* **118**, 529–537.

Omoto, C.K. and Nakamaye, K. (1989). ATP analogs substituted on the 2-position as substrates for dynein ATPase activity. *Biochim. Biophys. Acta* **999**, 221–224.

Omoto, C.K., Yagi, T., Kurimoto, E., and Kamiya, R. (1996). Ability of paralyzed flagella mutants of *Chlamydomonas* to move. *Cell Motil. Cytoskeleton* **33**, 88–94.

Padma, P., Hozumi, A., Ogawa, K., and Inaba, K. (2001). Molecular cloning and characterization of a thioredoxin/nucleoside diphosphate kinase related dynein intermediate chain from the ascidian, *Ciona intestinalis*. *Gene* **275**, 177–183.

Paschal, B.M., Shpetner, H.S., and Vallee, R.B. (1987). MAP 1C is a microtubule-activated ATPase which translocates microtubules *in vitro* and has dynein-like properties. *J. Cell Biol.* **105**, 1273–1282.

Patel, S. and Latterich, M. (1998). The AAA team: related ATPases with diverse functions. *Trends Cell Biol.* **8**, 65–71.

Patel-King, R.S., Benashki, S.E., Harrison, A., and King, S.M. (1996). Two functional thioredoxins containing redox-sensitive vicinal dithiols from the *Chlamydomonas* outer dynein arm. *J. Biol. Chem.* **271**, 6283–6291.

Patel-King, R.S., Benashski, S.E., Harrison, A., and King, S.M. (1997). A *Chlamydomonas* homologue of the putative murine *t* complex distorter Tctex-2 is an outer arm dynein light chain. *J. Cell Biol.* **137**, 1081–1090.

Patel-King, R.S., Gorbatyuk, O., Takebe, S., and King, S.M. (2004). Flagellar radial spokes contain a Ca^{2+}-sensitive nucleoside diphosphate kinase. *Mol. Biol. Cell* **15**, 3891–3902.

Pazour, G.J. and Witman, G.B. (2000). Forward and reverse genetic analysis of microtubule motors in *Chlamydomonas*. *Methods* **22**, 285–298.

Pazour, G.J., Sineschekov, O.A., and Witman, G.B. (1995). Mutational analysis of the phototransduction pathway of *Chlamydomonas reinhardtii*. *J. Cell Biol.* **131**, 427–440.

Pazour, G.J., Wilkerson, C.G., and Witman, G.B. (1998). A dynein light chain is essential for the retrograde particle movement of intraflagellar transport (IFT). *J. Cell Biol.* **141**, 979–992.

Pazour, G.J., Koutoulis, A., Benashski, S.E., Dickert, B.L., Sheng, H., Patel-King, R.S., King, S.M., and Witman, G.B. (1999). LC2, the *Chlamydomonas* homologue of the *t* complex-encoded protein Tctex2, is essential for outer dynein arm assembly. *Mol. Biol. Cell* **10**, 3507–3520.

Pedersen, L., Rompolas, P., Christensen, S., Rosenbaum, J.L., and King, S.M. (2007). The lissencephaly protein Lis1 is present in motile mammalian cilia and requires outer dynein arm for targeting to *Chlamydomonas* flagella. *J. Cell Sci.* **120**, 858–867.

Peitsch, M.C. (1996). ProMod and Swiss-Model: internet-based tools for automated comparative protein modeling. *Biochem. Soc. Trans.* **24**, 274–279.

Penningroth, S.M., Cheung, A., Olehnik, K., and Koslosky, R. (1982). Mechanochemical coupling in the relaxation of rigor-wave sea urchin sperm flagella. *J. Cell Biol.* **92**, 733–741.

Perrone, C.A., Yang, P., O'Toole, E., Sale, W.S., and Porter, M.E. (1998). The *Chlamydomonas IDA7* locus encodes a 140-kDa dynein intermediate chain required to assemble the I1 inner arm complex. *Mol. Biol. Cell* **9**, 3351–3365.

Perrone, C.A., Myster, S.H., Bower, R., O'Toole, E.T., and Porter, M.E. (2000). Insights into the structural organization of the I1 inner arm dynein from a domain analysis of the 1β dynein heavy chain. *Mol. Biol. Cell* **11**, 2297–2313.

Pfister, K.K. and Witman, G.B. (1984). Subfractionation of *Chlamydomonas* 18 S dynein into two unique subunits containing ATPase activity. *J. Biol. Chem.* **259**, 12072–12080.

Pfister, K.K., Fay, R.B., and Witman, G.B. (1982). Purification and polypeptide composition of dynein ATPases from *Chlamydomonas* flagella. *Cell Motil.* **2**, 525–547.

Pfister, K.K., Haley, B.E., and Witman, G.B. (1984). The photoaffinity probe 8-azidoadenosine 5'-triphosphate selectively labels the heavy chain of *Chlamydomonas* 12 S dynein. *J. Biol. Chem.* **259**, 8499–8504.

Pfister, K.K., Haley, B.E., and Witman, G.B. (1985). Labeling of *Chlamydomonas* 18 S dynein polypeptides by 8-azidoadenosine 5'-triphosphate, a photoaffinity analog of ATP. *J. Biol. Chem.* **260**, 12844–12850.

Pfister, K.K., Fisher, E.M.C., Gibbons, I.R., Hays, T.S., Holzbaur, E.L.F., McIntosh, J.R., Porter, M.E., Schroer, T.A., Vaughan, K.T., Witman, G.B., King, S.M., and Vallee, R.B. (2005). Cytoplasmic dynein nomenclature. *J. Cell Biol.* **171**, 411–413.

Pfister, K.K., Shah, P.R., Hummerich, H., Russ, A., Cotton, J., Annuar, A.A., King, S.M., and Fisher, E.M.C. (2006). Genetic analysis of the cytoplasmic dynein subunit families. *PLoS Genet.* **2**, e1.

Piperno, G. and Luck, D.J. (1979). Axonemal adenosine triphosphatases from flagella of *Chlamydomonas reinhardtii*. Purification of two dyneins. *J. Biol. Chem.* **254**, 3084–3090.

Piperno, G. and Luck, D.J. (1981). Inner arm dyneins from flagella of *Chlamydomonas reinhardtii*. *Cell* **27**, 331–340.

Piperno, G. and Ramanis, Z. (1991). The proximal portion of *Chlamydomonas* flagella contains a distinct set of inner dynein arms. *J. Cell Biol.* **112**, 701–709.

Piperno, G., Ramanis, Z., Smith, E.F., and Sale, W.S. (1990). Three distinct inner dynein arms in *Chlamydomonas* flagella: molecular composition and location in the axoneme. *J. Cell Biol.* **110**, 379–389.

Piperno, G., Mead, K., and Shestak, W. (1992). The inner dynein arms I2 interact with a "dynein regulatory complex" in *Chlamydomonas* flagella. *J. Cell Biol.* **118**, 1455–1463.

Piperno, G., Mead, K., LeDizet, M., and Moscatelli, A. (1994). Mutations in the "dynein regulatory complex" alter the ATP-insensitive binding sites for inner arm dyneins in *Chlamydomonas* axonemes. *J. Cell Biol.* **125**, 1109–1117.

Porter, M.E., Power, J., and Dutcher, S.K. (1992). Extragenic suppressors of paralyzed flagellar mutations in *Chlamydomonas reinhardtii* identify loci that alter the inner dynein arms. *J. Cell Biol.* **118**, 1163–1176.

Porter, M.E., Knott, J.A., Gardner, L.C., Mitchell, D.R., and Dutcher, S.K. (1994). Mutations in the *SUP-PF-1* locus of *Chlamydomonas reinhardtii* identify a regulatory domain in the β-dynein heavy chain. *J. Cell Biol.* **126**, 1495–1507.

Porter, M.E., Knott, J.A., Myster, S.H., and Farlow, S.J. (1996). The dynein gene family in *Chlamydomonas reinhardtii*. *Genetics* **144**, 569–585.

Prag, S. and Adams, J. (2003). Molecular phylogeny of the kelch-repeat superfamily reveals and expansion of BTB/kelch proteins in animals. *BMC Bioinformatics* **4**, 42.

Rhoads, A. and Friedberg, F. (1997). Sequence motifs for calmodulin recognition. *FASEB J.* **11**, 331–340.

Rittinger, K., Walker, P.A., Eccleston, J.F., Smerdon, S.J., and Gamblin, S.J. (1997). Structure at 1.65Å of RhoA and its GTPase-activating protein in complex with a transition-state analogue. *Nature* **389**, 758–762.

Roberts, A.J., Numata, N., Walker, M.L., Kon, T., Ohkura, R., Burgess, S.A., Knight, P.J., and Sutoh, K. (2007). Sub-domain mapping with the dynein motor domain. *Mol. Biol. Cell* **18**. Abstract #2553. (CD-ROM)

Rompolas, P., Patel-King, R.S., Gorbatyuk, O., Wakabayashi, K., Pazour, G.J., and King, S.M. (2005). Identification of a third member of the LC8 family of dynein light chains in *Chlamydomonas* flagella. *Mol. Biol. Cell* **16**. Abstract #969. (CD-ROM)

Rompolas, P., Pedersen, L., Patel-King, R.S., and King, S.M. (2007). *Chlamydomonas* FAP133 is a dynein intermediate chain associated with the retrograde intraflagellar transport motor. *J. Cell Sci.* **120**, 3653–3665.

Rüffer, U. and Nultsch, W. (1987). Comparison of the beating of *cis*- and *trans*-flagella of *Chlamydomonas* cells held on micropipettes. *Cell Motil. Cytoskeleton* **7**, 87–93.

Rupp, G., O'Toole, E., Gardner, L.C., Mitchell, B.F., and Porter, M.E. (1996). The *sup-pf-2* mutations of *Chlamydomonas* alter the activity of the outer dynein arms by modification of the γ-dynein heavy chain. *J. Cell Biol.* **135**, 1853–1865.

Sadek, C.M., Damdimopoulos, A.E., Pelto-Huikko, M., Gustafsson, J.A., Spyrou, G., and Miranda-Vizuete, A. (2001). Sptrx-2, a fusion protein composed of one thioredoxin and three tandemly repeated NDP-kinase domains is expressed in human testis germ cells. *Genes Cells* **6**, 1077–1090.

Sakakibara, H. and Nakayama, H. (1998). Translocation of microtubules caused by the αβ, β and γ outer arm dynein subparticles of *Chlamydomonas*. *J. Cell Sci.* **111**, 1155–1164.

Sakakibara, H., Mitchell, D.R., and Kamiya, R. (1991). A *Chlamydomonas* outer arm dynein mutant missing the α heavy chain. *J. Cell Biol.* **113**, 615–622.

Sakakibara, H., Takada, S., King, S.M., Witman, G.B., and Kamiya, R. (1993). A *Chlamydomonas* outer arm dynein mutant with a truncated β heavy chain. *J. Cell Biol.* **122**, 653–661.

Sakakibara, H., Kojima, H., Sakai, Y., Katayama, E., and Oiwa, K. (1999). Inner-arm dynein c of *Chlamydomonas* flagella is a single-headed processive motor. *Nature* **400**, 586–590.

Sakato, M. and King, S.M. (2003). Calcium regulates ATP-sensitive microtubule binding by *Chlamydomonas* outer arm dynein. *J. Biol. Chem.* **278**, 43571–43579.

Sakato, M. and King, S.M. (2004). Design and regulation of the AAA$^+$ microtubule motor dynein. *J. Struct. Biol.* **146**, 58–71.

Sakato, M., Sakakibara, H., and King, S.M. (2007). *Chlamydomonas* outer arm dynein alters conformation in response to Ca^{2+}. *Mol. Biol. Cell* **18**, 3620–3634.

Sale, W.S. and Satir, P. (1977). Direction of active sliding of microtubules in *Tetrahymena* cilia. *Proc. Natl. Acad. Sci. U.S.A.* **74**, 2045–2049.

Sale, W.S., Goodenough, U.W., and Heuser, J.E. (1985). The substructure of isolated and *in situ* outer dynein arms of sea urchin sperm flagella. *J. Cell Biol.* **101**, 1400–1412.

Salisbury, J.L., Baron, A.T., and Sanders, M.A. (1988). The centrin-based cytoskeleton of *Chlamydomonas reinhardtii*: distribution in interphase and mitotic cells. *J. Cell Biol.* **107**, 635–641.

Samant, S.A., Ogunkua, O., Hui, L., Fossella, J., and Pilder, S.H. (2002). The *t* complex distorter 2 candidate gene, Dnahc8, encodes at least two testis-specific axonemal dynein heavy chains that differ extensively at their amino and carboxyl termini. *Dev. Biol.* **250**, 24–43.

Samsó, M. and Koonce, M.P. (2004). 25 Å-resolution structure of a cytoplasmic dynein motor reveals a seven-member planar ring. *J. Mol. Biol.* **340**, 1059–1072.

Samsó, M., Radermacher, M., Frank, J., and Koonce, M.P. (1998). Structural characterization of a dynein motor domain. *J. Mol. Biol.* **276**, 927–937.

Sanders, M.A. and Salisbury, J.L. (1989). Centrin-mediated microtubule severing during flagellar excision in *Chlamydomonas reinhardtii*. *J. Cell Biol.* **108**, 1751–1760.

Sasaki, S., Shionoya, A., Ishida, M., Gambello, M., Yingling, J., Wynshaw-Boris, A., and Hirotsone, S. (2000). A LIS1/NUDEL/cytoplasmic dynein heavy chain complex in the developing and adult nervous system. *Neuron* **28**, 681–696.

Satir, P. (1968). Studies on cilia: III. Further studies on the cilium tip and a "sliding filament" model of ciliary motility. *J. Cell Biol.* **39**, 77–94.

Scheffzek, K., Ahmadian, M.R., Kabsch, W., Weismuller, L., Lautwein, A., Schmitz, F., and Wittinghofer, A. (1997). The Ras-RasGAP complex: structural basis for GTPase activation and its loss in oncogenic Ras mutants. *Science* **277**, 333–338.

Shimizu, T. and Kimura, I. (1974). Effects of *N*-ethylmaleimide on dynein adenosinetriphosphatase activity and its recombining ability with outer fibers. *J. Biochem. (Tokyo)* **76**, 1001–1008.

Silver, L.M. (1993). The peculiar journey of a selfish chromosome. *Trends Genet.* **9**, 250–254.

Sloan, D. and Mildvan, A. (1976). Nuclear magnetic relaxation studies of the conformation of adenosine 5′-triphosphate on pyruvate kinase from rabbit muscle. *J. Biol. Chem.* **251**, 2412–2420.

Smith, E.F. and Sale, W.S. (1991). Microtubule binding and translocation by inner dynein arm subtype I1. *Cell Motil. Cytoskeleton* **18**, 258–268.

Smith, E.F. and Sale, W.S. (1992). Structural and functional reconstitution of inner dynein arms in *Chlamydomonas* flagellar axonemes. *J. Cell Biol.* **117**, 573–581.

Sondek, J., Bohm, A., Lambright, D.G., Hamm, H.E., and Sigler, P.B. (1996). Crystal structure of a G-protein β/γ dimer at 2.1 Å resolution. *Nature* **379**, 369–374.

Song, J., Tyler, R.C., Lee, M.S., Tyler, E.M., and Markley, J.L. (2005). Solution structure of isoform 1 of Roadblock/LC7, a light chain in the dynein complex. *J. Mol. Biol.* **354**, 1043–1051.

Spormann, A.M. and Kaiser, D. (1999). Gliding mutants of *Myxococcus xanthus* with high reversal frequencies and small displacements. *J. Bacteriol.* **181**, 2593–2601.

Stolc, V., Samanta, M.P., Tongprasit, W., and Marshall, W.F. (2005). Genome-wide transcriptional analysis of flagellar regeneration in *Chlamydomons reinhardtii* identifies orthologs of ciliary disease genes. *Proc. Natl. Acad. Sci. U.S.A.* **102**, 3703–3707.

Summers, K. and Gibbons, I.R. (1971). Adenosine triphosphate-induced sliding of tubules in trypsin-treated flagella of sea urchin sperm. *Proc. Natl. Acad. Sci. U.S.A.* **68**, 3092–3096.

Susalka, S.J., Nikulina, K., Salata, M.W., Vaughan, P.S., King, S.M., Vaughan, K.T., and Pfister, K.K. (2002). The roadblock light chain binds a novel region of the cytoplasmic dynein intermediate chain. *J. Biol. Chem.* **277**, 32939–32946.

Tai, A.W., Chuang, J.Z., Bode, C., Wolfrum, U., and Sung, C.H. (1999). Rhodopsin's carboxy-terminal cytoplasmic tail acts as a membrane receptor for cytoplasmic dynein by binding to the dynein light chain Tctex-1. *Cell* **97**, 877–887.

Tai, C.Y., Dujardin, D.L., Faulkner, N.E., and Vallee, R.B. (2002). Role of dynein, dynactin, and CLIP-170 interactions in LIS1 kinetochore function. *J. Cell Biol.* **156**, 959–968.

Taillon, B.E., Adler, S.A., Suhan, J.P., and Jarvik, J.W. (1992). Mutational analysis of centrin: an EF-hand protein associated with three distinct contractile fibers in the basal body apparatus of *Chlamydomonas*. *J. Cell Biol.* **119**, 1613–1624.

Takada, S. and Kamiya, R. (1994). Functional reconstitution of *Chlamydomonas* outer dynein arms from α, β and γ subunits: requirement of a third factor. *J. Cell Biol.* **126**, 737–745.

Takada, S. and Kamiya, R. (1997). Beat frequency difference between the two flagella of *Chlamydomonas* depends on the attachment site of outer dynein arms on the outer-doublet microtubules. *Cell Motil. Cytoskeleton* **36**, 68–75.

Takada, S., Sakakibara, H., and Kamiya, R. (1992). Three-headed outer arm dynein from *Chlamydomonas* that can functionally combine with outer-arm-missing axonemes. *J. Biochem. (Tokyo)* **111**, 758–762.

Takada, S., Wilkerson, C.G., K, W., Kamiya, R., and Witman, G.B. (2002). The outer dynein arm-docking complex: composition and characterization of a sub-unit (Oda1) necessary for outer arm assembly. *Mol. Biol. Cell* **13**, 1015–1029.

Tang, W.Y. and Gibbons, I.R. (1987). Photosensitized cleavage of dynein heavy chains. Cleavage at the V2 site by irradiation at 365 nm in the presence of oligo-vanadate. *J. Biol. Chem.* **262**, 17728–17734.

Toba, S., Watanabe, T.M., Yamaguchi-Okimoto, L., Toyoshima, Y.Y., and Higuchi, H. (2006). Overlapping hand-over-hand mechanism of single molecular motility of cytoplasmic dynein. *Proc. Natl. Acad. Sci. U.S.A.* **103**, 5741–5745.

Tynan, S.H., Gee, M.A., and Vallee, R.B. (2000). Distinct but overlapping sites within the cytoplasmic dynein heavy chain for dimerization and for intermediate chain and light intermediate chain binding. *J. Biol. Chem.* **275**, 32769–32774.

Vale, R.D. and Toyoshima, Y.Y. (1988). Rotation and translocation of microtubules *in vitro* induced by dyneins from *Tetrahymena* cilia. *Cell* **52**, 459–469.

Wagner, V., Gessner, G., Heiland, I., Kaminski, M., Hawat, S., Scheffler, K., and Mittag, M. (2006). Analysis of the phosphoproteome of *Chlamydomonas reinhardtii* provides new insights into various cellular pathways. *Eukaryot. Cell* **5**, 457–468.

Wakabayashi, K. and King, S.M. (2006). Modulation of *Chlamydomonas reinhardtii* flagellar motility by redox poise. *J. Cell Biol.* **173**, 743–754.

Wakabayashi, K., Takada, S., Witman, G.B., and Kamiya, R. (2001). Transport and arrangement of the outer-dynein-arm docking complex in the flagella of *Chlamydomonas* mutants that lack outer dynein arms. *Cell Motil. Cytoskeleton* **48**, 277–286.

Wang, J., Song, J.J., Franklin, M.C., Kamtekar, S., Im, Y.J., Rho, I.S., Seong, C. S., Chung, C.H., and Eom, S.H. (2001). Crystal structures of the HS1VU peptidase-ATPase complex reveal an ATP-dependent proteolysis mechanism. *Structure* **9**, 177–184.

Watanabe, T. and Flavin, M. (1976). Nucleotide-metabolizing enzymes in *Chlamydomonas* flagella. *J. Biol. Chem.* **251**, 182–192.

Weber, C., Lee, V.D., Chazin, W.J., and Huang, B. (1994). High level expression in *Escherichia coli* and characterization of the EF-hand calcium-binding protein caltractin. *J. Biol. Chem.* **269**, 15795–15802.

Wilkerson, C.G., King, S.M., and Witman, G.B. (1994). Molecular analysis of the γ heavy chain of *Chlamydomonas* flagellar outer-arm dynein. *J. Cell Sci.* **107**, 497–506.

Wilkerson, C.G., King, S.M., Koutoulis, A., Pazour, G.J., and Witman, G.B. (1995). The 78,000 $M_{(r)}$ intermediate chain of *Chlamydomonas* outer arm dynein is a WD-repeat protein required for arm assembly. *J. Cell Biol.* **129**, 169–178.

Williams, J.C., Xie, H., and Hendrickson, W.A. (2005). Crystal structure of dynein light chain Tctex1. *J. Biol. Chem.* **280**, 21981–21986.

Wirschell, M., Pazour, G., Yoda, A., Hirono, M., Kamiya, R., and Witman, G.B. (2004). Oda5p, a novel axonemal protein required for assembly of the outer dynein arm and an associated adenylate kinase. *Mol. Biol. Cell* **15**, 2729–2741.

Wirschell, M., Hendrickson, T.W., Fox, L., Haas, N.A., Silflow, C.D., Witman, G.B., and Sale, W.S. (2005). IC97, a novel dynein intermediate chain from flagellar inner arm dynein I1, interacts with tubulin *in situ*. *Mol. Biol. Cell* **16**. Abstract #1023. (CD-ROM).

Witman, G.B. and Minervini, N. (1982). Dynein arm conformation and mechanochemical transduction in the eukaryotic flagellum. *Symp. Soc. Exp. Biol.* **35**, 203–223.

Witman, G.B., Plummer, J., and Sander, G. (1978). *Chlamydomonas* flagellar mutants lacking radial spokes and central tubules. Structure, composition, and function of specific axonemal components. *J. Cell Biol.* **76**, 729–747.

Witman, G.B., Johnson, K.A., Pfister, K.K., and Wall, J.S. (1983). Fine structure and molecular weight of the outer arm dyneins of *Chlamydomonas*. *J. Submicrosc. Cytol.* **15**, 193–197.

Wright, R.L., Adler, S.A., Spanier, J.G., and Jarvik, J.W. (1989). Nucleus-basal body connector in *Chlamydomonas*: evidence for a role in basal body segregation and against essential roles in mitosis or in determining cell polarity. *Cell Motil. Cytoskeleton* **14**, 516–526.

Wu, H. and King, S.M. (2003). Backbone dynamics of dynein light chains. *Cell Motil. Cytoskeleton* **54**, 267–273.

Wu, H., Maciejewski, M.W., Marintchev, A., Benashski, S.E., Mullen, G.P., and King, S.M. (2000). Solution structure of a dynein motor domain associated light chain. *Nat. Struct. Biol.* **7**, 575–579.

Wu, H., Blackledge, M., Maciejewski, M.W., Mullen, G.P., and King, S.M. (2003). Relaxation-based structure refinement and backbone molecular dynamics of the dynein motor domain-associated light chain. *Biochemistry* **42**, 57–71.

Wu, H., Maciejewski, M.W., Takebe, S., and King, S.M. (2005). Solution structure of the Tctex1 dimer reveals a mechanism for dynein–cargo interactions. *Structure* **13**, 213–223.

Wynshaw-Boris, A. and Gambello, M.J. (2001). LIS1 and dynein motor function in neuronal migration and development. *Genes Dev.* **15**, 639–651.

Xiang, X., Osmani, A.H., Osmani, S.A., Xin, M., and Morris, N.R. (1995). NUDF, a nuclear migration gene in *Aspergillus nidulans*, is similar to the human LIS-1 gene required for neuronal migration. *Mol. Biol. Cell* **6**, 297–310.

Yagi, T. (2000). ADP-dependent microtubule translocation by flagellar inner-arm dyneins. *Cell Struct. Funct.* **25**, 263–267.

Yagi, T., Minoura, I., Fujiwara, A., Saito, R., Yasunaga, T., Hirono, M., and Kamiya, R. (2005). An axonemal dynein particularly important for flagellar movement at high viscosity: implications from a new *Chlamydomonas* mutant deficient in the dynein heavy chain gene DHC9. *J. Biol. Chem.* **280**, 41412–41420.

Yamamoto, R., Yanagisawa, H., Yagi, T., and Kamiya, R. (2006). A novel subunit of axonemal dynein conserved among lower and higher eukaryotes. *FEBS Lett.* **580**, 6357–6360.

Yamamoto, R., Yanagisawa, H., Yagi, T., and Kamiya, R. (2008). Novel 44-kilodalton subunit of axonemal dynein conserved from *Chlamydomonas* to mammals. *Eukaryot. Cell* **7**, 154–161.

Yanagisawa, H.A. and Kamiya, R. (2001). Association between actin and light chains in *Chlamydomonas* flagellar inner-arm dyneins. *Biochem. Biophys. Res. Commun.* **288**, 443–447.

Yanagisawa, H. and Kamiya, R. (2004). A tektin homologue is decreased in *Chlamydomonas* mutants lacking an axonemal inner-arm dynein. *Mol. Biol. Cell* **15**, 2105–2115.

Yang, P. and Sale, W.S. (1998). The M_r 140,000 intermediate chain of *Chlamydomonas* flagellar inner arm dynein is a WD-repeat protein implicated in dynein arm anchoring. *Mol. Biol. Cell* **9**, 3335–3349.

Yang, P., Diener, D.R., Rosenbaum, J.L., and Sale, W.S. (2001). Localization of calmodulin and dynein light chain LC8 in flagellar radial spokes. *J. Cell Biol.* **153**, 1315–1326.

Yu, J., Hu, S., Wang, J., Wong, G.K.-S., Li, S., Liu, B., Deng, Y., Dai, L., Zhou, Y., Zhang, X., Cao, M., Liu, J., Sun, J., Tang, J., Chen, Y., Huang, X., Lin, W., Ye, C., Tong, W., Cong, L., Geng, J., Han, Y., Li, L., Li, W., Hu, G., Huang, X., Li, W., Li, J., Liu, Z., Li, L., Liu, J., Qi, Q., Liu, J., Li, L., Li, T., Wang, X., Lu, H., Wu, T., Zhu, M., Ni, P., Han, H., Dong, W., Ren, X., Feng, X., Cui, P., Li, X., Wang, H., Xu, X., Zhai, W., Xu, Z., Zhang, J., He, S., Zhang, J., Xu, J., Zhang, K., Zheng, X., Dong, J., Zeng, W., Tao, L., Ye, J., Tan, J., Ren, X., Chen, X., He, J., Liu, D., Tian, W., Tian, C., Xia, H., Bao, Q., Li, G., Gao, H., Cao, T., Wang, J., Zhao, W., Li, P., Chen, W., Wang, X., Zhang, Y., Hu, J., Wang, J., Liu, S., Yang, J., Zhang, G., Xiong, Y., Li, Z., Mao, L., Zhou, C., Zhu, Z., Chen, R., Hao, B., Zheng, W., Chen, S., Guo, W., Li, G., Liu, S., Tao, M., Wang, J., Zhu, L., Yuan, L., and Yang, H. (2002). A draft sequence of the rice genome (*Oryza sativa* L. ssp. *indica*). *Science* **296**, 79–92.

Yu, X., West, S.C., Jahn, R., and Brunger, A.T. (1998). Structure of the ATP-dependent oligomerization domain of *N*-ethylmaleimide sensitive factor complexed with ATP. *Nat. Struct. Biol.* **5**, 803–811.

Zhang, Z., Futamura, M., Vikis, H., Wang, M., Li, J., Wang, Y., Guan, K., and You, M. (2003). Positional cloning of the major quantitative trait locus underlying lung tumor susceptibility in mice. *Proc. Natl. Acad. Sci. U.S.A.* **100**, 12642–12647.

Zhou, T., Zimmerman, W., Liu, X., and Erikson, R.L. (2006). A mammalian NudC-like protein essential for dynein stability and cell viability. *Proc. Natl. Acad. Sci. U.S.A.* **103**, 9039–9044.

CHAPTER 7

The Flagellar Radial Spokes

Pinfen Yang[1] and Elizabeth F. Smith[2]
[1]Department of Biology, Marquette University, Milwaukee, Wisconsin, USA
[2]Department of Biological Sciences, Dartmouth College, Hanover, New Hampshire, USA

CHAPTER CONTENTS

I. INTRODUCTION

The radial spokes are a major structural feature of 9 + 2 axonemes and are crucial for flagellar beating. Each spoke consists of a thin stalk attached to the A-tubule of the axonemal doublet microtubules and a bulbous head projecting toward the central apparatus (reviewed in Curry and Rosenbaum, 1993). In *Chlamydomonas* flagella, the spokes repeat in pairs every 96 nm and maintain a right-handed helix along the length of the axoneme (Goodenough and Heuser, 1985) (Figure 7.1). The heads of the spoke pair

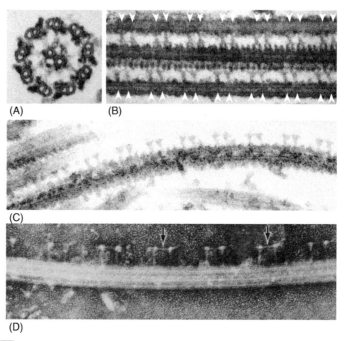

(A) (B)

(C)

(D)

FIGURE 7.1 *Electron micrographs of* Chlamydomonas *axonemes and outer doublets showing the radial spokes. (A) A cross-section shows that radial spokes are anchored to the A-tubule of each outer doublet and project toward the central pair of microtubules. (B) A longitudinal section shows that the radial spokes repeat in pairs every 96 nm. The spoke pairs on opposing outer doublets are not aligned because of the helical arrangement of spokes (white arrowheads). (C) A longitudinal section of an outer doublet that has separated from the rest of the axoneme yields an unobstructed view of the radial spokes. (D) A negatively stained outer doublet from a disrupted axoneme reveals a thin fiber (arrows) connecting the bulbous heads of each member of a spoke pair. (Modified from Witman et al., 1978. Reproduced from* The Journal of Cell Biology. *1978, 76, 729–747, by copyright permission of The Rockefeller University Press.)*

appear to be connected by a thin fiber (Witman et al., 1978). Combined biochemical and genetic studies have revealed that these structures contain 23 different proteins, termed radial spoke proteins or RSPs. Among them, five proteins (RSP1, 4, 6, 9, and 10) are located in the spoke heads and 18 proteins are located in the spoke stalks. The enormous variety of research tools that *Chlamydomonas* offers as a model system, such as radial spoke-defective mutants and *in vitro* functional assays, as well as the unique biochemical properties of spoke components, have all contributed to our understanding of the radial spokes. This chapter reviews the composition, assembly, and possible functions of this macromolecular complex.

II. GENETIC DISSECTION OF THE RADIAL SPOKES

Chlamydomonas mutants that are defective in radial spoke assembly have played a central role in studies of spoke composition and function. Paralyzed

Table 7.1   Mutants with defects in radial spokes

Mutant	Gene product	Motility phenotype	Morphological defect	Proteins missing	Proteins in reduced amount	References
pf1	RSP4	Paralyzed	Headless, stalks only	RSP1, 4, 6, 9, 10	None	Ebersold et al. (1962), Huang et al. (1981)
pf5	? (see text)	Paralyzed	?	RSP13, RSP15, an inner-arm dynein component, and a <15-kD protein	?	Ebersold et al. (1962), Huang et al. (1981)
pf14	RSP3	Paralyzed	Spokeless	All spoke proteins	None	Piperno et al. (1981), Ebersold et al. (1962)
pf17	RSP9	Paralyzed	Headless, stalks only	RSP1, 4, 6, 9, 10	None	Huang et al. (1981)
pf24	RSP2	Paralyzed	Missing some spokeheads, and parts of some stalks[c]	None	RSP1, 4, 6, 9, 10, 2, 16, 23	Huang et al. (1981), Patel-King et al. (2004), Yang et al. (2004)
pf25	RSP11	Paralyzed/ swimming	None	RSP11	8	Huang et al. (1981), Yang and Yang (2006)
pf26[ts]	RSP6	Paralyzed at restrictive temperature	None	None	RSP1, 4, 6[a], 9, 10	Huang et al. (1981)
pf27	?	Paralyzed	Missing some spokeheads and stalks[c]	None	All spoke proteins[b]	Huang et al. (1981), Yang and Yang (2006)
fla14	RSP22/ LC8	Paralyzed, short flagella	Radial spokes missing or defective, other defects[d]	RSP3, 22 (others)[d]	Multiple[d]	Pazour et al., 1998 Yang et al., submitted

[a]Mutated RSP6 from pf26[ts] mutant migrated as a smaller and more basic protein than wild-type RSP6 in two-dimensional gels.
[b]Reduced phosphorylation levels of RSP2, 3, 13 as well.
[c]See Huang et al. (1981) for detailed descriptions.
[d]Morphological and biochemical deficiencies vary with allele.

flagellar mutants that lack all or part of the radial spokes are easily identified by electron microscopy (Witman et al., 1978; Huang et al., 1981). Early screens for paralyzed *Chlamydomonas* strains generated by chemical mutagens or irradiation led to the identification of eight mutants with radia spoke defects (Ebersold et al., 1962; Luck et al., 1977; Huang et al., 1981). More recently, the mutant *fla14* also was shown to be defective in radial spokes (Pazour et al., 1998).

The initial characterization of radial spoke mutants relied heavily on two-dimensional gel electrophoresis. The majority of spoke proteins are highly acidic, and thus their absence in spoke mutants is easily detected by comparison with two-dimensional gels of wild-type axonemes (Piperno et al., 1977). With the exception of *pf5* and *fla14*, defects in all of these mutants are limited to the radial spokes (Huang et al., 1981). The motility phenotypes and radial spoke components that are lacking for each of these strains have been described in detail (Huang et al., 1981, Piperno et al., 1981; Yang et al., 2005; Yang and Yang, 2006) and are summarized in Table 7.1. Additional noteworthy characteristics of these mutants are discussed below.

Immotile cells with paralyzed flagella are the signature phenotype for mutants with defects in either the central apparatus or radial spokes. However, the paralysis of these two groups of mutants differs. In contrast to the rigid flagella of mutants lacking the entire central apparatus, mutants lacking all radial spokes and spoke components (*pf14*), some radial spokes (*pf27*), the spoke heads (*pf1* and *pf17*), or some spoke heads and part of some of the spoke stalks (*pf24*), have flaccid flagella that only occasionally beat. The common structural defect for each of these mutants is the lack of spoke head components. This observation supports the hypothesis that the direct interaction between the spoke heads and central pair projections is crucial for the motility of 9 + 2 cilia and flagella (see Chapters 8 and 9 for a more complete discussion).

In contrast to the spoke mutants with paralyzed flagella, the motility defect of *pf25* is less severe. Interestingly, the cells of five available *pf25* alleles can swim; however, their ability to swim is sensitive to culture conditions. When the cells are initially suspended in liquid culture or have exhausted the media of nutrients, the cells have flagella that are either entirely paralyzed or beat with a jerky motion (Yang and Yang, 2006). Yet, at mid-log phase the cells are motile and indistinguishable from wild-type cells. Axonemes isolated from *pf25* cells only lack the *PF25* gene product, RSP11, and have a reduced amount of RSP8. In structural analyses of *pf25* axonemes, the morphology of the radial spokes appears normal. The cause of the dramatic change in *pf25* motility under different culture conditions remains to be elucidated.

pf26ts is a conditional, temperature-sensitive mutant, i.e., its phenotype depends on the temperature of the cell culture during flagellar growth. When assembled at the restrictive temperature, the flagella are paralyzed; these flagella contain normal or reduced amounts of spoke head proteins, including the *PF26* gene product RSP6, which has altered mobility in two-dimensional gels (Huang et al., 1981). Interestingly, if the flagella assemble at the permissive temperature and then the cells are shifted to the restrictive temperature, the flagella remain motile. Most likely, at the restrictive temperature the mutation in RSP6 causes changes in conformation that affect protein–protein interactions. Following assembly into the radial spoke complex, RSP6 may be stabilized against such conformational changes.

The *pf27* mutant is intriguing in two ways. First, dikaryon rescue experiments indicate that the *PF27* gene product is not a component of the radial spokes. Second, the phosphorylation of three spoke stalk components (RSP2, 3, and 13) that are normally phosphorylated is inhibited (Huang et al., 1981) and the overall amount of spoke proteins is significantly reduced (Yang and Yang, 2006). Based on this unusual phenotype, the *pf27* mutant may provide a unique opportunity to understand the role of phosphorylation in axoneme assembly and potentially in the control of flagellar beating.

The *FLA14* gene encodes LC8, an 8-kD (91-amino-acid) component of radial spokes, outer dynein arms, inner dynein arm I1, and cytoplasmic dynein 1b, the retrograde motor for intraflagellar transport (see Chapters 4 and 6). Null mutants of LC8 (*fla14-1, fla14-2*) have reduced numbers of radial spokes, outer arms, and inner arms, as well as deficiencies in intraflagellar transport, leading to short, paralyzed flagella (Pazour et al., 1998). The *fla14-3* allele (see next paragraph) has a similar but slightly less severe flagellar phenotype with milder defects in axonemal structure (Yang et al., manuscript submitted).

Huang et al. (1981) briefly reported that two alleles at the *pf5* locus were deficient for RSPs, including RSP13 and 15, and also were missing a component of the inner arm dynein now called I1 or f-dynein, and a small protein (<15 kD). Whether both these alleles are extant is not known. The *Chlamydomonas* Center currently has two strain lineages with *pf5* mutations represented by CC-1028 and CC-1331. CC-1331 has reduced levels of radial spokes and is lacking IC97 of inner arm dynein I1 (Yang et al., manuscript submitted), consistent with the molecular phenotype of *pf5* as described by Huang and colleagues. Detailed analysis of strain CC-1331 indicates that it has a read-through mutation in the *FLA14* gene, leading to expression of an LC8 protein with an additional 23 amino acids at its C-terminus. That LC8 is associated with both inner arms and radial spokes readily explains why a defect in this protein could affect the assembly of both structures; that the larger LC8 would migrate in a different place than normal LC8 in two-dimensional gels would explain the absence of a <15-kD protein as reported by Huang and colleagues. *PF5* has been mapped to linkage group III between *AC208* and *NIT2*; this region also contains *FLA14*, but definitive crosses have not been done to show whether *PF5* and *FLA14* are two separate loci or represent alternative phenotypes at a single locus. However, the phenotype of CC-1331 is rescued by transformation with the wild-type LC8 gene (Yang et al., manuscript submitted). In light of this evidence, the *pf5* allele in CC-1331 has been renamed *fla14-3*. Further work will be required to determine the defective gene in CC-1028.

Considering that at least 23 proteins comprise the isolated radial spokes and that only eight or nine mutants have been identified (Yang et al., 2001; Patel-King et al., 2004), new spoke mutants remain to be discovered. The currently available mutants were recovered by screening for obvious motility defects. Perhaps, defects in some radial spoke proteins do not cause paralysis and thus have eluded detection in motility screens. The identification of additional mutants may require alternative approaches, such as conducting systematic insertional mutagenesis screens, screening with modified strategies, or reducing the expression of particular spoke components by RNA-mediated interference.

III. BIOCHEMICAL CHARACTERIZATION OF RADIAL SPOKES AND SPOKE PROTEINS

A. Extraction of the unextractable

For decades, the axonemal dynein arms have been the target of biochemical and functional analysis due in large part to the ease with which they can be extracted and purified from isolated flagella. However, under the conditions used to solubilize the dynein arms (0.6 M KCl or NaCl), the radial spokes remain associated with the doublet microtubules. A significant advance in characterizing radial spoke components was the discovery that the radial spokes could be efficiently extracted from the axoneme using buffers containing 0.5–0.6 M KI or 0.6 M NaBr (Yang et al., 2001). Using an extraction buffer with high ionic strength was inspired by the work of Linck (1976), who used chaotropic reagents to dissociate the doublet microtubule framework, but preserved the protofilaments of sea urchin sperm flagella. This extraction strategy also has been successfully used to identify components of the central pair projections (Mitchell and Sale, 1999) and the dynein regulatory complex (Rupp and Porter, 2003).

Importantly, despite the dissociation of the microtubule framework using these extraction conditions, the extracted radial spokes remain intact and can be further separated by sucrose gradient sedimentation. The spoke complexes sediment as approximately 20S particles, toward the bottom of a 5–20% or 10–30% sucrose gradient. This sedimentation coefficient is similar to the components of the outer dynein arms as well as the inner dynein arm I1. However, a relatively pure fraction of intact radial spokes can be obtained by extracting the spokes from the double mutant *pf28pf30*, which lacks the ~20S dyneins. This approach allowed for the unequivocal purification and characterization of individual spoke components by subsequent two-dimensional gel electrophoresis and mass spectrometry, and facilitated the discovery of new proteins that associate with the extracted spokes (Yang et al., 2001, 2006). Using isolated radial spokes, Patel-King et al. (2004) demonstrated that the extracted spokes accounted for 45% of the nucleoside diphosphate kinase (NDK) activity in axonemes (see below).

Despite their stability in high-ionic strength solutions, it is possible to subfractionate radial spokes. Piperno et al. (1981) demonstrated that spoke head proteins and RSP5 can be extracted from axonemes by prolonged dialysis against low ionic strength buffers, and then cosediment at the top of

5–20% sucrose gradients. In addition, the extracted radial spoke particles become partially dissociated if the axonemes are extracted with 0.6 M KI at low protein concentration, such as 0.5 mg/ml. This property may allow further dissection of the spoke macromolecular complex.

B. Composition of the radial spokes

Using a combination of available spoke defective mutants and spoke purification methods, 23 radial spoke proteins have been identified (Table 7.2). The first 17 axonemal proteins to be assigned to the radial spokes (RSP1–17) were identified based on their absence in two-dimensional gels of spokeless axonemes (pf14) compared with that of wild-type (Piperno et al., 1977, 1981). Of these 17 polypeptides, Piperno et al. (1981) reported that five (RSP2, 3, 5, 13, 17) were among the >80 axonemal proteins that are phospholabeled when cells are grown on agar plates containing ^{32}P-phosphoric acid. All of these phosphorylated spoke proteins are located in the stalk, and RSP2 appears to be most heavily labeled. RSP18–22 were subsequently identified based on their cosedimentation with extracted spokes as well as spoke stalks. RSP23, which contains an NDK domain, was discovered independently in an investigation of axonemal NDK activity (Patel-King et al., 2004).

The localization of individual polypeptides to specific spoke subdomains was deduced by correlating defects in spoke morphology with defects in the assembly of specific polypeptides (Table 7.1). For example, RSP1, 4, 6, 9, and 10 are presumed to be components of the spoke head based on their absence in the headless mutants pf1 and pf17 (Piperno et al., 1977; Huang et al., 1981). The rest of the RSPs were considered to form the spoke stalk. Within the spoke stalk, RSP2, 5, 16, and 23 are postulated to reside adjacent to the spoke head. This localization is based in part on the observation that RSP5 is coextracted and cosediments with the five spoke head proteins in low-salt buffer (Piperno et al., 1981), and in part on analysis of the pf24 mutant. In pf24 axonemes, RSP2, 16, and 23 as well as the spoke head proteins are diminished (Huang et al., 1981; Patel-King et al., 2004; Yang et al., 2005), whereas the remaining stalk proteins are present in normal amounts. Thus, RSP2, 5, 16, and 23 are most likely localized at the junction of the spoke stalk with the spoke head, while the remaining stalk proteins are located toward the base of the spoke stalk.

RSP3 is believed to be located at the very base of the spoke stalk based on several independent lines of evidence. The entire radial spoke fails to assemble in the RSP3 mutant, pf14. In addition, localization studies using indirect immunoelectron microscopy and anti-RSP3 antibodies revealed the presence of gold particles near the base of the spokes with an approximate periodicity similar to that of the radial spokes (Johnson, 1995). Finally, in reconstitution experiments, recombinant RSP3 binds to pf14 axonemes but not to wild-type axonemes or to purified microtubules assembled in vitro

Table 7.2 *Chlamydomonas* RSPs and some predicted homologues

RSP[a]	Reference	pI/Mr[b] (2D-NEPHGE)	pI/Mr (Theoretical)	Accession number	Linkage group (and mutant)	Similar to H.s. I.D.	H.s. Entrez Gene I.D., chromosomal location	Non mammalian homologues I.D.
RSP1	Yang et al. (2006)	5.2/123	4.55/78.6	ABC02025	III	NP_543136 (meichroacidin/TSGA2)	89765, 21q22.3	
RSP2	Yang et al. (2004)	5.0/118	4.48/77.4	AAQ92371	X (pf24)	NI		
RSP3	Williams, et al. (1989)	5.5/86	4.85/56.8	P12759	VI (pf14)	NP_114130 (RSHL2)	83861, 6q25.3	
RSP4	Curry et al. (1992)	5.1/76	4.58/49.8	A44498	V (pf1)	XP_294004 (RSHL3) NP_110412 (RSHL1)	345895, 6q22 81492, 19q13	
RSP5	Yang et al. (2006)	5.1/69	4.68/55.9	ABC02018	XII/XIII	NI		NP_565656 (Arabidopsis)
RSP6	Curry et al. (1992)	5.1/67	4.53/48.8	B44498	V (pf26)	NP_110412 (RSHL1) XP_294004 (RSHL3)	81492, 19q13 345895, 6q22	
RSP7	Yang et al. (2006)	5.1/58	4.54/55	ABC02026	VII	NI		XP_821614 (Trypanosome)
RSP8	Yang et al. (2006)	6.5/40	5.76/40.5	ABC02019	II	NI		XP_001033533 (Tetrahymena)
RSP9	Yang et al. (2006)	5.7/26	5.02/29.5	ABC02020	VII (pf17)	NP_689945 (C6orf206)	221421, 6p21.1	XP_780230 (Sea urchin)
RSP10	Yang et al. (2006)	5.6/24	5.02/23.5	ABC02021	I	NP_543136 (TSGA2) NP_775836	89765, 21q22.3 222967, 7p22.2	XP_768260 (Giardia)
RSP11	Yang et al. (2006)	4.8/22	4.53/21.5	ABC02022	X (pf25)	NP_114122 (ASP)	83853, 5p15.31	XP_001311110 (Trichomonas)
RSP12	Yang et al. (2006)	6.3/20	5.46/19.7	ABC02023	II	NP_775943 (PPIL6)	285755, 6q21	

RSP13		6.3/98	—	—	—		
RSP14	Yang et al. (2006)	6.8/41	5.84/28.3	ABC02024	XV	NP_055248 (RTDR1)	27156, 22q11.2
RSP15[c]	Yang et al. (2006)	5.7/38	—	—	—	NI	—
RSP16	Yang et al. (2005)	7.1/34	6.78/39.0	EDP05321	XII/XIII	NP_705842 (TSARG6) NP_006136 (DNAJB1)	374407, 11q13.3 3337, 19p13.2
RSP17	Yang et al. (2006)	6.2/124	5.56/98.5	ABC02027	VII	NI	
RSP18[c]	Yang et al. (2006)	5.4/210	—	—	—	NI	
RSP19 (β-tubulin)	Yang et al. (2006)	5.5/140	4.82/49.6	P04690	XII	Multiple	Multiple
RSP20 (calmodulin)	Zimmer et al. (1988), Yang et al. (2001)	4.3/18	4.3/18.3	AAA33083	III	Multiple	Multiple
RSP21		6.2/16	—	—	—		
RSP22 (LC8)	King and Patel-King (1995), Yang et al. (2001)	6.8/8	6.89/10.3	Q39580	III (fla14)	NP_542408 (Dlc2)	140735, 13q23.2
RSP23	Patel-King et al. (2004)	5.4/102	4.67/61	AY452667	XVIII	NP_003542 (NME5)	8382, 5q31

Note: NI, none identified.

[a] Radial spoke head proteins are indicated by white characters on a black background.

[b] pI and M_r are from Piperno et al. (1981), Yang et al. (2001), and Patel-King et al. (2004).

[c] Single peptide sequences were obtained from mass spectrometry analyses of RSP15 and RSP18, but the genes encoding these peptides have not been completely characterized. See Yang et al. (2006) for details.

(Diener et al., 1993). These combined results suggest that RSP3 may function to link the radial spoke to the outer doublet microtubules. The failure of recombinant RSP3 to bind to wild-type axonemes or to microtubules assembled *in vitro* suggests that some other protein or structure, perhaps a radial spoke docking complex, is necessary for RSP3 binding, and by extension, radial spoke binding.

The genes encoding 19 of the 23 radial spoke proteins have been cloned and their sequences analyzed (William et al., 1989; Curry et al., 1992; Patel-King et al., 2004; Yang et al., 2004, 2005, 2006) (Table 7.2). Because the majority of spoke proteins are very acidic, their positions on two-dimensional gels allowed for the systematic isolation of some of the spoke proteins for the production of specific antibodies, which could then be used to clone the corresponding genes by screening expression libraries (Williams et al., 1986, 1989; Curry et al., 1992). The sequences for additional spoke proteins were obtained by determining peptide sequences from isolated proteins using mass spectrometry, after which the complete coding sequences were deduced from ESTs or obtained by amplification of cDNAs followed by sequencing of the products (Yang et al., 2004, 2006). The assignment of RSP2, 3, 4, 6, 9, and 11 to the genes *PF24*, *PF14*, *PF1*, *PF26*, *PF17*, and *PF25*, respectively, was made possible by analysis of two-dimensional gels of dikaryons and intragenic revertants of the respective mutants (Luck et al., 1977; Huang et al., 1981). That RSP3 is the gene product of *PF14* and RSP11 the gene product of *PF25* was confirmed by transformation rescue of the *pf14* and *pf25* mutants with the corresponding wild-type genes (Diener et al., 1990; Yang and Yang, 2006). Specific mutations in the *pf14* and *pf25* genes were then identified by sequencing (Williams et al., 1989; Yang et al., 2004).

C. Functional domains in radial spoke proteins

Four interesting observations have emerged from a comprehensive analysis of the predicted amino acid sequences of spoke proteins. First, domains characteristic of proteins involved in signal transduction pathways are abundant and located exclusively in the spoke stalk. This observation suggests that the spoke heads contain structural proteins specialized for contact with the central apparatus, while the stalk contains regulatory elements for modulating flagellar beating. Second, two stalk proteins, HSP40 (RSP16) and peptidyl prolyl isomerase (RSP12), are members of the molecular chaperone family of proteins (see Chapter 19 in Volume 2), suggesting a sophisticated mechanism for assembly of the radial spokes. Third, there are at least five instances where two spoke proteins share the same molecular domain or motif, possibly indicating similar function. These proteins are illustrated in Figure 7.2, which indicates the approximate spoke location for individual proteins and also groups together proteins with the same

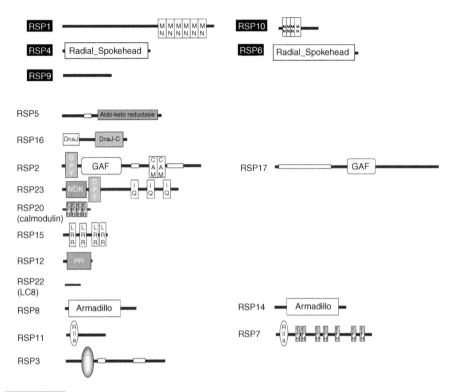

FIGURE 7.2 *Diagram representing molecular domains and approximate locations of those* Chlamydomonas *radial spoke proteins that have been characterized at the molecular level. The proteins are listed sequentially from the spoke head (top of figure) to the base of the spoke stalk (bottom of figure) based on their predicted positions in the radial spokes. For example, the first five spoke proteins with their names on a black background are spoke head proteins, while RSP3, located at the base of the spoke stalk, is placed at the bottom of the figure. For pairs of spoke proteins that share similar domains or sequence homology, the similar protein in the pair is placed in the second column. MN, MORN domain; DPY, Dpy-30 motif; GAF, cyclic GMP, Adenylyl cyclase, FhlA domain; CAM, 1-8-14 calmodulin-binding motif; AKAP, A-Kinase Anchoring Protein motif; RIIa, RII alpha motif; EFH, EF-hand domain; PPI, peptidyl prolyl isomerase motif; LRR, leucine-rich repeat; DnaJ and DnaJ-C, DnaJ-J and DnaJ-C molecular chaperone homology domains; NDK, NDK domain; IQ, IQ calmodulin-binding motif. Coiled-coil domains are indicated by an open bar.*

molecular motifs, and in Figure 7.3, which shows the predicted locations in the spoke for the functional domains of the proteins. Fourth, Blastp searches revealed candidate orthologues for several spoke proteins in mammals and ciliated organisms. The most conserved proteins are listed in Table 7.2 and most likely provide the essential core functions of the radial spokes. The less-conserved proteins may be involved in producing specialized flagellar beat parameters unique to the motility of *Chlamydomonas*. The individual spoke proteins are discussed in detail in Yang et al. (2006). Here, we focus discussion on proteins with signature domains suggesting

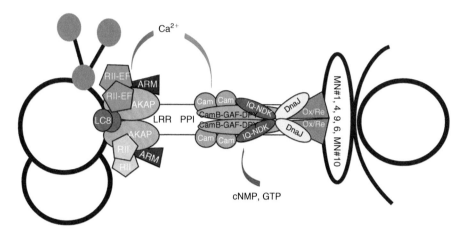

FIGURE 7.3 *Schematic diagram depicting the functional motifs in the predicted subdomains of the radial spoke. The base of the radial spoke is attached to an outer doublet microtubule (indicated in outline form on left); its head is in close contact with a central pair microtubule and its projections (outlined, right). LC8, dynein light chain LC8; ARM, armadillo repeats; Ox/Re, redox domains of RSP5; see Figure 7.2 for explanation of other motifs.*

that they function in the calcium and cyclic nucleotide signal transduction pathways that modulate cilia and flagellar beating (reviewed by Smith and Yang, 2004; also see Chapter 9).

Among the numerous axonemal proteins containing the calcium-binding EF-hand domain (Pazour et al., 2005) are the radial spoke proteins RSP20 and RSP7 (Yang et al., 2001; Yang et al., 2006). RSP20 was identified as calmodulin based on its migration in two-dimensional gels and by western blot analysis using anti-calmodulin antibodies. Calmodulin is an abundant flagellar protein found in the "membrane-plus-matrix" fraction as well as associated with the central apparatus and radial spokes; several studies have implicated calmodulin in the calcium-sensitive motility changes in cilia and flagella of a variety of organisms including *Chlamydomonas* (Witman and Minervini, 1982; Yang et al., 2001; Smith 2002a; Wargo et al., 2005).

Two radial spoke proteins, RSP2 and RSP23, have been identified as calmodulin-binding proteins. As noted above, both of these components are localized to the spoke stalk just below the spoke head. RSP2 contains 1-8-14 calmodulin-binding motifs and binds calmodulin in a calcium-dependent manner *in vitro* (Yang et al., 2004). RSP23 contains three calmodulin-binding IQ motifs. Interestingly, in *in vitro* calmodulin-binding experiments, one of these IQ domains exhibits calcium-independent calmodulin binding, whereas the other two exhibit calcium-dependent calmodulin binding (Patel-King et al., 2004).

As noted above, RSP23 also contains an NDK domain that converts NDP to NTP. Presumably, RSP23 is responsible for the calcium-stimulated NDK activity in isolated radial spokes (Patel-King et al., 2004). In addition to its calmodulin-binding motifs, RSP2 contains a GAF (cyclic <u>G</u>MP, <u>A</u>denylyl Cyclase, <u>F</u>hlA) domain, which is thought to bind small ligands including cNMP. Thus, the RSP2 and RSP23 components of the radial spoke stalk may mediate cross-talk between calcium, nucleotide, and cNMP-mediated signal transduction pathways.

Additional clues that the radial spokes play a role in mediating signaling pathways are found in the motifs of other radial spoke components. Gel overlay assays with the recombinant regulatory subunit (RII) of cAMP-dependent protein kinase (PKA) revealed that RSP3 is an A-kinase anchoring protein (AKAP) (Gaillard et al., 2001). Using recombinant proteins in *in vitro* binding assays, Gaillard et al. (2001) identified the amphipathic regions and key peptides in RSP3 required for binding RII. To assess the function of the RSP3 RII-binding domain *in vivo*, Gaillard et al. (2006) transformed *pf14* with a gene encoding RSP3 in which the RII-binding site was abolished. The resulting *pf14* transformants contained two populations of cells, those with motile and those with paralyzed flagella, even after single colony isolation (Gaillard et al., 2006). It is not clear why this mutation results in two different populations of cells, but these results demonstrate a role for the RII-binding domain of RSP3 in modulating motility.

The motility phenotype of the above *pf14* transformants is reminiscent of the phenotype of *pf25*, the RSP11 mutant. Importantly, RSP11 contains an RIIa domain. The RIIa domain is responsible for the binding of PKA RII to its AKAP, and indeed recombinant RSP11 binds to recombinant RSP3 (Yang and Yang, 2006). However, RSP11 does not contain the cAMP-binding domains of the RII regulatory subunit. The sensitivity of *pf25* motility to medium conditions suggests that RSP11 does play a role in regulating flagellar movement, but, given the lack of the cAMP-binding domain, RSP11 is unlikely to directly mediate cAMP-dependent activation or inactivation.

Similar to RSP11, RSP7 possesses an RIIa domain but lacks the cAMP-binding motifs. However, RSP7 also contains seven calcium-binding EF-hands. It remains to be demonstrated whether RSP7 also binds to RSP3 via its RIIa domain, and whether the EF-hands are functionally significant. The presence of an AKAP and two proteins containing RIIa domains in the radial spoke is consistent with the well-established principle that AKAPs anchor multiple signal transduction pathways (Beene and Scott, 2007). The future challenge will be to understand the details of how these proteins participate in the signal transduction mechanisms that mediate Ca^{2+} and cAMP-induced changes in flagellar motility (see below).

IV. ASSEMBLY OF THE RADIAL SPOKE COMPLEX

It is well established that axonemal proteins synthesized in the cell body are largely added onto the axoneme at the plus (distal) ends of the microtubules (Witman, 1975; Johnson and Rosenbaum, 1992). The polarity of assembly was elegantly demonstrated in dikaryon rescue experiments where unlabeled cells with half-length flagella were mated with cells having full-length flagella and expressing HA-tagged tubulin. As predicted, immunofluorescence microscopy revealed that the tagged tubulins first appeared at the tips of the elongating, unlabeled flagella of the quadriflagellate dikaryons. Similar results were obtained during spoke assembly in dikaryons of wild-type and *pf14* cells. Immunofluorescence microscopy revealed that the RSP3 protein contributed by the wild-type cell appeared first at the tip of full-length *pf14* flagella and then gradually extended toward the base (Johnson and Rosenbaum, 1992). This study conclusively demonstrated that radial spokes are transported to the tip of the flagellum for incorporation into the axonemal superstructure. We now know that the tipward transport of radial spoke components during flagellar elongation involves intraflagellar transport (IFT) by a kinesin motor (for complete discussion of IFT, see Chapter 4).

To determine whether the spokes are transported as individual subunits or as a complex, the membrane-plus-matrix fraction was prepared from regenerating wild-type flagella and fractionated by sucrose gradient sedimentation. Radial spoke complexes were present in peaks at 12S and 20S, with the former being predominant. However, in a similar analysis of the membrane-plus-matrix fraction from disassembling flagella (from the mutant *fla10*, which has a temperature-sensitive defect in the IFT anterograde motor kinesin-2 that causes the flagella to shorten under restrictive conditions – see Chapter 4), the 12S peak was predominant. The 20S peak, normally present in the cytoplasm of the wild-type cell body, was absent from the cytoplasm of mutants in which retrograde IFT was blocked. From these comparisons and related experiments, it was concluded that the 12S particles are spoke precursors being transported by IFT to the flagellar tip for assembly, whereas the 20S particles are mature spokes disassembled from the axoneme and being transported back to the cell body by retrograde IFT (Qin et al., 2004).

It is not clear why the sedimentation coefficient of spoke precursors and mature spokes differs, or how spoke precursors are converted into mature spokes. Biochemical analysis of the membrane-plus-matrix fraction indicates that RSP16, a type II HSP40 of DnaJ co-chaperones, is not in the precursor complex but in a separate 5S particle (Yang et al., 2005). Like many DnaJ-domain containing proteins, RSP16 forms a homodimer. In *Chlamydomonas* cells in which expression of RSP16 is knocked down by RNA-mediated interference, the cells have jerky, full-length flagella despite the fact that the mutant axonemes contain all of the remaining spoke proteins, which sediment as a complex at 20S (Yang et al., 2008). This result indicates that the spoke HSP40 (RSP16) is not required for the conversion of spoke precursors to mature 20S spokes, but is essential for motility.

Thus, there must be alternative mechanisms for the maturation of the 12S spoke particles.

V. RADIAL SPOKE FUNCTION: PART OF THE CONTROL SYSTEM FOR REGULATING FLAGELLAR BEATING

A. Genetic evidence: suppressor mutations

A variety of approaches including genetic, structural, and functional studies have provided significant insights into understanding the function of the radial spokes. As noted above, mutants with defects in radial spoke components have paralyzed flagella (e.g., Witman et al., 1978). In *Chlamydomonas*, extragenic suppressor mutations have been identified that restore beating to these paralyzed mutants without restoring the missing structures (Huang et al., 1982). Importantly, the flagella from the suppressed mutant cells produce only symmetric waveforms (Brokaw et al., 1982), indicating that the radial spokes may be involved in converting symmetric bends into asymmetric waveforms for forward swimming *in vivo*. However, *in vitro* studies indicate that the radial spokes are not absolutely essential for waveform conversion (see below). The suppressor mutations were later found to be mutations in outer (Huang et al., 1982; Porter et al., 1994) or inner (Porter et al., 1992) dynein arm components, or in polypeptides comprising a dynein regulatory complex (Huang et al., 1982; Piperno et al., 1992, 1994). These observations provide strong evidence that regulation of dynein-driven microtubule sliding involves a signal transduction pathway that includes the radial spokes. For a more complete discussion of the dynein regulatory complex and its involvement in the central pair/radial spoke control system, see Chapter 9.

B. The radial spokes transduce mechanical forces

Based on the observation that the radial spoke heads are in close apposition with the projections of the central microtubules and that the radial spoke stalk is anchored to the A-tubule of the doublet microtubules, the radial spokes appear to provide a structural linkage between the central apparatus and the dynein arms. It is not yet known if there are direct interactions between the radial spokes and dynein arm components. However, structural analyses indicate that the bases of the spokes may be in a position to interact with subsets of inner dynein arms or with components of the dynein regulatory complex (Piperno et al., 1990, 1992; Kagami and Kamiya, 1992; Gardner et al., 1994; Nicastro et al., 2006; also see Chapter 9).

Structural studies of cilia in other organisms have revealed that as the axoneme bends, the radial spokes are not extensible and undergo cycles of detachment and reattachment with the central pair projections (Warner and Satir, 1974). These contacts cause the radial spokes to tilt relative to the longitudinal axis of the axoneme as the microtubules slide. In *Chlamydomonas*, it has not been established whether the spokes make similar contacts with the central apparatus. However, it is known that in *Chlamydomonas* the central apparatus rotates once per beat (reviewed in Omoto et al., 1999; see also Mitchell, 2003; Wargo and Smith, 2003; Mitchell and Nakatsugawa, 2004; and Chapter 8 for discussion of central pair function). If the central pair projections interact with the radial spoke heads during rotation to induce tilt in the spokes relative to the circumferential axis of the axoneme, then the radial spokes would potentially experience distortion in two directions: longitudinally as the microtubules slide, and circumferentially as the central apparatus rotates. The key question is: What is the relationship of radial spoke distortion to changes in dynein activity?

One approach to addressing this question has been the application of mathematical models to simulate beating flagella (e.g., Lindemann, 1994, 2003, 2004; Brokaw, 2002; Cibert, 2003). Lindemann (1994) has proposed a "geometric clutch" model for flagellar beating in which dynein cross-bridge formation on specific doublets is limited by interdoublet spacing. Once dynein arms induce sliding between doublets, a transverse force is generated that is sufficiently strong to exceed the force developed by the dynein arms and results in the detachment of the dynein cross-bridges. At this switch-point, dynein arms on the opposite side of the axoneme are now in a position to form cross-bridges. Lindemann proposes that the radial spokes and central apparatus play a role in redistributing this transverse force during the switching event, and that movable attachments between the radial spokes and central pair projections are required for mediating the transmission of this force (Lindemann, 2003). This model has been used to simulate the beating of a *Chlamydomonas* flagellum in both the forward and reverse beating modes (Lindemann and Mitchell, 2007). Testing the prediction that strain in the radial spoke stalk locally modulates dynein activity will most likely require the development of new *in vitro* assays combined with micromanipulation technology (e.g., Shingyoji et al., 1998; Lindemann, 2004).

C. The radial spokes transduce chemical signals

While mechanical interactions between the central apparatus and radial spokes may provide an intrinsic mechanism for control of ciliary and flagellar beating, substantial evidence has emerged to indicate that cells utilize chemical signals to alter ciliary and flagellar motility (reviewed in Porter and Sale, 2000). In *Chlamydomonas*, a combination of physiological, pharmacological, and biochemical analyses of isolated wild-type and mutant axonemes has been used to elucidate these signal transduction pathways, and has provided evidence that the radial spokes play an important role in mediating motility responses to these chemical signals.

For all eukaryotic cilia and flagella, motility is regulated by second messengers such as cyclic nucleotide monophosphates and calcium (reviewed in Brokaw, 1987; Tash, 1989; Satir, 1995, 1999). In *Chlamydomonas*, cAMP has been shown to inhibit the motility of demembranated, reactivated axonemes (Hasegawa et al., 1987), and changes in intraflagellar calcium concentration are involved in modulating motility in response to light. *Chlamydomonas* cells normally swim forward with an asymmetric waveform; this waveform can be reproduced *in vitro* by reactivating isolated axonemes in low calcium conditions (~pCa8). *Chlamydomonas* cells also positively and negatively phototax. This behavior is mediated by small increases in calcium (pCa7–pCa6), which induce a shift in flagellar dominance, differentially activating one or the other flagellum and causing the cell to turn toward or away from light (Kamiya and Witman, 1984). A larger increase in calcium (pCa5–pCa4) causes a momentary cessation of motility followed by a complete switch from an asymmetric to a symmetric waveform (Bessen et al., 1980). The switch from asymmetric to symmetric waveform can also be reproduced *in vitro* simply by adjusting the calcium concentration of the reactivation medium (Bessen et al., 1980; Omoto and Brokaw, 1985). (For complete discussion of the ability of *Chlamydomonas* to sense and respond to light, see Chapter 13.)

Based on the fact that modulation of motility by second messengers such as cAMP and calcium can be reproduced *in vitro* using isolated axonemes, regulation must involve interaction of these molecules with flagellar components that are anchored to the axoneme. These components likely include cyclic nucleotide dependent kinases, kinase-independent cNMP-binding proteins, and a variety of calcium-binding proteins. As noted above and in Figure 7.2, several radial spoke proteins contain domains, such as EF-hands, calmodulin-binding domains, NDK domains, and GAF domains, which suggest a role in mediating second–messenger-stimulated motility responses.

The use of isolated axonemes in a microtubule sliding assay has proven to be an extremely powerful approach to reveal that the radial spokes are part of a control system involved in integrating chemical signals. *Chlamydomonas* radial spoke mutants have paralyzed flagella. However, axonemes isolated from these mutants undergo dynein-driven microtubule sliding *in vitro* (Witman et al., 1978), albeit at significantly reduced velocity compared to wild-type axonemes (Smith and Sale, 1992, 1994; Habermacher and Sale, 1997; Smith, 2002b). By reconstituting dynein arms isolated from wild-type axonemes onto extracted spokeless axonemes, wild-type dynein activity was restored (Smith and Sale, 1992); these results led to the prediction that the radial spokes controlled dynein motors by a stable posttranslational modification (Smith and Sale, 1992).

Consistent with this hypothesis, PKA inhibitors restored dynein activity of radial spoke-defective axonemes to wild-type levels (Howard et al., 1994). Furthermore, as noted above, cAMP, which stimulates PKA activity, inhibits reactivation of wild-type *Chlamydomonas* axonemes (Hasegawa et al., 1987). These experiments suggest that PKA is a component of the axoneme and plays a role in regulating dynein activity. Predictably, when the radial spokes are defective, PKA activity is not regulated and leads to inhibition of flagellar dynein activity. The radial spokes may be involved in control of cAMP-stimulated PKA activity. Additional experiments using similar approaches led to the discovery that radial spoke-mediated regulation of dynein activity also includes modulating the activity of second-messenger-independent kinases and phosphatases, such as casein kinase 1 (CK1) and protein phosphatase 1 and protein phosphatase 2A (Habermacher and Sale, 1996; Yang and Sale, 2000; also see Chapter 9).

Like the cAMP-signaling pathway, modulation of motility by changes in intraflagellar calcium concentration must involve calcium sensors anchored to the axoneme. It has long been suspected that calmodulin is involved in calcium signaling (reviewed in Otter, 1989), and we now know that calmodulin is a structural component of both the radial spokes (RSP20) and central apparatus (Yang et al., 2001; Wargo et al., 2005). Using the microtubule-sliding assay, Smith (2002a) discovered that wild-type dynein activity is restored to central apparatus defective mutants, but not radial spokeless mutants, in the presence of high calcium. The application of calmodulin inhibitors blocked the calcium-induced increase in dynein activity in central pairless axonemes (Smith, 2002a). These results indicate that calcium regulation of flagellar motility in *Chlamydomonas* involves regulation of dynein-driven microtubule sliding, that calmodulin may mediate the calcium signal, and that the central apparatus and radial spokes are among the key components of the calcium signaling pathway (Smith, 2002a).

Investigations of flagellar calmodulin-binding proteins have revealed the presence of a third calmodulin-binding complex that modulates dynein activity and appears to include an AKAP-binding protein associated with RSP3 (Dymek and Smith, 2007). This complex may represent an important functional and structural link between the radial spokes and dynein arms or dynein regulatory complex.

A key question that resulted from these studies is: Are there specific dynein subforms that are specifically regulated by the spoke/central pair control system? While the outer dynein arms are homogeneous in organization and composition, there are numerous inner-arm dyneins with specific locations relative to and in close proximity with the radial spokes (for a detailed discussion of these dyneins, see Chapter 6). In *Chlamydomonas*, the activity of inner-arm dynein I1, located proximal to spoke 1 (Piperno et al., 1990), appears to be regulated by the control system. Inhibitors of PKA and CK1 do not restore dynein activity in axonemes from double mutants with radial spoke defects and lacking I1 (Habermacher and Sale, 1997; Yang and Sale, 2000; also see Chapter 9). In contrast, mutations which affect I1 assembly can suppress paralysis in mutants lacking the C1 microtubule of the central apparatus (Porter et al., 1992). These observations reveal a distinct functional role for the I1 dynein and provide a functional link between the central apparatus, radial spokes, and inner dynein arm I1.

It was subsequently demonstrated that phosphorylation of IC138, an intermediate chain subunit of I1, is correlated with the inhibition of dynein activity (Habermacher and Sale, 1997; Yang and Sale, 2000; also see Chapter 9 for a more complete discussion of IC138 phosphorylation). The role of IC138 phosphorylation in modulating motility has also been suggested by the phenotype of mutants lacking subunits of I1 (Perrone et al., 2000) or with hyperphosphorylated IC138 (King and Dutcher, 1997). In both cases, these mutants are defective for phototactic behavior. Therefore, the phosphorylation state of I1 appears to be important for light-induced motility changes, a process that is dependent on changes in intraflagellar calcium concentration (Hyams and Borisy, 1978; Bessen et al., 1980). It is not yet known if the phosphorylation state of I1 is modulated in response to calcium, or if the phosphorylation state of I1 is specific to particular doublet microtubules. Nevertheless, the functional assays described above suggest that regulation of I1 phosphorylation includes the integration of second-messenger signals using the network of enzymes associated with the central apparatus and radial spokes.

Despite the fact that radial spoke-defective *Chlamydomonas* mutants have paralyzed flagella, these structures are not absolutely required for reactivation of beating *in vitro*. By using ATP analogs or low concentrations of ATP ($<50\,\mu M$ ATP), axonemes isolated from radial spoke- or central-apparatus-defective mutants can be reactivated to produce modest waveforms at very low beat frequency ($<5\,Hz$) (Omoto et al., 1996). These results suggest that ATP plays an inhibitory role (in the paralyzed mutants) and that the function of the radial spoke/central pair system is to release the ATP inhibition in a controlled manner.

Wakabayashi et al. (1997) and Frey et al. (1997) have also demonstrated that isolated axonemes from central pairless and radial spokeless mutants reactivated at low ATP concentration undergo modest waveform conversion in response to changes in calcium concentration (Frey et al., 1997; Wakabayashi et al., 1997). Since the reactivated axonemes have reduced beat frequency and irregular waveforms compared to wild-type axonemes, the authors propose that these structures are important for rapid movements and regular beating with proper waveforms. In similar studies using $200\,\mu M$ ATP, various salts ($MgSO_4$ in particular), and organic compounds, Yagi and Kamiya (2000) were able to reactivate central pairless and radial spokeless mutant axonemes with increased beat frequencies ($\sim20\,Hz$) and larger amplitude waveforms. These studies indicate that the central pair/radial spokes are not absolutely required for high-frequency oscillatory beating, a hallmark of 9 + 2 axonemes, and are consistent with the observation that some naturally occurring 9 + 0 cilia and flagella can beat with frequencies of up to 90 Hz.

This series of reactivation studies demonstrates that the central pair/radial spoke system is not an absolute requirement to produce asymmetric waveforms, high-frequency beating, or calcium-induced changes in waveform, since under certain conditions these beat parameters can be induced in the absence of the radial spokes or the central pair. Based on the large number of axonemal proteins that appear to have putative calcium-binding domains, including dynein arm subunits (Pazour et al., 2005), the axoneme most likely possesses both a spoke-dependent and spoke-independent mechanism for sensing and modulating motility in response to changes in calcium concentration. In addition, these reactivation studies do not address the differential changes in motility of the *cis* and *trans* flagellum that are required for phototaxis (Kamiya and Witman, 1984, King and Dutcher, 1997). One possibility is that as intraflagellar calcium concentration increases, different subsets of calcium-binding proteins are engaged to produce the desired motility response. We are only beginning to assess the calcium-binding properties of specific axonemal proteins and determine what protein interactions are affected by calcium binding.

It is possible that the central pair/radial spoke system contributes to the production of planar waveforms. Cilia and flagella with 9 + 2 axonemes generally have large-amplitude, planar waveforms during at least one phase of the beat cycle, if not the entire beat cycle. The control system may facilitate the more powerful planar effective stroke as the intrinsic waveform for 9 + 2 axonemes beating under physiological conditions.

VI. FUTURE DIRECTIONS

An enormous amount of evidence indicates that the radial spokes play an important role in modulating flagellar motility. The discovery of second-messenger-dependent and -independent signaling pathways revealed by functional analysis of spoke mutants, the presence of structural domains within spoke proteins that are related to signal transduction, the abnormal motility of mutants with defects in some of these structural domains, and the analysis of suppressor strains, are all consistent with a model in which the radial spokes function as a transducer of mechanochemical signals. Most likely, the physical interactions between the radial spokes and central apparatus modulate the on–off switch for dynein arms on specific subsets of doublet microtubules to establish an intrinsic mode of oscillatory, planar beating in 9 + 2 axonemes. Control of flagellar movement during behavioral responses, such as phototaxis, photoshock, and quiescence, probably involves specific modification (such as phosphorylation or calcium binding) of particular radial spoke components to alter this intrinsic beat pattern.

The continued molecular dissection of the radial spokes is essential for elucidating a molecular mechanism for control of flagellar motility. Significant progress has been made in identifying radial spoke components, but many questions remain to be answered. For example, we still do not know the precise stoichiometry and structural organization of all spoke components. We also do not know which components of the radial spokes interact with specific central apparatus components, and the functional consequences of these interactions. Similarly, while substantial evidence indicates that the spokes modulate dynein activity (and the activity of specific dynein subforms) on specific doublets, we do not know the molecular mechanism for how this change in activity is achieved. High-resolution images of axonemal components and their localization will continue to provide important insights into the structural organization of these regulatory components (see Nicastro et al., 2006; Chapter 9).

Finally, the radial spokes present a unique opportunity to study macromolecular assembly. It is still not known what occurs to convert 12S spoke precursors into 20S mature spokes in flagella and what prevents that from happening in the cytoplasm, which subunits associate for transport into the flagella by IFT, how subunits recognize the correct binding site on the axonemal microtubules, and how radial spokes are recycled to the cytoplasm. Most likely, investigations of radial spoke assembly in *Chlamydomonas* will reveal important principles that are generally applicable to the assembly of these components in all motile eukaryotic cilia and flagella.

REFERENCES

Beene, D.L. and Scott, J.D. (2007). A-kinase anchoring proteins take shape. *Curr. Opin. Cell Biol.* **19**, 192–198.

Bessen, M., Fay, R.B., and Witman, G.B. (1980). Calcium control of waveform in isolated flagellar axonemes of *Chlamydomonas*. *J. Cell Biol.* **86**, 446–455.

Brokaw, C.J. (1987). Regulation of sperm flagellar motility by calcium and cAMP-dependent phosphorylation. *J. Cell Biochem.* **35**, 175–184.

Brokaw, C.J. (2002). Computer simulation of flagellar movement VIII: coordination of dynein by local curvature control can generate helical bending waves. *Cell Motil. Cytoskeleton* **53**, 103–124.

Brokaw, C.J., Luck, D.J., and Huang, B. (1982). Analysis of the movement of *Chlamydomonas* flagella: the function of the radial spoke system is revealed by comparison of wild-type and mutant flagella. *J. Cell Biol.* **92**, 722–732.

Cibert, C. (2003). Entropy and information in flagellar axoneme cybernetics: a radial spokes integrative function. *Cell Motil. Cytoskeleton* **54**, 296–316.

Curry, A.M. and Rosenbaum, J.L. (1993). Flagellar radial spoke: a model molecular genetic system for studying organelle assembly. *Cell Motil. Cytoskeleton* **24**, 224–232.

Curry, A.M., Williams, B.D., and Rosenbaum, J.L. (1992). Sequence analysis reveals homology between two proteins of the flagellar radial spoke. *Mol. Cell Biol.* **12**, 3967–3977.

Diener, D.R., Ang, L.H., and Rosenbaum, J.L. (1993). Assembly of flagellar radial spoke proteins in *Chlamydomonas*: identification of the axoneme binding domain of radial spoke protein 3. *J. Cell Biol.* **123**, 183–190.

Diener, D.R., Curry, A.M., Johnson, K.A., Dentler, W.D., Lefebvre, P.A., Kindle, K.L., and Rosenbaum, J.L. (1990). Rescue of a paralyzed-flagella mutant of *Chlamydomonas* by transformation. *Proc. Natl. Acad. Sci. U.S.A.* **87**, 5739–5743.

Dymek, E.E. and Smith, E.F. (2007). A conserved CaM- and radial spoke-associated complex mediates regulation of flagellar dynein activity. *J. Cell Biol.* **179**, 515–526.

Ebersold, W.T., Levine, R.P., Levine, E.E., and Olmsted, M.A. (1962). Linkage maps in *Chlamydomonas reinhardi*. *Genetics* **47**, 531–543.

Frey, E., Brokaw, C.J., and Omoto, C.K. (1997). Reactivation at low ATP distinguishes among classes of paralyzed flagella mutants. *Cell Motil. Cytoskeleton* **38**, 91–99.

Gaillard, A.R., Diener, D.R., Rosenbaum, J.L., and Sale, W.S. (2001). Flagellar radial spoke protein 3 is an A-kinase anchoring protein (AKAP). *J. Cell Biol.* **153**, 443–448.

Gaillard, A.R., Fox, L.A., Rhea, J.M., Craige, B., and Sale, W.S. (2006). Disruption of the A-kinase anchoring domain in flagellar radial spoke protein 3 results in unregulated axonemal cAMP-dependent protein kinase activity and abnormal flagellar motility. *Mol. Biol. Cell.* **17**, 2626–2635.

Gardner, L.C., O'Toole, E., Perrone, C.A., Giddings, T., and Porter, M.E. (1994). Components of a "dynein regulatory complex" are located at the junction between the radial spokes and the dynein arms in *Chlamydomonas* flagella. *J. Cell Biol.* **127**, 1311–1325.

Goodenough, U.W. and Heuser, J.E. (1985). Substructure of inner dynein arms, radial spokes, and the central pair projection complex. *J. Cell Biol.* **100**, 2008–2018.

Habermacher, G. and Sale, W.S. (1996). Regulation of flagellar dynein by an axonemal type-1 phosphatase in *Chlamydomonas*. *J. Cell Sci.* **109**, 1899–1907.

Habermacher, G. and Sale, W.S. (1997). Regulation of flagellar dynein by phosphorylation of a 138-kD inner arm dynein intermediate chain. *J. Cell Biol.* **136**, 167–176.

Hasegawa, E., Hayashi, H., Asakura, S., and Kamiya, R. (1987). Stimulation of *in vitro* motility of *Chlamydomonas* axonemes by inhibition of cAMP-dependent phosphorylation. *Cell Motil. Cytoskeleton* **8**, 302–311.

Howard, D.R., Habermacher, G., Glass, D.B., Smith, E.F., and Sale, W.S. (1994). Regulation of *Chlamydomonas* flagellar dynein by an axonemal protein kinase. *J. Cell Biol.* **127**, 1683–1692.

Huang, B., Piperno, G., Ramanis, Z., and Luck, D.J. (1981). Radial spokes of *Chlamydomonas* flagella: genetic analysis of assembly and function. *J. Cell Biol.* **88**, 80–88.

Huang, B., Ramanis, Z., and Luck, D.J. (1982). Suppressor mutations in *Chlamydomonas* reveal a regulatory mechanism for flagellar function. *Cell* **28**, 115–124.

Hyams, J.S. and Borisy, G.G. (1978). Isolated flagellar apparatus of *Chlamydomonas*: characterization of forward swimming and alteration of waveform and reversal of motion by calcium ions *in vitro*. *J. Cell Sci.* **33**, 235–253.

Johnson, K.A. (1995). Immunoelectron microscopy. *Methods Cell Biol.* **47**, 153–162.

Kagami, O. and Kamiya, R. (1992). Translocation and rotation of microtubules caused by multiple species of *Chlamydomonas* inner-arm dynein. *J. Cell Sci.* **103**, 653–664.

Johnson, K.A. and Rosenbaum, J.L. (1992). Polarity of flagellar assembly in *Chlamydomonas*. *J. Cell Biol.* **119**, 1605–1611.

Kamiya, R. and Witman, G.B. (1984). Submicromolar levels of calcium control the balance of beating between the two flagella in demembranated models of *Chlamydomonas*. *J. Cell Biol.* **98**, 97–107.

King, S.J. and Dutcher, S.K. (1997). Phosphoregulation of an inner dynein arm complex in *Chlamydomonas reinhardtii* is altered in phototactic mutant strains. *J. Cell Biol.* **136**, 177–191.

King, S.M. and Patel-King, R.S. (1995). Identification of a Ca^{2+}-binding light chain within *Chlamydomonas* outer arm dynein. *J. Cell Sci.* **108**, 3757–3764.

Linck, R.W. (1976). Flagellar doublet microtubules: fractionation of minor components and alpha-tubulin from specific regions of the A-tubule. *J. Cell Sci.* **20**, 405–439.

Lindemann, C.B. (1994). A model of flagellar and ciliary functioning which uses the forces transverse to the axoneme as the regulator of dynein activation. *Cell Motil. Cytoskeleton* **29**, 141–154.

Lindemann, C.B. (2003). Structural–functional relationships of the dynein, spokes, and central-pair projections predicted from an analysis of the forces acting within a flagellum. *Biophys. J.* **84**, 4115–4126.

Lindemann, C.B. (2004). Testing the geometric clutch hypothesis. *Biol. Cell* **96**, 681–690.

Lindemann, C.B. and Mitchell, D.R. (2007). Evidence for axonemal distortion during the flagellar beat of *Chlamydomonas*. *Cell Motil. Cytoskeleton* **64**, 580–589.

Luck, D.J., Piperno, G., Ramahis, Z., and Huang, B. (1977). Flagellar mutants of *Chlamydomonas*: studies of radial spoke-defective strains by dikaryon and revertant analysis. *Proc. Natl. Acad. Sci. U. S. A.* **74**, 3456–3460.

Mitchell, D.R. (2003). Orientation of the central pair complex during flagellar bend formation in *Chlamydomonas*. *Cell Motil. Cytoskeleton* **56**, 120–129.

Mitchell, D.R. and Nakatsugawa, M. (2004). Bend propagation drives central pair rotation in *Chlamydomonas reinhardtii* flagella. *J. Cell Biol.* **166**, 709–715.

Mitchell, D.R. and Sale, W.S. (1999). Characterization of a *Chlamydomonas* insertional mutant that disrupts flagellar central pair microtubule-associated structures. *J. Cell Biol.* **144**, 293–304.

Nicastro, D., Schwartz, C., Pierson, J., Gaudette, R., Porter, M.E., and McIntosh, J.R. (2006). The molecular architecture of axonemes revealed by cryoelectron tomography. *Science* **18**, 944–948.

Omoto, C.K. and Brokaw, C.J. (1985). Bending patterns of *Chlamydomonas* flagella: II. Calcium effects on reactivated *Chlamydomonas* flagella. *Cell Motil.* **5**, 53–60.

Omoto, C.K., Yagi, T., Kurimoto, E., and Kamiya, R. (1996). Ability of paralyzed flagella mutants of *Chlamydomonas* to move. *Cell Motil. Cytoskeleton* **33**, 88–94.

Omoto, C.K., Gibbons, I.R., Kamiya, R., Shingyoji, C., Takahashi, K., and Witman, G.B. (1999). Rotation of the central pair microtubules in eukaryotic flagella. *Mol. Biol. Cell* **10**, 1–4.

Otter, T. (1989). Calmodulin and the control of flagellar movement. In: *Cell Movement, Vol. 1* (F.D. Warner, P. Satir, and I.R. Gibbons, Eds.) pp. 281–298. Alan R. Liss, New York, NY.

Patel-King, R.S., Gorbatyuk, O., Takebe, S., and King, S.M. (2004). Flagellar radial spokes contain a Ca^{2+}-stimulated nucleoside diphosphate kinase. *Mol. Biol. Cell* **15**, 3891–3902.

Pazour, G.J., Wilkerson, C.G., and Witman, G.B. (1998). A dynein light chain is essential for the retrograde particle movement of intraflagellar transport (IFT). *J. Cell Biol.* **141**, 979–992.

Pazour, G.J., Agrin, N., Leszyk, J., and Witman, G.B. (2005). Proteomic analysis of a eukaryotic cilium. *J. Cell Biol.* **170**, 103–113.

Perrone, C.A., Myster, S.H., Bower, R., O'Toole, E.T., and Porter, M.E. (2000). Insights into the structural organization of the I1 inner arm dynein from a domain analysis of the 1beta dynein heavy chain. *Mol. Biol. Cell* **11**, 2297–2313.

Piperno, G., Huang, B., and Luck, D.J. (1977). Two-dimensional analysis of flagellar proteins from wild-type and paralyzed mutants of *Chlamydomonas reinhardtii*. *Proc. Natl. Acad. Sci. U.S.A.* **74**, 1600–1604.

Piperno, G., Huang, B., Ramanis, Z., and Luck, D.J. (1981). Radial spokes of *Chlamydomonas* flagella: polypeptide composition and phosphorylation of stalk components. *J. Cell Biol.* **88**, 73–79.

Piperno, G., Ramanis, Z., Smith, E.F., and Sale, W.S. (1990). Three distinct inner dynein arms in *Chlamydomonas* flagella: molecular composition and location in the axoneme. *J. Cell Biol.* **110**, 379–389.

Piperno, G., Mead, K., and Shestak, W. (1992). The inner dynein arms I2 interact with a "dynein regulatory complex" in *Chlamydomonas* flagella. *J. Cell Biol.* **118**, 1455–1463.

Piperno, G., Mead, K., LeDizet, M., and Moscatelli, A. (1994). Mutations in the "dynein regulatory complex" alter the ATP-insensitive binding sites for inner arm dyneins in *Chlamydomonas* axonemes. *J. Cell Biol.* **125**, 1109–1117.

Porter, M.E. and Sale, W.S. (2000). The 9 + 2 axoneme anchors multiple inner arm dyneins and a network of kinases and phosphatases that control motility. *J. Cell Biol.* **151**, F37–F42.

Porter, M.E., Power, J., and Dutcher, S.K. (1992). Extragenic suppressors of paralyzed flagellar mutations in *Chlamydomonas reinhardtii* identify loci that alter the inner dynein arms. *J. Cell Biol.* **118**, 1163–1176.

Porter, M.E., Knott, J.A., Gardner, L.C., Mitchell, D.R., and Dutcher, S.K. (1994). Mutations in the *SUP-PF-1* locus of *Chlamydomonas reinhardtii* identify a regulatory domain in the beta-dynein heavy chain. *J. Cell Biol.* **126**, 1495–1507.

Qin, H., Diener, D.R., Geimer, S., Cole, D.G., and Rosenbaum, J.L. (2004). Intraflagellar transport (IFT) cargo: IFT transports flagellar precursors to the tip and turnover products to the cell body. *J. Cell Biol.* **164**, 255–266.

Rupp, G. and Porter, M.E. (2003). A subunit of the dynein regulatory complex in *Chlamydomonas* is a homologue of a growth arrest-specific gene product. *J. Cell Biol.* **7**, 47–57.

Satir, P. (1995). Landmarks in cilia research from Leeuwenhoek to us. *Cell Motil. Cytoskeleton* **32**, 90–94.

Satir, P. (1999). The cilium as a biological nanomachine. *FASEB J.* **13**, S235–S237.

Shingyoji, C., Higuchi, H., Yoshimura, M., Katayama, E., and Yanagida, T. (1998). Dynein arms are oscillating force generators. *Nature* **393**, 711–714.

Smith, E.F. (2002a). Regulation of flagellar dynein by calcium and a role for an axonemal calmodulin and calmodulin-dependent kinase. *Mol. Biol. Cell* **13**, 3303–3313.

Smith, E.F. (2002b). Regulation of flagellar dynein by the axonemal central apparatus. *Cell Motil. Cytoskeleton* **52**, 33–42.

Smith, E.F. and Sale, W.S. (1992). Regulation of dynein-driven microtubule sliding by the radial spokes in flagella. *Science* **257**, 1557–1559.

Smith, E.F., and Sale, W.S. (1994). Mechanisms of flagellar movement: functional interactions between dynein arms and radial spoke-central apparatus complex. In: *Microtubules* (J.S. Hyams and C. Lloyd, Eds.), pp. 381–392. Wiley Liss Inc., New York, NY.

Smith, E.F. and Yang, P. (2004). The radial spokes and central apparatus: mechano-chemical sensors that regulate flagellar motility. *Cell Motil. Cytoskeleton* **57**, 8–17.

Tash, J.S. (1989). Protein phosphorylation: the second messenger signal transducer of flagellar motility. *Cell Motil. Cytoskeleton* **14**, 332–339.

Wakabayashi, K., Yagi, T., and Kamiya, R. (1997). Ca^{2+}-dependent waveform conversion in the flagellar axoneme of *Chlamydomonas* mutants lacking the central-pair/radial spoke system. *Cell Motil. Cytoskeleton* **38**, 22–28.

Wargo, M.J. and Smith, E.F. (2003). Asymmetry of the central apparatus defines the location of active microtubule sliding in *Chlamydomonas* flagella. *Proc. Natl. Acad. Sci. U.S.A.* **100**, 137–142.

Wargo, M.J., Dymek, E.E., and Smith, E.F. (2005). Calmodulin and PF6 are components of a complex that localizes to the C1 microtubule of the flagellar central apparatus. *J. Cell Sci.* **118**, 4655–4665.

Warner, F.D. and Satir, P. (1974). The structural basis of ciliary bend formation. *J. Cell Biol.* **63**, 35–63.

Williams, B.D., Mitchell, D.R., and Rosenbaum, J.L. (1986). Molecular cloning and expression of flagellar radial spoke and dynein genes of *Chlamydomonas*. *J. Cell Biol.* **103**, 1–11.

Williams, B.D., Velleca, M.A., Curry, A.M., and Rosenbaum, J.L. (1989). Molecular cloning and sequence analysis of the *Chlamydomonas* gene coding for radial spoke protein 3: flagellar mutation *pf-14* is an ochre allele. *J. Cell Biol.* **109**, 235–245.

Witman, G.B. (1975). The site of *in vivo* assembly of flagellar microtubules. *Ann. N.Y. Acad. Sci.* **30**, 178–191.

Witman, G.B. and Minervini, N. (1982). Role of calmodulin in the flagellar axoneme: effect of phenothiazines on reactivated axonemes of *Chlamydomonas*. *Prog. Clin. Biol. Res.* **80**, 199–204.

Witman, G.B., Plummer, J., and Sander, G. (1978). *Chlamydomonas* flagellar mutants lacking radial spokes and central tubules. *J. Cell Biol.* **76**, 729–747.

Yagi, T. and Kamiya, R. (2000). Vigorous beating of *Chlamydomonas* axonemes lacking central pair/radial spoke structures in the presence of salts and organic compounds. *Cell Motil. Cytoskeleton* **46**, 190–199.

Yang, C. and Yang, P. (2006). The flagellar motility of *Chlamydomonas pf25* mutant lacking an AKAP-binding protein is overtly sensitive to medium conditions. *Mol. Biol. Cell* **17**, 227–238.

Yang, C., Compton, M.M., and Yang, P. (2005). Dimeric novel HSP40 is incorporated into the radial spoke complex during the assembly process in flagella. *Mol. Biol. Cell* **16**, 637–648.

Yang, C., Owen, H.A., and Yang, P. (2008). Dimeric heat shock protein 40 binds radial spokes for generating coupled power strokes and recovery strokes of 9 + 2 flagella. *J. Cell Biol.* **180**, 403–415.

Yang, P. and Sale, W.S. (2000). Casein kinase I is anchored on axonemal doublet microtubules and regulates flagellar dynein phosphorylation and activity. *J. Biol. Chem.* **275**, 18905–18912.

Yang, P., Diener, D.R., Rosenbaum, J.L., and Sale, W.S. (2001). Localization of calmodulin and dynein light chain LC8 in flagellar radial spokes. *J. Cell Biol.* **153**, 1315–1326.

Yang, P., Yang, C., and Sale, W.S. (2004). Flagellar radial spoke protein 2 is a calmodulin binding protein required for motility in *Chlamydomonas reinhardtii*. *Eukaryot. Cell* **3**, 72–81.

Yang, P., Diener, D.R., Yang, C., Kohno, T., Pazour, J.G., Dienes, J.M., Agrin, N.S., King, S.M., Sale, W.S., Kamiya, R., Rosenbaum, J.L., and Witman, G.B. (2006). Radial spoke proteins of *Chlamydomonas* flagella. *J. Cell Sci.* **15**, 1165–1174.

Zimmer, W.E., Schloss, J.A., Silflow, C.D., Youngblom, J., and Watterson, D.M. (1988). Structural organization, DNA sequence, and expression of the calmodulin gene. *J. Biol. Chem.* **263**, 19370–19383.

CHAPTER 8

The Flagellar Central Pair Apparatus

David R. Mitchell

Department of Cell and Developmental Biology,
SUNY Upstate Medical University, Syracuse, New York, USA

CHAPTER CONTENTS

I. INTRODUCTION

The central pair of singlet microtubules and their associated structures (the central pair apparatus, central pair complex, or simply the central pair) are essential for flagellar motility, as revealed by the paralyzed flagella phenotype of central pair assembly mutants. Analysis of *Chlamydomonas* mutations that disrupt different central pair structures has provided a basis for detailed structural characterization of the central pair and aided in biochemical identification of proteins located in specific central pair associated complexes. Because radial spoke heads form intimate contacts with central pair projections, regulation is thought to work through central pair modulation of spokes, which in turn alter the activity of doublet-associated dynein motors. Several proteins with known signaling properties, including a protein phosphatase, an A kinase anchoring protein (AKAP), and calmodulin, have been localized to the central pair and may participate in these

235

regulatory pathways. In addition, the central pair harbors motor proteins of the kinesin family, and proteins that may play important roles in flagellar nucleotide metabolism. This structure appears to have evolved prior to the divergence of eukaryotes from a common ancestor (Mitchell, 2004), and hence extensive work on the *Chlamydomonas* central pair has provided a basis for studies of central pair function in systems as distantly related as mammals (Zhang et al., 2004) and trypanosomes (Branche et al., 2006). As genome sequences have become available from all branches of eukaryotic organisms, the conservation of central pair proteins, as well as central pair structure, has been broadly confirmed. Because of its dominant role in studies of central pair biology, much of what we know about the *Chlamydomonas* central pair has been discussed in two general reviews of this suborganellar structure that cover all work up to that time (Smith and Lefebvre, 1997b; Smith and Yang, 2004).

II. CENTRAL PAIR STRUCTURE

Our current understanding of central pair structure derives primarily from work in *Chlamydomonas* using electron microscopy of freeze-etch replicas, negative stain preparations, and thin sections. Early work identified differences in the length of prominent projections from the two microtubules, with C1 defined as the microtubule with longer projections (Hopkins, 1970; Witman et al., 1978; Dutcher et al., 1984). The two longer projections from C1 are designated 1a and 1b, and those from C2 are 2a and 2b. Additional densities were characterized following image averaging of central pair cross sections (Figure 8.1A) (Mitchell and Sale, 1999), as summarized in Figure 8.1B. The axial periodicity of some C1 projections is 32 nm (four tubulin dimers), whereas others repeat at 16 nm (two tubulin dimers) (Figure 8.1D, summarized diagrammatically in Figure 8.2); in contrast, all C2 projections appear to repeat every 16 nm (Goodenough and Heuser, 1985; Mitchell and Sale, 1999; Mitchell, 2003b). The central pair microtubules are held together along most of their length by bridging structures that also have a 16-nm repeat period. Additional connections between the two microtubules as seen in cross section include a diagonal bridge between C1 projection 1b and the C2 microtubule wall, and an interaction between the tips of projections 1b and 2b. Thin circumferential arcs or sheath elements (*sh* in Figure 8.2) link the tips of some projections and give the entire central pair a circular cross section. These arcs may be lacking from central pair structure in metazoans.

The two microtubules of the central pair both start from a region above the basal body and transition zone (Ringo, 1967; Rosenbaum et al., 1969) (see Figure 10.1A and F in Chapter 10), and end in a membrane-associated cap structure at the flagellar tip (Ringo, 1967; Dentler and Rosenbaum, 1977) (see Figure 10.3C in Chapter 10). Work in other organisms indicates

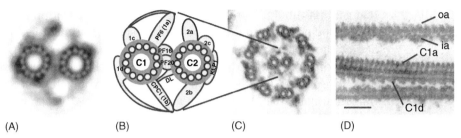

(A)   (B)   (C)   (D)

FIGURE 8.1 *(A) TEM image average of the central region of wild-type* Chlamydomonas *axonemes, with central pair orientation maintained constant. Because the central pair does not maintain a fixed orientation relative to the surrounding doublet microtubules and associated radial spokes, spoke head images blur into a ring around the central pair complex. Adapted from Figure 8.4 of Yokoyama et al., 2004, Proc. Natl. Acad. Sci. 101:17398–17403, by copyright permission of the National Academy of Sciences. (B) Diagram of the central pair in which the two microtubules (C1 and C2) and prominent cross sectional densities (1a–1d and 2a–2c) are labeled. The locations of complexes containing identified central pair proteins (PF6, PF16, PF20, CPC1, and KLP1) are also indicated; in addition, hydin is located in 2b. DL, diagonal link. (C) Electron micrograph of a cross section through a wild-type axoneme, showing the relationship between outer doublets, radial spokes, and the central pair complex. (D) A longitudinal ultrathin section through a pf14 cpc1 axoneme reveals the periodicity of C1a projections (16 nm), C1d projections (32 nm), outer row dyneins (oa; 24 nm), and inner row dyneins (ia, 96 nm). The central pair does not remain centered in the axoneme when spokes are absent, as in this mutant. Proximal end to the left. Scale bar 100 nm.*

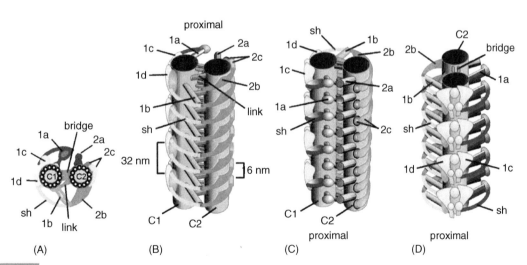

(A)   (B)   (C)   (D)

FIGURE 8.2 *Color-coded diagrams summarizing the relationships among structural elements of the central pair. (A) Cross sectional view of the central pair as seen from the proximal toward the distal end. (B) View of the 1b and 2b projections and associated structures with the proximal end toward the top of the page. Note the 32-nm spacing of projection 1d (yellow), and an associated attachment point of the 1b sheath (blue ovals), whereas most other structures visible from this view repeat every 16 nm. The topmost 1b projection and its associated sheath element (sh) have been omitted to provide a clearer view of the diagonal links (green) that connect 1b projections to the C2 microtubule. (C) The view in B has been flipped end-for-end to expose projections 1a, 2a, and 2c. The sheath connects every other 1a projection to a 1c density. (D) The view in C has been rotated by 90° to reveal the unusually dense particle distribution on the surface of C1, and the tilt of projections toward the distal end of the axoneme. Modified from Mitchell, 2003b, Cell Motil. Cytoskeleton 55:188–199, by copyright permission of Wiley-Liss, Inc.*

that gamma-tubulin is important for central pair nucleation (McKean et al., 2003), but no structural or biochemical specializations have been identified at the proximal end of the *Chlamydomonas* central pair, which can appear coincident with the distal end of the stellate transition zone structure (Ringo, 1967), or begin after a short gap (Mitchell et al., 2005). The transition zone may function as a physical barrier that prevents the central pair from sliding into the basal body cavity, as this defect occurs frequently in transition zone mutants (Jarvik and Suhan, 1991). Addition of new tubulin subunits to the central pair occurs at the distal end (Johnson and Rosenbaum, 1992), consistent with the view that these microtubules are nucleated by gamma-tubulin in the transition zone (Silflow et al., 1999; McKean et al., 2003) and with the observation that mutations affecting central pair microtubule assembly, such as beta-tubulin modifications in *Tetrahymena* (Thazhath et al., 2004) or *Drosophila* (Nielsen et al., 2001), can result in complete absence of the central pair, or in a central pair that is only present in proximal regions of the flagellum, but never in a central pair that is only assembled in distal regions. Assembly may be space-limited since a second, parallel central pair apparatus will assemble within the cage of outer doublets when space permits. This occurs when radial spokes are missing and the central pair cross sectional area is reduced by projection assembly mutations, such as in the triple mutant *pf14 cpc1 pf6* (Figure 8.3), or when the cage of doublets is enlarged from 9 to 10 in the basal body defective mutant strain *bld12* (Nakazawa et al., 2007; Nakazawa et al., personal communication).

The central pair extends distally beyond the end of the outer doublet microtubules by as much as 0.5 μm. This distal extension appears free of associated projections and is distinguished by an electron-dense sheet between the two microtubules (Ringo, 1967) (and see Figure 10.3B and C in Chapter 10). Because the central pair complex rotates relative to the outer doublet microtubules (see section V), a shear zone must exist within the cap structure itself

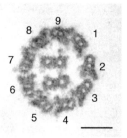

FIGURE 8.3 *A cross sectional micrograph of a* pf14 pf6 cpc1 *axoneme in which two central pair complexes have assembled (9 + 2 + 2). The absence of radial spokes (pf14) and a reduced cross sectional area of the central pair from loss of projections 1a (pf6) and 1b (cpc1) provides room for two central pair complexes to assemble within the normal nine doublets. Both central pair complexes are viewed from their proximal ends with C1 to the left. Doublet numbers are based on identification of doublet one, which lacks outer row dyneins. Scale bar 100 nm.*

or within the adjacent membrane. Although the flagellar tip region also plays an important role as the site of reversal of IFT motility (see Chapter 4), the central pair (and by inference the cap) are not required for this process, as IFT appears unaltered in strains carrying central pair assembly mutations. Central pair tip structures have not been characterized at the biochemical level.

III. GENETIC DISSECTION OF CENTRAL PAIR STRUCTURE AND FUNCTION

Mutations at four loci destabilize the entire central pair complex: *pf15*, *pf18*, *pf19*, and *pf20* (Table 8.1). These loci were some of the earliest to be mapped in *Chlamydomonas* (Ebersold et al., 1962) and analyzed for structural defects by electron microscopy (Warr et al., 1966), and were later studied in greater biochemical and structural detail (Witman et al., 1978; Adams et al., 1981). Phenotypically, these mutations all result in primarily nonmotile cells with flagella that do not beat, although a few cells in populations of these mutants display limited motility (Warr et al., 1966). Unlike the paralyzed flagella of radial spoke mutants (see Chapter 7), which are curved (most likely due to the intrinsic helical curvature of the central pair; see below), flagella of central pair mutants are straight. Although the central pair microtubules fail to assemble in these mutants, many central pair associated proteins are still present in the flagellar compartment and may interact to form a higher-order dense core structure (Witman et al., 1978; Adams et al., 1981). These remaining proteins and structures partition into the soluble phase when flagellar membranes are removed with detergents. Two of these four loci have been identified at the molecular level. The *PF15* locus encodes a p80 (noncatalytic) subunit of the microtubule-severing protein

Table 8.1   Mutations affecting central pair structure

Strain	Structural defect in flagella	Structural defect in axonemes	Gene product	References[a]
pf6	9 + 2, 1a missing	9 + 2, 1a missing	PF6	[1], [2]
pf15	9 + dense core	9 + 0	Katanin p80	[3], [4], [5], [6]
pf16	9 + 2	9 + 1, C1 missing	PF16	[1], [7]
pf18	9 + dense core	9 + 0		[3], [5]
pf19	9 + dense core	9 + 0		[3], [4], [5]
pf20	9 + 2, 9 + 1, 9 + 0	9 + 0	PF20	[3], [5], [8]
cpc1	1b missing	1b missing	CPC1	[9], [10]

[a][1] Dutcher et al. (1984); [2] Rupp et al. (2001); [3] Starling and Randall (1971); [4] Witman et al. (1978); [5] Adams et al. (1981); [6] Dymek et al. (2004); [7] Smith and Lefebvre (1996); [8] Smith and Lefebvre (1997a); [9] Mitchell and Sale (1999); [10] Zhang and Mitchell (2004).

katanin (Dymek et al., 2004), supporting a role for katanin in regulation of central pair microtubule assembly. The PF15 protein itself, although associated with the axoneme, does not appear to reside within the central pair structure. *PF20* encodes a WD-repeat protein that localizes to the bridge between C1 and C2 (Smith and Lefebvre, 1997a). The mammalian orthologue of PF20, SPAG16L, is also localized to the central pair (Zhang et al., 2002), and is essential for normal sperm motility, but not for central pair assembly (Zhang et al., 2006). Gene products of *PF18* and *PF19* have not been identified.

Three additional mutations have been characterized that partially disrupt the C1 microtubule or its associated structures: *pf6*, *pf16*, and *cpc1*. Nonconditional mutations at *PF16* destabilize the entire C1 microtubule, which remains assembled in intact *pf16* flagella but is selectively lost upon flagellar demembranation (Dutcher et al., 1984; Mitchell and Sale, 1999). Phenotypically, *pf16* flagella twitch but fail to propagate bends, and cells are nonmotile. When isolated *pf16* axonemes are exposed to reactivation conditions in the presence of a protease, they display an ability to disintegrate by doublet sliding, indicating residual dynein activity. However, the sliding rate is considerably slower than that of wild-type axonemes, and is indistinguishable from the sliding rate of axonemes that completely lack the central pair such as *pf15* and *pf18* (Smith, 2002a, b). The *PF16* locus encodes an armadillo-repeat protein associated with the C1 microtubule (Smith and Lefebvre, 1996, 2000). The orthologue in mammals (SPAG6) is also a central pair protein essential for ciliary and flagellar motility (Sapiro et al., 2002), and yeast two-hybrid studies suggest that, in mammals, PF16 interacts directly with PF20 (Zhang et al., 2002) and is therefore likely localized at the interface between C1 and C2 (Figure 8.1B). A similar location is supported by the structural defects in *pf16* axonemes under conditions that partially stabilize the C1 microtubule (Mitchell and Sale, 1999). However, selection of extragenic suppressors of a temperature-sensitive *Chlamydomonas pf16* allele identified *pf9-2*, an inner row I1 dynein assembly mutation, as a bypass suppressor specific to the *pf16-BR3*[ts] allele (Porter et al., 1992), a result that is hard to reconcile with a positioning of PF16 in the bridge region. *In vitro* doublet microtubule sliding rate assays have also shown that I1 assembly mutations relieve the inhibition of doublet sliding caused by *pf16* (Smith, 2002b). These results suggest that some regulatory signals generated by the central pair complex and involving PF16-associated or C1-associated proteins may specifically control I1 dynein activity (see Chapter 9 for additional information on I1 dynein regulation).

Mutations at *PF6* disrupt assembly of the C1a projection, and result in flagella that may beat, but too slowly for effective swimming (Dutcher et al., 1984). The gene product is a large (~240 kD) structural protein (Rupp et al., 2001) associated with five other subunits (Table 8.2), including both calmodulin and an 86-kD calmodulin-binding protein that has limited

Table 8.2 Central pair complex proteins

Structure/ location	Protein/ mutant[a]	Mass[b] (kD)	pI[b]	Accession number	Properties	References[c]
C1a	**PF6**	240	4.84	AAK38270	Alanine-proline rich; ASH domain	[1], [2]
	C1a-86	86	5.54	AAZ31187	PKA RII-like LRR domain	[2]
	C1a-34	34	5.63	AAZ31186	Coiled-coil, dimerizes with C1a-32	[2]
	C1a-32	32	5.56	AAZ31185	Resembles C1a-34	[2]
	C1a-18	18	5.70	AAZ31184	MORN domains	[2]
	Calmodulin	18	4.30	AAA33083	Ca^{2+}-binding protein	[2]
C1b	C1b-350 (FAP42)	350	5.37	EDP00757	Five guanylate kinase domains, one adenylate kinase domain	[3]
	CPC1	265	5.45	AAT40992	CH, adenylate kinase domains	[3], [4]
	C1b-135 (FAP69)	135	5.61	EDP06190	Armadillo repeat protein	[3]
	HSP70	78	5.25	P25840	Chaperone	[3], [5]
	Enolase	56	5.26	P13683	Glycolytic enzyme	[3]
C1	**PF16**	57	6.0	AAC49169	Armadillo repeat protein	[6]
	PP1c	35	4.8	AAD38850	Protein phosphatase 1c	[7]
C2b	Hydin	540	5.80	EDP09735	ASH domains	[8]
C2c	KLP1	96	6.6	P46870	Kinesin-like protein	[9], [10]
C1–C2 bridge	**PF20**	63	7.7	AAB41727	WD-repeat protein	[11]

[a] Proteins listed in **bold face** are associated with loci at which Chlamydomonas mutations have been characterized; protein names correspond to locus names (Table 8.1).

[b] Mass as estimated by SDS-PAGE, pI as calculated from primary sequence.

[c] [1] Rupp et al. (2001); [2] Wargo et al. (2005); [3] Mitchell et al. (2005); [4] Zhang and Mitchell (2004); [5] Bloch and Johnson (1995); [6] Smith and Lefebvre (1996); [7] Yang et al. (2000); [8] Lechtreck and Witman (2007); [9] Bernstein et al. (1994); [10] Yokoyama et al. (2004); [11] Smith and Lefebvre (1997a).

homology to protein kinase A (PKA) RII subunits (Wargo et al., 2005). Analysis of dynein regulation by calcium suggests that the PF6-calmodulin complex contributes to calcium-dependent regulation of dynein activation patterns, as seen by the pattern of doublet microtubule sliding in protease-treated axonemes (Wargo et al., 2004). Immunolocalization studies in mice show that the mouse orthologue of PF6 is also a central pair protein (Zhang et al., 2005). One region of retained sequence similarity between the algal and vertebrate PF6 proteins also appears in hydin, another conserved central pair protein (see below and Table 8.2), and defines a domain of unknown function that has been termed the ASH domain (Ponting, 2006).

The *CPC1* locus, similar to the *PF6* locus, encodes a large protein essential for assembly of a C1 microtubule projection (C1b) (Zhang and Mitchell, 2004). The tip of C1 projection 1b appears to interact with the tip of C2 projection 2b (Figure 8.1), and the absence of 1b correlates with frequent absence of projection 2b in *cpc1* axonemes (Mitchell and Sale, 1999). Although a domain homologous to adenylate kinases appears within the

CPC1 sequence, the functional significance of this domain has not been determined and axonemal adenylate kinase activity appears unaltered by this mutation (Zhang and Mitchell, 2004; Mitchell et al., 2005). A calponin homology (CH) domain near the N-terminus of CPC1 resembles CH domains of known microtubule-binding proteins such as EB1 (Bu and Su, 2003) and Spef1/CLAMP (Dougherty et al., 2005; Chan et al., 2005). The CPC1 protein can be extracted as part of a complex that contains another large (350 kD) protein with homology to nucleoside diphosphate kinases, a novel 135-kD structural protein, the heat shock protein HSP70, and enolase (Mitchell et al., 2005). Mutations at *cpc1* disrupt assembly of these subunits and result in motility with a reduced beat frequency. This phenotype can be largely attributed to a reduction in intraflagellar ATP concentrations that result from the loss of glycolytic ATP synthesis supported by enolase. Other enzymes necessary (along with enolase) for one of the ATP-generating steps of glycolysis have also been localized to the flagellar compartment, but do not appear to interact with the central pair complex (Mitchell et al., 2005). KPL2, the mammalian orthologue of CPC1 (Ostrowski et al., 2002; Zhang and Mitchell, 2004), occurs in several splice variants, at least one of which is essential for normal central pair assembly and motility in mammalian spermatozoa (Sironen et al., 2006).

In addition to the *Chlamydomonas* mutations listed above, whose characterizations first identified the gene products as central pair components and led to their later identification in mammals, one mammalian mutation became the starting point for identification of a *Chlamydomonas* central pair protein. The mouse *hy3* mutation in the gene encoding Hydin, a large protein expressed in ciliated cells (Davy and Robinson, 2003), results in hydrocephalus. The knockdown of the *Trypanosoma brucei* (Broadhead et al., 2006) or *Chlamydomonas* (Lechtreck and Witman, 2007) orthologues results in reduced motility, and the *Chlamydomonas* protein was further localized to the C2b central pair projection, where it stabilizes the association of kinesin-like protein KLP1 with the C2 microtubule (Lechtreck and Witman, 2007). The presence of a conserved ASH domain in both PF6 and hydin (Table 8.2) together with their locations at rotationally symmetric positions on the C1 and C2 microtubules suggests that these proteins may play similar roles as scaffolds for other central pair components.

IV. BIOCHEMICAL IDENTIFICATION OF CENTRAL PAIR PROTEINS

Comparative gel analysis was used over 20 years ago to catalog 23 proteins that are reduced or missing from axonemes of *Chlamydomonas* central pair assembly mutants (Adams et al., 1981; Dutcher et al., 1984) and to further show that 11 of these are likely associated with C1, based on their depletion from *pf16* axonemes. Reassessment of those results based on data from the molecular studies of *pf6*, *pf16*, and *cpc1* suggest that fewer than half of these probable central pair proteins have been characterized at the molecular level (Table 8.2). However, in addition to the proteins identified through analysis of mutants, biochemical methods have been used to identify a few central pair proteins in two categories: kinesin-like proteins and signal transduction proteins.

Kinesin-like proteins were first identified in the *Chlamydomonas* central pair through western blot analysis with pan-kinesin antibodies (Fox et al., 1994; Johnson et al., 1994), and through amplification of *Chlamydomonas* kinesin-related sequences (Bernstein et al., 1994). Two proteins with kinesin-like properties were localized to the central pair in these studies, but only one has been verified as a *bona fide* kinesin-related protein at the molecular level. The KLP1 protein was first identified through PCR of kinesin-related sequences, and complete characterizations of the gene and cDNA show that KLP1 has a conserved amino-terminal motor domain, but a carboxy-terminal tail unrelated to other kinesins (Bernstein et al., 1994). Sequence comparisons show KLP1 to be a member of the kinesin-9 family (Miki et al., 2005) that appears widely in the genomes of organisms with motile 9 + 2 organelles (Wickstead and Gull, 2006; Richardson et al., 2006), including mammals, whose kinesin-9 homologues are highly represented in testis and lung cDNA libraries (Miki et al., 2003). An antibody to *Chlamydomonas* KLP1 recognizes a 96-kD central pair protein and decorates the C2 microtubule. RNAi depletion of KLP1 results in abnormal motility and disrupts structures on the outer margin of the C2 microtubule (Yokoyama et al., 2004) (Figure 8.1). KLP1 shows nucleotide-sensitive binding to reassembled microtubules *in vitro*, but its association with the central pair is not nucleotide sensitive (Bernstein, 1995; Yokoyama et al., 2004). Although KLP1 has been hypothesized to act as a motor for rotation of the central pair in *Chlamydomonas*, more recent studies conclude that central pair rotation can occur as a passive response to bend propagation (see section V), and the activity of a central pair-associated motor is not required. In addition, kinesin-9 family members appear in the genomes of organisms such as mammals and trypanosomes whose flagella have fixed (nonrotating) central pairs (Omoto et al., 1999; Branche et al., 2006; Gadelha et al., 2006). Instead, it may be of value to consider KLP1 as a conformational switch whose activation could be linked to a change in central pair curvature, or to a change in the conformation of associated proteins important for signal transduction.

The second central pair kinesin is a 110-kD protein that reacts with antibodies directed against two conserved kinesin sequences, HIPYR and LAGSE, and that is missing from axonemes of central pair mutants such as *pf18* and *pf19* (Fox et al., 1994; Johnson et al., 1994). The molecular identity and central pair location of this protein are not known. Immunolocalization with an antibody directed against conserved kinesin neck domain sequences (αHD antibody) identified the C1 microtubule as a site of kinesin-like protein attachment (Johnson et al., 1994). The αHD antibody also detects a 110-kD band on blots, but this band is retained in *pf18* axonemes, and its relationship to the HIPYR and LAGSE band has not been determined.

Two biochemical studies suggest that additional signaling events feature central pair proteins. Through blot overlays with the PKA RII subunit, Sale and colleagues identified two AKAP-like proteins in the *Chlamydomonas* flagellum (Gaillard et al., 2001). One of these is missing from the radial spoke assembly mutant *pf14* and was identified by Gaillard et al. as the *PF14* gene product (see Chapter 7 for additional details). The other AKAP is missing from central pair mutant axonemes. This 240-kD central pair AKAP is retained in *pf16* axonemes, and thus likely associates with C2, but its molecular identity and functional role have not been determined. Biochemical and western blot analysis of protein phosphatases in the *Chlamydomonas* flagellum showed that most of the flagellar serine/threonine phosphatase PP1c is linked to the C1 central pair microtubule (Yang et al., 2000). PP1 is known to regulate axonemal dynein activity in *Chlamydomonas* (see Chapter 9), so it is tempting to assume that the central pair-linked PP1c is involved. However, some PP1c immunoreactive signal was also found associated with outer doublet microtubules, where dyneins are anchored. Therefore, the central pair PP1c may not be involved directly in this dynein regulatory pathway, and downstream targets of the central pair enzyme remain unknown.

V. CENTRAL PAIR REGULATION OF FLAGELLAR MOTILITY

The mechanisms through which the central pair influences axonemal motility have apparently gone through one major evolutionary shift, since all metazoans examined to date have a central pair complex that remains fixed in its orientation relative to the surrounding doublet microtubules, whereas protists from several clades have a central pair that is twisted, and which rotates during bend propagation (Omoto et al., 1999). *Chlamydomonas* has long been recognized as an organism with a central pair complex that could rotate, based on an observed variation in central pair orientation (relative to doublets and basal bodies) in electron micrographs of fixed cells, and on rotation of the extruded tip of the central pair in disintegrating reactivated axonemes (Kamiya, 1982; Kamiya et al., 1982; see videos "Central pair rotating in demembranated *Chlamydomonas* cell model" and "Central pair rotating in *Chlamydomonas* axoneme" at http://www.elsevierdirect.com/companions/9780123708731). The entire central pair complex in *Chlamydomonas* has a helical conformation, best observed when the central pair has been released from the rest of the flagellum (Figure 8.4) (Kamiya et al., 1982; Mitchell

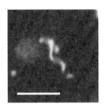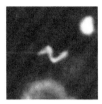

FIGURE 8.4 *Dark-field light micrographs of central pair complexes extruded from* in vitro-reactivated Chlamydomonas *axonemes. The central pair is helical when not constrained by surrounding doublet microtubules. Scale bar 5 μm.*

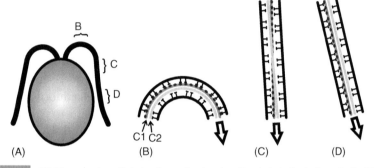

FIGURE 8.5 *(A) Diagram of a cell during forward swimming. Brackets indicate the principal bend (B), interbend (C), and reverse bend (D) regions illustrated in the diagrams to the right. (B) Central pair orientation in principal bends. The C1 microtubule is distinguished by a row of projections (dark ovals) repeating every 32 nm. (C) The central pair twists in interbend regions. (D) Central pair orientation in the reverse bend region. Arrows point toward distal end of flagellum. Adapted from Mitchell, 2005,* Am. Inst. Phys. Conf. Proc. *555:130–136, by copyright permission of the American Institute of Physics.*

and Nakatsugawa, 2004). Constraint of this helix within the cage of doublet microtubules results in a fixed relationship between central pair orientation and flagellar bends, such that the C1 microtubule faces the outside of every bend (Figure 8.5); the central pair twists by 180° between alternating bends to maintain the bend-dependent orientation (Mitchell, 2003a). As a result, during bend propagation, the twisted central pair rotates by 360° between successive beats.

The relationship between central pair rotation and regulation of bend formation has been partially resolved through experiments using suppressor mutations. As detailed in Chapters 7 and 9, the paralyzed phenotype that results from disruption of either radial spoke or central pair assembly can be partially overcome by bypass suppressors that restore bend propagation, without restoring the disrupted structures. When one of these suppressors, $sup_{pf}1$, was combined with $pf1$, which blocks radial spoke head assembly, the resulting flagella could beat in the absence of normal interactions between radial spokes and the central pair. In these flagella, the central pair maintained its twisted conformation and its bend-dependent orientation, showing that central pair rotation is independent of spoke/central pair interactions (Mitchell and Nakatsugawa, 2004). Bend propagation must be sufficient to drive central pair rotation, as first suggested by Kamiya (1982), without the requirement of a rotational motor linked to the central pair microtubules.

To determine the exact contribution of the central pair complex to motility regulation, one can start from the premise that the central pair complex must provide regulation for an aspect of motility that is specifically and consistently missing in central pair–defective mutants. When flagella from such mutants beat in the presence of suppressor mutations, they beat with a reduced frequency and an altered waveform (Brokaw et al., 1982; Brokaw and Luck, 1985), suggesting that both of these parameters may be modified through central pair/radial spoke interactions. During forward swimming, wild-type *Chlamydomonas* flagella beat with a highly asymmetric waveform in which bends are largely restricted to a single plane (rather than being helical) and principal bends have a much greater curvature than reverse ones. The waveform remains essentially planar in suppressed central pair mutants, indicating that the central pair is not essential for this form of beat asymmetry, but principal bend curvature is reduced and reverse bend curvature increased, suggesting that the central pair may be involved in regulation of the relative contributions of dyneins on different doublets to bend curvature at the low flagellar calcium concentrations typical of forward swimming. This would be consistent with a model in which the central pair regulates the underlying curvature-dependent activation of dynein arms on different doublets and modulates bending patterns.

When isolated *Chlamydomonas* axonemes from central pair assembly mutants are reactivated, they fail to beat except under highly selective buffer or nucleotide conditions. As was the case for suppression by mutations, these conditions generally only permit bend propagation with reduced frequency, reduced asymmetry of principal and reverse bends, and a tendency to a more three-dimensional (helical) waveform (Omoto et al., 1996; Frey et al., 1997; Wakabayashi et al., 1997; Yagi and Kamiya, 2000). Although reactivation in 10 mM $MgSO_4$ produced similar beat frequencies and waveforms in *pf18* and wild-type axonemes (Yagi and Kamiya, 2000), the beat frequency of the wild-type axonemes was reduced by these conditions compared with standard conditions (and compared with living cells). The calcium-dependent switch between asymmetric and symmetric (planar) waveforms (see Chapter 13) does not depend on the central pair (Wakabayashi et al., 1997), and thus must not be a function of the calmodulin complex associated with PF6.

Studies of doublet sliding in protease-treated axonemes have clearly shown a relationship between central pair/radial spoke interactions and regulation of I1 inner row dynein, as described in detail in Chapter 9. The central pair–specific aspects of that regulation are worth emphasizing in this chapter. Doublet sliding rates at low calcium concentrations are reduced in both central pair and radial spoke mutants compared to wild-type axonemes, and this reduced sliding rate is seen in *pf16* axonemes, which lack only C1, as well as in mutants such as *pf15* and *pf18* that lack the entire central pair, but not in *pf6* axonemes, which are only missing the 1a projection (Smith, 2002b). Doublet sliding rates in *pf16* and *pf18* axonemes return to wild-type levels when I1 dynein assembly is blocked (*ida1*), or when the serine/threonine protein kinase CK1 is inhibited. At high calcium concentrations, sliding rates are wild type in central pairless axonemes (e.g., *pf15, pf18*), but are reduced in C1-defective *pf16* axonemes (Smith, 2002a). This result suggests that the *pf16* central pair modulates radial spokes so that spoke-dependent calcium activation, seen in the absence of the central pair, is prevented. Because the pattern of doublets that slide in protease-treated axonemes is related to central pair orientation (Wargo and Smith, 2003) and calcium concentration (Wargo et al., 2004), these results can be interpreted to show that different central pair projections send operationally unique signals through radial spokes, and that modulation of dynein activity patterns is dependent on central pair orientation.

The mechanism of signal transduction between the central pair and radial spokes remains unclear. In metazoans, which lack central pair rotation, radial spokes tilt in the plane of the bend in response to bend-induced displacement of doublets relative to the central pair (Warner and Satir, 1974). The spoke tilt shows that spoke heads physically interact with central pair projections during bend formation. As each bend grows, tilt increases and reaches a maximum, after which spoke heads must translocate along the central pair without further tilt. Axonemal geometry requires that for a planar bend, the doublets in the bend plane (numbers 1 and 5–6) show little or no sliding displacement relative to their neighboring doublets, and therefore the dyneins on these doublets should contribute little or nothing to bend formation. Therefore, spokes tilt maximally along doublets that have largely inactive dyneins, which should require little regulation, and minimally along doublets with highly active dyneins, whose regulation would modify beat asymmetry (waveform). How then do spoke/central pair interactions regulate motility? Spoke tilt could function as a mechanism for the central pair/radial spoke system to locally sense both the direction and the extent of curvature. In this way, signals transmitted by spokes to dyneins could be modulated locally by the extent of tilt (as well as globally by physiological signal modulations such as changes in intraflagellar calcium concentration), so that dynein regulation would always be influenced by both the direction and the size of the bend. Whether similar longitudinal spoke tilt occurs in *Chlamydomonas*, or spoke tilt includes a radial displacement due to central pair rotation, has not been established.

VI. FUTURE DIRECTIONS

At this time the central pair field has at least as many questions as answers, but the many advances from work with *Chlamydomonas* show the continuing power of this organism for unraveling the complexities of flagellar mechanics and motility regulation. Especially needed is a more complete proteomic analysis coupled with a more precise three-dimensional structure of the central pair. The completion of a radial spoke proteome (Yang et al., 2006) and the discovery of methods to purify intact radial spokes (Yang et al., 2001) provide hope that some aspects of radial spoke/central pair interactions may yield to *in vitro* rebinding and biochemical approaches. Selection of additional mutants or depletion of proteins through RNAi technology will also be needed to clarify the *in vivo* role of specific central pair proteins in regulatory pathways. Finally, further comparative studies will ultimately be essential for understanding the evolution of this complex structure and its central importance in flagellar motility.

REFERENCES

Adams, G.M.W., Huang, B., Piperno, G., and Luck, D.J.L. (1981). Central-pair microtubular complex of *Chlamydomonas* flagella: polypeptide composition as revealed by analysis of mutants. *J. Cell Biol.* **91**, 69–76.

Bernstein, M. (1995). Flagellar kinesins: new moves with an old beat. *Cell Motil. Cytoskeleton* **32**, 123–128.

Bernstein, M., Beech, P.L., Katz, S.G., and Rosenbaum, J.L. (1994). A new kinesin-like protein (Klp1) localized to a single microtubule of the *Chlamydomonas* flagellum. *J. Cell Biol.* **125**, 1313–1326.

Bloch, M.A. and Johnson, K.A. (1995). Identification of a molecular chaperone in the eukaryotic flagellum and its localization to the site of microtubule assembly. *J. Cell Sci.* **108**, 3541–3545.

Branche, C., Kohl, L., Toutirais, G., Buisson, J., Cosson, J., and Bastin, P. (2006). Conserved and specific functions of axoneme components in trypanosome motility. *J. Cell Sci.* **119**, 3443–3455.

Broadhead, R., Dawe, H.R., Farr, H., Griffiths, S., Hart, S.R., Portman, N., Shaw, M.K., Ginger, M.L., Gaskell, S.J., McKean, P.G., and Gull, K. (2006). Flagellar motility is required for the viability of the bloodstream trypanosome. *Nature* **440**, 224–227.

Brokaw, C.J. and Luck, D.J.L. (1985). Bending patterns of *Chlamydomonas* flagella: III. A radial spoke head deficient mutant and a central pair deficient mutant. *Cell Motil.* **5**, 195–208.

Brokaw, C.J., Luck, D.J.L., and Huang, B. (1982). Analysis of the movement of *Chlamydomonas* flagella: the function of the radial-spoke system is revealed by comparison of wild-type and mutant flagella. *J. Cell Biol.* **92**, 722–732.

Bu, W. and Su, L.K. (2003). Characterization of functional domains of human EB1 family proteins. *J. Biol. Chem.* **278**, 49721–49731.

Chan, S.W., Fowler, K.J., Choo, K.H., and Kalitsis, P. (2005). Spef1, a conserved novel testis protein found in mouse sperm flagella. *Gene* **353**, 189–199.

Davy, B.E. and Robinson, M.L. (2003). Congenital hydrocephalus in hy3 mice is caused by a frameshift mutation in Hydin, a large novel gene. *Hum. Mol. Genet.* **12**, 1163–1170.

Dentler, W.L. and Rosenbaum, J.L. (1977). Flagellar elongation and shortening in *Chlamydomonas*. *J. Cell Biol.* **74**, 747–759.

Dougherty, G.W., Adler, H.J., Rzadzinska, A., Gimona, M., Tomita, Y., Lattig, M.C., Merritt, R.C., Jr., and Kachar, B. (2005). CLAMP, a novel microtubule-associated protein with EB-type calponin homology. *Cell Motil. Cytoskeleton* **62**, 141–156.

Dutcher, S.K., Huang, B., and Luck, D.J.L. (1984). Genetic dissection of the central pair microtubules of the flagella of *Chlamydomonas reinhardtii*. *J. Cell Biol.* **98**, 229–236.

Dymek, E.E., Lefebvre, P.A., and Smith, E.F. (2004). PF15p is the *Chlamydomonas* homologue of the Katanin p80 subunit and is required for assembly of flagellar central microtubules. *Eukaryotic Cell* **3**, 870–879.

Ebersold, W.T., Levine, R.P., Levine, E.E., and Olmsted, M.A. (1962). Linkage maps in *Chlamydomonas reinhardi*. *Genetics* **47**, 531–543.

Fox, L.A., Sawin, K.E., and Sale, W.S. (1994). Kinesin-related proteins in eukaryotic flagella. *J. Cell Sci.* **107**, 1545–1550.

Frey, E., Brokaw, C.J., and Omoto, C.K. (1997). Reactivation at low ATP distinguishes among classes of *paralyzed flagella* mutants. *Cell Motil. Cytoskeleton* **38**, 91–99.

Gadelha, C., Wickstead, B., McKean, P.G., and Gull, K. (2006). Basal body and flagellum mutants reveal a rotational constraint of the central pair microtubules in the axonemes of trypanosomes. *J. Cell Sci.* **119**, 2405–2413.

Gaillard, A.R., Diener, D.R., Rosenbaum, J.L., and Sale, W.S. (2001). Flagellar radial spoke protein 3 is an A-kinase anchoring protein (AKAP). *J. Cell Biol.* **153**, 443–448.

Goodenough, U.W. and Heuser, J.E. (1985). Substructure of inner dynein arms, radial spokes, and the central pair/projection complex. *J. Cell Biol.* **100**, 2008–2018.

Hopkins, J.M. (1970). Subsidiary components of the flagella of *Chlamydomonas reinhardii*. *J. Cell Sci.* **7**, 823–839.

Jarvik, J.W. and Suhan, J.P. (1991). The role of the flagellar transition region: inferences from the analysis of a *Chlamydomonas* mutant with defective transition region structures. *J. Cell Sci.* **99**, 731–740.

Johnson, K.A. and Rosenbaum, J.L. (1992). Polarity of flagellar assembly in *Chlamydomonas*. *J. Cell Biol.* **119**, 1605–1611.

Johnson, K.A., Haas, M.A., and Rosenbaum, J.L. (1994). Localization of a kinesin-related protein to the central pair apparatus of the *Chlamydomonas reinhardtii* flagellum. *J. Cell Sci.* **107**, 1551–1556.

Kamiya, R. (1982). Extrusion and rotation of the central-pair microtubules in detergent-treated *Chlamydomonas* flagella. *Cell Motil. [Suppl.]* **1**, 169–173.

Kamiya, R., Nagai, R., and Nakamura, S. (1982). Rotation of the central-pair microtubules in *Chlamydomonas* flagella. In: *Biological Functions of Microtubules and Related Structures* (H. Sakai, H. Nohri, and G.G. Borisy, Eds.), pp. 189–198. Academic Press, New York.

Lechtreck, K.F. and Witman, G.B. (2007). *Chlamydomonas reinhardtii* hydin is a central pair protein required for flagellar motility. *J. Cell Biol.* **176**, 473–482.

McKean, P.G., Baines, A., Vaughan, S., and Gull, K. (2003). Gamma-tubulin functions in the nucleation of a discrete subset of microtubules in the eukaryotic flagellum. *Curr. Biol.* **13**, 598–602.

Miki, H., Setou, M., and Hirokawa, N. (2003). Kinesin superfamily proteins (KIFs) in the mouse transcriptome. *Genome Res.* **13**, 1455–1465.

Miki, H., Okada, Y., and Hirokawa, N. (2005). Analysis of the kinesin superfamily: insights into structure and function. *Trends Cell Biol.* **15**, 467–476.

Mitchell, B.F., Pedersen, L.B., Feely, M., Rosenbaum, J.L., and Mitchell, D.R. (2005). ATP production in *Chlamydomonas reinhardtii* flagella by glycolytic enzymes. *Mol. Biol. Cell* **16**, 4509–4518.

Mitchell, D.R. (2003a). Orientation of the central pair complex during flagellar bend formation in *Chlamydomonas*. *Cell Motil. Cytoskeleton* **56**, 120–129.

Mitchell, D.R. (2003b). Reconstruction of the projection periodicity and surface architecture of the flagellar central pair complex. *Cell Motil. Cytoskeleton* **55**, 188–199.

Mitchell, D.R. (2004). Speculations on the evolution of 9 + 2 organelles and the role of central pair microtubules. *Biol. Cell* **96**, 691–696.

Mitchell, D.R. (2005). Regulation of eukaryotic flagellar motility. *Am. Inst. Phys. Conf. Proc.* **555**, 130–136.

Mitchell, D.R. and Nakatsugawa, M. (2004). Bend propagation drives central pair rotation in *Chlamydomonas reinhardtii* flagella. *J. Cell Biol.* **166**, 709–715.

Mitchell, D.R. and Sale, W.S. (1999). Characterization of a *Chlamydomonas* insertional mutant that disrupts flagellar central pair microtubule-associated structures. *J. Cell Biol.* **144**, 293–304.

Nakazawa, Y., Hiraki, M., Kamiya, R., and Hirono, M. (2007). SAS-6 is a cartwheel protein that establishes the 9-fold symmetry of the centriole. *Curr. Biol.* **17**, 2169–2174.

Nielsen, M.G., Turner, F.R., Hutchens, J.A., and Raff, E.C. (2001). Axoneme-specific beta-tubulin specialization: a conserved C-terminal motif specifies the central pair. *Curr. Biol.* **11**, 529–533.

Omoto, C.K., Yagi, T., Kurimoto, E., and Kamiya, R. (1996). Ability of paralyzed flagella mutants of *Chlamydomonas* to move. *Cell Motil. Cytoskeleton* **33**, 88–94.

Omoto, C.K., Gibbons, I.R., Kamiya, R., Shingyoji, C., Takahashi, K., and Witman, G.B. (1999). Rotation of the central pair microtubules in eukaryotic flagella. *Mol. Biol. Cell* **10**, 1–4.

Ostrowski, L.E., Blackburn, K., Radde, K.M., Moyer, M.B., Schlatzer, D.M., Moseley, A., and Boucher, R.C. (2002). A proteomic analysis of human cilia: identification of novel components. *Mol. Cell Proteomics* **1**, 451–465.

Ponting, C.P. (2006). A novel domain suggests a ciliary function for ASPM, a brain size determining gene. *Bioinformatics* **22**, 1031–1035.

Porter, M.E., Power, J., and Dutcher, S.K. (1992). Extragenic suppressors of paralyzed flagellar mutations in *Chlamydomonas reinhardtii* identify loci that alter the inner dynein arms. *J. Cell Biol.* **118**, 1163–1176.

Richardson, D.N., Simmons, M.P., and Reddy, A.S. (2006). Comprehensive comparative analysis of kinesins in photosynthetic eukaryotes. *B.M.C. Genomics* **7**, 18.

Ringo, D.L. (1967). Flagellar motion and fine structure of the flagellar apparatus in *Chlamydomonas*. *J. Cell Biol.* **33**, 543–571.

Rosenbaum, J.L., Moulder, J.E., and Ringo, D.L. (1969). Flagellar elongation and shortening in *Chlamydomonas*: The use of cycloheximide and colchicine to study the synthesis and assembly of flagellar proteins. *J. Cell Biol.* **41**, 600–619.

Rupp, G., O'Toole, E., and Porter, M.E. (2001). The *Chlamydomonas PF6* locus encodes a large alanine/proline-rich polypeptide that is required for assembly of a central pair projection and regulates flagellar motility. *Mol. Biol. Cell* **12**, 739–751.

Sapiro, R., Kostetskii, I., Olds-Clarke, P., Gerton, G.L., Radice, G.L., and Strauss, I.I.I. (2002). Male infertility, impaired sperm motility, and hydrocephalus in mice deficient in sperm-associated antigen 6. *Mol. Cell Biol.* **22**, 6298–6305.

Silflow, C.D., Liu, B., LaVoie, M., Richardson, E.A., and Palevitz, B.A. (1999). Gamma-tubulin in *Chlamydomonas*: characterization of the gene and localization of the gene product in cells. *Cell Motil. Cytoskeleton* **42**, 285–297.

Sironen, A., Thomsen, B., Andersson, M., Ahola, V., and Vilkki, J. (2006). An intronic insertion in KPL2 results in aberrant splicing and causes the immotile short-tail sperm defect in the pig. *Proc. Natl. Acad. Sci. U.S.A.* **103**, 5006–5011.

Smith, E.F. (2002a). Regulation of flagellar dynein by calcium and a role for an axonemal calmodulin and calmodulin-dependent kinase. *Mol. Biol. Cell* **13**, 3303–3313.

Smith, E.F. (2002b). Regulation of flagellar dynein by the axonemal central apparatus. *Cell Motil. Cytoskeleton* **52**, 33–42.

Smith, E.F. and Lefebvre, P.A. (1996). *PF16* encodes a protein with armadillo repeats and localizes to a single microtubule of the central apparatus in *Chlamydomonas* flagella. *J. Cell Biol.* **132**, 359–370.

Smith, E.F. and Lefebvre, P.A. (1997a). *PF20* gene product contains WD repeats and localizes to the intermicrotubule bridges in *Chlamydomonas* flagella. *Mol. Biol. Cell* **8**, 455–467.

Smith, E.F. and Lefebvre, P.A. (1997b). The role of central apparatus components in flagellar motility and microtubule assembly. *Cell Motil. Cytoskeleton* **38**, 1–8.

Smith, E.F. and Lefebvre, P.A. (2000). Defining functional domains within PF16: a central apparatus component required for flagellar motility. *Cell Motil. Cytoskeleton* **46**, 157–165.

Smith, E.F. and Yang, P. (2004). The radial spokes and central apparatus: mechanochemical transducers that regulate flagellar motility. *Cell Motil. Cytoskeleton* **57**, 8–17.

Starling, D. and Randall, J. (1971). The flagella of temporary dikaryons of *Chlamydomonas reinhardii*. *Genet. Res. Camb.* **18**, 107–113.

Thazhath, R., Jerka-Dziadosz, M., Duan, J., Wloga, D., Gorovsky, M.A., Frankel, J., and Gaertig, J. (2004). Cell context-specific effects of the beta-tubulin glycylation domain on assembly and size of microtubular organelles. *Mol. Biol. Cell* **15**, 4136–4147.

Wakabayashi, K., Yagi, T., and Kamiya, R. (1997). Ca^{2+}-dependent waveform conversion in the flagellar axoneme of *Chlamydomonas* mutants lacking the central-pair radial spoke system. *Cell Motil. Cytoskeleton* **38**, 22–28.

Wargo, M.J. and Smith, E.F. (2003). Asymmetry of the central apparatus defines the location of active microtubule sliding in *Chlamydomonas* flagella. *Proc. Natl. Acad. Sci. U.S.A.* **100**, 137–142.

Wargo, M.J., McPeek, M.A., and Smith, E.F. (2004). Analysis of microtubule sliding patterns in *Chlamydomonas* flagellar axonemes reveals dynein activity on specific doublet microtubules. *J. Cell Sci.* **117**, 2533–2544.

Wargo, M.J., Dymek, E.E., and Smith, E.F. (2005). Calmodulin and PF6 are components of a complex that localizes to the C1 microtubule of the flagellar central apparatus. *J. Cell Sci.* **118**, 4655–4665.

Warner, F.D. and Satir, P. (1974). The structural basis of ciliary bend formation. Radial spoke positional changes accompanying microtubule sliding. *J. Cell Biol.* **63**, 35–63.

Warr, J.R., McVittie, A., Randall, J., and Hopkins, J.M. (1966). Genetic control of flagellar structure in *Chlamydomonas reinhardii*. *Genet. Res. Camb.* **7**, 335–351.

Wickstead, B. and Gull, K. (2006). A "holistic" kinesin phylogeny reveals new kinesin families and predicts protein functions. *Mol. Biol. Cell* **17**, 1734–1743.

Witman, G.B., Plummer, J., and Sander, G. (1978). *Chlamydomonas* flagellar mutants lacking radial spokes and central tubules. *J. Cell Biol.* **76**, 729–747.

Yagi, T. and Kamiya, R. (2000). Vigorous beating of *Chlamydomonas* axonemes lacking central pair/radial spoke structures in the presence of salts and organic compounds. *Cell Motil. Cytoskeleton* **46**, 190–199.

Yang, P., Fox, L., Colbran, R.J., and Sale, W.S. (2000). Protein phosphatases PP1 and PP2A are located in distinct positions in the *Chlamydomonas* flagellar axoneme. *J. Cell Sci.* **113**, 91–102.

Yang, P., Diener, D.R., Rosenbaum, J.L., and Sale, W.S. (2001). Localization of calmodulin and dynein light chain LC8 in flagellar radial spokes. *J. Cell Biol.* **153**, 1315–1326.

Yang, P., Diener, D.R., Yang, C., Kohno, T., Pazour, G.J., Dienes, J.M., Agrin, N.S., King, S.M., Sale, W.S., Kamiya, R., Rosenbaum, J.L., and Witman, G.B. (2006). Radial spoke proteins of *Chlamydomonas* flagella. *J. Cell Sci.* **119**, 1165–1174.

Yokoyama, R., O'Toole, E., Ghosh, S., and Mitchell, D.R. (2004). Regulation of flagellar dynein by a central pair kinesin. *Proc. Natl. Acad. Sci. U.S.A.* **101**, 17398–17403.

Zhang, H. and Mitchell, D.R. (2004). Cpc1, a *Chlamydomonas* central pair protein with an adenylate kinase domain. *J. Cell Sci.* **117**, 4179–4188.

Zhang, Z., Sapiro, R., Kapfhamer, D., Bucan, M., Bray, J., Chennathukuzhi, V., McNamara, P., Curtis, A., Zhang, M., Blanchette-Mackie, E.J., and Strauss, J.F., III (2002). A sperm-associated WD repeat protein orthologous to *Chlamydomonas* PF20 associates with Spag6, the mammalian orthologue of *Chlamydomonas* PF16. *Mol. Cell Biol.* **22**, 7993–8004.

Zhang, Z., Kostetskii, I., Moss, S.B., Jones, B.H., Ho, C., Wang, H.B., Kishida, T., Gerton, G.L., Radice, G.L., and Strauss, J.F., III (2004). Haploinsufficiency for the murine orthologue of *Chlamydomonas* PF20 disrupts spermatogenesis. *Proc. Natl. Acad. Sci. U.S.A.* **101**, 12946–12951.

Zhang, Z., Jones, B.H., Tang, W., Moss, S.B., Wei, Z., Ho, C., Pollack, M., Horowitz, E., Bennett, J., Baker, M.E., and Strauss, J.F., III (2005). Dissecting the axoneme interactome: the mammalian orthologue of *Chlamydomonas* PF6 interacts with sperm-associated antigen 6, the mammalian orthologue of *Chlamydomonas* PF16. *Mol. Cell Proteomics* **4**, 914–923.

Zhang, Z., Kostetskii, I., Tang, W., Haig-Ladewig, L., Sapiro, R., Wei, Z., Patel, A.M., Bennett, J., Gerton, G.L., Moss, S.B., Radice, G.L., and Strauss, J.F., III (2006). Deficiency of SPAG16L causes male infertility associated with impaired sperm motility. *Biol. Reprod.* **74**, 751–759.

CHAPTER 9

The Regulation of Axonemal Bending

Maureen Wirschell,[1] Daniela Nicastro,[2] Mary E. Porter[3] and Winfield S. Sale[1]

[1]Department of Cell Biology, Emory University School of Medicine, Atlanta, Georgia, USA
[2]Department of Biology, Brandeis University, Waltham, Massachusetts, USA
[3]Department of Genetics, Cell Biology and Development, University of Minnesota, Minneapolis, Minnesota, USA

CHAPTER CONTENTS

I. INTRODUCTION

This chapter reviews an emerging view for how ciliary and flagellar motility is regulated. The mechanisms are complex and are only just beginning to be understood, but include a physically integrated network of axonemal structures and signaling proteins required for regulation of the dynein motors. The axonemal components include the "dynein regulatory complex" or DRC (Piperno et al., 1992), linkers connecting the dynein arms to each other and the DRC (Goodenough and Heuser, 1982, 1984; Nicastro et al., 2006), the nexin links (Goodenough and Heuser, 1985b; Woolley, 1997), and a group of kinases and phosphatases localized to the axonemal microtubules (Figure 9.1; see Porter and Sale, 2000; Smith and Yang, 2004). Genetic, structural, and biochemical studies of *Chlamydomonas* have demonstrated that these conserved structural linkers, the DRC, and the axonemal signaling proteins play a central role in the regulatory pathway that integrates the function of the central pair apparatus and the radial spokes with the activity of the dynein arms. The topics of this chapter are complemented by Chapters 6–8 on the dyneins, radial spokes, and central pair apparatus, respectively.

Although we have a good understanding of the axonemal dyneins and their role in powering microtubule sliding, we do not know how dynein

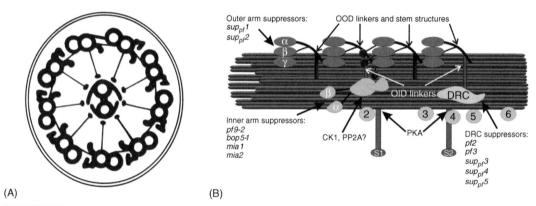

(A)　　　　　　　　　　　　　　(B)

FIGURE 9.1 *The* Chlamydomonas *flagellar axoneme. (A) Schematic diagram of the flagellar 9 + 2 axoneme in cross-section, showing the inner and outer dynein arms attached to the A-tubules of the outer doublet microtubules, and the radial spokes extending toward the center of the axoneme, where they interact with projections from the central pair of microtubules (from Pazour et al., 2005, with modification.) (B) Arrangement of the inner dynein arms and other structures on the A-tubule of the outer doublet as viewed from the B-tubule of the adjacent outer doublet. The structures repeat along the length of the outer doublet microtubule with a 96-nm periodicity, referred to as the 96-nm repeat; the structures in a single repeat are shown here. The proximal end (adjacent to the cell body) is to the left. Inner and outer dynein arms are multisubunit ATPases that generate relative sliding movements between outer doublet microtubules. Interdoublet sliding is normally constrained by the interdoublet linkages and attachment to the basal body, leading to flagellar bending. The outer arms (blue) are on the top and the radial spokes (S1 and S2, dark green) are on the bottom. The I1 dynein is the trilobed structure (light green) proximal to the first radial spoke (S1) in each repeat; the positions of the single-headed inner arm dyneins are indicated (2–6). The crescent-shaped structure above the second radial spoke (S2) is the DRC. The postulated locations of regulatory signaling molecules (PKA, CK1, and PP2A) are indicated. Chlamydomonas suppressor mutations that affect specific axonemal structures (outer arms, inner arm I1, and the DRC) also are indicated. The OOD linkers are labeled, and the orange rods represent the OID linkers (connecting the outer dynein arms to I1 and the DRC) that have been identified by cryoelectron microscopy (Nicastro et al., 2006; Ishikawa et al., 2007).*

activity is locally regulated such that microtubule sliding is converted into the bending motion that characterizes ciliary and flagellar motility. The basic oscillatory bending of the axoneme likely involves a mechanical feedback mechanism that locally activates, or inhibits, the dynein motors (Satir, 1980; Brokaw and Kamiya, 1987; Shingyoji et al., 1998; Lindemann, 2003). Specifically, since dyneins are minus-end motors (Sale and Satir, 1977; Fox and Sale, 1987), and all microtubules in the 9 + 2 axoneme are uniformly oriented, the generation and propagation of alternating bends must involve switching of dynein activity across the axis and along the length of the axoneme (Satir, 1984, 1985; see also Lechtreck and Witman, 2007). Little is known about the mechanisms responsible for oscillatory bending, but structural studies have revealed linkers between the dynein arms and details about organization of the interdoublet nexin links that may be critical for mechanical feedback control and coordination of axonemal dynein (Nicastro et al., 2006; see Figures 9.1 and 9.2). Further advances in understanding how microtubule sliding is converted to oscillatory bending will require more information about the biochemical composition of these dynein-associated linkers as well as a better understanding of the mechanical/physical basis for dynein motor activity (Burgess et al., 2003; Oiwa and Sakakibara, 2005; Koonce, 2006).

Additional regulation is required for the behavioral responses exhibited by cilia and flagella. Cilia and flagella from various sources can dramatically alter the pattern, shape, and size of the bending motion. For example, *Chlamydomonas* can precisely control the waveform of each flagellum for accurate control of cell turning and phototaxis or chemotaxis, functions common to most cells with motile cilia or flagella (Witman, 1993; also see Chapter 13). Alteration in the size, shape, or direction of propagation of the flagellar bend involves changes in the local regulation of dynein-driven microtubule sliding – the same process that is responsible for the basic oscillatory mechanism described above. Based on diverse studies in *Chlamydomonas*, modulation of axonemal bending involves the central pair apparatus, the radial spokes, and the structures discussed in this chapter, including the DRC and the axonemal kinases and phosphatases. The DRC and signaling proteins are thought to regulate dynein-driven microtubule sliding by coupling the central pair apparatus/radial spoke mechanism to the local control of dynein activity.

Diverse studies in *Chlamydomonas* have implicated the central pair and radial spoke structures as key regulators of dynein activity. The radial spokes are located in close proximity to the inner arms (Figure 9.1), and in some species, including *Chlamydomonas*, the central pair rotates during the beat cycle (Omoto et al., 1999; Smith and Yang, 2004). The central pair microtubules are structurally asymmetric and biochemically distinct (see Chapter 8). One model is that the central pair projections function like a distributor to send a signal to a specific subset of radial spokes for selective activation of dynein arms (Omoto et al., 1999; Smith and Yang, 2004). Importantly, *Chlamydomonas* flagellar mutants lacking central pair/radial spoke structures are paralyzed under physiological conditions (Witman et al., 1978).

The coupling of the central pair/radial spoke system to regulation of dynein also involves control of protein phosphorylation (Porter and Sale, 2000; Smith and Yang, 2004; Salathe, 2006). The kinases and phosphatases that mediate control of phosphorylation are physically anchored within the axoneme (Howard et al., 1994; Habermacher and Sale, 1995, 1996; Yang et al., 2000; Yang and Sale, 2000; Gaillard et al., 2001, 2006). Continued progress on the identification of axonemal signaling molecules has been possible due to functional and biochemical studies of isolated axonemes from wild-type and mutant *Chlamydomonas* cells and development of the flagellar proteome (Pazour et al., 2005). These signaling proteins are localized within the axoneme by anchoring proteins that target and dock the kinases and phosphatases in position to control the dyneins (Gaillard et al., 2001, 2006). The axonemal kinases and phosphatases, together with the DRC, are thought to locally modulate dynein and alter flagellar bending to achieve cell turning during phototaxis and chemotaxis.

This chapter primarily focuses on the regulatory mechanisms, signalling molecules, and the discovery and structural organization of axonemal components, revealed in studies of *Chlamydomonas*, which control axonemal bending. We also review the use of cryoelectron microscopy for elucidating details of axonemal structure, with particular emphasis on how structural components of the axoneme may contribute to control of flagellar motility.

II. SUPPRESSOR MUTANTS AND THE DRC

A. Discovery of the DRC

The first evidence for a control system that inhibits dynein activity came from the characterization of a series of suppressor mutations that restore motility (i.e., dynein activity) to paralyzed central pair/radial spoke mutants without restoring the missing structures (Huang et al., 1982). These second-site mutations permit modified motility in the central pair/radial spoke mutants, revealing a previously unrecognized dynein regulatory control system that globally inhibits dynein activity when the central pair/radial spokes are defective (Huang et al., 1982; Piperno, 1994, 1995). In the absence of the radial spokes or central pair structures, this inhibitory system is active, and dyneins are inactive. When the system is mutated, it disrupts the inhibition and allows for dynein activity. Many of these mutations reside in subunits of the dynein arms, or in components of the DRC. For the purpose of this discussion, it is useful to categorize the suppressors into groups that include: (a) suppressor mutations that reside within the dynein

Table 9.1	Paralyzed-flagella suppressor mutants			
Mutant	**Gene product**	**Phenotype**	**Suppresses**	**References**
$sup_{pf}1$	ODA4 (β DHC)	reduced flagellar beat frequency	CP/RS	Huang et al. (1982), Porter et al. (1994)
$sup_{pf}2$	ODA2 (γ DHC)	reduced flagellar beat frequency	CP/RS	Huang et al. (1982), Rupp et al. (1996)
pf2	DRC4	slow swimming, abnormal waveform	CP/RS[a]	Huang et al. (1982), Piperno et al. (1994), Rupp and Porter (2003)
pf3	DRC1	slow swimming, abnormal waveform	CP/RS[a]	Huang et al. (1982), Gardner et al. (1994), Piperno et al. (1994)
pf9	1α HC (f dynein)	slow swimming, abnormal waveform	CP	Porter et al. (1992)
$sup_{pf}3$	[b]	wild-type motility	CP/RS[a]	Huang et al. (1982), Gardner et al. (1994), Piperno et al. (1994)
$sup_{pf}4$	DRC5	wild-type motility	CP/RS[a]	Huang et al. (1982), Gardner et al. (1994), Piperno et al. (1994)
$sup_{pf}5$	?	slow swimming, abnormal waveform	CP/RS	Piperno et al. (1992), Piperno et al. (1994)
bop2-1	?		pf10	Dutcher et al. (1988), King et al. (1994)
bop5-1	IC138		pf10	Dutcher et al. (1988), Hendrickson et al. (2004)

Note: CP, central pair; RS, radial spokes.

[a]*Huang et al. (1982) originally found these four mutants suppressed only radial spoke defects. Piperno et al. (1994) suggests these four mutants also suppress the central pair motility mutants pf15 and pf18. This discrepancy is likely due to the criteria used to score suppression. Piperno et al. (1994) describe DRC-central pair double mutants that show regular flagellar beating. However, the flagellar movement is not sufficient to generate forward swimming.*

[b]*Piperno et al. (1994) suggest that sup_{pf}3 is allelic to pf2.*

heavy chains (HCs) ($sup_{pf}1$, $sup_{pf}2$, and pf9-2); (b) extragenic mutations, including those that affect dynein intermediate chains (ICs), that suppress paralysis in the *pf10* mutant (*bop2* and *bop5*); and (c) suppressor mutations that reside in and define the DRC (*pf2*, *pf3*, $sup_{pf}3$, $sup_{pf}4$, $sup_{pf}5$). These mutants are listed in Tables 9.1 and 9.2 and will be discussed in turn.

Table 9.2 DRC components, mutants, phenotypes, and biochemical defects

Protein	Accession	Mr[a]	Missing in:	Phenotype	Mutant	Biochemical defect	References
DRC1		83/108	pf3, sup$_{pf}$5	abnormal waveform	pf3	missing DRC1, 2, 5, and 6; missing inner-arm dynein e; missing RSP13; reduced tektin levels	Gardner et al. (1994), Piperno et al. (1994), Piperno (1995), Yanagisawa and Kamiya (2004)
DRC2		70/83	pf3, sup$_{pf}$5	?	?	Not determined	Piperno et al. (1994)
DRC3		62/65	pf2, sup$_{pf}$3	?	?	Not determined	Piperno et al. (1994)
DRC4	AAP57169	55/60	pf2, sup$_{pf}$3	abnormal waveform	pf2, sup$_{pf}$3	missing DRC3, 4, 5, 6, and 7; reduced inner-arm dyneins e and b	Gardner et al. (1994), Piperno et al. (1994), Rupp and Porter (2003)
DRC5		40/40	pf2, pf3, sup$_{pf}$3, sup$_{pf}$4, sup$_{pf}$5	wild-type motility	sup$_{pf}$4	missing DRC5 and 6	Huang et al. (1982), Piperno et al. (1994)
DRC6		29/29	pf2, pf3, sup$_{pf}$3, sup$_{pf}$4, sup$_{pf}$5	?	?	Not determined	Piperno et al. (1994)
DRC7		192	pf2	?	?	Not determined	Piperno et al. (1994)
?		?	?	slow swimming	sup$_{pf}$5	missing DRC1, 2, 5, and 6; missing RSP13; missing inner-arm dynein e	Piperno et al. (1992, 1994)
?		?	?	slow swimming	ida6	missing inner-arm dynein e	Yanagisawa and Kamiya (2004)

[a]Published apparent molecular weights vary (Huang et al., 1982; Piperno et al., 1992, 1994).

B. The dynein suppressors

Table 9.1 lists the suppressor mutants and their known gene products that either reside in the dyneins, or were isolated by reversion analysis of *pf10* (see next subsection). Included are mutations in both the outer arm dyneins and the inner arm dynein I1. Among the best studied suppressors are *sup$_{pf}$1* and *sup$_{pf}$2*, mutations that alter the activity of the outer arm dynein HCs (Brokaw et al., 1982; Huang et al., 1982; Porter et al., 1994; Rupp et al., 1996). The mutants *sup$_{pf}$1* and *sup$_{pf}$2* affect the β HC and γ HC of the outer arm, respectively. Although several *sup$_{pf}$1* and *sup$_{pf}$2* alleles have been isolated, the mutant lesions have only been identified in two *sup$_{pf}$1* alleles. Both are in-frame deletions in the coiled-coil domains of the microtubule-binding stalk of the β HC (Porter et al., 1994; Rupp et al., 1996; also see Chapter 6). Interestingly, all of the *sup$_{pf}$1* and *sup$_{pf}$2* mutations appear to reduce the activity and/or assembly of the outer dynein arms *in situ*. The detailed mechanism of suppression is unknown, but may involve changes in the regulatory sites that otherwise inhibit dynein activity at physiological levels of ATP (Rupp et al., 1996). These mutations imply that there are physical linkages that couple the radial spoke structures and the outer arm dyneins. Such linkages could include the DRC and/or inner arm dyneins together with a series of inter-dynein linkers (Nicastro et al., 2006) described in detail in section III.

The inner arm dynein I1 (also known as the f dynein) appears to play an important role in control of microtubule sliding by receiving signals that originate from the central pair and radial spoke structures (Porter and Sale, 2000; Smith and Yang, 2004; Wirschell et al., 2007). These conclusions are founded on both genetic analysis of I1 dynein mutants (Porter et al., 1992) and *in vitro* functional studies using isolated axonemes (Habermacher and Sale, 1997; King and Dutcher, 1997; Yang and Sale, 2000). The first evidence that the I1 dynein is a critical component of the central pair/radial spoke signal transduction pathway was the observation that a mutation (*pf9-2*) that disrupts the assembly of the I1 dynein also suppresses paralysis of the temperature-sensitive, central pair motility mutant *pf16BR3* (Porter et al., 1992). Later work demonstrated that the *pf9/ida1* mutations fall within the gene encoding the 1α HC subunit of the I1 dynein (Myster et al., 1997, 1999). Mutations in I1 dynein also result in loss of control of flagellar waveform (Brokaw and Kamiya, 1987) and phototaxis (King and Dutcher, 1997; Okita et al., 2005), further indicating a central role for I1 in control of microtubule sliding and axonemal waveform.

C. Mutations in the I1 intermediate chain IC138

Mutations (*mia1, mia2,* and *bop5-1*) that affect another I1 component, IC138, were identified in screens for phototaxis mutants or recovered as suppressors of paralyzed flagellar mutant *pf10* (Dutcher et al., 1988; King and Dutcher, 1997; Hendrickson et al., 2004). These studies revealed that changes in phosphorylation of the IC138 subunit are linked to the central pair/radial spoke signal transduction mechanism (Habermacher and Sale, 1997; King and Dutcher, 1997; Yang and Sale, 2000; Hendrickson et al., 2004). For example, *in vitro* studies reveal that increased phosphorylation of IC138 correlates with decreased dynein-driven microtubule sliding velocity (inactive dynein), while dephosphorylation of IC138 correlates with increased sliding velocities (active dynein) (Habermacher and Sale, 1997; King and Dutcher, 1997; Yang and Sale, 2000; see section IV.B). Furthermore, hyperphosphorylation of IC138 in the *mia1* and *mia2* mutants results in inhibition of dynein-driven microtubule sliding and altered phototaxis phenotypes (King and Dutcher, 1997), and IC138 has been found to be hyperphosphorylated in radial spoke and central pair mutants (Hendrickson et al., 2004). These findings provide evidence of a functional link between I1 dynein activity, IC138 phosphorylation, and the central pair/radial spoke mechanism that controls microtubule sliding.

In the paralyzed flagellar mutant *pf10* (Lewin, 1954), the asymmetric flagellar waveform of wild-type cells is altered to a nearly symmetric waveform. Dutcher et al. (1988) described the isolation of a number of *pf10* suppressors, 11 of which are intragenic and 70 of which are extragenic. Extragenic suppressors are potentially useful for understanding the structural and functional links that regulate flagellar beating. An example is the *bop5-1* (bypass of paralysis) allele, cited above, one of the *pf10* suppressors identified in this screen (Dutcher et al., 1988). The *bop5-1* mutation results in a truncated IC138 missing the last WD-repeat motif (Hendrickson et al., 2004; see Chapter 6), which is required for assembly of the LC7b subunit of I1dynein. Another example is *bop2-1*, a mutation also recovered as an extragenic suppressor of *pf10* and apparently involved in control of flagellar waveform (Dutcher et al., 1988; King et al., 1994). The *bop2-1* mutation may provide evidence of radial asymmetry in inner arm dyneins that could be fundamental to the formation of distinct, asymmetric effective and recovery flagellar bending patterns (King et al., 1994).

D. DRC suppressors

Screens for suppressors of paralyzed central pair/radial spoke mutant strains have also identified several mutations (*pf2*, *pf3*, *sup$_{pf}$3*, *sup$_{pf}$4*, and *sup$_{pf}$5*, see Table 9.2) that alter the assembly of a subset of closely associated polypeptides known as the DRC (Huang et al., 1982; Piperno et al., 1992, 1994). The DRC is a complex of at least seven axonemal proteins ranging in molecular mass from ~29 to ~192 kD (Table 9.2) that are tightly associated with the outer doublets in wild-type axonemes, but are missing to varying degrees in the DRC mutants. The seven DRC proteins are predicted to form a complex with a minimal molecular mass of ~530–572 kD, approximately the size of a dynein HC (Piperno et al., 1994). Several of these mutants are also associated with defects in the assembly of inner arm dynein HCs (Gardner et al., 1994; Piperno et al., 1994), suggesting that the DRC has a role in inner dynein arm assembly. Electron microscopy has identified a crescent-shaped structure at the base of the second radial spoke in each 96-nm axoneme repeat (see section III) that is altered or missing in the most severely defective DRC mutant strains (Mastronarde et al., 1992; Gardner et al., 1994; Nicastro et al., 2006). This structure, predicted to be the DRC, appears to be ideally positioned to mediate local signals, either mechanical, chemical, or both, between the radial spokes, interdoublet linkages, and the inner and outer dynein arms (Mastronarde et al., 1992; Gardner et al., 1994; Woolley, 1997; Nicastro et al., 2006).

DRC mutants display motility phenotypes ranging from abnormal waveform and slow swimming to nearly wild-type motility, indicating that this structure modifies dynein arm activity, hence the acronym. Drawing on genetic and biochemical characterizations, the current model is that the DRC acts as a reversible inhibitor of axonemal dyneins that is regulated by signals delivered by the central pair/radial spoke mechanosensory signal transduction machinery to mediate local control of dynein activity (Piperno, 1995; Porter and Sale, 2000; Rupp and Porter, 2003).

Based on dikaryon rescue, three DRC mutations have been assigned putative gene products (*pf2*, *pf3*, and *sup$_{pf}$4*; Piperno et al., 1994). To date, only DRC4, the protein encoded by the *PF2* gene, has been characterized at the molecular level. DRC4 is a highly coiled-coil polypeptide that is tightly associated with the outer doublets (Rupp and Porter, 2003). The *pf2* mutations are null alleles that completely disrupt DRC4, and the loss of DRC4 also blocks the assembly of at least four other DRC subunits into the axoneme (DRC3, DRC5, DRC6, and DRC7). As a result, the motility of the *pf2* strain is severely impaired (Piperno et al., 1994; Rupp and Porter, 2003). Chemical cross-linking of proteins in wild-type axonemes indicate that DRC4 interacts with at least three axonemal polypeptides ranging in size from 65 to 85 kD (Rupp and Porter, 2003). The identities of the three polypeptides are unknown, but their apparent sizes are similar to those reported for three DRC subunits, DRC1, DRC2, and DRC3 (Table 9.2). These observations are consistent with the hypothesis that the DRC subunits form a discrete polypeptide complex. Comparison with other DRC mutants suggest that *pf2* is most closely related to *sup$_{pf}$3* in terms of its biochemical and structural defects, and these mutations may even be allelic (Gardner et al., 1994; Piperno et al., 1994). Further work is needed to understand the relationship between these two mutants.

Interestingly, orthologues of DRC4 are expressed in several cell types, including tissues not associated with motile cilia (Rupp and Porter, 2003; Colantonio et al., 2006). DRC4 encodes a protein showing sequence homology to trypanin in *Trypanosoma bruceii* and to the mammalian growth-arrest-specific protein known as Gas8 (mouse) or GAS11 (human) (Hill et al., 2000; Rupp and Porter, 2003). Knockdown of trypanin expression by RNAi can suppress *pf16* mutations in trypanosomes, suggesting a genetic interaction between trypanin and PF16. As PF16 is a central pair protein (see Chapter 8), this result is similar to the extragenic suppression of central pair mutations observed in *Chlamydomonas* (Piperno et al., 1994; Ralston et al., 2006).

One hypothesis, founded upon studies of *Chlamydomonas* DRC4, is that the DRC functions as a scaffold for the attachment of regulatory enzymes that modify dynein activity (Rupp and Porter, 2003). A second hypothesis is that the DRC interacts with the interdoublet linkages and/or radial spokes to sense tension or strain within the axoneme, providing a mechanical feedback to the dynein arms (Rupp and Porter, 2003). As illustrated in the next section, ultrastructural studies have revealed links between the DRC and the outer dynein arms and, importantly, have revealed details of DRC–nexin interactions (Nicastro et al., 2006). Thus, the DRC likely plays multiple functional roles, including roles as a sensor of microtubule sliding strain and as a transducer of signals from the radial spokes to the outer and inner dynein arms. Further advances in understanding the DRC will require comprehensive analyses of each DRC component, how they interact with each other and the axoneme, more detailed structural studies of axonemes in each bend position, and evaluation of whether the DRC may regulate dynein activity elsewhere in the cell.

III. THE 96-NM AXONEME REPEAT

A. Complexity of inner arm region

Both the stability and highly ordered, repetitive arrangement of the axoneme components have made the ciliary/flagellar axoneme an ideal specimen for structural analysis. Ciliary and flagellar axonemes from many different organisms have been studied by conventional electron microscopy, using thin sections, negative staining, and/or metal replicas of quick-freeze, deep-etch samples. Such studies have demonstrated the 24-nm periodicity of the outer dynein arms as well as the complex arrangements of inner dynein arms and radial spokes within a 96-nm axoneme repeat (Goodenough and Heuser, 1985a, b; Burgess et al., 1991; Mastronarde et al., 1992; Porter et al., 1992). The isolation and characterization of motility mutations in *Chlamydomonas* that disrupt the assembly of specific inner arm dynein isoforms further revealed the heterogeneity of structures within the 96-nm repeat (Figures 9.1 and 9.2; see Brokaw and Kamiya, 1987; Piperno, 1990; Kamiya, 1991). Comparison of wild-type and mutant axonemes using computed difference maps of averaged electron microscopic images permitted the identification of many inner arm structures, but also showed that there is still much to be learned about the organization of axoneme components in the inner arm region (Mastronarde et al., 1992; O'Toole et al., 1995; Porter and Sale, 2000). Details of dynein organization are included in Chapter 6; however, key features of structural organization critical to this chapter are also briefly reviewed here.

The *Chlamydomonas* genome contains 11 distinct dynein HC genes whose expression patterns suggest that they are likely to encode inner dynein arm isoforms (Porter, 1996; Porter et al., 1996); of these, at least eight inner arm HCs have been resolved biochemically by high-salt extraction and FPLC ion-exchange chromatography (Kagami and Kamiya, 1992, 1995; Yagi et al., 2005; Yagi and Kamiya, 2006). These are organized with various IC and light chain (LC) subunits into seven distinct complexes: one two-headed dynein (I1 or species f) and six single-headed isoforms (I2 and I3, or species a, b, c, d, e, and g). Structures corresponding to six of the eight dynein HC subunits have thus far been identified in the 96-nm repeat (Mastronarde et al., 1992; Nicastro et al., 2006).

The two-headed I1 dynein is composed of the 1α and 1β HCs, three ICs (IC140, IC138, IC97), and several LCs (LC8, LC7a, LC7b, tctex1, tctex2b) (Myster et al., 1997; Yang and Sale, 1998; Perrone et al., 1998, 2000; DiBella et al., 2001, 2004a, b; Hendrickson et al., 2004; Wirschell et al., 2006) and forms a trilobed structure proximal to the first radial spoke in each 96-nm repeat (Goodenough and Heuser, 1985b; Mastronarde et al., 1992). In general, mutations in *DHC1* (encoding the 1α HC), *DHC10* (encoding the 1β HC), or *ida7* (encoding IC140) prevent the assembly of the I1 dynein into the flagellar axoneme (Myster et al., 1997; Perrone et al., 1998, 2000). However, expression of amino-terminal dynein HC constructs in the appropriate I1 mutant background (*pf9/ida1* or *ida2*) results in the reassembly of an I1-dynein complex lacking one of the HC motor domains. The 1α HC motor domain is located in the innermost lobe, whereas the 1β HC motor domain is located closer to the outer dynein arms (Figures 9.1 and 9.2; see Myster et al., 1997; Perrone et al., 2000).

The IC/LC complex is thought to comprise the third lobe at the base of the I1 dynein, closest to the first radial spoke in the 96-nm repeat, a position consistent with its proposed role in dynein regulation (Porter and Sale, 2000). Images obtained by cryoelectron tomography have revealed considerable substructure within the IC/LC domain as well as the presence of linkers between this domain and neighboring structures in both the outer and inner dynein arm regions (Nicastro et al., 2006). The kinases and phosphatases that modify the phosphorylation state of IC138 and I1 activity must also be located close to the base of the I1 dynein, although their specific locations are still unknown (King and Dutcher, 1997; Yang et al., 2000; Yang and Sale, 2000). The presence of linkers connecting the I1 dynein to both the outer dynein arms and adjacent inner dynein arms suggests a mechanism for coordinating motor activity between multiple dynein isoforms that may operate at different intrinsic velocities (Kurimoto and Kamiya, 1991; Kagami and Kamiya, 1992).

Four of the single-headed inner arm dyneins have also been identified within the 96-nm axoneme repeat. Mutations in the *IDA4* gene encoding the LC subunit p28 result in the failure to assemble three inner arm dynein isoforms (species a, c, and d, each of which contains p28 as a subunit – see Chapter 6) (Kagami and Kamiya, 1992; Mastronarde et al., 1992; LeDizet and Piperno, 1995a, b). Interestingly, these three isoforms do not coassemble *in situ*, but are instead located at three distinct sites within the 96-nm repeat, one at the base of each radial spoke (S1 and S2) and a third at the distal end of the repeat (Mastronarde et al., 1992). The densities missing in *ida4* axonemes correspond to the inner arm structures 2, 4, and 6 seen by cryoelectron tomography (Nicastro et al., 2006; see Figure 9.2, panels A and B). The p28 LC associated with these inner arm dyneins may recognize a structural motif that forms three unique binding sites within the 96-nm repeat. The nature of this proposed motif is not well understood, but it may be related to the tektin subunits that coassemble with tubulin protofilaments and are thought to form a distinct microtubule lattice on the surface of the A-tubule (Setter et al., 2006; see also Yanagisawa and Kamiya, 2004). Alternatively, these three dyneins may be destabilized and degraded in the cytoplasm in the absence of p28, thus preventing assembly in the axoneme.

The fourth single-headed dynein isoform (species e) identified within the 96-nm repeat is a centrin-associated dynein HC that appears to be closely associated with the DRC (Gardner et al., 1994). Assembly of this isoform is disrupted in the inner arm motility mutant *ida6* and the DRC mutant *pf3* (Kato et al., 1993; Gardner et al., 1994). Loss of this dynein also correlates with reduced levels of tektin in the isolated axonemes (Yanagisawa and Kamiya, 2004). Efforts to identify the *IDA6* and *PF3* gene products are ongoing, but based on the mutant phenotypes, it seems likely that these gene products are involved in anchoring at least some of the centrin-associated dynein HCs (DHC5, DHC8; see Chapter 6) to the DRC (Gardner et al., 1994; Piperno et al., 1994). Cryoelectron tomography of this region indicates that several of the single-headed dyneins attach to the A-tubule by long tails that insert close to the base of the radial spokes and/or below the DRC (Figure 9.2; see Nicastro et al., 2006).

B. The DRC, nexin links, and dynein linkers

As introduced in section II.D, image averaging of DRC mutant axonemes revealed that the DRC forms a discrete structural domain within the inner arm region, at the junction between the second radial spoke and the outer dynein arms (Mastronarde et al., 1992; Gardner et al., 1994; Rupp and

Porter, 2003). These early studies also recognized that the DRC is located in close proximity to the site where the nexin link attaches to the A-tubule (Goodenough and Heuser, 1985a, b; Burgess et al., 1991; Mastronarde et al., 1992; Woolley, 1997). However, the more recent three-dimensional (3D) images of the DRC and nexin link obtained by cryoelectron tomography have provided a much clearer view of the complexities of these structures and their relationship to other axoneme components (Nicastro et al., 2006).

The nexin linker was identified in early EM studies of the axoneme as a flexible filament that connects neighboring microtubules to form the ring of nine outer doublets (Warner, 1976; Goodenough, 1989; Stephens et al., 1989; Bozkurt and Woolley, 1993). It is thought to contribute to the elastic resistance that converts doublet sliding to axoneme bending (Summers and Gibbons, 1973; Brokaw, 1980; Woolley, 1997; Lindemann et al., 2005) and to be a primary target of the proteolytic enzymes used in sliding disintegration assays of outer doublet microtubules (Summers and Gibbons, 1973; Brokaw, 1980; Ikeda et al., 2003; Lindemann et al., 2005). By cryoelectron tomography, one end of the *Chlamydomonas* nexin linkage appears to form a bifurcated attachment to the B-tubule of the outer doublet (Nicastro et al., 2006). The other end is anchored to the A-tubule close to the base of the second radial spoke and the DRC (Figure 9.2, panels F and G). The part that connects the A-tubule anchor to the B-tubule bifurcation point appears to be a zigzag structure. The total length of the nexin linkage is on the order of ~40 nm, which is longer than the shortest distance between adjacent outer doublets (~30 nm) in straight axonemes. This arrangement could accommodate the local microtubule sliding that occurs during axonemal bending and may also contribute to the elastic restoring forces that nexin is proposed to provide (Ikeda et al., 2003; Lindemann et al., 2005). The protein subunits that form the nexin linkage are still unknown.

The DRC appears as a crescent-shaped density when viewed in thin-sectioned material, but cryoelectron tomography demonstrated that much of the mass associated with the DRC extends several nanometers away from the surface of the A-tubule and appears to be closely associated with the nexin link (Nicastro et al., 2006; see Figure 9.2, panel I). Ongoing efforts to identify nexin subunits are expected to provide further insights into the interactions between these two key axoneme structures.

The DRC also makes contact with the outer dynein arms through the distal outer arm–inner arm dynein (OID) linker (Figures 9.1B and 9.2B–D, J). The subunits that mediate interactions between the outer dynein arms and the DRC are still unknown, but are the subject of intense investigation. The linker connecting I1 to the outer dynein arm may contain the ODA7 protein (Freshour et al., 2006). The presence of the OID linkers suggests a mechanism by which signals from the central pair and radial spokes can be relayed through the I1 dynein and the DRC to the outer dynein arms. These structures could serve as mechanical sensors and feedback loops

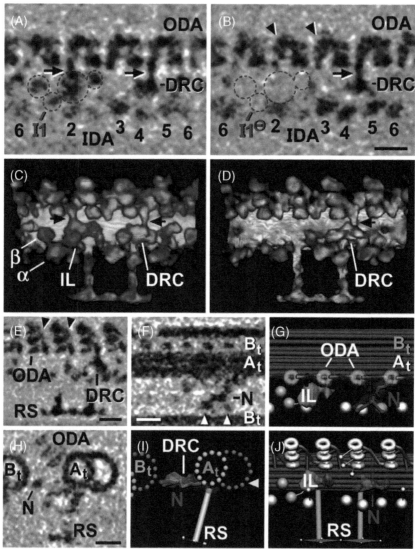

FIGURE 9.2 *The organization of the dynein arms and associated structures in the 96-nm axoneme repeat of* Chlamydomonas *flagella as revealed by cryoelectron tomography. Wild-type and I1 mutant (pf9) axonemes were rapidly frozen in liquid ethane. Multiple images of the frozen hydrated samples were collected using a cryotilt holder over a range of −65° to +65°, aligned, and averaged to generate 3D reconstructions. To increase the signal-to-noise ratio, multiple copies of the 96-nm repeat were aligned and averaged. Tomographic slices of (A) wild-type and (B) pf9 axonemes containing the 96-nm axoneme repeat in longitudinal view. (C) and (D) show the volume-rendering representations of the corresponding 3D volume reconstructions. Structures on the A-tubule are viewed from the perspective of the B-tubule of the adjacent outer doublet, with the proximal end of the A-tubule on the left. The outer dynein arms appear as three-headed structures on the top of the images, and there are four outer dynein arms per 96-nm repeat. Connections between the outer dynein arms and structures in the inner dynein arm region can be seen in both the tomographic slice and 3D images; these are the OID linkers (arrows). The I1 inner arm dynein is indicated by the dashed red circles in panels (A) and (B) and the red areas in panel (C). This structure is missing in the I1 mutant (B and D). The positions of the I1 1α and 1β dynein HC motor domains are indicated in panel (C) (α, β). The IC/LC complex of the I1 dynein is also indicated (IL); it connects to the first OID linker (left arrow in panels A and C) and also to other structures in the inner dynein arm row. The positions (2, 3, 4, 5, and 6) of the*

which sense tension during axonemal bending, and thereby influence the activity of nearby dynein isoforms (Nicastro et al., 2006). Alternatively, the linkers may represent part of a scaffold of tightly bound axonemal enzymes that regulates the phosphorylation state of key axonemal polypeptides (Nicastro et al., 2006; see section IV for details of the phosphorylation pathway). Given the large number of highly conserved axoneme components that have been identified in ciliary and flagellar proteomes, but not yet assigned to specific axoneme structures (Li et al., 2004; Pazour et al., 2005), further studies are clearly needed to address these questions.

The structural studies of Nicastro et al. (2006) also confirmed and extended the understanding of the outer arm–outer arm dynein (OOD) linkers (Figure 9.2; the structure of these linkers is discussed in more detail in Chapter 6). Although the composition and function of the OOD linkers are not understood, these linkers may play a role in cooperative assembly of the outer dynein arms (Goodenough and Heuser, 1984; Oda et al., 2007) or in coordination of outer arm dynein activity along each doublet microtubule. They also may support the structure of multiple motor domains in the outer arm (Nicastro et al., 2006; Ishikawa et al., 2007).

IV. REGULATION OF FLAGELLAR MOTILITY BY PHOSPHORYLATION

A. Overview/model

The evidence described above based on genetic, biochemical, and structural approaches indicates that the central pair, radial spokes, DRC, and inner arm dynein I1 (the f-dynein) form a conserved network of regulators that control dynein activity and flagellar bending. Consistent with this interpretation, the DRC is located near the base of the S2 radial spoke and inner

---

single-headed inner dynein arms are indicated in panels (A) and (B). The two radial spokes are the T-shaped structures located below the inner dynein arms in panels (C) and (D). The dynein regulatory complex (DRC) is located at the distal end of the repeat, between the inner dynein arms and the outer dynein arms. It connects to the outer dynein arms through the second OID linker (right arrow in panels A–D). Panel (E) is a tomographic slice of a wild-type axoneme, similar to panel (B), but oriented to include all three motor domains of the outer dynein arms (ODA), and the radial spoke heads (RS). The OOD linkers (arrowheads) connect to the α and β HC motor domains of the outer dynein arms on their proximal sides in panel (E), but to the β and γ HC motor domains on their distal sides in panel (B). Panel (F) is a longitudinal slice through the 3D reconstruction at a plane indicated by the yellow arrowhead in panel I; panel (G) is the corresponding graphic model in similar orientation. They contain an average of 160 nexin links (shown in red in panel G). Visible here is the attachment of nexin (N) to the A-tubule (A_t), the zigzag appearance of the nexin linkage, its bifurcation point, and its attachment to the B-tubule (B_t) of the adjacent doublet (white arrowheads in panel F). Panels (H) and (I) show the same region in cross-section, and reveal the close association between the nexin link (shown in red) and the DRC. (J) The 96-nm repeat showing the four outer dynein arms, the inner dynein arms, the OOD and OID linkers, the DRC, and nexin structures. Scale bars, 20 nm. This figure is reprinted with permission from Nicastro et al. (2006).

dynein arms in position to mediate and transmit signals from radial spoke S2 to the dynein motors (Figures 9.1B and 9.2). Predictably, a complex analogous to the DRC would be located at the base of radial spoke S1 to mediate signals to inner arm dynein I1, but such a complex has not yet been identified (for additional discussion of radial-spoke-associated complexes, see Dymek and Smith, 2007, and also Chapter 7). Phenotypic analysis of various mutants, including the radial spoke suppressor mutants, has led to the idea that the combination of the central pair, radial spokes, DRC, and inner arm dyneins regulates the size and shape of the flagellar bends by direct, local control of dynein-driven microtubule sliding (Brokaw et al., 1982; Brokaw and Kamiya, 1987). The question is: What is the nature of the signals that locally regulate dynein-driven microtubule sliding?

Diverse evidence indicates that part of the mechanism for regulating the axonemal bending configuration is the localized control of microtubule sliding by the dynein arms that in turn adjusts an otherwise relatively symmetric, intrinsic, and oscillatory bending configuration (Satir, 1980; Brokaw et al., 1982; Lindemann, 2003; Aoyama and Kamiya, 2005). Analysis of central pair and radial spoke mutant axonemes indicate that they have the intrinsic ability to beat under nonphysiological conditions, as long as the outer dynein arms are present (Omoto et al., 1996; Yagi and Kamiya, 2000). Thus, the dyneins retain the ability to generate and propagate bends, but controlling those bends in a normal physiological state requires the central pair and radial spoke structures.

For example, asymmetry of the central pair structure likely sends signals to specific doublet microtubules that alter dynein-driven microtubule sliding on one half of the axis of the axoneme (see Chapters 7 and 8). The mechanism may include mechanical interaction between the radial spoke heads and specific projections from the central pair apparatus (Warner and Satir, 1974). The mechanism could also involve chemical signals from second messengers and control of axonemal protein phosphorylation. For example, *in vitro* functional, pharmacological, and biochemical evidence using isolated axonemes from wild-type and mutant *Chlamydomonas* axonemes indicates that dynein-driven microtubule sliding is regulated by phosphorylation (Porter and Sale, 2000; Smith and Yang, 2004). The key kinases and phosphatases are physically anchored and localized in the axonemal structure (Figures 9.1B and 9.3A). To date, these include cAMP-dependent protein kinase (PKA), casein kinase 1 (CK1), protein phosphatase 1 (PP1), and protein phosphatase 2A (PP2A) (Howard et al., 1994; Habermacher and Sale, 1995, 1996; Yang et al., 2000; Yang and Sale, 2000; Gaillard et al., 2001, 2006; Smith, 2002a). Furthermore, the *Chlamydomonas* flagellar/axonemal proteome predicts additional kinases and phosphatases not yet characterized (Pazour et al., 2005). In this final section, we will describe the *in vitro* functional evidence and biochemical analysis, using isolated

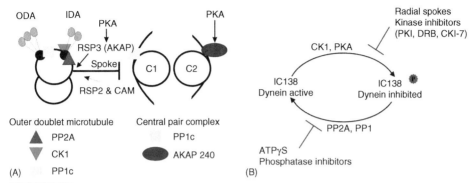

FIGURE 9.3 *Schematic of the axonemal signaling mechanism. (A) The central pair/radial spoke mechanism for regulating axonemal dynein involves changes in phosphorylation. The key kinases and phosphatases identified in this pathway and their putative locations are indicated. ODA, outer dynein arm; IDA, inner dynein arm; C1 and C2, central microtubules 1 and 2; CAM, calmodulin. (Modified from Figure 2A in Porter and Sale, 2000.) (B) IC138 is a critical phosphoprotein in I1 dynein that is modulated by the signal transduction network. The cycle of phosphorylation and dephosphorylation and how it alters I1 dynein activity is shown.*

axonemes, which revealed that phosphorylation, particularly of IC138 of inner arm dynein I1 (Figure 9.3B), plays a role in the control of microtubule sliding and that the key signaling proteins – kinases and phosphatases – are physically anchored and localized in the structure of the axoneme.

B. Axonemal kinases and phosphatases regulate interdoublet sliding

Pharmacological and biochemical evidence from several experimental systems have shown that ciliary and flagellar motility is regulated in part by phosphorylation (Tash and Bracho, 1994; Walczak and Nelson, 1994; Porter and Sale, 2000; Salathe, 2006). For example, cAMP-dependent phosphorylation of axonemal proteins activates sperm motility (Inaba et al., 1999), increases beat frequency in *Paramecium* cilia (Hamasaki et al., 1991), and inhibits flagellar motility in *Chlamydomonas* (Hasegawa et al., 1987). Moreover, several kinases and phosphatases appear to be anchored in the axoneme (Hasegawa et al., 1987; Hamasaki et al., 1989; San Agustin and Witman, 1994; Yang et al., 2000; Yang and Sale, 2000). The challenge has been to identify the enzymes most directly involved in the control of motility, to determine how each enzyme is anchored in the axoneme, and to identify the relevant phosphoprotein targets in the axoneme (Piperno et al., 1981; Hamasaki et al., 1991; Wagner et al., 2006).

One successful strategy for unraveling this complexity has been to combine a pharmacological approach with microtubule-sliding assays on isolated axonemes from specific *Chlamydomonas* flagellar mutants. The methods are reviewed in Sale and Howard (1995), and are based on the pioneering studies of Summers and Gibbons (1971), and further developed by Okagaki and Kamiya (1986) and Kagami and Kamiya (1992). One advantage of the microtubule-sliding assay is the ability to measure dynein activity in both wild-type and paralyzed flagellar axonemes (Witman et al., 1978).

The primary observation for study of regulation was that microtubule sliding is greatly inhibited in axonemes from radial spoke mutants, despite the complete assembly and presence of the dynein arm structures (Smith and Sale, 1992a, b). The observation indicated that dynein activity is globally inhibited in the paralyzed flagellar mutants lacking the radial spokes, as was predicted from the studies of suppressor mutations (Huang et al., 1982; Porter et al., 1992; Piperno, 1995). The mechanism was predicted to involve inhibitory posttranslational modifications of dynein. To test this, an *in vitro* reconstitution approach was developed in which dynein, derived from either wild-type or mutant axonemes, was combined with dynein-depleted wild-type or mutant axonemes and microtubule sliding subsequently measured (Smith and Sale, 1992a, b; see Yamamoto et al., 2006). For example, it was predicted that reconstituting dynein derived from radial spoke mutants with axonemes from wild-type cells would result in rescue of wild-type dynein activity. As predicted, the "inhibited" dynein from radial spoke mutants, when reconstituted on wild-type axonemes, was restored to wild-type activity. In contrast, dynein derived from wild-type axonemes, when reconstituted on axonemes from the radial spoke mutant, retained wild-type dynein activity.

Although the results of these studies could be interpreted in a couple of ways, the simplest interpretation is that dynein is inhibited in radial spoke mutants by an inhibitory posttranslational modification that is reversed when the dynein is reconstituted on wild-type axonemes. The most obvious possibility is that dynein is inappropriately phosphorylated, or dephosphorylated, through a failure in regulation of axonemal kinases or phosphatases. To test this, Howard et al. (1994) took advantage of kinase inhibitors to examine the idea that dynein-driven microtubule sliding in radial spoke mutants is inhibited by inappropriately regulated axonemal kinase activity. As predicted, inhibition of the PKA by PKI or the purified regulatory subunit RII, both highly specific PKA inhibitors, restored wild-type microtubule sliding in axonemes from paralyzed cells lacking the radial spokes (Howard et al., 1994; Habermacher and Sale, 1997; Smith, 2002b). A similar strategy revealed that the axoneme contains a CK1-type kinase that also inhibits dynein in paralyzed flagellar mutants (Yang and Sale, 2000). In this case, addition of the CK1 inhibitors, DRB and CKI–7, rescued dynein activity in axonemes derived from radial spoke or central pair mutant cells (Smith, 2002b).

Based on these data, the simplest model was that the kinases, PKA and CK1, are anchored in the axoneme and that in radial spoke or central pair mutants, kinase activity is not properly regulated, leading to inappropriate phosphorylation and global inhibition of dynein. Predictably, the target regulatory substrates include dynein subunits that, when phosphorylated, inhibit microtubule sliding. Consistent with this, and as described in section II.C above, one of the key target phosphoprotein substrates is IC138 of I1 dynein (Habermacher and Sale, 1997; King and Dutcher, 1997; Yang and Sale, 2000; Hendrickson et al., 2004). Furthermore, IC138, which appears to be a direct substrate for CK1 (Yang and Sale, 2000), is hyperphosphorylated in paralyzed flagellar mutants defective in radial spoke and central pair assembly, and in phototactic mutants (King and Dutcher, 1997; Hendrickson et al., 2004). Therefore, paralysis in the radial spoke and central pair mutants results in part through unregulated phosphorylation of IC138 and the general inhibition of dynein. Detailed tests of this model will require identification of the key phosphorylated residues in IC138, site-directed mutagenesis, and analysis of the resulting phenotypes to determine the molecular mechanism of I1 dynein function. Additional immediate questions include (1) whether PKA and CK1 each work independently in separate pathways, or in a cascade to control dynein; and (2) whether the control mechanism also includes phosphatases located in the axoneme and required to reverse phosphorylation and rescue dynein activity (Figure 9.3B).

These results indicated the presence of axonemal phosphatases that are required for reversal of IC138 phosphorylation and rescue of dynein-driven microtubule sliding (Figure 9.3B). For example, if dephosphorylation of IC138 (phosphorylated *in vivo* by unregulated axonemal kinases) is required for rescue of dynein-driven microtubule sliding, then the axoneme must contain these protein phosphatases. Therefore, the addition of phosphatase inhibitors would predictably block kinase inhibitor-induced rescue of dynein activity (Figure 9.3B). As predicted, addition of the phosphatase inhibitors okadaic acid, microcystin, and inhibitor-2 blocked kinase inhibitor-mediated rescue of dynein activity in radial spoke and central pair mutant axonemes (Habermacher and Sale, 1996, 1997). These data are consistent with the presence of PP1 and PP2A in the axoneme, located on the outer doublet microtubules adjacent to IC138 of I1-dynein. To test this, further biochemical and structural approaches were taken (Yang et al., 2000).

C. Locations of axonemal kinases and phosphatases

Direct biochemical and enzymatic characterization, as well as proteomic analysis of isolated axonemes, has confirmed the presence of CK1, PP1, and PP2A in the *Chlamydomonas* axoneme (Yang et al., 2000; Yang and Sale, 2000; Pazour et al., 2005). However, although pharmacological analysis using PKI and the RII regulatory subunits has definitively defined an axonemal PKA activity, and at least two axonemal A-kinase anchoring proteins (AKAPs) have been identified (Gaillard et al., 2001, 2006), PKA subunits have not been defined in *Chlamydomonas*. It is possible that *Chlamydomonas* expresses a PKA that has an unconventional structure.

CK1 has been directly isolated from axonemes, confirming that CK1 is a structural protein in the axoneme (Yang and Sale, 2000; Pazour et al., 2005). CK1 is likely to be localized to the outer doublet microtubules, as would be expected for a kinase that operates directly on dynein subunits. More specifically, based on observations that the CK1 inhibitors DRB and CKI–7 block IC138 phosphorylation, CK1 is predicted to be anchored near I1 dynein and to repeat every 96-nm along each doublet microtubule (Figures 9.1B and 9.3A). Consistent with this model, CK1 is a relatively abundant axonemal protein. However, CK1 has not yet been directly localized in the axoneme, and the mechanisms for targeting and anchoring CK1 are not known. Based on the flagellar proteome (Pazour et al., 2005), the gene for the catalytic subunit of the axonemal CK1 (EDP02935) has been identified and cloned, revealing that it is highly conserved in amino acid sequence compared to CK1 from mammalian sources (Gokhale et al., 2006). Current challenges include the direct localization of CK1 in the axoneme and identifying the mechanism for targeting and anchoring CK1 in the cell.

PP1 and PP2A have also been identified by direct biochemical analysis of isolated axonemes (Yang et al., 2000). Therefore, as predicted from the functional analysis, the phosphatases are targeted and localized to the axoneme. The highly conserved axonemal PP1 catalytic subunit (AAD38856) is primarily localized to the C1 microtubule of the central pair apparatus (Yang et al., 2000) where it is predicted to operate in a pathway that regulates flagellar bending (see Chapter 8). The localization studies also indicate that a small fraction of PP1 is associated with the outer doublet microtubules. However, further work is required to determine how PP1 is anchored in the axoneme and to identify relevant PP1 substrates. The axonemal PP2A C and A (EDP00240) subunits have been isolated, revealing that they are relatively abundant axonemal proteins and that the catalytic subunit of PP2A is anchored on the outer doublet microtubule (Yang et al., 2000). These results are consistent with localization of PP2A near I1 dynein, directly in position to mediate IC138 dephosphorylation (Figures 9.1B and 9.3A). Further tests will require the direct localization of PP2A in the axoneme, identification of the mechanism for PP2A anchoring, and determination of whether PP2A acts directly to control the phosphorylation state of IC138.

D. Anchoring of signaling proteins in the axoneme

The evidence cited in this chapter provides a relatively new view of the 9 + 2 axoneme, not only as a conserved motility machine, but also as a conserved scaffold that localizes equally conserved signaling proteins that regulate motility and possibly participate in other sensory functions (Satir and Christensen, 2007). For example, here we cite evidence that otherwise ubiquitous and conserved kinases (PKA, CK1) and phosphatases (PP1, PP2A) are highly localized in the axoneme. Since the catalytic and regulatory subunits of the signaling proteins are used for multiple cell functions – e.g., CK1 is also localized at the eyespot (Schmidt et al., 2006) – then the challenge is to determine the flagellar and axonemal proteins that are responsible for targeting and anchoring PKA, CK1, PP1, and PP2A in the axoneme. The molecular mechanisms for localizing CK1, PP1, and PP2A are not known. However, the axoneme and the *Chlamydomonas* experimental system (genetic, molecular, structural, and biochemical advantages) offer exceptional opportunities to define the principles for targeting and anchoring the enzymes in the cell.

A well-established model for targeting and regulating protein kinases and phosphatases is the localization and control of PKA by the class of proteins called AKAPs (Wong and Scott, 2004). As described in section IV.B, physiological and pharmacological experiments demonstrated that the *Chlamydomonas* flagellar axoneme contains a PKA that regulates axonemal motility and dynein activity (Hasegawa et al., 1987; Howard et al., 1994), but the mechanism for localizing this enzyme in the axoneme was unknown. Gaillard et al. (2001) tested the idea that the axoneme contains members of the AKAP family of proteins designed to anchor otherwise ubiquitous PKA specifically in the axoneme. The approach used RII blot overlays on isolated axonemes from wild-type and mutant axonemes. Based on the high affinity of RII for AKAPs, the blot overlay is a standard starting procedure to identify candidate AKAPs. Two axonemal AKAPs were identified: a 240-kD AKAP associated with the central pair apparatus, and a 97-kD AKAP located in the radial spoke stalk.

Based on a detailed analysis, including western blots and expression of recombinant protein, it was determined that AKAP97 is radial spoke protein 3 (RSP3), the most proximal radial spoke protein required for spoke assembly (Diener et al., 1993; Smith and Yang, 2004; see Chapter 7). Further detailed *in vitro* and *in vivo* molecular analysis of RSP3 revealed a single RII-binding domain, required for PKA localization, located at amino acids 161–180 with sequence homology to other AKAP family members (Vijayaraghavan et al., 1999). Furthermore, PKA localization by RSP3 is required for control of normal flagellar motility (Gaillard et al., 2001, 2006). Importantly, RSP3 is predicted to be located near inner arm dyneins where an anchored PKA would be in direct position to modify dynein activity and regulate motility. As a further test of the physiological role of PKA anchoring in the axoneme, an RSP3 mutant, *pf14*, was transformed with an RSP3 gene containing a mutation in the PKA-binding domain. Analysis of the transformants confirmed that flagellar RSP3 is an AKAP and that a mutation in the PKA-binding domain results in unregulated axonemal PKA activity and inhibition of normal motility (Gaillard et al., 2006). Thus, the PKA-binding domain of RSP3 is required for proper regulation of axonemal

PKA, as well as for regulation of normal flagellar motility. The inhibitory effect of axonemal PKA is consistent with earlier reports on *in vitro* reactivation of motility (Hasegawa et al., 1987) and with previous studies on global inhibition of dynein-driven microtubule sliding in axonemes from paralyzed flagellar mutants also having unregulated PKA (Howard et al., 1994; Habermacher and Sale, 1996, 1997; Yang and Sale, 2000; Smith, 2002b; Hendrickson et al., 2004).

In wild-type cells, RSP3 may play a central role in regulating axonemal PKA in a pathway involving the radial spokes, by ultimately impinging on individual outer doublet microtubules to locally control inner arm dynein activity (Smith and Yang, 2004). Regulation of axonemal PKA activity by the PKA-binding domain of RSP3 is probably most important in local control of microtubule sliding and the modification of flagellar waveform, but not for control of flagellar beat frequency or initiation of oscillation (Brokaw et al., 1982; Brokaw and Kamiya, 1987; Smith and Yang, 2004; Aoyama and Kamiya, 2005). These studies are consistent with a model in which the radial spokes are required for proper regulation of axonemal PKA which, based on analysis of microtubule sliding, is required for regulation of I1 dynein activity (Smith and Sale, 1992a; Howard et al., 1994; Habermacher and Sale, 1996, 1997; King and Dutcher, 1997; Hendrickson et al., 2004). Specifically, this model predicts that a defect in the radial spokes would result in a failure to suppress axonemal PKA activity, thereby resulting in phosphorylation of the IC138 IC of I1 dynein, inappropriate inhibition of microtubule sliding, and global inhibition of motility. Direct tests of this model will require characterization of PKA in *Chlamydomonas* with the production of useful antibodies and the recovery of additional informative mutant cells.

The AKAP model described here provides a general plan for targeting all of the kinases and phosphatases in the axoneme. As stated above, the catalytic functions of CK1, PP1, and PP2A are not likely to be restricted to only the flagellar axoneme. Rather, in each case, as in other eukaryotic cells, these conserved and ubiquitous proteins are required for several functions in the cell. Therefore, one challenge is to define new classes of anchoring proteins for CK1 and for the phosphatases. As illustrated by the RSP3–AKAP story, the *Chlamydomonas* axonemal experimental system offers an outstanding opportunity for discovery of new classes of interacting/anchoring proteins used to localize ubiquitous enzymes to specific cellular sites.

V. FUTURE DIRECTIONS

In this chapter, we have summarized diverse experimental studies in *Chlamydomonas* that have revealed conserved structures and proteins responsible for control of the dynein motors and regulation of axonemal bending. The combined results provide a view of the 9 + 2 axoneme as a highly conserved scaffold that localizes each of the dynein arms, the radial spokes, the DRC, the nexin links, and signaling proteins including protein kinases, phosphatases, and calcium-binding proteins, within a 96-nm repeat unit. Major unsolved questions include: What is the structural basis for the axonemal 96-nm repeat and the molecular basis for targeting each component and signaling protein to the outer doublet structure? What are the protein interactions in the DRC and the mechanism of the DRC in control of dynein motor function? What is the precise role of each dynein motor for generation and control of axonemal bending? How is the activity of each dynein coordinated with the activity of other dynein motors, and do the OID and OOD linkers play a role in coordination of activities? What is the mechanism of the I1 dynein (the f-dynein) for control of microtubule sliding? How are signals transmitted between the central pair apparatus, the radial spokes, and dynein motors? Finally, does the central pair/radial spoke system regulate axonemal kinases and phosphatases located on the outer doublet microtubules? Answers will require continued genetic analysis in *Chlamydomonas,* a comprehensive protein–protein interaction map, the application of EM tomography to the study of mutant axonemes, and further high-resolution analysis of axonemal motility.

REFERENCES

Aoyama, S. and Kamiya, R. (2005). Cyclical interactions between two outer doublet microtubules in split flagellar axonemes. *Biophys. J.* **89**, 3261–3268.

Bozkurt, H.H. and Woolley, D.M. (1993). Morphology of nexin links in relation to interdoublet sliding in the sperm flagellum. *Cell Motil. Cytoskeleton* **24**, 109–118.

Brokaw, C.J. (1980). Elastase digestion of demembranated sperm flagella. *Science* **207**, 1365–1367.

Brokaw, C.J. and Kamiya, R. (1987). Bending patterns of *Chlamydomonas* flagella: IV. Mutants with defects in inner and outer dynein arms indicate differences in dynein arm function. *Cell Motil. Cytoskeleton* **8**, 68–75.

Brokaw, C.J., Luck, D.J., and Huang, B. (1982). Analysis of the movement of *Chlamydomonas* flagella: the function of the radial-spoke system is revealed by comparison of wild-type and mutant flagella. *J. Cell Biol.* **92**, 722–732.

Burgess, S.A., Carter, D.A., Dover, S.D., and Woolley, D.M. (1991). The inner dynein arm complex: compatible images from freeze-etch and thin section methods of microscopy. *J. Cell Sci.* **100**, 319–328.

Burgess, S.A., Walker, M.L., Sakakibara, H., Knight, P.J., and Oiwa, K. (2003). Dynein structure and power stroke. *Nature* **421**, 715–718.

Colantonio, J.R., Bekker, J.M., Kim, S.J., Morrissey, K.M., Crosbie, R.H., and Hill, K.L. (2006). Expanding the role of the dynein regulatory complex to nonaxonemal functions: association of GAS11 with the golgi apparatus. *Traffic* **7**, 538–548.

DiBella, L.M., Benashski, S.E., Tedford, H.W., Harrison, A., Patel-King, R.S., and King, S.M. (2001). The Tctex1/Tctex2 class of dynein light chains. Dimerization, differential expression, and interaction with the LC8 protein family. *J. Biol. Chem.* **276**, 14366–14373.

DiBella, L.M., Sakato, M., Patel-King, R.S., Pazour, G.J., and King, S.M. (2004a). The LC7 light chains of *Chlamydomonas* flagellar dyneins interact with components required for both motor assembly and regulation. *Mol. Biol. Cell* **15**, 4633–4646.

DiBella, L.M., Smith, E.F., Patel-King, R.S., Wakabayashi, K., and King, S.M. (2004b). A novel Tctex2-related light chain is required for stability of inner dynein arm I1 and motor function in the *Chlamydomonas* flagellum. *J. Biol. Chem.* **279**, 21666–21676.

Diener, D.R., Ang, L.H., and Rosenbaum, J.L. (1993). Assembly of flagellar radial spoke proteins in *Chlamydomonas*: identification of the axoneme binding domain of radial spoke protein 3. *J. Cell Biol.* **123**, 183–190.

Dutcher, S.K., Gibbons, W., and Inwood, W.B. (1988). A genetic analysis of suppressors of the *PF10* mutation in *Chlamydomonas reinhardtii*. *Genetics* **120**, 965–976.

Dymek, E.E. and Smith, E.F. (2007). A conserved CaM- and radial spoke-associated complex mediates regulation of flagellar dynein activity. *J. Cell Biol.* **179**, 515–526.

Freshour, J., Yokoyama, R., and Mitchell, D.R. (2006). *Chlamydomonas* flagellar outer row dynein assembly protein ODA7 interacts with both outer row and I1 inner row dyneins. *J. Biol. Chem.* **282**, 5404–5412.

Fox, L.A. and Sale, W.S. (1987). Direction of force generated by the inner row of dynein arms on flagellar microtubules. *J. Cell Biol.* **105**, 1781–1787.

Gaillard, A.R., Diener, D.R., Rosenbaum, J.L., and Sale, W.S. (2001). Flagellar radial spoke protein 3 is an A-kinase anchoring protein (AKAP). *J. Cell Biol.* **153**, 443–448.

Gaillard, A.R., Fox, L.A., Rhea, J.M., Craige, B., and Sale, W.S. (2006). Disruption of the A-kinase anchoring domain in flagellar radial spoke protein 3 results in unregulated axonemal cAMP-dependent protein kinase activity and abnormal flagellar motility. *Mol. Biol. Cell* **17**, 2626–2635.

Gardner, L.C., O'Toole, E., Perrone, C.A., Giddings, T., and Porter, M.E. (1994). Components of a "dynein regulatory complex" are located at the junction between the radial spokes and the dynein arms in *Chlamydomonas* flagella. *J. Cell Biol.* **127**, 1311–1325.

Gokhale, A., Wirschell, M., and Sale, W.S. (2006). Characterization of the protein kinase CK1 that regulates dynein-driven microtubule sliding and is anchored in *Chlamydomonas* flagellar axonemes. *Mol. Biol. Cell* **17**, 1. abstract #975 (CD-ROM)

Goodenough, U.W. (1989). Cilia, flagella and the basal apparatus. *Curr. Opin. Cell Biol.* **1**, 58–62.

Goodenough, U.W. and Heuser, J. (1982). Structure of the outer dynein arm. *J. Cell Biol.* **95**, 798–815.

Goodenough, U. and Heuser, J. (1984). Structural comparison of purified dynein proteins with *in situ* dynein arms. *J. Mol. Biol.* **180**, 1083–1118.

Goodenough, U.W. and Heuser, J.E. (1985a). Outer and inner dynein arms of cilia and flagella. *Cell* **41**, 341–342.

Goodenough, U.W. and Heuser, J.E. (1985b). Substructure of inner dynein arms, radial spokes, and the central pair/projection complex of cilia and flagella. *J. Cell Biol.* **100**, 2008–2018.

Habermacher, G. and Sale, W.S. (1995). Regulation of dynein-driven microtubule sliding by an axonemal kinase and phosphatase in *Chlamydomonas* flagella. *Cell Motil. Cytoskeleton* **32**, 106–109.

Habermacher, G. and Sale, W.S. (1996). Regulation of flagellar dynein by an axonemal type-1 phosphatase in *Chlamydomonas*. *J. Cell Sci.* **109**, 1899–1907.

Habermacher, G. and Sale, W.S. (1997). Regulation of flagellar dynein by phosphorylation of a 138-kD inner arm dynein intermediate chain. *J. Cell Biol.* **136**, 167–176.

Hamasaki, T., Murtaugh, T.J., Satir, B.H., and Satir, P. (1989). *In vitro* phosphorylation of *Paramecium* axonemes and permeabilized cells. *Cell Motil. Cytoskeleton* **12**, 1–11.

Hamasaki, T., Barkalow, K., Richmond, J., and Satir, P. (1991). cAMP-stimulated phosphorylation of an axonemal polypeptide that copurifies with the 22S dynein arm regulates microtubule translocation velocity and swimming speed in *Paramecium*. *Proc. Natl. Acad. Sci. U.S.A.* **88**, 7918–7922.

Hasegawa, E., Hayashi, H., Asakura, S., and Kamiya, R. (1987). Stimulation of *in vitro* motility of *Chlamydomonas* axonemes by inhibition of cAMP-dependent phosphorylation. *Cell Motil. Cytoskeleton* **8**, 302–311.

Hendrickson, T.W., Perrone, C.A., Griffin, P., Wuichet, K., Mueller, J., Yang, P., Porter, M.E., and Sale, W.S. (2004). IC138 is a WD-repeat dynein intermediate chain required for light chain assembly and regulation of flagellar bending. *Mol. Biol. Cell* **15**, 5431–5442.

Hill, K.L., Hutchings, N.R., Grandgenett, P.M., and Donelson, J.E. (2000). T lymphocyte-triggering factor of African trypanosomes is associated with the flagellar fraction of the cytoskeleton and represents a new family of proteins that are present in several divergent eukaryotes. *J. Biol. Chem.* **275**, 39369–39378.

Howard, D.R., Habermacher, G., Glass, D.B., Smith, E.F., and Sale, W.S. (1994). Regulation of *Chlamydomonas* flagellar dynein by an axonemal protein kinase. *J. Cell Biol.* **127**, 1683–1692.

Huang, B., Ramanis, Z., and Luck, D.J. (1982). Suppressor mutations in *Chlamydomonas* reveal a regulatory mechanism for flagellar function. *Cell* **28**, 115–124.

Ikeda, K., Brown, J.A., Yagi, T., Norrander, J.M., Hirono, M., Eccleston, E., Kamiya, R., and Linck, R.W. (2003). Rib72, a conserved protein associated with the ribbon compartment of flagellar A-microtubules and potentially involved in the linkage between outer doublet microtubules. *J. Biol. Chem.* **278**, 7725–7734.

Inaba, K., Kagami, O., and Ogawa, K. (1999). Tctex2-related outer arm dynein light chain is phosphorylated at activation of sperm motility. *Biochem. Biophys. Res. Commun.* **256**, 177–183.

Ishikawa, T., Sakakibara, H., and Oiwa, K. (2007). The architecture of outer dynein arms *in situ*. *J. Mol. Biol.* **368**, 1249–1258.

Kagami, O. and Kamiya, R. (1992). Translocation and rotation of microtubules caused by multiple species of *Chlamydomonas* inner-arm dynein. *J. Cell Sci.* **103**, 653–664.

Kagami, O. and Kamiya, R. (1995). Separation of dynein species by high-pressure liquid chromatography. *Methods Cell Biol.* **47**, 487–489.

Kamiya, R. (1991). Selection of *Chlamydomonas* dynein mutants. *Methods Enzymol.* **196**, 348–355.

Kato, T., Kagami, O., Yagi, T., and Kamiya, R. (1993). Isolation of two species of *Chlamydomonas reinhardtii* flagellar mutants, *ida5* and *ida6*, that lack a newly identified heavy chain of the inner dynein arm. *Cell Struct. Funct.* **18**, 371–377.

King, S.J. and Dutcher, S.K. (1997). Phosphoregulation of an inner dynein arm complex in *Chlamydomonas reinhardtii* is altered in phototactic mutant strains. *J. Cell Biol.* **136**, 177–191.

King, S.J., Inwood, W.B., O'Toole, E.T., Power, J., and Dutcher, S.K. (1994). The *bop2-1* mutation reveals radial asymmetry in the inner dynein arm region of *Chlamydomonas reinhardtii*. *J. Cell Biol.* **126**, 1255–1266.

Koonce, M.P. (2006). Dynein shifts into second gear. *Proc. Natl. Acad. Sci. U.S.A.* **103**, 17587–17588.

Kurimoto, E. and Kamiya, R. (1991). Microtubule sliding in flagellar axonemes of *Chlamydomonas* mutants missing inner- or outer-arm dynein: velocity measurements on new types of mutants by an improved method. *Cell Motil. Cytoskeleton* **19**, 275–281.

Lechtreck, K. and Witman, G.B. (2007). *Chlamydomonas reinhardtii* hydin is a central pair protein required for flagellar motility. *J. Cell Biol.* **176**, 473–482.

LeDizet, M. and Piperno, G. (1995a). *ida4-1*, *ida4-2*, and *ida4-3* are intron splicing mutations affecting the locus encoding p28, a light chain of *Chlamydomonas* axonemal inner dynein arms. *Mol. Biol. Cell* **6**, 713–723.

LeDizet, M. and Piperno, G. (1995b). The light chain p28 associates with a subset of inner dynein arm heavy chains in *Chlamydomonas* axonemes. *Mol. Biol. Cell* **6**, 697–711.

Lewin, R.A. (1954). Mutants of *Chlamydomonas moewusii* with impaired motility. *J. Gen. Microbiol.* **11**, 358–363.

Li, J.B., Gerdes, J.M., Haycraft, C.J., Fan, Y., Teslovich, T.M., May-Simera, H., Li, H., Blacque, O.E., Li, L., Leitch, C.C., Lewis, R.A., Green, J.S., Parfrey, P.S., Leroux, M.R., Davidson, W.S., Beales, P.L., Guay-Woodford, L.M., Yoder, B.K., Stormo, G.D., Katsanis, N., and Dutcher, S.K. (2004). Comparative genomics identifies a flagellar and basal body proteome that includes the BBS5 human disease gene. *Cell* **117**, 541–552.

Lindemann, C.B. (2003). Structural–functional relationships of the dynein, spokes, and central-pair projections predicted from an analysis of the forces acting within a flagellum. *Biophys. J.* **84**, 4115–4126.

Lindemann, C.B., Macauley, L.J., and Lesich, K.A. (2005). The counterbend phenomenon in dynein-disabled rat sperm flagella and what it reveals about the interdoublet elasticity. *Biophys. J.* **89**, 1165–1174.

Mastronarde, D.N., O'Toole, E.T., McDonald, K.L., McIntosh, J.R., and Porter, M.E. (1992). Arrangement of inner dynein arms in wild-type and mutant flagella of *Chlamydomonas*. *J. Cell Biol.* **118**, 1145–1162.

Myster, S.H., Knott, J.A., O'Toole, E., and Porter, M.E. (1997). The *Chlamydomonas Dhc1* gene encodes a dynein heavy chain subunit required for assembly of the I1 inner arm complex. *Mol. Biol. Cell* **8**, 607–620.

Myster, S.H., Knott, J.A., Wysocki, K.M., O'Toole, E., and Porter, M.E. (1999). Domains in the 1alpha dynein heavy chain required for inner arm assembly and flagellar motility in *Chlamydomonas*. *J. Cell Biol.* **146**, 801–818.

Nicastro, D., Schwartz, C., Pierson, J., Gaudette, R., Porter, M., and McIntosh, J. (2006). The molecular architecture of axonemes revealed by cryoelectron tomography. *Science* **313**, 944–948.

Oda, T., Hirokawa, N., and Kikkawa, M. (2007). Three-dimensional structures of the flagellar dynein-microtubule complex by cryoelectron microscopy. *J. Cell Biol.* **177**, 243–252.

Oiwa, K. and Sakakibara, H. (2005). Recent progress in dynein structure and mechanism. *Curr. Opin. Cell Biol.* **17**, 98–103.

Okagaki, T. and Kamiya, R. (1986). Microtubule sliding in mutant *Chlamydomonas* axonemes devoid of outer or inner dynein arms. *J. Cell Biol.* **103**, 1895–1902.

Okita, N., Isogai, N., Hirono, M., Kamiya, R., and Yoshimura, K. (2005). Phototactic activity in *Chlamydomonas* 'non-phototactic' mutants deficient in Ca^{2+}-dependent control of flagellar dominance or in inner-arm dynein. *J. Cell Sci.* **118**, 529–537.

Omoto, C.K., Yagi, T., Kurimoto, E., and Kamiya, R. (1996). Ability of paralyzed flagella mutants of *Chlamydomonas* to move. *Cell Motil. Cytoskeleton* **33**, 88–94.

Omoto, C.K., Gibbons, I.R., Kamiya, R., Shingyoji, C., Takahashi, K., and Witman, G.B. (1999). Rotation of the central pair microtubules in eukaryotic flagella. *Mol. Biol. Cell* **10**, 1–4.

O'Toole, E., Mastronarde, D., McIntosh, J.R., and Porter, M.E. (1995). Computer-assisted analysis of flagellar structure. *Methods Cell Biol.* **47**, 183–191.

Pazour, G.J., Agrin, N., Leszyk, J., and Witman, G.B. (2005). Proteomic analysis of a eukaryotic cilium. *J. Cell Biol.* **170**, 103–113.

Perrone, C.A., Yang, P., O'Toole, E., Sale, W.S., and Porter, M.E. (1998). The *Chlamydomonas IDA7* locus encodes a 140-kDa dynein intermediate chain required to assemble the I1 inner arm complex. *Mol. Biol. Cell* **9**, 3351–3365.

Perrone, C.A., Myster, S.H., Bower, R., O'Toole, E.T., and Porter, M.E. (2000). Insights into the structural organization of the I1 inner arm dynein from a domain analysis of the 1beta dynein heavy chain. *Mol. Biol. Cell* **11**, 2297–2313.

Piperno, G. (1990). Functional diversity of dyneins. *Cell Motil. Cytoskeleton* **17**, 147–149.

Piperno, G. (1995). Regulation of dynein activity within *Chlamydomonas* flagella. *Cell Motil. Cytoskeleton* **32**, 103–105.

Piperno, G., Huang, B., Ramanis, Z., and Luck, D.J. (1981). Radial spokes of *Chlamydomonas* flagella: polypeptide composition and phosphorylation of stalk components. *J. Cell Biol.* **88**, 73–79.

Piperno, G., Mead, K., and Shestak, W. (1992). The inner dynein arms I2 interact with a "dynein regulatory complex" in *Chlamydomonas* flagella. *J. Cell Biol.* **118**, 1455–1463.

Piperno, G., Mead, K., LeDizet, M., and Moscatelli, A. (1994). Mutations in the "dynein regulatory complex" alter the ATP-insensitive binding sites for inner arm dyneins in *Chlamydomonas* axonemes. *J. Cell Biol.* **125**, 1109–1117.

Porter, M.E. (1996). Axonemal dyneins: assembly, organization, and regulation. *Curr. Opin. Cell Biol.* **8**, 10–17.

Porter, M.E. and Sale, W.S. (2000). The 9 + 2 axoneme anchors multiple inner arm dyneins and a network of kinases and phosphatases that control motility. *J. Cell Biol.* **151**, F37–F42.

Porter, M.E., Power, J., and Dutcher, S.K. (1992). Extragenic suppressors of paralyzed flagellar mutations in *Chlamydomonas reinhardtii* identify loci that alter the inner dynein arms. *J. Cell Biol.* **118**, 1163–1176.

Porter, M.E., Knott, J.A., Gardner, L.C., Mitchell, D.R., and Dutcher, S.K. (1994). Mutations in the *SUP-PF-1* locus of *Chlamydomonas reinhardtii* identify a regulatory domain in the beta-dynein heavy chain. *J. Cell Biol.* **126**, 1495–1507.

Porter, M.E., Knott, J.A., Myster, S.H., and Farlow, S.J. (1996). The dynein gene family in *Chlamydomonas reinhardtii*. *Genetics* **144**, 569–585.

Ralston, K.S., Lerner, A.G., Diener, D.R., and Hill, K.L. (2006). Flagellar motility contributes to cytokinesis in *Trypanosoma brucei* and is modulated by an evolutionarily conserved dynein regulatory system. *Eukaryotic Cell* **5**, 696–711.

Rupp, G. and Porter, M.E. (2003). A subunit of the dynein regulatory complex in *Chlamydomonas* is a homologue of a growth arrest-specific gene product. *J. Cell Biol.* **162**, 47–57.

Rupp, G., O'Toole, E., Gardner, L.C., Mitchell, B.F., and Porter, M.E. (1996). The *sup-pf-2* mutations of *Chlamydomonas* alter the activity of the outer dynein arms by modification of the gamma-dynein heavy chain. *J. Cell Biol.* **135**, 1853–1865.

Salathe, M. (2006). Regulation of mammalian ciliary beating. *Annu. Rev. Physiol.* **69**, 401–422.

Sale, W.S. and Howard, D.R. (1995). Microscopic assays of flagellar dynein activity. *Methods Cell Biol.* **47**, 257–262.

Sale, W.S. and Satir, P. (1977). Direction of active sliding of microtubules in *Tetrahymena* cilia. *Proc. Natl. Acad. Sci. U.S.A.* **74**, 2045–2049.

San Agustin, J.T. and Witman, G.B. (1994). Role of cAMP in the reactivation of demembranated ram spermatozoa. *Cell Motil. Cytoskeleton* **27**, 206–218.

Satir, P. (1980). Structural basis of ciliary movement. *Environ. Health Perspect.* **35**, 77–82.

Satir, P. (1984). The generation of ciliary motion. *J. Protozool.* **31**, 8–12.

Satir, P. (1985). Switching mechanisms in the control of ciliary motility. In: *Modern Cell Biology*, Vol. 4 (B. Satir, Ed.), pp. 1–46. Alan R. Liss, Inc, New York, NY.

Satir, P. and Christensen, S.T. (2007). Overview of structure and function of mammalian cilia. *Annu. Rev. Physiol.* **69**, 377–400.

Schmidt, M., Gessner, G., Luff, M., Heiland, I., Wagner, V., Kaminski, M., Geimer, S., Eitzinger, N., Reissenweber, T., Voytsekh, O., Fiedler, M., Mittag, M., and Kreimer, G. (2006). Proteomic analysis of the eyespot of *Chlamydomonas reinhardtii* provides novel insights into its components and tactic movements. *Plant Cell* **18**, 1908–1930.

Setter, P.W., Malvey-Dorn, E., Steffen, W., Stephens, R.E., and Linck, R.W. (2006). Tektin interactions and a model for molecular functions. *Exp. Cell Res.* **312**, 2880–2896.

Shingyoji, C., Higuchi, H., Yoshimura, M., Katayama, E., and Yanagida, T. (1998). Dynein arms are oscillating force generators. *Nature* **393**, 711–714.

Smith, E.F. (2002a). Regulation of flagellar dynein by calcium and a role for an axonemal calmodulin and calmodulin-dependent kinase. *Mol. Biol. Cell* **13**, 3303–3313.

Smith, E.F. (2002b). Regulation of flagellar dynein by the axonemal central apparatus. *Cell Motil. Cytoskeleton* **52**, 33–42.

Smith, E.F. and Sale, W.S. (1992a). Regulation of dynein-driven microtubule sliding by the radial spokes in flagella. *Science* **257**, 1557–1559.

Smith, E.F. and Sale, W.S. (1992b). Structural and functional reconstitution of inner dynein arms in *Chlamydomonas* flagellar axonemes. *J. Cell Biol.* **117**, 573–581.

Smith, E.F. and Yang, P. (2004). The radial spokes and central apparatus: mechano-chemical transducers that regulate flagellar motility. *Cell Motil. Cytoskeleton* **57**, 8–17.

Stephens, R.E., Oleszko-Szuts, S., and Linck, R.W. (1989). Retention of ciliary nine-fold structure after removal of microtubules. *J. Cell Sci.* **92**, 391–402.

Summers, K.E. and Gibbons, I.R. (1971). Adenosine triphosphate-induced sliding of tubules in trypsin-treated flagella of sea-urchin sperm. *Proc. Natl. Acad. Sci. U.S.A.* **68**, 3092–3096.

Summers, K.E. and Gibbons, I.R. (1973). Effects of trypsin digestion on flagellar structures and their relationship to motility. *J. Cell Biol.* **58**, 618–629.

Tash, J.S. and Bracho, G.E. (1994). Regulation of sperm motility: emerging evidence for a major role for protein phosphatases. *J. Androl.* **15**, 505–509.

Vijayaraghavan, S., Liberty, G.A., Mohan, J., Winfrey, V.P., Olson, G.E., and Carr, D.W. (1999). Isolation and molecular characterization of AKAP110, a novel, sperm-specific protein kinase A-anchoring protein. *Mol. Endocrinol.* **13**, 705–717.

Wagner, V., Gessner, G., Heiland, I., Kaminski, M., Hawat, S., Scheffler, K., and Mittag, M. (2006). Analysis of the phosphoproteome of *Chlamydomonas reinhardtii* provides new insights into various cellular pathways. *Eukaryotic Cell* **5**, 457–468.

Walczak, C.E. and Nelson, D.L. (1994). Regulation of dynein-driven motility in cilia and flagella. *Cell Motil. Cytoskeleton* **27**, 101–107.

Warner, F.D. (1976). Ciliary inter-microtubule bridges. *J. Cell Sci.* **20**, 101–114.

Warner, F.D. and Satir, P. (1974). The structural basis of ciliary bend formation. Radial spoke positional changes accompanying microtubule sliding. *J. Cell Biol.* **63**, 35–63.

Wirschell, M., Yanagisawa, H.A., Kamiya, R., Witman, G.B., Porter, M.E., and Sale, W.S. (2006). The role of the IC97 in I1-dynein assembly and axonemal anchoring. *Mol. Biol. Cell* **17**. abstract #1602 (CD-ROM)

Wirschell, M., Hendrickson, T., and Sale, W.S. (2007). Keeping an eye on I1: I1 dynein as a model for flagellar dynein assembly and regulation. *Cell Motil. Cytoskeleton* **64**, 569–579.

Witman, G.B. (1993). *Chlamydomonas* phototaxis. *Trends Cell Biol.* **3**, 403–408.

Witman, G.B., Plummer, J., and Sander, G. (1978). *Chlamydomonas* flagellar mutants lacking radial spokes and central tubules. Structure, composition, and function of specific axonemal components. *J. Cell Biol.* **76**, 729–747.

Wong, W. and Scott, J.D. (2004). AKAP signalling complexes: focal points in space and time. *Nat. Rev. Mol. Cell Biol.* **5**, 959–970.

Woolley, D.M. (1997). Studies on the eel sperm flagellum. I. The structure of the inner dynein arm complex. *J. Cell Sci.* **110**, 85–94.

Yagi, T. and Kamiya, R. (2000). Vigorous beating of *Chlamydomonas* axonemes lacking central pair/radial spoke structures in the presence of salts and organic compounds. *Cell Motil. Cytoskeleton* **46**, 190–199.

Yagi, T. and Kamiya, R. (2006). Identification of the genes of all known *Chlamydomonas* inner-arm dynein heavy chains. *Mol. Biol. Cell* **17**, L47. (Late Abstracts, The American Society for Cell Biology 46th Annual Meeting)

Yagi, T., Minoura, I., Fujiwara, A., Saito, R., Yasunaga, T., Hirono, M., and Kamiya, R. (2005). An axonemal dynein particularly important for flagellar movement at high viscosity. Implications from a new *Chlamydomonas* mutant deficient in the dynein heavy chain gene *DHC9*. *J. Biol. Chem.* **280**, 41412–41420.

Yamamoto, R., Yagi, T., and Kamiya, R. (2006). Functional binding of inner-arm dyneins with demembranated flagella of *Chlamydomonas* mutants. *Cell Motil. Cytoskeleton* **63**, 258–265.

Yanagisawa, H.-A. and Kamiya, R. (2004). A tektin homologue is decreased in *Chlamydomonas* mutants lacking an axonemal inner-arm dynein. *Mol. Biol. Cell* **15**, 2105–2115.

Yang, P. and Sale, W.S. (1998). The Mr 140,000 intermediate chain of *Chlamydomonas* flagellar inner arm dynein is a WD-repeat protein implicated in dynein arm anchoring. *Mol. Biol. Cell* **9**, 3335–3349.

Yang, P. and Sale, W.S. (2000). Casein kinase I is anchored on axonemal doublet microtubules and regulates flagellar dynein phosphorylation and activity. *J. Biol. Chem.* **275**, 18905–18912.

Yang, P., Fox, L., Colbran, R.J., and Sale, W.S. (2000). Protein phosphatases PP1 and PP2A are located in distinct positions in the *Chlamydomonas* flagellar axoneme. *J. Cell Sci.* **113**, 91–102.

CHAPTER 10

Microtubule–Membrane Interactions in *Chlamydomonas* Flagella

William Dentler

Department of Molecular Biosciences,
University of Kansas, Lawrence, Kansas, USA

CHAPTER CONTENTS

I. INTRODUCTION

The flagellar membrane forms a tube around the microtubular axoneme as the microtubules grow from the basal body. The membrane is linked to the basal bodies, to the doublet microtubules along the flagellar axis, and to the tips of the central and A-microtubules by a variety of bridge structures. Connections between the membrane and basal body/transition region likely maintain the flagellum at the surface of the cell and may create unique flagellar membrane and matrix domains. Connections along the flagellar axis may stabilize the membrane and attached mastigonemes as the flagellum beats, facilitate flagellar surface motility, position signaling complexes, and, in gametes, stabilize contact between mating cells. Capping structures that link the flagellar membrane to the distal tips of microtubules are the only known structures that bind to the inner wall of microtubules and that are exclusively located at the microtubule plus-ends. The caps may directly regulate microtubule assembly and disassembly or provide a scaffold to which regulatory factors can bind. The caps also may cover the microtubule ends and protect them from physical damage that would occur as they rub along the membrane during flagellar beating.

Each of the classes of microtubule–membrane bridges described in this review has been found in all cilia and flagella examined, although structural details and specific signaling complexes often are species- and cilia-specific. Their structures and their occurrences in many different cilia have been reviewed elsewhere (Dentler, 1981, 1987, 1990), so this review will focus principally on the microtubule–membrane bridges found in *Chlamydomonas*. Given their importance in flagellar maintenance, assembly, and signal transduction, it is hoped that this review will stimulate further study of these poorly understood structures.

II. CILIARY BASES AND TRANSITION REGIONS

A. Introduction to the transition region

The transition region serves as a structural attachment site for flagellar basal bodies and, when basal bodies are removed, as in the green alga *Chlorogonium* during cell division (Hoops and Witman, 1985), for the attachment of the flagellum to the cell body. It also forms the boundary that separates flagellar membrane components, soluble and insoluble components of the flagellar axoneme, and signaling complexes located on the flagellum from those that are located in the cytoplasm or on the plasma membrane. Chapter 2 provides additional details regarding basal body and transition region structure.

B. Structural links between the basal body/transition region and the membrane

Basal bodies determine the site at which flagella will assemble, establish the orientation of the two flagella, and maintain flagellar attachment to the cell body. To assemble flagella, basal bodies first attach to the plasma membrane by transition fibers associated with the triplet microtubules (Ringo, 1967; Weiss et al., 1977; Gaffal, 1988; Silflow et al., 2001; O'Toole et al., 2003). We have little detailed understanding of the mechanisms by which basal bodies dock to the plasma membrane in *Chlamydomonas*, but clues may come from studies in mammalian cells. In these cells, basal bodies often are centrioles that move to the cell surface and attach to the membrane. A membrane vesicle first attaches to the distal end of the centriole and forms a cap over the end of the centriole, and the complex then migrates to the cortex, where it initiates the formation of a primary cilium (Sorokin, 1968). A number of centrosomal proteins that help anchor centrioles to the centrosomal matrix have been identified (reviewed by Bornens, 2002). A centriole-associated protein, Odf2, or cenexin, forms a discrete appendage on centrioles that appears to be essential for the formation of primary cilia. In $odf2^{-/-}$ mouse cells, ciliogenesis, but not cell division, is completely suppressed (Ishikawa et al., 2005), which indicates that Odf2 may be an important docking protein.

Once basal bodies attach to the plasma membrane and to the cytoskeleton (Wright et al., 1983; Hoops et al., 1984; Silflow et al., 2001; O'Toole et al., 2003), flagellar doublet microtubules assemble onto the ends of the A- and B-tubules of each triplet, and a transition region, which has prominent links to the newly forming flagellar membrane and contains an inner H-shaped cup region, distal to which the central microtubules grow, is formed (Figure 10.1C–G). The polarity of the central microtubules is the same as that of the doublet microtubules (the plus-end is at the distal flagellar tip) (Euteneur and McIntosh, 1981a,b) and it is generally accepted that central microtubules, like the doublets, assemble at the distal tip (Johnson and Rosenbaum, 1992). However central microtubules have been found to grow proximally, into the basal body, in living *Chlamydomonas* transition region mutants (Jarvik and Suhan, 1991) and in isolated respiratory cilia (Dentler and LeCluyse, 1982a). The site at which central microtubules assemble, therefore, is not completely understood.

Basal bodies and transition regions are linked to the plasma membrane by three different structures, each of whose attachment to the membrane is associated with intramembrane particles seen in freeze-fractured cells (Weiss et al., 1977 and see Chapter 11). Transition fibers, or basal body struts (Ringo, 1967; O'Toole et al., 2003), radiate from the distal end of the basal body and anchor to particles in the plasma membrane (Weiss et al., 1977). The base of the transition region, where the distal end of the basal body attaches to the membrane, is attached to a ring of particles called the "flagellar bracelet"

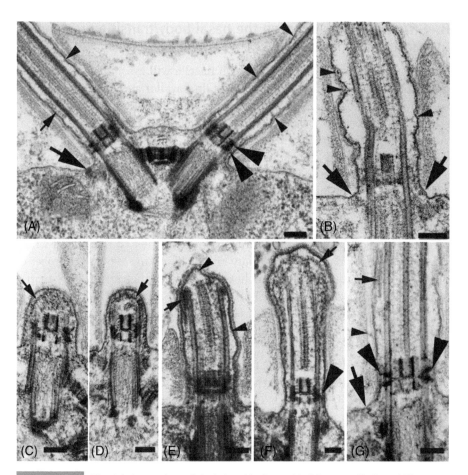

FIGURE 10.1 *Microtubule-membrane links in basal bodies and in fully-grown (A, B, and G) and in regenerating (C–F) flagella. Fully-grown flagella maintain a V-shaped orientation that is maintained by the distal striated fiber (the large structure linking basal bodies) and microtubule rootlets (not shown). Prominent Y-shaped links (large arrowheads) bind the transition regions to the membrane, and transition fibers (large arrows) link the distal end of the basal body to the membrane. Filamentous bridges (small arrowheads) and IFT particles (small arrows) bridge the membrane to doublet microtubules. Following deflagellation, new flagella assemble onto the distal tips of basal body microtubules (C–F). Initially, IFT particles line the flagellar membrane (C and F). As microtubules elongate, the IFT particles gradually spread out and are seen as discrete raft-like particle arrays linked together by filamentous structures (Dentler, 2005). Bars = 0.1 μm.*

(Weiss et al., 1977). As seen in freeze-fractured preparations (see Figure 11.3), the bracelet appears to be an important structure in *Chlamydomonas* and is composed of uniformly sized particles that touch one another (Weiss et al., 1977). Distal to the basal body, near the mid-transition region, flagellar microtubules are associated with particles that form the flagellar "necklace." These have been proposed to be attached to prominent bridges, or "connectives," that form pairs of links (Figure 10.1A, G) that appear as Y-shaped "champagne glass" structures in cross-section (Ringo, 1967; Gilula and Satir, 1972; Besharse and Horst, 1990) and hence are often referred to as "Y-shaped" connectors or links. These appear to tether the membrane to the doublet microtubules in the transition region. However, the number of necklace rows should match the number of connectives seen in thin-sectioned flagella; that is, there should be two rows of particles, but, in *Chlamydomonas*, Weiss et al. (1977) reported three rows of necklace particles, which would be more consistent with particles being excluded from the attachment sites of the connectives. Ciliary necklaces are commonly found in other cilia, including cilia of *Paramecium* (Watanabe, 1990), *Tetrahymena* (Sattler and Staehelin, 1974), and avian reproductive tracts (Chailley et al., 1990).

The ciliary necklace region of the membrane is likely protein rich, because it resists solubilization by non-ionic detergents (Snell et al., 1974; Goodenough, 1983; Kamiya and Witman, 1984). This region is similar to that in the connecting cilium in photoreceptor cells (Lechtreck et al., 1999), in which the connective bridges clearly link the doublet microtubules to the plasma membrane even after non-ionic detergent extraction (Besharse and Horst, 1990).

Another set of intramembrane particles found in ciliated protozoans is the ciliary granule plaques (Gilula and Satir, 1972; Sattler and Staehelin, 1974; Watanabe, 1990). These intramembrane particles form rectangular arrays positioned adjacent to doublet microtubules and appear to be linked to the doublets by bridging structures. Their function is unknown, but they may be involved in calcium ion transport (reviewed in Dentler, 1981, 1990). These plaques, however, do not appear to be present in *Chlamydomonas* or avian respiratory cilia (Weiss et al., 1977; Chailley et al., 1990).

Distal to the transition region, the matrix of the newly formed flagellar compartment initially fills with granular structures that line the inner face of the flagellar membrane (Figure 10.1C, D; Dentler, 2005). Some of these are likely to be intraflagellar transport (IFT) components (see Chapter 4), but others form the microtubule capping structures (section IV) and other flagellar structures. As the microtubules grow, the granular material collects at the distal tip, where tubulin is added to the growing microtubules (Figure 10.1F; Johnson and Rosenbaum, 1992).

C. Proteins associated with the transition region microtubule–membrane links

Specific glycoproteins associated with the Y-shaped linkers have been identified on the extraflagellar surface of photoreceptor connecting cilia (Horst et al., 1987), but similar glycoproteins have not been identified in *Chlamydomonas*. The best-characterized linker protein is p210, a protein associated with the Y-shaped linkers in *Spermatozopsis similis* that remains associated with the distal end of basal bodies after flagellar amputation (Lechtreck et al., 1999). Antibodies against the *S. similis* p210 protein also stained the basal body/transition region of *Chlamydomonas*, but, when examined by immuno-gold electron microscopy, the proximal portion of the axoneme and a structure surrounding the distal connecting fiber were labeled; additionally, the antibody recognized 65- and 85-kD polypeptides on western blots but not any high-molecular-weight proteins (Schoppmeier and Lechtreck, 2002). Therefore, the relationship between the p210 *S. similis* linker protein and proteins associated with *Chlamydomonas* transition region linkers, if any, is unclear.

D. Functions associated with basal body and transition region bridges

In *Chlamydomonas* and in most mammalian cilia, the assembly of flagellar microtubules occurs after basal bodies dock with the plasma membrane, so a primary function of the basal body and transition region appears to be to initiate flagellar assembly. Once assembled, the basal body and transition region become the sites at which proteins and membrane components, including IFT particles, are selected for transport into the flagellum. The role of the membrane in the initiation of flagellar assembly is not well understood. In *Drosophila* sperm (Tokuyasu et al., 1972; Han et al., 2003; Witman, 2003), basal bodies nucleate microtubules in the cytoplasm but a small cap of membrane is associated with the distal tips of the growing microtubules. However, a membrane cap has not been identified in *Giardia*, *Cryptosporidium*, two *Theileria* species, and seven *Plasmodium* species (Sinden et al., 1976; Briggs et al., 2004), where axonemes grow from basal bodies in the cytoplasm and later attach to the plasma membrane to protrude from the cell. Since a major portion of the flagellum is assembled while in contact with the cytoplasm, some or all IFT-particle proteins and associated machinery that are essential for *Chlamydomonas* flagella and sensory cilia are absent and unnecessary (Avidor-Reiss et al., 2004; Briggs et al., 2004).

Attachments of basal bodies and transition regions to the membrane are permanent during interphase, and the attachments remain after flagellar amputation and after extensive washes with non-ionic detergent (Snell et al., 1974). The attachments do not appear to restrict the rotational movement of basal bodies or of flagella, because basal bodies can freely rotate in the plane of the membrane until interactions among the basal bodies and cytoplasmic rootlets are established (Dentler, 1981, 1987; Boisvieux-Ulrich et al., 1988; Sandoz et al., 1988; Silflow et al., 2001). Prior to cell division, *Chlamydomonas* flagella are disassembled; the basal bodies remain attached to the plasma membrane and are associated with spindle poles (Coss, 1974 and see Chapter 14). In the related algae *Chlorogonium* and *Polytoma*, basal bodies detach from the flagellum without flagellar disassembly and without affecting flagellar motility (Hoops and Witman, 1985).

In *Chlamydomonas* and other organisms, the transition region appears to form a boundary that separates flagellar and plasma membrane glycoproteins (Musgrave et al., 1986). For example, the mastigonemes, fine filamentous structures that project from the surface of the flagellum, are limited to the flagellar membrane. The site at which *Chlamydomonas* mastigonemes are attached is not understood, but, based on studies of *Euglena* (Bouck et al., 1990), it is reasonable to consider that the mastigonemes are secreted near the base of the flagellum and then moved onto the flagellum proper as a result of attachment to a transmembrane anchoring protein that is transported through the transition region (see Chapter 11). All integral membrane proteins found in the flagellar membrane must pass through the transition region. Some of these proteins may be modified or activated by enzymes in the transition region. For example, sexual agglutinins are activated as they pass through the transition region or shortly thereafter (Hunnicutt et al., 1990), although an active role for transition region enzymes has not been discovered.

The transition region also is the site at which signaling components for flagellar autotomy are located (Lewin and Lee, 1985; Lohret et al., 1998; Mahjoub et al., 2004; see Chapter 3), and is proposed to be the site at which flagellar proteins are selected for transport into the flagellar compartment (Rosenbaum and Witman, 2002). Targeting sequences for *Chlamydomonas* flagellar proteins or protein complexes have not been identified, but eighteen flagellar "membrane-plus-matrix" proteins contain a myristoylation motif (Pazour et al., 2005), which has been shown to direct membrane-spanning proteins into *Leishmania* and *Trypanosoma* flagella (Godsel and Engman, 1999; Tull et al., 2004).

The transition region or the flagellar base also is the site at which IFT particles enter the flagellum (Kozminski et al., 1993; Piperno et al., 1998; Iomini et al., 2001; see Chapter 4). The transport of these particles up and down the flagellum is essential for flagellar growth and maintenance. Although we have relatively little understanding of what these particles are transporting, some flagellar proteins are reported to enter the flagellar compartment and assemble onto the axonemes independently of IFT (Piperno et al., 1996; Song and Dentler, 2001). The transition region, or a region close to the basal bodies, must be a site for the loading of cargo onto IFT particles and for the regulation of IFT motors. For example, particles containing IFT52 proteins dock at the basal body transition fibers (Deane et al., 2001). Proteins that dock at the transition region in other cilia include two IFT particle-associated Bardet-Biedl syndrome proteins (BBS-7 and -8) in *Caenorhabditis elegans* (Blacque et al., 2004), IFT20 in mammalian cells (Follit et al., 2006), and nephrocystin in renal respiratory cilia and photoreceptor connecting cilia (Fliegauf et al., 2006).

III. LATERAL INTERACTIONS IN THE AXONEME

A. Two types of bridges mediate lateral interactions

Along the flagellum, filamentous structures (bridges) link the doublet microtubules to the surrounding membrane. In some flagella, bridges also

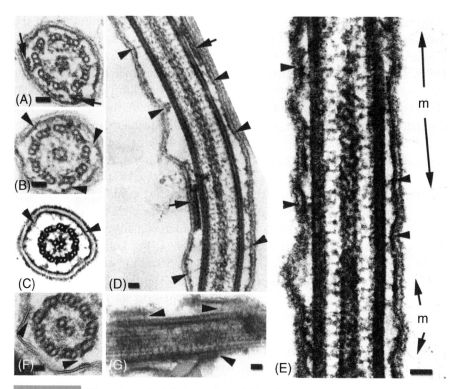

FIGURE 10.2 *Filamentous microtubule-membrane bridges (small arrowheads) and IFT particles (small arrows) link microtubules to membranes along the axoneme. Flagella on fixed cells are shown in (A)–(E). Isolated flagella extracted with 1% Nonidet P-40 are shown in (F) and (G). Membranes are swollen and separated from the axonemes by the fixation procedure, but where IFT particles and filamentous bridges are present, the membrane is held more tightly to the microtubules. The membranes are linked to microtubules by filamentous bridges even after most membrane is released by detergent-extraction (F and G). The IFT particles are linked primarily to the B-tubules of each doublet but the filamentous bridges appear to be linked to each doublet microtubule at the A–B junctions or to a portion of the outer dynein arms. Mastigonemes (m in E) project from the flagellar surface but do not appear to be associated with filamentous bridges. Panel C, courtesy of R.A. Bloodgood. Bars = 50nm.*

attach doublets to paraflagellar rods or to large mastigoneme complexes that extend through the flagellar membrane (Dentler, 1981; Bouck et al., 1990). In *Chlamydomonas*, two types of bridges appear to link the flagellar membrane to the microtubules (Figure 10.2). Transient or moving bridges may be formed by IFT particles that are linked to the membrane and to the doublet microtubules but which continuously move up and down the flagellum (Kozminski et al., 1995; Dentler, 2005). Evidence that these bridges link the membrane to the doublet microtubules is based on morphological observations of short filaments between the IFT particles and the membrane (Kozminski et al., 1995; Dentler, 2005), and observations that the flagellar membrane appears to adhere to the doublet microtubules more tightly in the presence of the IFT particles. This interpretation should be regarded cautiously, however, because it does not prove that the IFT particles are attached to the membrane.

Filamentous bridge structures may be more permanent and are most easily recognized when the membrane is slightly pulled away from the axoneme (arrows, Figures 10.1B and 10.2D, E). Both types of bridges bring the membrane in close apposition to the doublet microtubule. The filamentous bridges also maintain attachment of the membrane to the microtubules after isolated flagella are extracted with non-ionic detergent (Figure 10.2F,G) and may maintain the integrity of the flagellar membrane on motile flagella (Dentler, 1981, 1990).

B. Structure of bridges that link doublets to the membrane

Unlike most flagellar structures, the finer filamentous bridges appear to be irregularly spaced and difficult to preserve. They also appear to be flexible, because they maintain attachments between the membrane and doublets even when the membrane has partially pulled away from the axoneme. The separation of membranes from the axoneme may be due to poor membrane fixation, because cryopreservation maintains smooth membranes that are more closely opposed to the microtubules (Nicastro et al., 2006). Slight swelling of the membrane, however, reveals the bridges and IFT particles, so the micrographs of thin sections presented in this chapter are of conventionally fixed *Chlamydomonas* cells and flagella. Additionally, some bridges that link membranes and microtubules may not be preserved or stained by known procedures, as discussed below. Despite their poor preservation, bridges clearly serve to link the membrane to the axonemes, and with the discovery of signaling complexes arrayed along flagella, it is likely that the bridges are important scaffolds to position and anchor membrane-associated proteins along cilia and flagella.

The sites to which the linkers attach to the doublets are not well understood. IFT particles bind to the B-tubules of each doublet (Kozminski et al., 1995; Pedersen et al., 2006), although they also appear to interact with the outer row of dynein arms in some images (arrow, Figure 10.2A). Since IFT particles can extend to the distal tips of flagella, distal to the ends of the B-tubules and dynein arms (Pedersen et al., 2006), it is possible that the IFT particle is a linear aggregate with a motor at only one end, allowing the remainder of the particle to "swing" up into the distal tip of the flagellum (Dentler, 2005).

The thinner filamentous bridges appear to attach either to the junction of the A- and B-tubules or to the outer row of dynein arms (arrowheads, Figure 10.2B–F; see Figure 2 in Bloodgood, 1990). The bridges may bind to the A-tubules as well, since the A-tubule extensions of the doublets are clearly bridged to the membrane near the distal tip of the axoneme (Figure 10.3C). The bridges that appear to link the membrane to the outer arm could be artifacts created by superimposition of the outer arms and membranes in cross-sections, since the thickness of cross-sections are greater than the 24 nm periodicity of dynein arms. A careful comparison of the doublet-microtubule–membrane bridge structures in flagella from wild-type *Chlamydomonas* with those in *oda* mutants, which lack outer dynein arms, should reveal whether or not these bridges are associated with the outer arms.

Based on studies of isolated cilia from *Tetrahymena* and scallop gills (Dentler et al., 1980), structures that maintain tight association between membranes and doublet microtubules are difficult to stain and observe in sectioned material, so it is possible that important links between membranes and the microtubules in *Chlamydomonas* as well as in other cilia and flagella remain to be discovered.

C. Biochemical analysis of bridge proteins

The only biochemical analysis of microtubule–membrane linkers in cilia was carried out with cilia isolated from scallop gills and from *Tetrahymena pyriformis* (Dentler et al., 1980). In these cilia, a cleavable photoactivatable cross-linker, *N,N'*-dithiobisphenylazide, was used to stabilize the bridges in isolated cilia. The cilia were then extracted with non-ionic detergent to release membrane that was not attached to the microtubules. When examined by electron microscopy, the cross-linker preserved patches of membrane tightly linked to doublet microtubules; when cross-linked cilia were treated with a reducing agent to cleave the cross-linker, membranes were released from the doublet microtubules. Activation of the cross-linker on living scallop gills and *Tetrahymena* cells also inhibited ciliary motility. This indicated that the bridges linking membranes to microtubules must not be permanent in motile cilia.

To identify proteins associated with the bridges, cross-linked polypeptides were isolated using non-reducing sodium dodecyl sulfate-polyacrylamide gel electrophoresis, and re-run under reducing conditions that cleaved the cross-linkers. In both scallop and *Tetrahymena* cilia, the major cross-linked proteins included a tubulin-like protein and a high-molecular-weight cytoplasmic dynein (Dentler et al., 1980). None of the axonemal proteins were cross-linked, confirming that the cross-linked polypeptides were specifically membrane-associated. These results were the first to indicate a role for cytoplasmic dynein in maintaining membrane and axonemal connections. Unfortunately, the bridge structures linking the membranes and microtubules could not be stained or resolved by transmission electron microscopy, so it was not possible to identify unambiguously the filamentous bridges as part of the cytoplasmic dynein complexes. In retrospect, the cross-linking reagent may have preserved the filamentous bridges and/or IFT particles and their retrograde motor.

The cross-linking experiments have not been repeated with *Chlamydomonas*, but it would be interesting to do so because they may reveal proteins linking IFT complexes to the membrane, motors associated with flagellar surface motility (Bloodgood, 1990), and other membrane-associated protein complexes that connect membranes to axonemes. Additional potential bridge proteins include Ca^{2+}-ATPases and a polycystin 2 homologue that, while predicted to be transmembrane proteins, fractionate with the axoneme after non-ionic detergent extraction (Pazour et al., 2005).

Finally, hair-like mastigonemes (Figure 10.2E) and a surface-exposed glycoprotein also remain attached to axonemes after detergent extraction (Dentler, 1980; Reinhart and Bloodgood, 1988), but, since these proteins have no known transmembrane domains, it is unclear if they bind directly to the microtubules or, more likely, bind to other proteins associated with microtubules and the membrane.

D. Functions of the bridges

Without links between the membrane and microtubules, is it likely that the tubular structure would be lost and the membrane would form a bulbous extension of the plasma membrane or would "flap" as flagella beat. Some indication of this can be seen in flagella fixed in low osmotic strength media, in which the membranes often balloon away from the axoneme even though most of the membrane remains linked to the microtubules (Figure 10.2; Dentler et al., 1980; Dentler, 1990). At the flagellar tips, the membrane often can swell, carrying the attached doublet microtubules with the membrane (Figure 10.3C). A *Chlamydomonas* mutant, *spon1*, in which the membrane ballooned away from the axonemes, was described by Randall (1969), but this mutant is not available in any known *Chlamydomonas* mutant collections. The isolation of one mutant, however, indicates that bridge mutants may be isolated in future screens. Clearly, the bridges maintain tight associations between the membrane and the microtubules.

The bridges may function similar to myosin-1a or ezrin family proteins that link microvillar membrane proteins to actin filaments (Bretscher et al., 2002; Tyska et al., 2005). In myosin-1a deficient mice, brush border membranes are partially distended and disorganized, consistent with disruptions in a membrane-cytoskeleton linker protein (Tyska et al., 2005). Based on proteomic analysis, both ezrin- and myo1-like proteins are in the detergent-solubilized membrane-plus-matrix fraction of *Chlamydomonas* flagella (Pazour et al., 2005), even though flagellar actin filaments, their presumed binding partners, have not been identified (Detmers et al., 1985). Actin is associated with *Chlamydomonas* inner arm dyneins (Watanabe et al., 2004) and with the detergent-solubilized membrane-plus-matrix (Kozminski et al., 1993; Pazour et al., 2005), but its role, if any, in the linkage of the membrane to the microtubules is not known. Curiously, cytochalasin D, which disrupts actin filaments, induces flagella to undergo periods of dynamic growth and shortening (Dentler and Adams, 1992), suggesting a role for actin filaments in flagellar maintenance even though flagellar actin filaments have not been identified.

Some of the microtubule–membrane linkers may be involved with flagellar surface motility, discussed elsewhere in this volume (see Chapter 11). The movements of microspheres along flagella may be associated with the movements of IFT particles, because inhibition of IFT in the *fla10* mutant at non-permissive temperature also is associated with inhibition of the movements of microspheres along flagellar surfaces (Kozminski et al., 1995). However, the IFT particles and the proteins associated with flagellar surface motility are unlikely to be permanently linked to one another, because microsphere movements gradually are inhibited by protein synthesis inhibitors without inducing flagellar shortening (Bloodgood et al., 1979), whereas continuous IFT appears to be required to maintain flagellar length. Thus, some of the filaments that appear to project from the membrane towards the microtubules may transiently attach to IFT particles and be involved with surface motility.

Until the proteins comprising the bridges are identified, it will be difficult to determine the function of the bridges. When they are identified, the sites of their binding and their effects on the movement of IFT particles will be important to examine, particularly in light of the effects of various microtubule-associated proteins on kinesin-based organelle transport in mammalian cells (Mandelkow et al., 2004; Baas and Qiang, 2005). One possibility is that the bridges help stabilize the membranes against forces generated by rapidly moving flagella, but the bridges also may affect the rates of IFT or pauses observed in IFT movements (Dentler, 2005; Mueller et al., 2005). Another possibility is that the bridges link the various signaling systems that are associated with flagella (Pazour and Witman, 2003; Davis et al., 2006; Benzing and Walz, 2006; Inglis et al., 2006) to the doublet microtubules and the membrane. Such linkage may be important for transport, positioning, and anchorage of the signaling complexes along the flagellum.

IV. CAPPING STRUCTURES LINK THE MICROTUBULE ENDS TO THE MEMBRANE

A. Dynamic processes occur at the flagellar tip

The assembly, disassembly, and turnover of flagellar microtubules and most axoneme-associated proteins occur at the plus-ends of flagellar microtubules, at the flagellar distal tip (Johnson and Rosenbaum, 1992). This requires not only the addition and removal of tubulins, but also of the other flagellar axonemal proteins (Song and Dentler, 2001). The distal tips also are the sites to which IFT particles are targeted in growing, non-growing, and shortening flagella, the sites at which IFT particles are remodeled, the sites at which cargo is unloaded or loaded, and at which anterograde kinesin-2 motors are deactivated and cytoplasmic dynein 1b motors are activated (Kozminski et al., 1995; Iomini et al., 2001; Dentler, 2005; Pedersen et al.,

2006 and see Chapter 4). Therefore, proteins and/or structures localized at the membrane tips and those that interact with the microtubules are likely to have important roles in the assembly, disassembly, and maintenance of flagellar microtubules.

B. Microtubule capping structures at flagellar tips

The distal tips of the central microtubules and the A-microtubules of each of the nine doublets are linked to the tip of the flagellar membrane by morphologically distinct capping structures (Figure 10.3; Dentler and Rosenbaum, 1977; Dentler, 1980). Although a variety of proteins have been identified that localize at the plus-ends of flagellar and cytoplasmic microtubules (Miller et al., 1990; Howard and Hyman, 2003; Mimori-Kiyosue and Tsukita, 2003; Pedersen et al., 2003), most appear to be transported along the surface of microtubules to the distal tip, where they accumulate. In contrast to these proteins, the caps described here are the only structures known to be exclusively located at the microtubule plus-ends and are the only structures that bind to the microtubules via a plug-shaped structure that inserts into the center, or lumen, of each microtubule.

The most prominent capping structure is the central microtubule cap that links the plus-ends of the central microtubules to the flagellar membrane (Figure 10.3A–E; Dentler, 1980). The central microtubule cap in *Chlamydomonas* is similar to that found at the tips of *Tetrahymena* cilia (Figure 10.3A; Dentler, 1980, 1984). Each of the two central microtubules is attached to the most proximal of two plates by short plugs that insert into the microtubule lumen (Dentler, 1980, 1984). The distal plate attaches to the membrane by a spherical bead (c in Figure 10.3A–C). The membrane is closely associated with the bead and appears to wrap around it when the flagellum is extracted with detergent (c in Figure 10.3E). This small tip of the membrane may contain specific docking sites for the cap, since normal-appearing central microtubule cap structures have been found in *Chlamydomonas* mutants that completely lack central microtubules (Dentler, unpublished results).

Each A-microtubule plus-end is attached to the tapering tip of the membrane by filamentous structures, called distal filaments (Dentler and Rosenbaum. 1977; d in Figure 10.3). The filaments lie along the inner surface of the membrane and are linked to the membrane by short bridge-like structures (Dentler, 1980). The filaments are bound together to form a compact carrot-shaped plug that inserts into the A-microtubule and is attached to the inner wall a short distance from the microtubule tip (Dentler, 1980; Figure 10.3A–D).

The caps are found at the tips of most cilia and flagella (Dentler, 1990). Two families of cap structures, based on ciliary function, include those found in "water-moving cilia" and in "mucus-moving cilia." In "water-moving

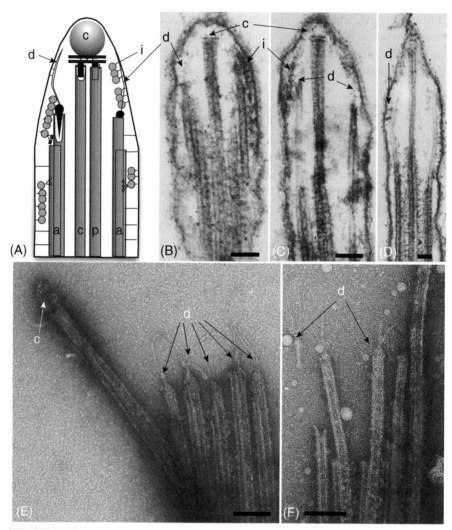

FIGURE 10.3 *Diagram and electron micrographs showing the distal filament and central microtubule capping structures at the distal tips of flagellar A- and central microtubules. (A) The central microtubule cap (c) is attached to the membrane and is formed by a spherical structure, two plates, and two plugs that insert into the distal ends of the central pair of microtubules (cp). Distal filaments (d) are attached to the flagellar membrane by short bridges and form a plug structure that inserts into the distal end of each of the A-microtubules (a). Attachment of the distal filaments and central microtubule cap to the membrane is best seen in thin sections (B–D). The spherical bead of membrane that remains attached to the central microtubule cap after detergent extraction is seen in (E). Distal filaments are best seen in negatively stained flagella (E and F). In (E), the distal filaments can be seen to extend from a spherical end on each A-microtubule. In (F), the filaments can be seen to be attached to a plug structure that has been released from the A-microtubule tips. Plug structures that link the central microtubules to the central microtubule cap are shown in Dentler (1984). IFT particles (i) travel along the doublet microtubules and extend into the flagellar tips, often in association with the distal filaments (B and C). Bars = 100nm.*

cilia," the entire cilium is in contact with the surrounding fluid as the cell swims. In these, the central pair caps and the A-microtubule distal filament caps attach independently to the membrane. The doublet caps often appear to lie along the membrane and be linked to it by short bridges (Dentler, 1980; Figure 10.3A–C). In cilia on cells that transport mucus across tissue surfaces, the plug structures in each of the A-microtubules are linked to a large cap that also is linked to plug structures inserted into the central microtubules (Dentler and LeCluyse, 1982b; Dentler, 1990). This complex may provide structural support to the tips as they push mucus across the epithelium (reviewed in Dentler, 1981, 1987, 1990).

The capping structures first appear during the first 1–2 μm of ciliary growth and link the membrane to the microtubule tips as flagella grow (Dentler and Rosenbaum, 1977; Portman et al., 1987; Dentler, 1990). In regenerating *Chlamydomonas* flagella, the tips first fill with an array of IFT-like particles that line the membrane; the caps gradually appear as the IFT particles distribute (Figure 10.1C–F). Once they appear, caps are attached exclusively to the plus-ends as the flagellum assembles, is maintained at steady length, and eventually shortens.

C. Composition of capping structures

All of the proteins so far identified at the tips of both cytoplasmic and ciliary microtubules appear to bind to the outer surface of the microtubules. Some end-binding proteins found in mammalian and yeast cells, including APC, Tea1P, and CLIP170 (Mimori-Kiyosue and Tsukita, 2003), and in protozoan cilia (Wang et al., 1994) can bind along microtubules but are concentrated at the microtubule plus-ends and are plus-end tracking proteins. Other proteins, including CLIPs, p150[glued], and EB1 (end-binding protein1) are exclusively concentrated at the ends of microtubules (Howard and Hyman, 2003; Mimori-Kiyosue and Tsukita, 2003). Of these, only EB1 (AAO62368) has been found in *Chlamydomonas*, where it is concentrated at the basal bodies and distal tips of the flagella. Depletion of EB1 at flagellar tips occurs concomitant with flagellar disassembly (Pedersen et al., 2003), suggesting that it plays a role in microtubule assembly and/or stability. The relationship between the capping structures seen at the tips of ciliary and flagellar microtubules and these end-binding proteins is not known, and none of the end-binding proteins associated with cytoplasmic microtubules has been shown to be associated with the caps.

Attempts to purify intact capping structures from *Chlamydomonas* have been unsuccessful, probably because of the low abundance of the capping structures and their instability after release from microtubule tips (Dentler, unpublished). Each flagellum contains a single central microtubule cap and nine outer doublet caps and, based on structural studies, each cap is composed of numerous proteins. The proportion of caps and cap proteins relative to all flagellar proteins is extremely small and may be below the detection limits of current flagellar proteomic studies (Pazour et al., 2005). Attempts to purify caps from *Tetrahymena* cilia were more successful, because each *Tetrahymena* cell has several hundred cilia that can be isolated with intact caps (Suprenant and Dentler, 1988). A *Tetrahymena* 97-kD ciliary tip polypeptide was recognized by an antibody to mammalian kinetochores (which also bind microtubule plus-ends) (Miller et al., 1990) but the antibody did not recognize tip proteins in *Chlamydomonas* (Dentler, unpublished results). Other approaches to identify cap proteins include comparing proteins isolated from long vs. short cilia (short cilia should be enriched in caps but not other major axonemal structures) and cross-linking cap fractions with proteins that appear to be enriched at tips. For example, one might identify proteins that co-purify with the end-binding protein EB1, which is enriched at flagellar tips (Pedersen et al., 2003). To date, the plugs and caps in cilia and flagella remain the only proteins exclusively found at microtubule ends, and proteins that are exclusively located at microtubule tips remain to be identified.

D. Functions of microtubule caps

1. Overview

The functions of the caps are unknown but possible functions include (1) protecting the microtubule tips from the membrane, (2) regulating flagellar growth and shortening, (3) regulating IFT motors and cargo loading and unloading, and (4) serving as a scaffold to position flagellar-signaling components. The regulation of microtubule assembly, of IFT motors and remodeling, and the positioning of signaling components are all known to occur at the flagellar tips, and the caps are in the proper position to serve as a scaffold to which the regulatory agents bind and would be maintained exclusively at the microtubule ends.

2. Caps may protect microtubule ends

Flagellar motility is caused by microtubule sliding, which extends to the tips of the flagellum. Thus, the tip of each of the doublet microtubules, actually the A-microtubule of each doublet, which is longer than the B-tubule, moves along the membrane, in the flagellar tip compartment. The central microtubules rotate during flagellar beating (Omoto et al., 1999) and their distal tips directly face the flagellar membrane. One might predict that the tips of these microtubules would be damaged as the flagellum rapidly beats and the membrane is dragged over the microtubule plus-ends. The central microtubule cap separates the microtubule tips from the membrane (Figures 10.1E and 10.3A–D). Since the microtubules are firmly attached to the plates by plugs, the plates must rotate in relation to the surrounding membrane. The spherical cap must be tightly bound to the membrane, because the membrane wraps around it when flagella are extracted with detergent. A membrane shear zone, perhaps similar to that found in a termite gut flagellate (Tamm, 1979), may form to permit the central microtubules and caps to rotate while maintaining the integrity of the membrane and minimizing any restraint on the ability of the central microtubules to rotate.

The longer distal filaments on the A-microtubules are attached along the membrane and similarly may protect doublets from shear forces generated by rubbing along the membrane surface (Figure 10.3A–C). The umbrella-shaped caps on mucus-transporting respiratory cilia may mediate the greatest force on ciliary tips, because only the tips of these cilia contact the mucus layer (see Dentler, 1981, 1987). Thus, one function of the elaborate capping structures may be to protect the microtubule tips and/or the surrounding membrane from frictional damage caused during flagellar beating. Similar frictional damage would not occur at the tips of cytoplasmic microtubules or microtubules associated with the mitotic apparatus, which may explain why prominent capping structures have not been found in cytoplasmic or mitotic microtubules.

3. Do caps regulate microtubule length?

Do caps, or microtubule plus-end-binding proteins associated with caps, regulate microtubule assembly and disassembly? They certainly are at the right place – the plus-ends of flagellar microtubules, where microtubule assembly and disassembly occurs (Johnson and Rosenbaum, 1992), they become assembled on the flagellar microtubule plus-ends shortly after flagella begin to assemble, and they remain attached to the microtubules as flagella grow, maintain steady-state length, and disassemble (see Dentler, 1987, 1990). The caps block tubulin addition and microtubule growth *in vitro* (Dentler and Rosenbaum, 1977; Dentler and LeCluyse, 1982a), so flagella must be able to regulate the associations of caps with microtubules to facilitate the addition or removal of tubulin from capped microtubules. The caps may regulate growth directly, or they may be the targets of other regulators.

One of the most interesting aspects of ciliary and flagellar microtubule assembly is that, within a single cilium, the lengths of individual microtubules are precisely regulated, and structurally distinct caps are found at the ends of specific microtubules (see Dentler, 1987, 1990). For example, in water-moving cilia and flagella, the central microtubules are significantly longer than the doublet microtubules, and structurally distinct caps – the distal filament caps on A-tubules and central microtubule cap on the central microtubules – are found on these different microtubules. In mucus-transporting cilia, the lengths of central microtubules and each of the doublet A-microtubules are nearly identical, and the caps attached to each of the doublets and the central microtubules appear to be identical in structure. In certain flatworm cilia, one set of doublet microtubules are nearly the same length as the central microtubules but another set of doublet microtubules (in the same cilium) are approximately one micron shorter than the other doublets. In elegant micrographs, Tyler (1979) showed that the caps attached to each of the microtubule sets had distinctly different structures (also see Dentler, 1981, 1987). The observations of structurally different caps in cilia and of similar caps being associated with similar length microtubules certainly are consistent with caps being involved in ciliary and flagellar length regulation.

4. Caps and IFT

The caps are the most obvious structural features of the flagellar tip compartment, so it would be surprising if the caps or cap-associated proteins did not have a role in the regulation of IFT motor activity, IFT-particle remodeling, and/or the unloading or loading of cargo carried by IFT (Qin et al., 2004; Pedersen et al., 2006; see Chapter 4).

Once IFT particles enter the flagellum, most move continuously from the base to the tip and from the tip to the base. Although occasional pauses in particle movement are observed, most particles do not reverse direction along the flagellum (Iomini et al., 2001; Dentler, 2005; Mueller et al., 2005). Therefore the distal tips must be the sites at which kinesin-2 motors are silenced and cytoplasmic dynein 1b motors are activated to carry particles to the flagellar base. Based on the size of anterograde and retrograde IFT particles, IFT-particle remodeling also must occur at the flagellar tips (Piperno et al., 1998). Several proteins, in addition to EB1 (Pedersen et al., 2003), are concentrated at flagellar tips. These include NIMA-related kinases associated with ciliary disassembly in *Tetrahymena* (Wloga et al., 2006). In *Chlamydomonas*, NIMA-related kinases were found in the flagellar proteome, but have not yet been localized within the flagellum. Also present at the ciliary tip in *Tetrahymena* is a 97-kD plus-end-binding protein whose interaction with microtubules is regulated by phosphorylation (Wang et al., 1994). Each of these proteins, while not exclusively located at the tips, likely has a role in the regulation of flagellar length and IFT activity.

5. Caps may serve as scaffolds for regulatory factors

The numbers of activities that occur at the flagellar tips suggest that the caps are scaffolds to which a number of different regulatory factors can attach. While attempts to purify caps or to identify flagellar assembly mutants specifically associated with caps have not been successful to date, the eventual characterization of these structures will likely reveal essential regulatory components for flagellar length as well as interesting mechanisms by which these components are delivered to microtubule tips and maintained at the tips throughout the life of specific ciliary and flagellar microtubules. Although most of our studies have been focused on the role of structures or proteins that bind to microtubules, the flagellar membrane also may play an important role in length regulation. Membrane proteins turn over rapidly in steady-state flagella (Song and Dentler, 2001), flagellar growth is inhibited by brefeldin A, which blocks Golgi trafficking (Haller and Fabray, 1998; Dentler, 2006), and an IFT-particle protein required for ciliary assembly is associated with the Golgi (Follit et al, 2006). Further studies of the interactions between microtubules and membranes will reveal the mechanisms and importance of the coordinated growth of flagellar microtubules and the surrounding membrane.

V. SUMMARY

Interactions between the flagellar membrane and doublet microtubules represent an important and potentially very fertile area for future studies. We currently have little understanding of the bridges that link the membranes to the doublets along the flagellar shaft, yet it is likely that these bridges will be key to understanding how signaling molecules and complexes can be anchored along the flagellum. We are only now beginning to identify the proteins that make up the connections between the basal body/transition region and the membrane, and still do not know the composition of the specialized cap structures that connect microtubules to the membrane at the tips of cilia and flagella. However, these structures are likely to play critical roles in the regulation of IFT and in the control of flagellar length, and defects in them may have devastating consequences for human health. Although many of these structures appear to be universally present in cilia and flagella, *Chlamydomonas* is still one of the best experimental systems with which to study them, because *Chlamydomonas* provides an easy and inexpensive cell system with access to biochemical, genetical, molecular biological, and morphological approaches for analyzing the composition, assembly and functioning of flagella.

ACKNOWLEDGMENT

I would like to thank Bob Bloodgood and George Witman for critically reading this chapter and for many stimulating conversations as the chapter was being written.

REFERENCES

Avidor-Reiss, T., Maer, A.M., Koundakjian, E., Polyanovsky, A., Keil, T., Subramaniam, S., and Zuker, C.S. (2004). Decoding cilia function: Defining specialized genes required for compartmentalized cilia biogenesis. *Cell* **117**, 527–539.

Baas, P.W. and Qiang, L. (2005). Neuronal microtubules: When the MAP is the roadblock. *Trends Cell Biol.* **15**, 183–187.

Benzing, T. and Walz, G. (2006). Cilium-generated signaling: A cellular GPS? *Curr. Opin. Nephrol. Hypertens.* **12**, 245–249.

Besharse, J.C. and Horst, C.J. (1990). The photoreceptor connecting cilium: A model for the transition zone. In: *Ciliary and Flagellar Membranes* (R.A. Bloodgood, Ed.), pp. 389–417. Plenum Press, New York, London.

Blacque, O.E., Reardon, M.J., Li, C., McCarthy, J., Mahjoub, M.R., Ansley, S.J., Badano, J.L., Mah, A.K., Beales, P.L., Davidson, W.S., Johnsen, R.C., Audeh, M., Plasterk, R.H.A., Baillie, D.L., Katsanis, N., Quamby, L.M., Wicks, S.R., and Leroux, M.R. (2004). Loss of *C. elegans* BBS-7 and BBS-8 protein function results in cilia defects and compromised intraflagellar transport. *Genes. Dev.* **18**, 1630–1642.

Bloodgood, R.A. (1990). Gliding motility and flagellar glycoprotein dynamics in *Chlamydomonas.* In: *Ciliary and Flagellar Membranes* (R.A. Bloodgood, Ed.), pp. 91–124. Plenum Press, New York and London.

Bloodgood, R.A., Leffler, E.M., and Bojczuk, A.T. (1979). Reversible inhibition of *Chlamydomonas* flagellar surface motility. *J. Cell Biol.* **83**, 664–674.

Briggs, L.J., Davidge, J.A., Wickstead, B., Gingerm, M.L., and Gull, K. (2004). More than one way to build a flagellum: Comparative genomics of parasitic protozoa. *Curr. Biol.* **14**, R611–R612.

Bornens, M. (2002). Centrosome composition and microtubule anchoring mechanisms. *Curr. Opin. Cell Biol.* **14**, 25–34.

Boisvieux-Ulrich, E., Laine, M.C., and Sandoz, D. (1988). *In vitro* effects of taxol on ciliogenesis in quail oviduct. *J. Cell Sci.* **92**, 9–20.

Bouck, G.B., Rosiere, T.K., and Levasseur, P.J. (1990). *Euglena gracilis*: A model for flagellar surface assembly, with reference to other cells that bear flagellar mastigonemes and scales. In: *Ciliary and Flagellar Membranes* (R.A. Bloodgood, Ed.), pp. 65–90. Plenum Press, New York and London.

Bretscher, A., Edwards, K., and Fehon, R.G. (2002). ERM proteins and merlin: Integrators at the cell cortex. *Nat. Rev. Mol. Cell Biol.* **3**, 586–599.

Chailley, B., Boisvieux-Ulrich, E., and Sandoz, D. (1990). Structure and assembly of the oviduct ciliary membrane. In: *Ciliary and Flagellar Membranes* (R.A. Bloodgood, Ed.), pp. 337–362. Plenum Press, New York and London.

Coss, R.A. (1974). Mitosis in *Chlamydomonas reinhardtii* basal bodies and the mitotic apparatus. *J. Cell Biol.* **63**, 325–329.

Davis, E.E., Brueckner, M., and Katsanis, N. (2006). The emerging complexity of the vertebrate cilium: New functional roles for an ancient organelle. *Dev. Cell* **11**, 9–19.

Deane, J.A., Cole, D.G., Seely, E.S., Diener, D.F., and Rosenbaum, J.L. (2001). Localization of the intraflagellar transport protein IFT52 identifies the transitional fibers of the basal bodies as the docking site for IFT particles. *Curr. Biol.* **11**, 1586–1590.

Dentler, W.L. (1980). Structures linking the tips of ciliary and flagellar microtubules to the membrane. *J. Cell Sci.* **42**, 207–220.

Dentler, W.L. (1981). Microtubule–membrane interactions in cilia and flagella. *Int. Rev. Cytol.* **72**, 1–47.

Dentler, W.L. (1984). Attachment of the cap to the central microtubules of *Tetrahymena* cilia. *J. Cell Sci.* **66**, 167–173.

Dentler, W.L. (1987). Cilia and flagella. *Int. Rev. Cytol.* **Suppl.17**, 391–456.

Dentler, W.L. (1990). Linkages between microtubules and membranes in cilia and flagella. In: *Ciliary and Flagellar Membranes* (R.A. Bloodgood, Ed.), pp. 31–64. Plenum Press, New York and London.

Dentler, W.L. (2005). Intraflagellar transport (IFT) during assembly and disassembly of *Chlamydomonas* flagella. *J. Cell Biol.* **170**, 649–659.

Dentler, W.L. (2006). Importance of Golgi for the regulation of flagellar length in *Chlamydomonas*. *Mol. Biol. Cell* **17**. abstract #1596 (CD-ROM).

Dentler, W.L. and Adams, C. (1992). Flagellar microtubule dynamics in *Chlamydomonas*: Cytochalasin D induces periods of microtubule shortening and elongation; and colchicine induces disassembly of the distal, but not proximal, half of the flagellum. *J. Cell Biol.* **117**, 1289–1298.

Dentler, W.L. and LeCluyse, E.L. (1982a). The effects of structures attached to the tips of tracheal ciliary microtubules on the nucleation of microtubule assembly *in vitro*. *Cell Motil. Cytoskel.* **Suppl. 1**, 13–18.

Dentler, W.L. and LeCluyse, E.L. (1982b). Microtubule capping structures at the tips of tracheal cilia: Evidence for their firm attachment during ciliary bend formation and the restriction of microtubule sliding. *Cell Motil.* **2**, 549–572.

Dentler, W.L. and Rosenbaum, J.L. (1977). Flagellar elongation and shortening in *Chlamydomonas* III. Structures attached to the tips of flagellar microtubules and their relationship to the directionality of microtubule assembly. *J. Cell Biol.* **74**, 747–759.

Dentler, W.L., Pratt, M.M., and Stephens, R.E. (1980). Microtubule–membrane interactions in cilia. II. Photochemical cross-linking of bridge structures and the identification of a membrane-associated ATPase. *J. Cell Biol.* **84**, 381–403.

Euteneur, U. and McIntosh, J.R. (1981a). Polarity of some motility-related microtubules. *Proc. Natl. Acad. Sci. U. S. A.* **78**, 372–376.

Euteneur, U. and McIntosh, J.R. (1981b). Structural polarity of kinetochore microtubules in PtK1 cells. *J. Cell Biol.* **89**, 338–345.

Detmers, P.A., Carboni, J.M., and Condeelis, J. (1985). Localization of actin in *Chlamydomonas* using antiactin and NBD-phallacidin. *Cell Motil.* **5**, 415–430.

Fliegauf, M., Horvath, J., von Schnackenburg, C., Olbrich, H., Muller, D., Thumfart, J., Schermer, B., Pazour, G.J., Neumann, H.P.H., Zentgraf, H., Benzing, T., and Orman, H. (2006). Nephrocystin specifically localizes to the transition zone of renal and respiratory cilia and photoreceptor connecting cilia. *J. Am. Soc. Nephrol.* **17**, 2424–2433.

Follit, J.A., Tuft, R.A., Fogarty, K.E., and Pazour, G.J. (2006). The intraflagellar transport protein IFT20 is associated with the Golgi complex and is required for cilia assembly. *Mol. Biol. Cell* **17**, 3781–3792.

Gaffal, K.P. (1988). The basal body-root complex of *Chlamydomonas reinhardtii* during mitosis. *Protoplasma* **143**, 118–129.

Gilula, N.B. and Satir, P. (1972). The ciliary necklace: A ciliary membrane specialization. *J. Cell Biol.* **53**, 494–509.

Godsel, L.M. and Engman, D.M. (1999). Flagellar protein localization mediated by a calcium-myristoyl/palmitoyl switch mechanism. *EMBO J.* **18**, 2057–2065.

Goodenough, U.W. (1983). Motile detergent-extracted cells of *Tetrahymena* and *Chlamydomonas*. *J. Cell Biol.* **96**, 1610–1621.

Haller, K. and Fabray, S. (1998). Brefeldin A affects synthesis and integrity of a eukaryotic flagellum. *Biochem. Biophys. Res. Commun.* **242**, 597–601.

Han, Y.G., Kwok, B.H., and Kernan, M.J. (2003). Intraflagellar transport is required in *Drosophila* to differentiate sensory cilia but not sperm. *Curr. Biol.* **13**, 1679–1686.

Hoops, H.J. and Witman, G.B. (1985). Basal bodies and associated structures are not required for normal flagellar motion or phototaxis in the green alga *Chlorogonium elongatum*. *J. Cell Biol.* **100**, 297–309.

Hoops, H.J., Wright, R.L., Jarvik, J.W., and Witman, G.B. (1984). Flagellar waveform and rotational orientation in a *Chlamydomonas* mutant lacking normal striated fibers. *J. Cell Biol.* **98**, 818–824.

Horst, C.H., Forestner, D.M., and Besharse, J.C. (1987). Cytoskeletal–membrane interactions: A stable interaction between cell surface glycoconjugates and doublet microtubules of the photoreceptor connecting cilium. *J. Cell Biol.* **105**, 2973–2988.

Howard, J. and Hyman, A.A. (2003). Dynamics and mechanics of the microtubule plus end. *Nature* **422**, 753–758.

Hunnicutt, G.R., Kosfiszer, M.G., and Snell, W.J. (1990). Cell body and flagellar agglutinins in *Chlamydomonas reinhardtii*: The cell body plasma membrane is a reservoir for agglutinins whose migration to the flagella is regulated by a functional barrier. *J. Cell Biol.* **111**, 1605–1616.

Inglis, P.N., Boroevich, K.A., and Leroux, M.R. (2006). Piecing together a ciliome. *Trends Genet.* **22**, 491–500.

Iomini, C., Babaev-Khaimov, V., Sassaroli, M., and Piperno, G. (2001). Protein particles in *Chlamydomonas* flagella undergo a transport cycle consisting of four phases. *J. Cell Biol.* **153**, 13–24.

Ishikawa, H., Akiharu, K., Shoichiro, T., and Sachiko, T. (2005). Odf2-deficient mother centrioles lack distal/subdistal appendages and the ability to generate primary cilia. *Nat. Cell Biol.* **7**, 517–524.

Jarvik, J.W. and Suhan, J.P. (1991). The role of the flagellar transition region: Inferences from the analysis of a *Chlamydomonas* mutant with defective transition region structures. *J. Cell Sci.* **99**, 731–740.

Johnson, K.A. and Rosenbaum, J.L. (1992). Polarity of flagellar assembly in *Chlamydomonas*. *J. Cell Biol.* **119**, 1605–1611.

Kamiya, R. and Witman, G.B. (1984). Submicromolar levels of calcium control the balance of beating between the two flagella in demembranated models of *Chlamydomonas*. *J. Cell Biol.* **98**, 97–107.

Kozminski, K.G., Johnson, K.A., Forscher, P., and Rosenbaum, J.L. (1993). A motility in the eukaryotic flagellum unrelated to flagellar beating. *Proc. Natl. Acad. Sci. U. S. A.* **90**, 5519–5523.

Kozminski, K.G., Beech, P.L., and Rosenbaum, J.L. (1995). The *Chlamydomonas* kinesin-like protein FLA10 is involved in motility associated with the flagellar membrane. *J. Cell Biol.* **131**, 1517–1527.

Lechtreck, K.F., Teltenkotter, A., and Grunow, A. (1999). A 210 kDa protein is located in a membrane-microtubule linker at the distal end of mature and nascent basal bodies. *J. Cell Sci.* **112**, 1633–1644.

Lewin, R.A. and Lee, K.W. (1985). Autotomy of algal flagella: Electron microscope studies of *Chlamydomonas* (Chlorophyceae) and *Tetraselmis* (Prasinophyceae). *Phycologia* **24**, 311–316.

Lohret, T.A., McNally, F.J., and Quarmby, L.M. (1998). A role for katanin-mediated axonemal severing during *Chlamydomonas* deflagellation. *Mol. Biol. Cell* **9**, 1195–1207.

Mahjoub, M.R., Rasi, M.Q., and Quarmby, L.M. (2004). A NIMA-related kinase, Fa2p, localizes to a novel site in the proximal cilia of *Chlamydomonas* and mouse kidney cells. *Mol. Biol. Cell* **15**, 5172–5186.

Mandelkow, E.-M., Thies, E., Trinczek, B., Biernat, J., and Mandelkow, E. (2004). MARK/PAR1 kinase is a regulator of microtubule-dependent transport in axons. *J. Cell Biol.* **167**, 99–110.

Miller, J.M., Wang, W., Balczon, R., and Dentler, W.L. (1990). Ciliary microtubule capping structures contain a mammalian kinetochore antigen. *J. Cell Biol.* **110**, 703–714.

Mimori-Kiyosue, Y. and Tsukita, S. (2003). Search and capture of microtubules through plus-end-binding proteins (+TIPs). *J. Biochem.* **134**, 321–326.

Mueller, J., Perrone, C.A., Bower, R., Cole, D.G., and Porter, M.E. (2005). The FLA3 KAP subunit is required for localization of kinesin-2 to the site of flagellar assembly and processive anterograde intraflagellar transport. *Mol. Biol. Cell* **16**, 1341–1354.

Musgrave, A., DeWildt, P., van Etten, I., Pijst, H., Schholma, C., Kooyman, R., Homan, W., and van den Ende, H. (1986). Evidence for a functional membrane barrier in the transition zone between the flagellum and cell body of *Chlamydomonas eugametos* gametes. *Planta* **167**, 544–553.

Nicastro, D., Schwartz, C., Pierson, J., Gaudette, R., Porter, M.E., and McIntosh, J.R. (2006). The molecular architecture of axonemes revealed by cryoelectron tomography. *Science* **313**, 944–948.

Omoto, C.K., Gibbons, I.R., Kamiya, R., Shingyoji, C., Takahashi, K., and Witman, G.B. (1999). Rotation of the central pair microtubules in eukaryotic flagella. *Mol. Biol. Cell* **10**, 1–4.

O'Toole, E.T., Giddings, T.H., McIntosh, J.R., and Dutcher, S.K. (2003). Three-dimensional organization of basal bodies from wild-type and γ-tubulin deletion strains of *Chlamydomonas reinhardtii*. *Mol. Biol. Cell* **14**, 2999–3012.

Pazour, G.J. and Witman, G.B. (2003). The vertebrate primary cilium is a sensory organelle. *Curr. Opin. Cell Biol.* **15**, 105–110.

Pazour, G.J., Agrin, N., Leszyk, J., and Witman, G.B. (2005). Proteomic analysis of a eukaryotic cilium. *J. Cell Biol.* **170**, 103–113.

Pedersen, L.B., Geimer, S., Sloboda, R.D., and Rosenbaum, J.L. (2003). The microtubule plus end-tracking protein EB1 is localized to the flagellar tip and basal bodies in *Chlamydomonas reinhardtii*. *Curr. Biol.* **13**, 1969–1974.

Pedersen, L.B., Geimer, S., and Rosenbaum, J.L. (2006). Dissecting the molecular mechanisms of intraflagellar transport in *Chlamydomonas*. *Curr. Biol.* **16**, 450–459.

Piperno, G., Mead, K., and Henderson, S. (1996). Inner dynein arms but not outer dynein arms require the activity of kinesin homologue protein KHP[FLA10] to reach the distal part of flagella in *Chlamydomonas*. *J. Cell Biol.* **133**, 373–379.

Piperno, G., Siuda, E., Henderson, S., Segil, M., Vaananen, H., and Sassaroli, M. (1998). Distinct mutants of retrograde intraflagellar transport (IFT) share similar morphological and molecular defects. *J. Cell Biol.* **143**, 1591–1601.

Portman, R.W., LeCluyse, E.L., and Dentler, W.L. (1987). Development of microtubule capping structures in ciliated epithelial cells. *J. Cell Sci.* **87**, 85–94.

Qin, H., Diener, D.F., Geimer, S., Cole, D.G., and Rosenbaum, J.L. (2004). Intraflagellar transport (IFT) cargo: IFT transports flagellar precursors to the tip and turnover products to the cell body. *J. Cell Biol.* **164**, 255–266.

Randall, J. (1969). The flagellar apparatus as a model organelle for the study of growth and morphogenesis. *Proc. Roy. Soc. B* **173**, 31–62.

Ringo, D.L. (1967). Flagellar motion and fine structure of the flagellar apparatus in *Chlamydomonas*. *J. Cell Biol.* **3**, 543–571.

Reinhart, F.D. and Bloodgood, R.A. (1988). Membrane–cytoskeleton interactions in the flagellum: A 240,000 Mr surface-exposed glycoprotein is tightly associated with the axoneme in *Chlamydomonas moewusii*. *J. Cell Sci.* **89**, 521–531.

Rosenbaum, J.L. and Witman, G.B. (2002). Intraflagellar transport. *Nat. Rev. Mol. Cell Biol.* **3**, 813–825.

Sandoz, D., Chailley, B., Boisvieux-Ulrich, E., Lemullois, M., Laine, M.C., and Bautista-Harris, G. (1988). Organization and functions of cytoskeleton in metazoan ciliated cells. *Biol. Cell* **63**, 183–193.

Sattler, C.A. and Staehelin, L.A. (1974). Ciliary membrane differentiation in *Tetrahymena pyriformis*: *Tetrahymena* has four types of cilia. *J. Cell Biol.* **62**, 473–490.

Schoppmeier, J. and Lechtreck, K.-F. (2002). Localization of p210-related proteins in green flagellates and analysis of flagellar assembly in the green alga *Dunaliella bioculata* with monoclonal anti-p210. *Protoplasma* **220**, 29–38.

Silflow, C.A., LaVoie, M., Tam, L.-W., Tousey, S., Sanders, M., Wu, W., Borodovsky, M., and Lefebvre, P.A. (2001). The Vfl1 protein in *Chlamydomonas* localizes in a rotationally asymmetric pattern at the distal ends of basal bodies. *J. Cell Biol.* **153**, 63–74.

Sinden, R.E., Canning, E.U., and B Spain, B. (1976). Gametogenesis and fertilization in *Plasmodium yoelii nigeriensis*: A transmission electron microscope study. *Proc. R. Soc. Lond. B, Biol. Sci.* **193**, 55–76.

Snell, W.J., Dentler, W.L., Haimo, L., Binder, L.I., and Rosenbaum, J.L. (1974). Assembly of chick brain tubulin onto isolated basal bodies of *Chlamydomonas reinhardtii*. *Science* **185**, 357–360.

Song, L. and Dentler, W.L. (2001). Flagellar protein dynamics in *Chlamydomonas*. *J. Biol. Chem.* **276**, 29754–29763.

Sorokin, S.P. (1968). Reconstructions of centriole formation and ciliogenesis in mammalian lungs. *J. Cell Sci.* **3**, 207–230.

Suprenant, K.A. and Dentler, W.L. (1988). Release of intact microtubule capping structures from *Tetrahymena* cilia. *J. Cell Biol.* **107**, 2259–2270.

Tamm, S.L. (1979). Membrane movements and fluidity during rotational motility of a termite flagellate. *J. Cell Biol.* **80**, 141–149.

Tokuyasu, K.T., Peacock, W.J., and Hardy, R.W. (1972). Dynamics of spermiogenesis in *Drosophila melanogaster*. I. Individualization process. *Z. Zellforsch. Mikrosk. Anat.* **124**, 479–506.

Tull, D., Vince, J.E., Callaghan, J., Naderer, T., Spurck, T., McFadden, G.I., Currie, G., Ferguson, K., Bacic, A., and McConville, M.J. (2004). SMP-1, a member of a new family of small myristoylated proteins in kinetoplastid parasites, is targeted to the flagellum membrane in *Leishmania*. *Mol. Biol. Cell* **15**, 4775–4786.

Tyler, S. (1979). Distinctive features of cilia in metazoans and their significance for systematics. *Tissue Cell* **11**, 385–400.

Tyska, M.J., Mackey, A.T., Huang, J.D., Copeland, N.G., Jenkins, N.A., and Mooseker, M.S. (2005). Myosin-1a is critical for normal brush border structure and composition. *Mol. Biol. Cell* **16**, 2443–2457.

Wang, W., Himes, R., and Dentler, W.L. (1994). The binding of a ciliary microtubule plus-end binding protein complex to microtubules is regulated by ciliary protein kinase and phosphatase activities. *J. Biol. Chem.* **269**, 21460–21466.

Watanabe, T. (1990). The role of ciliary surfaces in mating in *Paramecium*. In: *Ciliary and Flagellar Membranes* (R.A. Bloodgood, Ed.), pp. 149–200. Plenum Press, New York and London.

Watanabe, Y., Hayashi, M., Yagi, T., and Kamiya, R. (2004). Turnover of actin in *Chlamydomonas* flagella detected by fluorescence recovery after photobleaching (FRAP). *Cell Struct. Funct.* **29**, 67–72.

Weiss, R.L., Goodenough, D.A., and Goodenough, U.W. (1977). Membrane particle arrays associated with the basal body and with contractile vacuole secretion in *Chlamydomonas*. *J. Cell Biol.* **72**, 133–143.

Witman, G.B. (2003). Cell motility: Deaf *Drosophila* keep the beat. *Curr. Biol.* **13**, R796–R798.

Wloga, D., Camba, A., Rogowski, K., Manning, G., Jerka-Dziadosz, M., and Gaertig, J. (2006). Members of the NIMA-related kinase family promote disassembly of cilia by multiple mechanisms. *Mol. Biol. Cell* **17**, 2799–2810.

Wright, R.L., Chojnacki, B., and Jarvik, J.W. (1983). Abnormal basal-body number, location, and orientation in a striated fiber-defective mutant of *Chlamydomonas reinhardtii*. *J. Cell Biol.* **96**, 1697–1707.

CHAPTER 11

The *Chlamydomonas* Flagellar Membrane and Its Dynamic Properties

Robert A. Bloodgood
Department of Cell Biology, University of Virginia School
of Medicine, Charlottesville, Virginia, USA

CHAPTER CONTENTS

I. THE FLAGELLUM AS A MEMBRANE-BOUNDED ORGANELLE

As a cellular organelle, the flagellum falls into a category distinct from either the true membrane-bounded organelles found within the cytoplasm (mitochondria, for instance) or the cytoskeleton. The flagellum is only partially enclosed within an extension of the plasma membrane; the rest of the compartmentalization is due to the complex structures of the transition zone at the base of the axoneme. The soluble protein compartment of the flagellum (the matrix) is distinct in protein composition from the cytosol of *Chlamydomonas*. In addition, the protein composition of the flagellar membrane appears to be distinct from that of the rest of the plasma membrane. Hence there must be barriers to the free exchange of macromolecules between the flagellum and the rest of the cell. The basal body complex at the base of the flagellum is thus analogous to the nuclear pore complex, providing a selective barrier and machinery for selective protein import into and export from the flagellum. Jekely and Arendt (2006) point out the similarity in domain composition (WD40 repeats, α-helical repeats, coiled-coils) of the intraflagellar transport (IFT) complex used for import of at least some proteins into the flagellum (Rosenbaum and Witman, 2002) and the nuclear pore complex. In addition to the basal body/transition zone complex at the base of the flagellum acting as a selective barrier to macromolecules located in the cytosol reaching the flagellar matrix, there must be a barrier within the plasma membrane that prevents free diffusional exchange between proteins in the flagellar membrane and the rest of the plasma membrane; evidence for this barrier is discussed later in this chapter. The existence of these barriers implies that there are active processes for translocation of proteins into and out of the flagellum and hence the need for targeting information (a flagellar zip code) for flagellar proteins synthesized in the cell body on free polyribosomes and the rough endoplasmic reticulum. There is some emerging evidence for targeting sequences in *Chlamydomonas* axonemal proteins but none yet for any flagellar membrane proteins. However, there is evidence for targeting sequences in flagellar membrane proteins in trypanosomatid flagellate protozoans (Bloodgood, 2000; Nasser and Landfear, 2004) and information on targeting sequences for membrane proteins in primary cilia is emerging.

In addition to transporting proteins along the length of the flagellum, IFT (Kozminski et al., 1993; Rosenbaum and Witman, 2002) may be associated with the process of identifying proteins destined for the flagellum and translocating them from the cytosol to the flagellum; this hypothesis is consistent with the observation that most IFT proteins are concentrated at the level of the basal body (Rosenbaum and Witman, 2002; see Chapter 4).

II. THE *CHLAMYDOMONAS* PLASMA MEMBRANE

A. Plasma membrane domains

As already implied, the plasma membrane of a *Chlamydomonas* cell must have at least two domains: (1) the membrane associated with each of the two flagella (Figure 11.1) and (2) the rest of the plasma membrane, the latter being associated with what is referred to as the "cell body". Using terminology developed for polarized epithelial cells, the flagellar membrane would be considered the apical plasma membrane of *Chlamydomonas* and the cell body plasma membrane would be considered the baso-lateral plasma membrane. Although cell body size and flagellar length vary with cell strain and growth conditions, the membrane associated with the two flagella makes up about 6–8% of the total surface area of the plasma membrane. It is far from clear whether these are the only two plasma membrane domains in *Chlamydomonas*. One can conceive of a plasma membrane domain at the base of the flagella that would be the site of targeting of membrane proteins

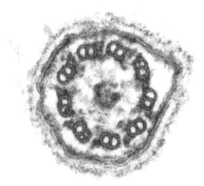

FIGURE 11.1 *Transmission electron micrograph showing a cross-section of a flagellum from* Chlamydomonas. *This is an image from the* pf18 *mutant and hence the structure of the central pair of microtubules in the axoneme is disrupted. Note the flagellar membrane, the prominent glycocalyx, and the proximal portion of a mastigoneme on the left side of the flagellum. Reproduced from Bloodgood and May (1982) with permission of the Rockefeller University Press, New York.*

to the flagellar membrane, since transport vesicles from the Golgi apparatus cannot enter the flagellum or reach the flagellar membrane proper, because of the physical barrier of the basal body/transition zone complex. This situation of a third plasma membrane domain would be analogous to the situation in the trypanosomatid flagellate protozoa which possess three clear plasma membrane domains: the flagellar membrane, the flagellar pocket membrane domain and the cell body plasma membrane domain. In trypanosomes, all endocytosis and exocytosis occurs at the flagellar pocket plasma membrane domain, which serves as a "sorting domain" for targeting newly synthesized plasma membrane proteins to the flagellar membrane and/or cell body membrane (Overath et al., 1997). *Euglena* also possesses a flagellar "reservoir" plasma membrane domain which appears to be the site where flagellar membrane glycoproteins initially appear on the cell surface. Pulse-chase studies with antibody Fab fragments show a direct transfer of at least some flagellar membrane proteins from the flagellar reservoir membrane to the flagellar membrane (Rogalski and Bouck, 1982). A similar situation exists in the alga *Pyramimonas* (McFadden and Wetherbee, 1985). There also is evidence for a periciliary sub-domain of the apical plasma membrane domain in polarized epithelial cells in mammals (Reiter and Mostov, 2006; Veira et al., 2006). This plasma membrane domain, located around the base of the cilium, is characterized by the presence of the protein galectin-3 and by the absence of two GPI-linked proteins that are found in the rest of the apical domain of these epithelial cells. This periciliary domain is hypothesized to be a site for targeting transport vesicles carrying membrane components destined for the cilium/flagellum. The only suggestion of a third plasma membrane domain in *Chlamydomonas*, which may be the homologue of the flagellar pocket membrane domain in trypanosomatid flagellate protozoa or the periciliary domain in polarized epithelial cells, comes from freeze-fracture electron micrographs that show a "flagellar bracelet" (Figures 11.2 and 11.3) of intramembrane particles located some distance basal to the "ciliary necklace", another circular array of intramembrane particles located at the point where the basal body is closely apposed to the plasma membrane (Weiss et al., 1977) (see section III.C). It is possible that the region of *Chlamydomonas* plasma membrane between the ciliary necklace (Figure 11.2) and the flagellar bracelet (Figures 11.2 and 11.3) is a periciliary domain for targeting and fusion of transport vesicles coming from the Golgi apparatus, perhaps via the pathway that has been visualized in mammalian cells using antibodies to IFT20 (Follit et al., 2006).

There is evidence for the existence of a barrier to free diffusion of membrane proteins between the flagellar membrane and the rest of the plasma membrane in vegetative cells of *C. reinhardtii* (Bloodgood, 1988), in gametic cells of *C. reinhardtii* (Hunnicutt et al., 1990), and in gametic cells of *C. eugametos* (Musgrave et al., 1986). Indeed, Hunnicutt et al. (1990) provided some data that this diffusion barrier can be regulated during mating, such that integral (e.g., FMG-1B) and peripheral (e.g., sexual

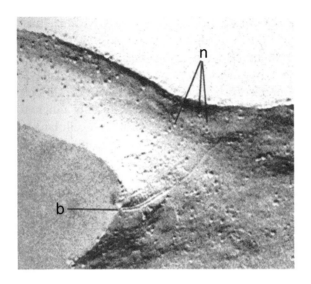

FIGURE 11.2

Image of the base of a flagellum from a freeze-fractured gametic cell of Chlamydomonas *showing both the flagellar necklace (n) and the flagellar bracelet (b). Reproduced from Weiss et al. (1977) with permission of the Rockefeller University Press, New York.*

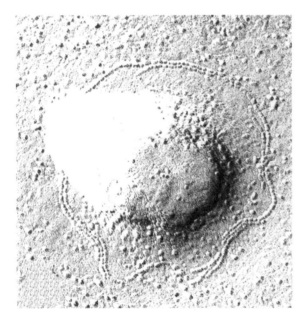

FIGURE 11.3

Image of the membrane of a freeze-fractured vegetative cell of Chlamydomonas *showing the disposition of the flagellar bracelet around the perimeter of a flagellum. Reproduced from Weiss et al. (1977) with permission of the Rockefeller University Press, New York.*

agglutinins) membrane proteins are recruited from the cell body plasma membrane to the flagellar membrane in response to a rise in cyclic AMP resulting from flagellar adhesion between gametes of opposite mating types. The morphological correlate of this diffusion barrier could be either the "ciliary necklace" or "flagellar bracelet" (Figures 11.2 and 11.3) observed using freeze-etch electron microscopy (Weiss et al., 1977). In mammalian (MDCK) polarized epithelial cells, the membrane at the base of the primary cilium has an unusually high degree of lipid ordering and this has been proposed as a possible barrier to diffusion of membrane proteins into and out of the ciliary membrane (Veira et al., 2006).

The cell body of *Chlamydomonas* is enclosed within a thick, multi-layered glycoprotein cell wall, limiting access to most of the plasma membrane. The flagellar membrane is the only plasma membrane domain exposed directly to the environment and hence normally available for cell–cell and cell–substrate interactions. One exception to this rule occurs during the *Chlamydomonas* sexual cell cycle. Gametes initially adhere to each other by their flagellar membrane surfaces; this initiates a cellular signaling pathway that leads to release of cell walls, allowing a second phase of cell–cell adhesion mediated by specialized domains on the cell body plasma membranes of the two gametes (see Chapter 12). Not surprisingly, we will see that the flagellar surface plays important roles in cell–cell recognition and adhesion (during the initial events of mating) as well as in cell–substrate adhesion and generation of force for whole-cell locomotion (gliding motility). There are cellular signaling pathways associated with both these processes.

B. Technical issues in studying the *Chlamydomonas* plasma membrane and its domains

Because the *Chlamydomonas* cell body is enclosed within a thick, multi-layered proteinaceous cell wall and the flagella are not, it is fairly routine to isolate highly purified preparations of flagella (Witman, 1986) from light-synchronized cultures, although care must be taken that one does not end up with green flagellar preparations, due to contamination with chloroplast components. Given that well-synchronized cells are used and flagellar preparations are isolated and purified with care (and are white, not green), one can generally deduce that any membrane or membrane proteins found within those preparations are likely to be derived from the flagellar membrane domain. The major source of contamination of "white" flagellar preparations comes from cell wall components (Musgrave et al., 1979, 1983; Monk et al., 1983). One characteristic expected of a putative flagellar membrane protein is that the mRNA encoding that protein is upregulated upon deflagellation (see Chapter 3). Identification of a flagellar protein as an integral membrane protein is based on features such as (1) glycosylation, (2) presence of signal sequence codons in the gene, (3) presence of one or more predicted transmembrane domains, (4) partitioning into Triton X-114 and (5) surface accessibility to vectorial labeling such as biotinylation or iodination (Monk et al., 1983; Bloodgood and Workman, 1984).

In contrast to the relative ease of isolation of flagella (and hence the separation of flagellar membrane from all other cellular membranes), the isolation of purified preparations of the cell body plasma membrane has proven to be much more difficult, in part because of the close association of the cell body plasma membrane with the glycoprotein cell wall and the presence of a single huge, membrane-rich chloroplast that dominates the cytoplasm and is easily disrupted when cells are lysed. Attempts at cell body plasma membrane isolation or labeling generally involve deflagellation and elimination of the cell wall (either using "cell wall-less" mutants or the natural cell wall lytic enzyme). It is not clear that either of these latter methods entirely removes all traces of the cell wall. Dolle (1988) was the first to prepare a chlorophyll-free preparation of *Chlamydomonas* plasma membrane using phase partitioning with an aqueous polymer two-phase system to separate the plasma membrane from the internal membranes of the cell. Using surface iodination of gametic cells from which both the flagella and the cell wall had been removed, Hunnicutt et al. (1990) showed that both the FMG-1B glycoprotein and the sexual agglutinin are associated with the cell body plasma membrane, albeit at much lower concentrations than in the flagellar membrane.

Purified *Chlamydomonas* flagella have been routinely separated into two fractions by extraction with non-ionic detergents such as Triton X-100 and Nonidet P-40. One fraction is the detergent-soluble fraction. This fraction, routinely referred to as the "membrane-plus-matrix" fraction, contains a mixture of proteins derived from the flagellar membrane and the flagellar matrix. In reality, this fraction contains small intact membrane vesicles as well as solubilized membrane proteins and matrix proteins. The second fraction is the detergent-insoluble fraction, which has been referred to as the "axonemal fraction". These are not necessarily clean and mutually exclusive fractions. In addition, their composition also depends upon the purity of the initial flagellar preparation (which can be contaminated with cell wall and chloroplast components) and the conditions (detergent type, detergent concentration, protein:detergent ratio, temperature, length of extraction) used for detergent extraction. The *Chlamydomonas* flagellar proteome (Pazour et al., 2005) contains many examples of peptides belonging to known flagellar membrane proteins (such as the major flagellar membrane glycoprotein [FMG-1B] and the mastigoneme protein) being found in the axoneme fraction as well as the membrane-plus-matrix fraction of the flagellum. This is consistent with the observation that pieces of flagellar membrane can be seen associated with axonemes after non-ionic detergent extraction, perhaps reflecting sites of tight association between the axoneme and the flagellar membrane (Dentler, 1980; see Chapter 10). However, such membrane was not observed in the axoneme preparations used for the proteomic analysis, and other predicted membrane proteins were found exclusively in the detergent-soluble fraction. Therefore, there may be flagellar components that truly belong to multiple flagellar compartments (for instance proteins that mechanically link the flagellar membrane to the axoneme (Dentler, 1990; see Chapter 10) or microtubule-associated motor proteins and IFT components whose cargo are membrane proteins.

One clever approach taken to obtain a fraction enriched in flagellar membrane proteins without truly purifying flagellar membrane involves freeze–thawing isolated flagella to create tears in the flagellar membrane and washing them to remove soluble matrix proteins (Wang and Snell, 2003; Huang et al., 2007). Subsequent extraction with non-ionic detergent yields solubilized membrane proteins and small membrane vesicles (after centrifugation to remove the axoneme).

A number of approaches have been taken to obtain "intact" flagellar membranes from isolated and purified flagella from vegetative and gametic *Chlamydomonas*: (1) rubbing membranes off isolated flagella using vortexing in a viscous sucrose solution (Snell, 1976) or shearing flagella through Pasteur pipettes (Witman et al., 1972); (2) cycles of freeze–thawing to release flagellar membrane vesicles (Wang and Snell, 2003; Iomini et al., 2006); (3) cycles of freeze–thaw followed by shearing of the flagella through a 23-gauge hypodermic needle (Pasquale and Goodenough, 1987); (4) extraction of flagella with 0.1% Sarkosyl (Witman et al., 1972); (5) extraction of whole flagella with dilute non-ionic detergents such as 0.04% Nonidet P-40 (Bloodgood and May, 1982); and (6) collection of membrane vesicles that are naturally shed into the *Chlamydomonas* growth medium (McLean et al., 1974; Bergman et al., 1975; Snell, 1976; Kalshoven et al., 1990). All of the methods mentioned above have the potential for contamination of the flagellar membrane preparations with non-membrane proteins or the possible loss of true flagellar membrane components. Any method that generates closed membrane vesicles can trap soluble flagellar matrix components within the lumen of the membrane vesicles. Generation of membrane vesicles using non-ionic detergent may extract legitimate membrane components from those vesicles. The membrane vesicles shed into the medium by cells may not have the same composition as the native flagellar membrane from which they were derived. Only one lab has attempted to subfractionate *Chlamydomonas* flagellar membranes; Iomini et al. (2006) reported that a fraction of flagellar membrane that is resistant to solubilization with 1% Triton X-100 is enriched in sphingolipids over the bulk flagellar membrane and contains an integral membrane protein (AGG2) that is localized to the base of the flagellum and is involved in photoresponses.

Clearly, no single, ideal method for selectively obtaining and characterizing components of the *Chlamydomonas* flagellar membrane currently exists, and the flagellar membrane composition will vary as a function of the isolation method used. As will be seen in the following sections, we are far from a complete biochemical characterization of the flagellar membrane.

III. STRUCTURE OF THE FLAGELLAR MEMBRANE

A. Introduction

The *Chlamydomonas* flagellar membrane is a sub-domain of and is morphologically continuous with the rest of the cell's plasma membrane. In stained thin sections prepared for transmission electron microscopy (TEM), the flagellar membrane shows a typical trilaminar image characteristic of plasma membranes. The major ultrastructural features of the flagellar membrane of vegetatively grown *Chlamydomonas* are a prominent glycocalyx and mastigonemes (both of which are absent from the cell body plasma membrane). In quick-freeze/deep-etch electron microscopy, the gametic flagellar membrane, in addition to these two features, exhibits extended structures which represent the sexual agglutinin molecules (section III.E) (see Chapter 12).

B. The glycocalyx

In thin-section electron micrographs, the *Chlamydomonas* flagellar membrane is covered by a darkly staining glycocalyx (sometimes called the flagellar sheath) of uniform diameter that extends about 16–18 nm from the surface of the membrane (Figure 11.1); there is a non-staining gap between the glycocalyx and the flagellar membrane surface. This glycocalyx is presumed to represent the very large and heavily glycosylated ectodomains of the principal flagellar membrane glycoprotein, FMG-1B. It is not clear whether the morphologically visible glycocalyx represents primarily the protein ectodomains or a combination of the protein ectodomains and the N-linked carbohydrate moieties. Evidence for a very compact layer of carbohydrate on the surface of the flagellum comes from several observations: (1) heavy staining of the flagellar surface with the lectin concanavalin A (McLean et al., 1981) and with antibodies to carbohydrate epitopes on FMG-1B (Bloodgood et al., 1986); (2) the inability of endoglycosidases to cleave the N-linked carbohydrates from the flagellar surface of intact cells (presumably because of steric issues) (Bloodgood, unpublished results); and (3) the inability of any monoclonal antibodies to protein epitopes on FMG-1B to label the flagellar surface, presumably due to masking by the extensive N-linked carbohydrate groups (Bloodgood et al., 1986). In TEM, the appearance of the glycocalyx is very similar on vegetative and gametic flagella, consistent with the presence of FMG-1B on both types of flagella (Goodenough et al., 1985) (see section IV.A.2). The glycocalyx is not present on the cell body plasma membrane.

C. Intramembrane particle arrays

The *Chlamydomonas* flagellar membrane has a number of distinctive intramembrane particle arrays observed by freeze-fracture TEM (Bergman et al., 1975; Snell, 1976; Weiss et al., 1977; McLean et al., 1981). Longitudinal rows of large intramembrane particles extend the length of the flagellar membrane in both vegetative and gametic cells of *C. reinhardtii* and *C. moewusii* (Bergman et al., 1975; Snell, 1976; McLean et al., 1981) (Figure 11.4). The spacing of the rows suggests that each row of intramembrane particles may be oriented over an axonemal doublet microtubule. These

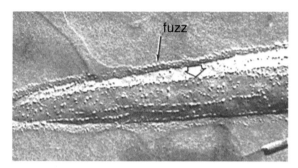

FIGURE 11.4 *Image of a freeze-fractured flagellar membrane from a vegetative cell of C. reinhardtii showing aligned rows of intramembranous particles (open arrow). Reproduced from Bergman et al. (1975) with permission of the Rockefeller University Press, New York.*

intramembrane particles may represent or be associated with the membrane-axonemal links that are observed by TEM (Bloodgood, 1987; Dentler, 1990; see Chapter 10); they could represent signaling molecules such as PKD2 tethered to the underlying doublets (Huang et al., 2007), or they could represent the location of axonemal anchored motor protein complexes responsible for the flagellar surface motility (glycoprotein movements; microsphere movements; gliding motility) described in section VI. Indeed, the observations that many small microspheres can be moving simultaneously on a single flagellum and microspheres can move past each other in opposite directions (Bloodgood, 1977) suggests that there are multiple independent domains in the flagellar membrane and multiple discrete "tracks" on the flagellar surface. In addition, the sexual agglutinin molecules (see section IV.A.7) on the gametic flagellar surface have been observed to be arrayed in parallel rows (Goodenough et al., 1985; Tomson et al., 1990).

The ciliary necklace (Figure 11.2) of *Chlamydomonas* is located in the membrane at the level of the basal body transition zone in both vegetative and gametic cells and consists of three rings of intramembrane particles (Weiss et al., 1977). This structure is a general feature of cilia and flagella (Gilula and Satir, 1972) and may function as the diffusion barrier to proteins and/or lipids that allows the flagellar membrane to maintain a composition distinct from the rest of the plasma membrane (see section II.A). Alternatively, the necklace may function primarily to anchor the base of the flagellar membrane to the underlying cytoskeleton (see Chapters 2 and 10).

As mentioned in section II.A, there is a third, unusual intramembrane particle array called the flagellar bracelet (Weiss et al., 1977) (Figures 11.2 and 11.3). The flagellar bracelet is a circular arrangement of small, tightly packed intramembranous particles organized into discontinuous strands. The bracelet is located more proximal to the cell body than the ciliary necklace. It is located at the base of the transition zone (see Chapter 2) at the point where the plasma membrane everts to form the flagellar membrane.

Weiss et al. (1977) raised the possibility that the tightly packed intramembranous particles form a diffusion barrier. Earlier in this chapter, it was speculated that the flagellar bracelet, in collaboration with the ciliary necklace, may define a third, periciliary, domain of the plasma membrane that could be a site of docking and fusion of transport vesicles containing membrane proteins and lipids destined for the flagellar membrane. Unlike the ubiquitous ciliary necklace, the flagellar bracelet has been described only in *Chlamydomonas* (Weiss et al., 1977) and green algae such as *Volvox* that are closely related to *Chlamydomonas* (Melkonian, 1982).

Robenek and Melkonian (1981) used filipin-labeling and freeze-etch electron microscopy to study the localization of 3-β-hydroxysterols in the plasma membrane of *Chlamydomonas* and found that these sterols were absent from sites of intramembrane particle arrays, including the flagellar necklace and the flagellar bracelet. However, this does not appear to be the case for the ciliary necklace in all cilia (Stephens and Good, 1990).

D. Mastigonemes

Mastigonemes are flagellar appendages (flagellar hairs; flimmer) that are only found on the flagella of protists (and are especially common among biflagellate algae). There are several categories of mastigonemes (Andersen et al., 1991). The non-tubular, fibrous type (such as those found on *Chlamydomonas*) is simple in design (long and slender glycoprotein polymers) and has been hypothesized (Bouck, 1972) to function by simply enlarging the effective surface area of the flagellum and hence increasing the effective thrust created by the breast-stroke action of the flagella. The second, more intensively studied, category of mastigonemes has a more complex architecture (thick tubular structures to which are attached much finer filaments) (Bouck, 1971). The tubular mastigonemes are found only on leading flagella (such as in the flagellate protozoan *Ochromonas*) that exhibit propagation of bends in the same direction as that of whole-cell movement. The tubular mastigonemes are thought to reverse the direction of thrust produced by these paradoxical bends, so that the cell is able to move in the same direction as bend propagation (Jahn et al., 1964; Holwill and Sleigh, 1967).

Chlamydomonas mastigonemes (Ringo, 1967; Witman et al., 1972; Bernstein and Rosenbaum, 1993; Nakamura et al., 1996) have a shaft of uniform length (0.9–1.0 μm) and diameter (16 nm) that appears to be a polymer of identical subunits with a center-to-center spacing of approximately 20 nm (in which case it would take as many as 50 subunits to assemble a single mastigoneme). The mastigoneme shaft is terminated by a short, thinner fibril (Figure 11.5). The mastigonemes appear to be organized into 2 rows on opposite sides of the flagellum; they are evenly spaced along the distal two-thirds of the flagellum but do not extend all the way to the

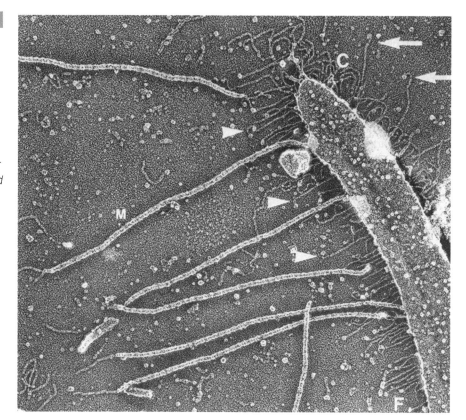

Quick-freeze/deep-etch image of a flagellum from a gametic cell of Chlamydomonas *showing the long mastigonemes (M) and the much shorter sexual agglutinin structures (arrows and arrowheads) associated with the flagellar surface. C: crescent-shaped fibrils, observed on surface of gametic flagella only, of unknown function but possibly sexual agglutinin molecules. Reproduced from Goodenough et al. (1985) with permission of the Rockefeller University Press, New York.*

base of the flagellum (Bergman et al., 1975; Snell, 1976; Nakamura et al., 1996) (Figures 11.5 and 11.6). This poses an interesting problem during flagellar assembly: how to position/anchor mastigonemes only over the distal two-thirds of the flagellar surface, given that mastigonemes must pass through the basal region of the flagellar membrane in order to gain access to other portions of the flagellar membrane.

Although mastigonemes appear to be very flexible in many TEM images, they appear to be quite stiff in negative-stain images (Witman et al., 1972) or rapid-freeze/deep-etch preparations (Goodenough et al., 1985) (Figure 11.5). When live cells are treated with a monoclonal antibody to the mastigoneme protein, cells lose their mastigonemes (Nakamura et al., 1996). Antibody-treated cells, lacking mastigonemes, exhibited a normal flagellar waveform and a slightly increased (10%) beat frequency, but their swimming velocity decreased to about 70–80% of control cells (Nakamura et al., 1996), consistent with the hypothesis (Bouck, 1971) that *Chlamydomonas* mastigonemes function to increase flagellar propulsive force by increasing the effective surface of the flagellum. *Chlamydomonas* mastigonemes are not involved in microsphere movement (Bloodgood, 1977), gliding motility (Nakamura et al., 1996) or flagellar adhesion during mating (Mclean et al., 1974;

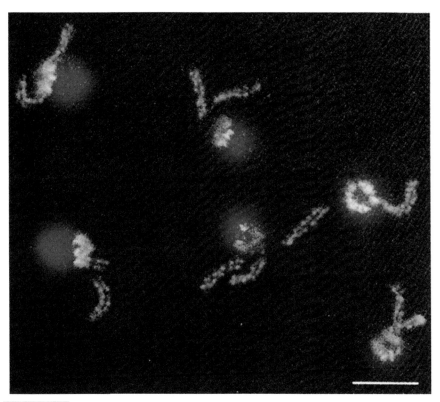

FIGURE 11.6 *Immunofluorescence staining of mastigonemes with an antibody to the mastigoneme protein. Mastigonemes are seen as a row of green fluorescent dots on each side of each flagellum on these vegetative cells of* Chlamydomonas. *In addition, there is a ring of green fluorescent particles at the anterior end of each of the cells; these may represent assembled mastigonemes inside cytoplasmic membranous organelles. The red is due to chlorophyll autofluorescence. Reproduced from Nakamura et al. (1996) with permission of The Company of Biologists, Ltd., Cambridge, UK.*

Bergman et al., 1975; Snell, 1976). Interestingly, in *Peranema*, mastigonemes are reported to be involved in both flagella-dependent gliding motility and microsphere movement (Saito et al., 2003).

Chlamydomonas mastigonemes can be easily purified to homogeneity (Witman et al., 1972; Bernstein, 1995; Nakamura et al. 1996). Mastigonemes are composed of a single 220- to 230-kD glycoprotein that binds concanavalin A. Removal of the carbohydrate yields an approximately 200-kD polypeptide (Bernstein, 1995). The *Chlamydomonas* mastigoneme cDNA was cloned and sequenced independently by Bernstein (unpublished results) and later by Song and Dentler (AF508983). The best evidence that this sequence encodes the mastigoneme protein is that Bernstein expressed the cDNA in bacteria, purified the expression product, and used it to raise a polyclonal antibody that reacted with the native mastigoneme protein (unpublished results). The mastigoneme gene is a single copy gene (8.6 kb) with 19 exons and 18 small introns yielding a 6.9 kb transcript; it maps to the left end of linkage group XVIII (Carolyn Silflow, personal communication). Expression of the gene encoding the mastigoneme protein is upregulated 1.8 fold after deflagellation (Pazour et al., 2005). Combining data from Bernstein, Song and Dentler, the *Chlamydomonas* genome, and ESTs, the best protein sequence prediction is a protein of 1956 amino acids (202 kD) plus a 31-amino-acid signal sequence. The available sequence information is consistent with a globular secretory glycoprotein that fits the size seen on SDS-PAGE as well as the subunit size seen in the TEM images (Witman et al., 1972). The protein is predicted to have an N-terminal signal sequence but no transmembrane domain. It has 2–3 predicted N-glycosylation sites and several predicted disulfide bonds. It has an 80-amino-acid proline-rich domain near the C-terminus that contains multiple copies of a PPSPX repeat that is also found in the *Chlamydomonas* GP1 cell wall protein (Ferris et al., 2001) and the *Chlamydomonas* sexual agglutinin proteins present on the gametic flagellar membrane surface (Ferris et al., 2005). The mastigoneme sequence also contains two Tumor Necrosis Factor Receptor domains as well as a subtilase serum protease motif (InterPro IPR000209). There are nineteen CXXCXXG repeats evenly spaced through the C-terminal half of the sequence. The mastigoneme protein is found in the *Chlamydomonas* phosphoproteome (Wagner et al., 2006) and is phosphorylated on serine, threonine, and tyrosine residues.

There remain many puzzles about the assembly and targeting of *Chlamydomonas* mastigonemes. The subunit protein is presumably synthesized in the rough endoplasmic reticulum and glycosylated in the Golgi apparatus. As observed for other types of mastigonemes (Bouck, 1972; Bouck et al., 1990), the assembly of the mastigonemes from the subunit protein must occur intracellularly because secreted subunits would be too dilute for assembly to occur outside the cell. Unlike the case with the tubular type of mastigoneme (Bouck, 1972), no one has reported, using TEM, the presence of mastigonemes within membranous organelles in the cytoplasm of *Chlamydomonas*. However, using fluorescence microscopy and an antibody to the *Chlamydomonas* mastigoneme glycoprotein, Nakamura et al. (1996) observed a fluorescent ring composed of more than 10 particles near the base of the flagella. This localization may represent fully assembled mastigonemes in secretory vesicles ready to be released by exocytosis (Figure 11.6). The cell appears to carefully regulate the length of the mastigonemes during their intracellular assembly (Marshall, 2004), but how this is accomplished is unknown.

One of the biggest puzzles about *Chlamydomonas* mastigonemes is how they are anchored to the flagellum/axoneme. Presumably, mastigonemes become associated by their ends with some membrane component, prior to exocytosis of transport vesicles at the base of the flagellum; indeed, it is possible that some transmembrane protein may nucleate the assembly of mastigoneme protein within the cell (at an as yet unknown site) and then function as the flagellar membrane anchoring protein after exocytosis and movement of mastigonemes onto the flagellar surface. Since mastigonemes are not observed on the cell body plasma membrane of *Chlamydomonas*, there must be a mechanism for directing mastigonemes onto the flagellar membrane after being externalized by exocytosis. *Chlamydomonas* may have the equivalent of a periciliary plasma membrane domain (see section II.A) which could prevent movement of mastigonemes onto the general plasma membrane while allowing entry onto the flagellar surface. While it is possible that a mastigoneme and its membrane protein anchor could diffuse onto the flagellar membrane and laterally within the plane of the flagellar membrane, it is more likely that mastigonemes are transported into the flagellum and along the flagellar surface by an active process, such as IFT (see Chapter 4) or the mechanism visualized by microsphere movement (section VI.C.4). Any active mechanism for movement of mastigonemes along the flagellar surface and anchoring into position would require a mechanical connection between the mastigoneme and the axoneme. Such a connection would be consistent with the finding of mastigoneme protein in the axonemal fraction of the flagellar proteome (Pazour et al., 2005). Indeed, sometimes one can see mastigonemes attached to what appears to be a membrane-free *Chlamydomonas* axoneme after non-ionic detergent extraction of flagella (Figure 2D in Bloodgood, 1987). It may be that some of the slender membrane-axoneme links observed in transmission electron micrographs of the *Chlamydomonas* flagellum (see Chapter 10) represent anchorage of individual mastigonemes to the axoneme.

Both the tubular mastigonemes in *Ochromonas* (Markey and Bouck, 1977) and the finer, fibrous mastigonemes of *Euglena* (Bouck et al., 1978) remain associated with the axoneme after treatment of flagella with non-ionic detergents. Markey and Bouck (1977) described an adapter structure that appears to mediate attachment of the mastigoneme to the axoneme in *Ochromonas*.

One interesting question is whether mastigonemes can turn over in the intact flagellum or whether they can only be added to the flagellar membrane during assembly. Because there is turnover of the flagellar membrane and its constituent membrane proteins (see section V.B) and continuous release of flagellar membrane vesicles containing mastigonemes (Bergman et al., 1975; Snell, 1976), mastigonemes must be continuously added to the surface of the intact flagellum. This turnover is dramatically illustrated by the work of Nakamura et al. (1996), who removed the mastigonemes from intact *Chlamydomonas* flagella through crosslinking with antibody. After removal of the antibody, the cells, now possessing mastigoneme-free flagella, replaced the mastigonemes on the flagellar surface in about 4 hours, which is the time observed by Bloodgood et al. (1979) for flagellar membrane turnover in *Chlamydomonas*.

E. Sexual agglutinin structures

A prominent structural feature of the flagellar surface of *Chlamydomonas* gametes, when viewed by freeze-etch procedures, are the large, extended or looped sexual agglutinin molecules that can extend out 225 nm from the flagellar surface, well beyond the glycocalyx, but not nearly as far as the mastigonemes, which are about 0.9–1.0 μm in length (Figure 11.5) (Adair et al., 1983; Goodenough et al., 1985). Both the *plus* and *minus* sexual agglutinins from *Chlamydomonas* have a very characteristic shape when the purified agglutinins are viewed in rapid-freeze/deep-etch preparations in TEM. The overall length of the isolated molecules is from 200 to 250 nm; each molecule is cane shaped with a large globular domain at the C-terminal end (called the head), a long shaft region that has a characteristic bend or kink (tail hook) and a smaller globular region at the other end (N-terminus) (Adair et al., 1983; Goodenough et al., 1985; Ferris et al., 2005). The tail-hook end (N-terminus) associates with receptor proteins in the flagellar membrane while the large head domain (C-terminus) is extended out from the flagellar surface in a position to interact with the flagellar surface of a gamete of the opposite mating type. As with the mastigonemes, the sexual agglutinins appear to be organized into linear rows (at least two) on the surface of the gametic flagellum (Goodenough et al., 1985; Tomson et al., 1990). The sexual agglutinins found on the flagellar surface of *plus* gametes interact with the sexual agglutinins found on the flagellar surface of *minus* gametes and this is the initial step in mating (Goodenough and Heuser, 1999; see Chapter 12). The biochemical characteristics of these large glycoproteins are described in section IV.A.7.

IV. COMPOSITION OF THE FLAGELLAR MEMBRANE

A. Flagellar membrane proteins

1. Overview

Most of the research focus on the *Chlamydomonas* flagellum has been on the structure, function, and biochemistry of the axoneme. Hence, it is not surprising that the protein composition of the flagellar membrane has not been well studied. While there are a number of observations suggesting that the protein composition of the flagellar membrane differs from that of the cell body plasma membrane (such as Musgrave et al., 1986, working with *C. eugametos*), rigorous biochemical comparison of the two plasma membrane domains has been thwarted by the technical difficulties in obtaining clean cell body plasma membrane fractions (see section II.B). There is some evidence (using vectorial labeling with ^{125}I) that at least some flagellar membrane proteins (in particular, FMG-1B and the sexual agglutinin) can be found on the cell body plasma membrane of *C. reinhardtii* (Hunnicutt et al., 1990).

Characteristics of some of the best studied *Chlamydomonas* flagellar membrane proteins are summarized in Table 11.1. At this time, the only

Table 11.1 *C. reinhardtii* flagellar membrane proteins that have been cloned and sequenced

Flagellar membrane protein	GenBank protein accession number	Gene size (kb)	Exon/intron	Predicted transcript size (kb)	Predicted protein size[a] (± signal sequence)	Signal sequence	Trans-membrane domains	Upregulated upon deflagellation	Comments
FMG-1A	AAO25117	14.5	14/13	11.6	3854 aa 382 kD 3824 aa 379 kD	Yes	1	No	Not expressed in vegetative cells
FMG-1B	AAO25118	14.5	9/8	13.2	4176 aa 413 kD 4149 aa 410 kD	Yes	1	Yes (5 fold)	
Mastigoneme	AAM33652	8.6	19/18	6.9	1987 aa 205 kD 1956 aa 202 kD	Yes	0	Yes (1.8 fold)	Found in phospho-proteome
Plus sexual agglutinin SAG1	AAS07044	14.8	15/14		3409 aa 3349 aa 330 kD	Yes	0		Expressed only in *plus* gametes
Minus sexual agglutinin SAD1	AAS07042	17.8	30/29		3889 aa 3853 aa 385 kD	Yes	0		Expressed only in *minus* gametes
AGG2	AAZ22341			1.2	183 aa 19.6 kD		2		
PKD2	ABR14113	8	12/11	5.8	1626 aa 181 kD	Yes	6	No Reduced (0.55 fold)	Increased expression in gametes

[a]aa, amino acids

Chlamydomonas flagellar membrane proteins that have been well characterized are: (1) the transmembrane glycoprotein FMG-1B, (2) the mastigoneme protein (see section III.D), (3) the sexual agglutinin proteins found only in gametic cells, (4) AGG2 and (5) PKD2. There is indirect evidence for the presence of a number of other proteins in the *Chlamydomonas* flagellar membrane, based on the *Chlamydomonas* flagellar proteome (Pazour et al., 2005), the *Chlamydomonas* phosphoproteome (Wagner et al., 2006), vectorial labeling of the flagellar surface using biotinylation and iodination, immunoblots of flagella using anti-carbohydrate monoclonal antibodies, blot overlays using lectins, and enzymatic activity assays of flagellar membrane fractions (Monk et al., 1983; Kooijman et al., 1989; Kalshoven et al., 1990; Bloodgood and Salomonsky, 1994; Wang et al., 2006).

2. FMG-1B – the major flagellar membrane protein

The protein composition of the *Chlamydomonas* flagellar membrane from both vegetative and gametic cells is dominated by a pair of high-molecular-weight polypeptides that migrate slower than the axonemal dyneins in SDS-PAGE gels (Figure 11.7). The source of these two electrophoretic isoforms (which always show equal staining intensity) could be two genes, alternative splice variants of a single gene, or post-translational modification. A wide variety of monoclonal, polyclonal, and peptide-antibodies (as well as

⇆ FMG1B

← Mast

● 205

● 116

● 97

● 66

FIGURE 11.7   *Proteins exposed at the flagellar surface. Live vegetative cells of Chlamydomonas were incubated in a reagent that biotinylates proteins exposed at the cell surface. Flagella were isolated and flagellar proteins were separated by SDS-PAGE. After transfer to nitrocellulose, biotinylated flagellar proteins were identified using peroxidase-labeled avidin. Arrows indicate the FMG-1B and mastigoneme glycoproteins. Numbers indicate molecular mass standards (in kD). Reproduced from Bloodgood (1990) with permission of Plenum Press, New York.*

the lectin concanavalin A) stain the two bands equally well. Trypsin digestion of the two bands yields an identical set of peptide sizes as determined by mass spectrometry. Although referred to in the early literature by various names such as the high-molecular-weight protein or the 350-kD protein, these two protein bands are now collectively referred to as FMG-1 (Flagellar Membrane Glycoprotein 1). Adair et al. (1983) estimated that there are approximately 90,000 copies of FMG-1 in the membrane of each gametic flagellum. In cells that are fixed prior to labeling with antibodies, FMG-1 is seen to be uniformly distributed along the flagellar surface (Figure 11.8).

A mixture of three different monoclonal antibodies (obtained by injecting mice with both gel isoforms) was used to screen a *Chlamydomonas* expression library (Bloodgood and Spano, 2000). One of the cDNAs identified was used to screen the *Chlamydomonas* genomic BAC library. BACs containing two related sequences were obtained and these two *FMG-1* genes were sequenced and designated *FMG-1A* (AY206771) and *FMG-1B* (AY208914). The proteins predicted to be encoded by these genes share 57% identity; there is 90% identity in the C-terminal 1000 amino acids of their sequences. Both genes map to the left end of linkage group IX; the *CNA47* marker is located within the *FMG-1B* sequence. Genomic sequencing showed that *FMG-1A* and *FMG-1B* are located very close together, separated

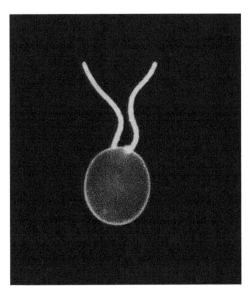

FIGURE 11.8 *Immunofluorescence image of a vegetative cell of* Chlamydomonas, *strain 21gr, labeled with an antibody to the carbohydrate portion of the major flagellar membrane glycoprotein, FMG-1B. Note the uniform, intense staining of the flagellar surface of this unpermeabilized cell. The thin rim of staining around the cell body is due to cross-reactivity of the antibody with carbohydrate epitopes on the 2BII cell wall glycoprotein. Reproduced from Bloodgood (1987) with permission of JAI Press, Greenwich, CT.*

FMG-1B   FMG-1A   FMG-1A
              fragment   10 kb

FIGURE 11.9   *Diagram illustrating the clustering of* FMG-1-*related genes within the* Chlamydomonas *genome; based on version 2.0 JGI genome assembly.*

by only 37 kb of sequence (Figure 11.9). The genomic DNA between the *FMG-1A* and *FMG-1B* genes contains a small sequence (less than 2 kb) having 69% identity to a limited region of the *FMG-1A* gene (Figure 11.9). The close spacing of these three sequences suggests endoduplication of an ancestral gene. Based on the total absence of any unique ESTs or unique peptides for FMG-1A, it appears that only FMG-1B is expressed in vegetative cells of *Chlamydomonas*. It is possible that FMG-1A may be expressed during the sexual cycle of *Chlamydomonas*. Expression of FMG-1B increases 5 fold after deflagellation (Pazour et al., 2005), as is expected for a major flagellar protein. Since only one *FMG-1* gene is expressed in vegetative *Chlamydomonas*, the two closely spaced bands consistently seen on SDS-PAGE of flagella, membrane-plus-matrix fractions, or purified flagellar membranes cannot be due to two different gene products but more likely result from alternative splicing or post-translational modification.

The *FMG-1B* gene contains an open reading frame predicted to encode a protein of 4149 amino acids (410 kD) plus a 27-amino-acid signal sequence. Three observations strongly suggest that this is the correct protein prediction for the *FMG-1B* gene and that the two bands seen on gels are encoded by the *FMG-1B* gene: (1) perfect matches to peptide sequences, obtained from the native protein by mass spectrometry (Pazour et al., 2005), are spread uniformly throughout this predicted sequence (indeed perfect matches cover 50% of the predicted sequence), (2) antibodies made to predicted peptide sequences react with the native protein (Bloodgood, unpublished data), and (3) the predicted amino acid composition is almost a perfect match to the amino acid composition of the combined bands excised from an SDS-polyacrylamide gel (Bloodgood, 1990).

The predicted amino acid sequence of FMG-1B has an N-terminal signal sequence, a single transmembrane domain, and a 17-amino-acid C-terminal cytoplasmic domain. This cytoplasmic domain, albeit short, is likely to play critical roles in signal transduction and the coupling of FMG-1B to the intraflagellar cytoskeletal machinery responsible for movement of FMG-1B within the plane of the flagellar membrane. Indeed, this is the likely binding site for a flagellar phosphoprotein known to bind FMG-1B and to play a role in the signal transduction pathway activated by crosslinking of FMG-1B (Bloodgood and Salomonsky, 1994, 1998). There is a surprising lack of clear protein motifs that would provide insight into the function of this membrane glycoprotein, for which there is abundant cell biological and biochemical evidence suggesting adhesive, signaling, and motility functions (see below) analogous to the functions performed by integrins in meta-zoans. The most obvious sequence domains/motifs in the ectodomain of FMG-1B are a 20-amino-acid proline-rich domain containing Pro–Ser–Pro–Pro repeats and an ATP/GTP-binding P-loop domain (InterPro IPR001687). There is a hydroxymethylglutaryl-coenzyme A reductase domain (InterPro IPR002202) in the C-terminal region. There are also some large threonine-rich and alanine-rich regions in the sequence, as well as a possible fibro-nectin type III repeat domain. Although *in vivo* labeling of cells with ^{32}P-phosphate followed by flagellar isolation, SDS-PAGE, and autoradiography failed to provide any evidence for phosphorylation of FMG-1B (Bloodgood and Salomonsky, 1994), mass spectrometry has identified several phospho-peptides in the FMG-1B ectodomain (Chinskey et al., 2006).

The cell wall of *Chlamydomonas* cross-reacts with anti-carbohydrate monoclonal antibodies made to FMG-1B (Bloodgood et al., 1986) and FMG-1B reacts with certain anti-carbohydrate monoclonal antibodies prepared to the 2BII cell wall glycoprotein (Smith et al., 1984; Roberts et al., 1985), sug-gesting that flagellar FMG-1B glycoproteins and the 2BII cell wall glycopro-tein share common carbohydrate epitopes. The large ectodomain of FMG-1B (over 4100 amino acids) contains 31 potential N-glycosylation sites; this is consistent with the biochemical evidence that FMG-1B is N-glycosylated on at least 15 peptides derived from FMG-1B by V-8 protease digestion (Bloodgood et al., 1986). The sugar composition of FMG-1B is consistent with N-linked glycosylation (Bloodgood, 1990); as expected, FMG-1B binds concanavalin A. FMG-1B is exposed at the flagellar surface and it is the major flagellar protein labeled by biotinylation (Bloodgood, 1990; Bloodgood and Salomonsky, 1995) (Figure 11.7) and iodination (Monk et al., 1983; Bloodgood and Workman, 1984) of live cells. FMG-1B is the major flagel-lar protein accessible to extracellular proteolysis and data suggest that the faster migrating of the two isoforms seen by SDS-PAGE is preferentially digested, perhaps due to a difference in glycosylation or tertiary conforma-tion (Bloodgood and May, 1982). Gel filtration data suggest that FMG-1B exists in the flagellar membrane as a dimer or tetramer (Bloodgood, unpub-lished data); the existence of a heterodimer would explain the equimolar iso-forms always observed by SDS-PAGE. The large extracellular domain, along with the extensive N-glycosylation, is thought to be responsible for the thick glycocalyx seen on the extracellular surface of the *Chlamydomonas* flagel-lum by TEM. Only anti-carbohydrate monoclonal antibodies can bind to the flagellar surface; anti-protein monoclonal antibodies cannot access protein epitopes on FMG-1B presumably because of the extensive N-glycosylation (Bloodgood et al., 1986).

3. Mastigoneme protein

The mastigonemes extending from the flagellar surface (Figure 11.5) are polymers of a single, 220-kD glycoprotein described in detail in section III.D.

4. Voltage-gated calcium channels

Chlamydomonas exhibits a number of photoresponses (positive phototaxis, negative phototaxis, and a photoshock/photophobic response) that involve the regulation of bending in the two flagella (see Chapter 13). High intensity pulses of light induce a graded, inward-directed, calcium-dependent photoreceptor current through the cell body plasma membrane overlying the eyespot, which then induces a calcium-dependent, all-or-none, inward flagellar membrane current. The latter is presumably mediated by voltage-dependent calcium channels within the flagellar membrane (Harz and Hegemann, 1991; Kateriya et al., 2004). Opening of these flagellar calcium channels allows influx of calcium from the medium and a rise in intraflagellar calcium concentration, triggering an alteration in the flagellar waveform (Holland et al., 1997). The influx is self-regulated in the sense that internal Ca^{2+} downregulates the current. If external Ca^{2+} is replaced by Ba^{2+}, extremely large and long lasting currents are observed. This also means that the Ca^{2+} influx must be compensated by a K^+ efflux to keep the membrane potential within an acceptable range (Holland et al. 1996). Flagellar regeneration studies showing the amplitude of the flagellar calcium current to be proportional to flagellar length suggest that the voltage-dependent flagellar calcium channels are distributed uniformly along the length of the flagella (Beck and Uhl, 1994).

Because of the evidence for voltage-regulated calcium channels in the flagellar membrane, it is presumed that the *Chlamydomonas* flagellar membrane must also possess a calcium pump that extrudes calcium from the flagellum. Although the *Chlamydomonas* flagellar proteome project identified several candidate calcium-transporting ATPases and one cation channel (Pazour et al., 2005), no voltage-dependent calcium channel or calcium-active transporter has yet been biochemically identified within the flagellar membrane. Pazour et al. (1995) isolated two insertional mutants of *Chlamydomonas* defective in both phototaxis and the photophobic response. These mutants (*ptx2* and *ptx8*) exhibit a normal photoreceptor current but an absence of the flagellar current, suggesting that one or both of the disrupted genes might encode for the hypothesized voltage-dependent calcium channel in the flagellar membrane. Yoshimura (1996) presented evidence for mechanosensitive calcium channels in the *Chlamydomonas* flagellar membrane, and the *Chlamydomonas* genome has several gene models for mechanosensitive ion channels.

5. AGG2

A 1% Triton X-100 detergent-resistant flagellar membrane fraction from *Chlamydomonas* is enriched both in sphingolipids and in a 20-kD protein containing two predicted transmembrane domains (Iomini et al., 2006). Antibodies specific for this flagellar membrane protein, named

AGG2 (AAZ22341), localize to the proximal region of the flagellum. RNAi knockdown of this protein results in a defect in phototaxis; cells exhibit negative phototaxis under conditions where wild-type cells exhibit positive phototaxis.

6. PKD2

Mammalian polycystin-2 is a calcium-channel protein (product of the *PKD2* gene) which is usually associated with the transmembrane receptor-like protein polycystin-1 (product of the *PKD1* gene) and is involved with a variety of sensory phenomena in primary and motile cilia (Kottgen and Walz, 2005). A *PKD2*-like gene (but not a *PKD1*-like gene) has been found in the *Chlamydomonas* genome and a polycystin-2 homologue was identified in the *Chlamydomonas* flagellar proteome (Pazour et al. 2005). Huang et al. (2007) have sequenced and characterized the *Chlamydomonas PKD2* gene and the 181-kD protein it encodes (ABR14113; 1626 amino acids). Using immunoblots, three forms of PKD2 (210, 120 and 90 kD) were found in vegetative and gametic *Chlamydomonas*; only the 120-kD and 90-kD forms (presumed to be cleavage products of the 210-kD protein) are found in the flagella (Huang et al., 2007). Although PKD2 was found within purified flagellar membrane vesicles and much of the flagellar PKD2 could be solubilized by non-ionic detergent, about a quarter of the flagellar PKD2 remained associated with the axoneme through repeated detergent extractions, suggesting that a subset of the protein is anchored to the doublet microtubules. RNAi knockdown resulted in a mating defect (Huang et al., 2007). This is not totally surprising, as gametic flagella have much more PKD2 than vegetative flagella and calcium is known to play a role in mating of *Chlamydomonas* gametes (Snell et al., 1982; Bloodgood and Levin, 1983; Goodenough et al., 1993). The *Chlamydomonas* genome is not predicted to encode a homologue of polycystin-1, raising the possibility that PKD2 interacts with some other receptor, such as a sexual agglutinin, in the *Chlamydomonas* flagellar membrane.

7. Sexual agglutinins

During gametogenesis, the flagellar membrane acquires new properties and new proteins. In particular, the gametic flagellar membrane contains a population of sexual agglutinin molecules (see Chapter 12). The sexual agglutinins are very large, species-specific, mating-type-specific, hydoxyproline-rich glycoproteins (Ferris et al., 2005); they are peripheral membrane proteins that bind to an integral membrane protein (Kooijman et al., 1989) that serves to anchor them to the gametic flagellar membrane. Limited electron microscopic (Goodenough and Heuser, 1999) and biochemical data (Versluis et al., 1992) indicate that the mating-type *plus* sexual agglutinin interacts directly with the mating-type *minus* sexual agglutinin as the initial event in gamete interaction during mating. The sexual agglutinins of *C. reinhardtii* are about 240 nm in length and have a unique cane shape (described in section III.E). The cell has the ability to regulate the state of activation (adhesiveness) of the sexual agglutinins on the gametic flagellar surface although this process is poorly understood (Saito et al., 1985). Among other signals, light appears to affect flagellar agglutinin activity (Kooijman et al., 1988); the blue-light receptor, phototropin, appears to be one mediator of agglutinin activation (Pan et al., 1997). At least in *C. reinhardtii*, the sexual agglutinins are also found associated with the cell body plasma membrane (Hunnicutt et al., 1990), but these sexual agglutinins are always kept in the inactive state (unable to bind to gametic flagella of the opposite mating type).

The *plus* and *minus* sexual agglutinins have been cloned and sequenced from *C. reinhardtii* and its close relative *C. incerta* (Table 11.1; see Chapter 12; Ferris et al., 2005). The *C. reinhardtii plus* agglutinin is coded by the *SAG1* gene (14.8 kb with 15 exons). The transcript encodes a protein of 3349 amino acids (330 kD) plus a 60-amino-acid signal sequence. The *minus* agglutinin is encoded by the *SAD1* gene (17.8 kb with 30 exons). The transcript encodes a protein of 3853 amino acids (385 kD) plus a 36-amino-acid signal sequence. They are both peripheral membrane glycoproteins that exhibit anomalous migration (1000 kD) on SDS-PAGE; they can be stripped from the external surface of the flagellar membrane by EDTA treatment. Each of the agglutinins has three domains: (1) an N-terminal domain (~500 amino acids) that is proposed to mediate binding to a receptor protein in the flagellar membrane; (2) a shaft domain (~1000 amino acids; 60% proline, mostly if not totally in the form of hydroxyproline; composed largely of PPSPX repeats) and (3) a large C-terminal globular head domain (~2000 amino acids; extends out from the flagellar surface and presumably interacts with the agglutinin of the opposite mating type). Both the N- and C-terminal domains possess multiple N-glycosylation sites; *in vivo* studies using the drug tunicamycin suggest that N-glycosylation is necessary for the adhesive function of the sexual agglutinins (Matsuda et al., 1982; Ray and Gibor, 1982; Wiese and Mayer, 1982). The sexual agglutinins in *C. eugametos* have been reported to also contain sulphated O-linked carbohydrates (Versluis et al., 1993). The sexual agglutinins are closely related to a hydroxyproline-rich glycoprotein, GP1, which has the same PPXPS repeats and is found in the *Chlamydomonas* cell wall (Ferris et al., 2001).

Sexual agglutinins are peripheral membrane proteins (lacking transmembrane domains) and must be anchored to integral flagellar membrane glycoproteins in order to perform their adhesive and signaling roles (see Chapter 12) and to be moved laterally along the flagellar surface (see section VI.B). In *C. eugametos*, a candidate sexual agglutinin receptor has been identified as a 125-kD flagellar membrane glycoprotein that binds the lectin wheat germ agglutinin (Kooijman et al., 1989; Kalshoven et al., 1990).

8. Additional flagellar membrane proteins

Although the protein composition of the vegetative flagellar surface is dominated by FMG-1B and the mastigoneme protein, and the gametic flagellar surface is dominated by FMG-1B, the mastigoneme protein, and the sexual agglutinins, there is good reason to believe that there are many additional *Chlamydomonas* flagellar membrane proteins yet to be identified and characterized, beyond the small number described above and in Table 11.1. If one takes all of the proteins from the flagellar proteome (Pazour et al., 2005) that were identified by a minimum of five peptides and contain at least one putative transmembrane domain, then one gets 20 presumptive vegetative *Chlamydomonas* flagellar membrane proteins, including FMG-1B, PKD2, the inositol 1,4,5,-tris-phosphate receptor, calcium-transporting ATPases, a TRP channel protein, fibrocystin and proteins with PAS sensory domains (Ponting and Aravind, 1997). Lectins and monoclonal antibodies have identified additional *Chlamydomonas* flagellar membrane glycoproteins that have not been fully characterized (Monk et al. 1983; Kooijman et al., 1989; Kalshoven et al., 1990; Bloodgood and Salomonsky, 1994). There is some evidence that components of the signaling pathway triggered by adhesion of sexual agglutinins (Pan and Snell, 2000), particularly a cGMP-activated protein kinase (Wang et al., 2006), associate with the gametic flagellar membrane.

B. Flagellar membrane lipids

Most of the literature on the lipid composition of ciliary and flagellar membranes comes from work on the cilia of *Tetrahymena*, and to a lesser extent, *Paramecium*. Regarding these studies, Kaneshiro (1990) stated that: "In all cases in which lipids of cilia and flagella have been compared with the rest of the cell, the lipid composition of the organelle has been found to be distinct." Most of the research on lipids in *Chlamydomonas* relates to the chloroplast or to the whole cell; there is only one study that has analyzed flagellar lipids (Gealt et al., 1981) and that study examined only sterols and fatty acids and compared flagella to whole cells. There has never been a study in *Chlamydomonas* comparing the lipid composition of flagella or flagellar membranes with that of the cell body plasma membrane, primarily because of the difficulty of obtaining cell body plasma membrane preparations (as discussed in section II.B). The major phospholipids of whole *Chlamydomonas* are phosphatidylglycerol, phosphatidylethanolamine, and phosphatidylinositol, and the major glycolipids are monogalactosyldiacylglycerol (MGDG), digalactosyldiacylglycerol (DGDG), and sulfoquinovosyldiacylglycerol (SQDG) (Eichenberger, 1976; Giroud et al., 1988; Arisz et al., 2000; see also Volume 2, Chapter 2). Janero and Barrnett (1981b, 1982) reported phosphatidylcholine as a non-thylakoid phospholipid in *Chlamydomonas*, but this appears to have been incorrect, as Giroud et al.

(1988) and Grenier et al. (1991) documented the total absence of phosphatidylcholine in *Chlamydomonas*. Among the glycerolipids of whole *Chlamydomonas*, glycolipids are much more abundant than phospholipids. Giroud et al. (1988) argue that the major lipids synthesized in the cytoplasm are diacylglyceryl(N,N,N-trimethyl)homoserine (DGTS) and phosphatidylethanolamine (with the rest being synthesized in the chloroplast). DGTS is thought to serve as a replacement for phosphatidylcholine in algal membranes and its synthesis is associated with microsomal fractions and not chloroplasts (Moore et al., 2001). Janero and Barrnett (1981a) reported that only 70% of the sulfolipid (SQDG) and 50% of the galactolipids (MGDG and DGDG) are found in the chloroplast. Assuming that flagellar lipids would be equivalent to or a subset of whole-cell non-thylakoid lipids, this leaves phosphatidylethanolamine, DGTS, MGDG, and DGDG as possible candidates for flagellar membrane lipids. Synthesis of the mRNA for the key regulatory enzyme (phosphoethanolamine cytidylyltransferase) in the pathway for synthesis of phosphatidylethanolamine is upregulated after deflagellation, suggesting an involvement in flagellar membrane synthesis (Yang et al., 2004). Supporting the possibility that galactolipids are present in the ciliary and flagellar membranes is the observation of a galactosyltransferase in the database of candidate ciliary and basal body proteins generated by comparing the genomes of ciliated vs. non-ciliated organisms (Li et al., 2004). Glycolipids have been reported in the flagellar membrane of *Euglena* (Chen and Bouck, 1984; Bouck et al., 1990; Kaneshiro, 1990). Chen and Bouck (1984) demonstrated that *Euglena* flagella possess a glycosyltransferase activity capable of glucosylating flagellar membrane lipids. Bloodgood et al. (1995) obtained a mouse monoclonal antibody specific for a carbohydrate epitope associated with a glycolipid in the *Chlamydomonas* flagellar membrane. Only a subset of cells in any strain exhibit flagellar surface staining with this antibody and a small number of cells exhibit differential staining on the two flagella of the same cell.

Gealt et al. (1981) found that the sterol composition of whole flagella (presumably representing the flagellar membrane sterol composition) is 55% ergosterol and 45% 7-dehydroporiferasterol (identical to that of whole cells); triterpenoid alcohols have also been found in *Tetrahymena* cilia (Kaneshiro, 1990) and *Ochromonas* flagella (Chen et al., 1976). However, Gealt et al. (1981) found that the fatty acid composition of *Chlamydomonas* flagella differed from that of whole cells. Flagella lack some fatty acids (16:2, 16:3) found in the whole cells and are enriched in others (16:1, 18:0, 18:2) relative to whole cells. In the whole cells, 19% of the 16-carbon fatty acids and 4% of the 18-carbon fatty acids are saturated; in contrast, in the flagella, 64% of the 16-carbon and 26% of the 18-carbon fatty acids are saturated. This raises the possibility that there may be regulation of the amount of unsaturated fatty acids in the flagella. Due to the lack of data on flagellar lipid composition in general and inconsistency in the data from different authors on fatty acid composition of various lipids and on the whole-cell vs. chloroplast thylakoid fatty acid data (Gealt et al., 1981; Janero and Barrnett, 1981a,b; Giroud et al., 1988), it is not possible to determine whether the differences in flagella vs. whole-cell-body fatty acid composition observed by Gealt et al. (1981) can be totally attributed to the chloroplasts (which tend to be enriched in unsaturated fatty acids) or if, indeed, the flagellar membrane lipid composition differs in any way from the rest of the plasma membrane.

There is some information suggesting that phosphoinositide signaling is involved in two flagella-related processes in *Chlamydomonas*: (1) mating (Musgrave et al., 1993) and (2) flagellar excision (Quarmby et al., 1992; see Chapter 3). While Brederoo et al. (1991) found phosphatidylinositol 4-phosphate (PtdInsP) and phosphatidylinositol 4,5-bisphosphate (PtdInsP$_2$) at low levels in the flagellar membrane of *C. eugametos* gametes, the concentration of PtdInsP$_2$ was 20 times less than in the cell body plasma membrane. The *C. reinhardtii* flagellar proteome contains a protein similar to the inositol 1,4,5,-tris-phosphate receptor (Pazour et al., 2005).

V. ASSEMBLY AND TURNOVER OF THE FLAGELLAR MEMBRANE

A. Assembly

Little is known about the assembly of the *Chlamydomonas* flagellar membrane. It must be assumed that most flagellar membrane proteins and lipids are synthesized in the rough endoplasmic reticulum, transported to and processed within the Golgi apparatus, and then transported to the cell surface by a vesicular pathway. Transport vesicles are assumed to be transported to a region of the plasma membrane close to the base of each flagellum (perhaps to a periciliary domain of the plasma membrane as discussed in section II.A). Jekely and Arendt (2006) pointed out that IFT-complex proteins have similarities to coatomer proteins (clathrin and COP1) and hypothesized that the primitive IFT complex mediated vesicular traffic to a specialized domain of the plasma membrane that later evolved into the cilium/flagellum, which then incorporated IFT for the entry of proteins into and transport within cilia and flagella. In *Caenorhabditis*, mutation of a subunit of the AP-1 clathrin adaptor complex prevents trafficking of transmembrane proteins to neuronal sensory cilia (Bae et al., 2006). Veira et al. (2006) have shown that FAPP2 (a Golgi apparatus protein with domains that bind phosphotidylinositol-4-phosphate and glycolipids) is necessary for ciliogenesis and for the targeting of membrane proteins and lipids to the primary cilium on MDCK cells; there was an accumulation of membrane vesicles near the centrosome in FAPP2 knockdown cells. Follit et al. (2006) showed that one IFT protein, IFT20, first identified in *Chlamydomonas*

(Cole et al., 1998), is associated with the Golgi apparatus as well as the base of the primary cilium and the cilium itself in mammalian cells. They described a track of IFT20 protein extending between the Golgi apparatus and the base of the cilium, suggesting a route for vesicular traffic carrying membrane components destined for the cilium. Moreover, moderate knock-down of IFT20 reduced the amount of polycystin-2, which is processed through the Golgi apparatus, on the ciliary membrane. This indicated that IFT20 has a role in ciliary and flagellar membrane assembly as well as axo-nemal assembly, although it is not certain if this reflects IFT20's function in the Golgi apparatus or in the cilium. Davidge et al. (2006) and Absalon et al. (2008) suggest that, in trypanosomes, the flagellar membrane can assemble in the absence of IFT and in the absence of axonemal assembly.

It is unclear whether the transport vesicles containing proteins destined for the flagellar membrane constitutively fuse with the plasma membrane or whether they can be stored at the apical end of the cell for future regulated exocytosis. The only evidence in *Chlamydomonas* for flagellar membrane components being stored in the apical end of the cell comes from Nakamura et al. (1996) who observed, using immunofluorescence microscopy, a ring of structures within the apical end of the cell that were labeled by the anti-mastigoneme antibody. In either case, exocytosis would be followed by movement of flagellar proteins and lipids into the flagellar membrane, either by diffusion (coupled to a diffusional barrier preventing diffusion into the cell body plasma membrane) or by directed transport utilizing IFT or perhaps the mechanism responsible for directed membrane protein move-ments within the flagellar membrane (see section VI.B). It is also possible that the cell body plasma membrane may be used as a reservoir for certain flagellar membrane glycoproteins; Hunnicutt et al. (1990) utilized vectorial labeling (iodination) to show that FMG-1B and sexual agglutinins can move from the plasma membrane into the flagellar membrane during flagellar regeneration.

The Rab family of small G proteins perform many roles in membrane trafficking in cells, and there is accumulating evidence for the role of cer-tain Rab proteins in the assembly of primary cilia (Shin-ichiro et al., 2007). The *Chlamydomonas* flagellar proteome contains five Rab proteins, two of which are IFT proteins (Pazour et al., 2005).

B. Turnover

There is abundant evidence for the turnover of *Chlamydomonas* flagellar membrane proteins in intact flagella that are maintaining a constant length (Bloodgood et al., 1979; Bloodgood and May, 1982; Remillard and Witman, 1982; Bloodgood, 1984; Song and Dentler, 2001). Indeed, flagellar mem-brane proteins turn over faster than axonemal proteins and may do so by a different mechanism.

Pronase treatment of live cells under conditions that have no effect on cell viability or cell motility results in a loss of adhesiveness of the flagellar surface for polystyrene microspheres and proteolysis of the FMG-1B flagellar membrane glycoprotein. After pronase removal, flagellar surface adhesiveness and FMG-1B recover; recovery can be blocked by inhibition of protein synthesis with cycloheximide (Bloodgood and May, 1982). If cells are pulse labeled for 1 hour with ^{35}S-sulfate under conditions where flagellar length is not changing, flagellar proteins become labeled. However, 60% of the trichloroacetic-acid-precipitable counts that were incorporated into the flagella during the pulse can be removed from the flagella by pronase treatment of live cells (Bloodgood and May, 1982). Because the pronase treatment did not affect viability or motility of the cells, these data demonstrate that flagellar membrane proteins were being imported into the full-length flagellum during the 1-hour pulse. Because the flagellar membrane contains only a small fraction of the total protein of the flagellum, this experiment also demonstrates that the flagellar membrane proteins are turning over more rapidly than other flagellar proteins.

Pulse labeling of cells with ^{35}S-sulfate, under conditions where the flagella were not changing length, demonstrated incorporation of label preferentially into the pair of FMG-1B bands (Bloodgood, 1984; Song and Dentler, 2001) and an approximately 65-kD membrane protein (Remillard and Witman, 1982; Bloodgood, 1984). To quantitate this differential labeling, a double label strategy was utilized; cells were uniformly labeled for 24 hours with ^{3}H-acetate and then pulse labeled for an hour with ^{35}S-sulfate. Duplicate gels were either stained for protein, dried, and subjected to autoradiography or cut into slices and the ratio of ^{35}S/^{3}H counts in each slice determined. By far the highest ratios of incorporation of the pulse label to the uniform label were found for FMG-1B and the 65-kD membrane protein (Bloodgood, 1984), suggesting that membrane proteins are turning over more rapidly than axonemal proteins. An alternative interpretation is that there are large differences in cellular pool sizes for flagellar membrane vs. axonemal proteins (Remillard and Witman, 1982). Inhibitors of protein synthesis and protein glycosylation result in the reversible loss of microsphere binding to the surface of the flagellum under conditions where length remains constant (Bloodgood et al., 1979). Because microspheres have been shown to interact primarily with FMG-1B (Bloodgood and Workman, 1984) and to cluster FMG-1B at the point of contact with the flagellar membrane (Bloodgood and Salomonsky, 1998), this suggests that the loss and recovery of microsphere binding on the intact flagellar surface reflects turnover of FMG-1B. Collectively, these observations suggest a rapid rate of turnover of flagellar membrane proteins and explains the ability of the flagellar membrane to recover from the effects of proteolytic treatment (Bloodgood and May, 1982), protein synthesis inhibitors (Bloodgood et al., 1979), and protein glycosylation inhibitors (Bloodgood et al., 1979).

This evidence for protein turnover in the flagellar membrane is also consistent with the rapid changes in flagellar membrane composition that occur during the early stages of mating (see below and Chapter 12).

Granted that membrane protein turnover occurs in the intact flagellum in the absence of net assembly or disassembly of the flagellum, there are three possible mechanisms by which membrane components could exit the flagellum:

1. Both vegetative and gametic *Chlamydomonas* exhibit a process of constitutive membrane vesicle release from the flagellum (Brown et al., 1968; McLean et al., 1974; Bergman et al., 1975; Snell, 1976; Huang et al., 2005). This process, of necessity, results in turnover of both flagellar membrane proteins and lipids and may be the principal means of turnover of flagellar membrane components. The major flagellar membrane components (including FMG-1B and mastigoneme protein, as well as sexual agglutinins in the case of gametes) are found in the vesicles that are shed continuously from the flagellar surface (McLean et al., 1974; Bergman et al., 1975; Snell, 1976). This constant loss of membrane from the flagellum into the medium requires a constant source of new membrane, presumably coming from the Golgi apparatus (directly or via the cell body plasma membrane). Consistent with this, Dentler (2006) showed that brefeldin A, a drug that disrupts the Golgi apparatus, results in shortening of the *Chlamydomonas* flagellum.

2. Membrane could be removed from the flagellum by endocytosis at the base of the flagellum. Relatively little is known about the process of endocytosis in *Chlamydomonas*. Clathrin-coated vesicles have been purified from *Chlamydomonas* (Denning and Fulton, 1989). One group of investigators has observed that (i) dextran is endocytosed in *Chlamydomonas*, (ii) clathrin immunolocalizes to punctate structures near the flagellar base, (iii) following resorption of flagella, clathrin no longer localizes to the flagellar base (J.Z. Rappoport, S. Kemal, R. Boursiquot and S.M. Simon, personal communication). Flagellar resorption does not appear to be associated with an accumulation of cytoplasmic vesicles near the base of the flagella.

3. Anterograde IFT transports axonemal proteins into the *Chlamydomonas* flagellum and retrograde IFT transports axonemal proteins from the flagellum to the cell body (Rosenbaum and Witman, 2002; see Chapter 4). However, there is, as yet, very limited evidence for the involvement of IFT in the transport of flagellar membrane proteins into or out of the *Chlamydomonas* flagellum. Qin et al. (2004) identified a variety of axonemal cargo

proteins associated with immunoprecipitated IFT complexes; however, they specifically looked for and did not find the FMG-1B flagellar membrane glycoprotein. On the other hand, Huang et al. (2007) reported that PKD2 accumulated in the flagella when IFT was shut down in the temperature-sensitive mutant *fla10* (defective in the anterograde IFT motor, kinesin-2), suggesting that IFT is required for removal of PKD2 from the flagellum. For further discussion about the role of IFT in moving flagellar membrane proteins, see section VI.B.

As previously noted, gametic cells possess sexual agglutinins on their flagellar surfaces and these species-specific and mating-type-specific glycoproteins mediate the initial contact of *plus* and *minus* gametic cells. Flagellar membrane turnover is important to the events of mating, a process that demands rapid changes in the properties of the gametic flagellar membrane (particularly in relation to the composition of active and inactive sexual agglutinins). Gametic flagella release membrane vesicles into the medium (Brown et al., 1968; McLean et al., 1974; Bergman et al., 1975; Snell, 1976). This release is certainly a manifestation of flagellar membrane turnover and a mechanism to allow rapid changes in the properties of the flagellar membrane. Membrane vesicles shed by gametic cells contain sexual agglutinins on their external surface and are capable of activating gametes of the opposite mating type by binding to their flagellar surface. This might have a physiological function in the sense that gametes of one mating type can pre-activate gametes of the opposite mating type even before making physical (flagella–flagella) contact. Flagellar membrane vesicles shed by gametes induce an increase in gametic flagellar surface adhesiveness on gametes of the opposite mating type (Demets et al., 1988). Flagellar membrane vesicles may be involved in the initial flagella–flagella adhesion of gametes during mating (see images in Goodenough and Heuser, 1999). If antibody crosslinking is utilized to induce the shedding of all of the sexual agglutinins from the flagellar surface, the sexual agglutinins are rapidly replaced in the flagellar membrane by a cycloheximide-sensitive process (Hunnicutt et al., 1990). Mesland et al. (1980) showed TEM images of membrane blebbing from the tips of flagella of gametes that had been activated by antibody crosslinking of the flagellar surface. Flagellar adhesion of gametes during mating initiates a rise in intracellular cyclic AMP that initiates a number of events in the flagellum (such as the tipping of membrane proteins described in section VI.B) as well as in the cell body (cell wall release and activation of fusion domains in the general cell body plasma membrane). In addition, this signaling pathway (dependent upon both adhesion and cyclic AMP) induces increased turnover of sexual agglutinins in the gametic flagellar membrane through shedding of existing flagellar agglutinin and movement of fresh sexual agglutinins from the cell body into the flagellar membrane (Snell and Moore, 1980;

Saito et al., 1985; Goodenough, 1989; Hunnicutt et al., 1990; Tomson et al., 1990). While the experiments of Hunnicutt et al. (1990) indicate that there is a pool of inactive sexual agglutinin on the cell body plasma membrane, there is disagreement about the source of the new sexual agglutinin molecules in the agglutinating flagellar membrane. Hunnicutt et al. (1990) argued that, in *C. reinhardtii*, sexual agglutinins newly recruited into the flagellum come, at least in part, from the plasma membrane over the cell body. Musgrave et al. (1986) argued that, in *C. eugametos*, newly recruited sexual agglutinins in the flagellum come entirely from cytoplasmic sources, presumably newly synthesized agglutinin delivered from the Golgi apparatus via vesicular transport directly to the plasma membrane near the base of the flagella.

VI. DYNAMIC PROPERTIES OF THE FLAGELLAR MEMBRANE

A. Overview

Because *Chlamydomonas* is enclosed in a thick, multi-layered glycoprotein cell wall, the only portion of the plasma membrane that is available for cell–cell and cell–substrate interactions is the flagellar membrane. The functions of the *Chlamydomonas* flagellar surface (plasma membrane and associated structures) can be classified as: (1) cell–cell adhesion during mating of gametes (see Chapter 12), (2) cell–substrate adhesion and associated signal transduction and force transduction during whole-cell gliding motility (see below), (3) amplification of thrust during swimming motility (using mastigonemes) and (4) regulation of flagellar beat through regulation of the intraflagellar ionic environment (particularly calcium), which is part of the entire photosensory system (see Chapter 13). There are other, less well defined, functions for the flagellar membrane; some of these are sensory functions associated with flagellar resorption (see Chapter 5), flagellar abscission (see Chapter 3), light responses, and mating. There may be some role for the flagellar membrane in flagellar length control (perhaps involving calcium channels or pumps). The flagellar membrane may play an important role in IFT (see Chapter 4) although it is equally possible that motor-driven movement of IFT particles along the axoneme could occur in the absence of the flagellar membrane (given the appropriate conditions). The process of flagellar membrane vesicle release plays a role in flagellar membrane turnover and possibly in events associated with mating. Underlying some of these flagellar membrane/flagellar surface functions are some unusual rapid, directed movements of flagellar membrane glycoproteins.

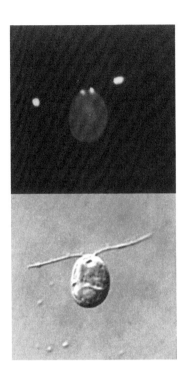

FIGURE 11.10 *Immunofluorescence and DIC images of a vegetative cell of the pf18 mutant, labeled with a monoclonal antibody to the carbohydrate portion of the major flagellar membrane glycoprotein, FMG-1B. The live cell was incubated with the antibody at room temperature for 5 minutes and then fixed. Most of the bound antibody has redistributed into a single patch towards the distal end of each flagellum, whose length is unchanged. With increased incubation times, the antibody cluster (presumably associated with a group of FMG-1B glycoproteins), will move to the base of the flagellum and then subsequently be shed from the flagellum into the medium.*

B. Directed movements of flagellar membrane proteins within the plane of the flagellar membrane

One of the most striking and dynamic phenomena associated with the flagellar membrane of *Chlamydomonas* is the regulated and directed movement of membrane glycoproteins within the plane of the flagellar membrane. This is a temperature-dependent process that requires crosslinking of membrane-glycoproteins and a signaling pathway induced by the crosslinking event. This phenomenon was first described as a response to monoclonal antibodies that crosslink FMG-1B or to the lectin concanavalin A that crosslinks the exposed portions of flagellar membrane glycoproteins (Bloodgood et al., 1986) (Figure 11.10). Crosslinking of FMG-1B is necessary; reagents that bind but do not crosslink (such as Fab fragments of the FMG-1B antibodies or succinyl-concanavalin A) do not induce the directed movements of flagellar membrane proteins. In wild-type cells, a high concentration of the crosslinking reagent induced a global crosslinking and "capping" of the FMG-1B (in the case of specific antibodies) or the concanavalin-A-binding membrane glycoproteins (in the case of concanavalin A) in the flagellar membrane into a single aggregate (Bloodgood et al., 1986) (Figure 11.10). At low concentrations, a number of smaller aggregates were formed (Huang et al., 2005). After concanavalin-A-induced clustering, most of the FMG-1B of the flagellum is found within the concanavalin A clusters; indeed it has been shown that FMG-1B represents a high proportion of the concanavalin-A-binding proteins of the flagellum. These crosslinking-induced membrane protein aggregates move along the length of the flagellum at rates comparable to the speed of microsphere movement or gliding motility (see below). An example of the directed movement of these aggregates along the flagellum at constant velocity can be seen at http://people.virginia.edu/~rab4m/RABResearch/Images/FMG1/WebPicts/pages/Picture1_jpg.htm and in the video "FMG1 Antibody Movement" at http://www.elsevierdirect.com/companions/9780123708731. FMG-1B is cleared from the flagellar surface from base to tip until it forms a round aggregate at the flagellar tip; this aggregate can then move at uniform velocity up and down the flagellum (Bloodgood et al., 1986). It usually ends up at the base of the flagellum; subsequently, it is shed into the medium (presumably associated with a membrane bleb but this has not been clearly demonstrated). Soon after the FMG-1B has been cleared from the flagellar surface, new copies of FMG-1B appear along the entire length of the flagellum (Bloodgood et al., 1986).

The crosslinking of FMG-1B (by antibodies, lectins, polystyrene microspheres and presumably by contact of the flagellar surface with a solid substrate) induces a transmembrane signaling pathway associated with calcium influx and calcium/calmodulin-dependent changes in levels of flagellar protein phosphorylation (see section VI.C.3). Crosslinking-induced movements of FMG-1B are inhibited by calcium-channel blockers, calmodulin antagonists, and serine/threonine protein kinase inhibitors (Bloodgood and Salomonsky, 1990, 1991). Presumably, the transmembrane signaling pathway induced by FMG-1B crosslinking activates a motor protein to engage the crosslinked membrane proteins and move them along the flagellar axoneme. It is likely that this process of directed membrane protein movement is responsible for both microsphere movements and gliding motility (section VI.C); however, in the case of microsphere movement and gliding motility, one imagines very local crosslinking and the creation of small patches of FMG-1B associated with microsphere binding (in the case of microsphere movement) or planar substrate binding (in the case of whole-cell gliding motility). Interference reflection microscopy shows local areas of contact of the flagellar surface with a solid substrate during gliding motility (Barbara Daniel and Conrad King as cited in Bloodgood, 1990). Indeed, it has been shown that microsphere binding induces a very local aggregation of FMG-1B at the site of microsphere contact and activation of the same signaling pathway (involving changes in flagellar protein phosphorylation) that is activated by global crosslinking with antibodies or lectins (Bloodgood and Salomonsky, 1998).

FMG-1B is not the only flagellar membrane glycoprotein that exhibits crosslinking-induced movement within the *Chlamydomonas* flagellar membrane. A monoclonal antibody that recognizes carbohydrate epitopes on a group of high-molecular-weight flagellar membrane glycoproteins (but not FMG-1B) also induces the directed movement of these glycoproteins within the flagellar membrane (Bloodgood et al., 1986).

Antibodies against the sexual agglutinin glycoproteins on the gametic flagellar surface of *C. eugametos* induce clustering and transport of these sexual agglutinins to the flagellar tip (Homan et al., 1987, 1988); this process requires crosslinking of sexual agglutinins (Fab antibody fragments are ineffective) and phenocopies a similar process of sexual agglutinin crosslinking, clustering, directed movement to the flagellar tip, and shedding from the flagellar surface that occurs during the early stages of the normal mating process in *C. reinhardtii* (see Chapter 12). The directed movement of the crosslinked sexual agglutinins to the flagellar tips is referred to as "tipping" (Goodenough and Jurivich, 1978; Goodenough, 1993) and requires a rise in intracellular cyclic AMP, which is initiated as a result of the initial crosslinking of sexual agglutinins, either by antibodies, the lectin concanavalin A, or during the normal process of adhesion of flagella from *plus* and *minus* gametes.

As described in section IV.A.7, the sexual agglutinins are large peripheral membrane proteins that are anchored to the flagellar membrane by association with an as yet poorly described integral membrane glycoprotein. In *C. eugametos*, a 125-kD glycoprotein in the flagellar membrane appears to anchor the sexual agglutinins to the flagellar membrane (Kooijman et al., 1989). Wheat germ agglutinin crosslinks this glycoprotein and induces its clustering and movement to the flagellar tip, inducing the same physiological responses that occur when sexual agglutinins are crosslinked with antibodies or when *plus* and *minus* gametic flagellar surfaces interact via the sexual agglutinin proteins during normal mating (Kooijman et al., 1989). Indeed, crosslinking with antibodies to either the sexual agglutinin or the agglutinin receptor protein result in co-redistribution of the two proteins to the flagellar tips and induction of a signaling pathway triggering later events in mating.

The process of flagellar tipping is critical to mating in *Chlamydomonas*; the real cargo being moved by this process is the gametic flagellum belonging to the mating partner. Initial interaction of the flagella of *plus* and *minus* gametes can occur at any point on their respective surfaces but it is necessary that the flagella become oriented such that the tips of the flagella of the *plus* gamete are aligned with the tips of the flagella of the *minus* gamete, at which point the flagellar tips are locked into place and a signal is sent to the cell body activating later events in mating (loss of the cell walls and activation of fusion sites called "mating structures" on the cell bodies of the two gametes). Aligning and locking the two sets of flagellar tips into the proper position ensures that subsequent cell body contact will bring the *plus* and *minus* mating structures into apposition allowing cell body fusion. The realignment of the flagella (movement of the flagellar contact points) during mating can be directly observed (Mesland, 1976; Homan et al., 1987); indeed one can observe isolated gametic flagella (which round up into spheres) binding to and then being moved to the distal tips of the flagella of gametes of the opposite mating type (Goodenough, 1993). Using this as an assay for the directed movement of sexual agglutinins, Goodenough (1993) showed that the same set of inhibitors that inhibit the crosslinking-induced movements of FMG-1B also inhibit the adhesion-induced movements of the sexual agglutinins along the flagellar surface. Hoffman and Goodenough (1980) observed that, during the tipping and tip-locking stages of normal mating, polystyrene microspheres accumulated at the flagellar tips, suggesting that the motors responsible for microsphere movement may well be involved in the directed movements of the sexual agglutinins and hence the realignment of the flagella that occurs during the process of tipping.

It is of much interest to know the identity of the motor proteins responsible for these directed movements of flagellar membrane glycoproteins in response to crosslinking and to know the role (if any) that IFT may be playing in this dynamic flagellar membrane process. Huang et al. (2005) showed that the movement of concanavalin-A-induced clusters within the *Chlamydomonas* flagellar membrane is inhibited at the non-permissive temperature in the *fla10* mutant, which has a temperature-sensitive defect in the anterograde IFT motor kinesin-2 (Kozminski et al., 1995, and see Chapter 4). However, this observation does not demonstrate that these flagellar membrane glycoproteins are transported by IFT, only that both processes have some dependence on kinesin-2. Indeed, Qin et al. (2004) found no evidence that FMG-1B is a cargo for IFT. Moreover, the crosslinking-dependent movements of FMG-1B (as well as microsphere movements and gliding motility) are dependent upon micromolar concentrations of free calcium in the medium (Bloodgood and Salomonsky, 1990) whereas IFT is not (Kozminski et al., 1993). While Qin et al. (2005) demonstrated IFT-dependent movement of TRPV channels in the membrane of sensory cilia of *Caenorhabditis*, they showed that another ciliary membrane protein, polycystin-2 (TRPP2), is not actively moved by IFT in these same sensory cilia. Huang et al. (2007) observed that most of the *Chlamydomonas* flagellar PKD2, as observed by immunofluorescence, was immobile. Using the *fla10* mutant defective in IFT coupled with kymograph and fluorescence-recovery-after-photobleaching analysis, they observed that only a small portion of the total flagellar PKD2 moves in an IFT-dependent manner. Further experiments will be necessary to identify the motor protein(s) involved in flagellar membrane protein movements in *Chlamydomonas*.

C. Cell–substrate adhesion and whole-cell gliding motility

1. Overview of gliding motility in Chlamydomonas

The major selective functions for motility in *Chlamydomonas* are probably the need to optimize light intensity for photosynthesis (see Volume 2) and the need to promote interaction of gametes of opposite mating types. *Chlamydomonas* has long been a model organism for the study of whole-cell swimming motility based on the propagation of bends along the length of the flagellar axoneme (see other chapters in this volume); *Chlamydomonas* can regulate the beating of its two flagella so as to exhibit both positive and negative phototaxis as well as transient backwards swimming (see Chapters 7 and 13). *Chlamydomonas* cells have traditionally been grown in liquid cultures and it has been assumed by generations of students that *Chlamydomonas* lives in bodies of freshwater, such as ponds, where they swim and orient themselves by this flagellar beating. However, species of *Chlamydomonas* have adapted to growth in a wide variety of ecological niches, including soils, sands, moss, glaciers, etc. (see Volume 1, Chapter 1). Most of the strains of *Chlamydomonas* used in laboratory studies have been obtained from soil samples, where motility is unlikely to be due to swimming through liquid water. An alternative form of flagella-based motility (called gliding motility) is likely the physiologically relevant form of whole-cell motility for many of these natural habitats. Gliding motility (Ulehla, 1911; Lewin, 1952; Bloodgood, 1981), like swimming motility, is dependent on machinery associated with the flagella but, unlike swimming motility, gliding motility is critically dependent upon substrate adhesion.

In the most general sense, gliding motility is defined as a form of whole-cell motility, observed among both prokaryotes and unicellular eukaryotes (protists), in which motility occurs without any observable deformation of the cell or any cell extension. Another universal characteristic of gliding motility is that it is dependent upon contact with a solid or semi-solid substrate. Eukaryotic gliding motility can be divided into flagella-dependent and flagella-independent categories. Many of the latter cases of eukaryotic gliding motility occur among the apicomplexan parasitic protozoa, such as *Gregarina*, *Plasmodium*, *Toxoplasma*, and *Eimeria*; in these organisms, gliding motility is dependent upon a unique class of myosins (myosin XIV) (Heintzelman, 2003, 2006). Other cases of eukaryotic gliding motility in non-flagellated protistan organisms (such as vegetative diatoms and *Labyrinthula*) also appear to be dependent upon actin and myosin (Heintzelman, 2006).

Flagella-dependent whole-cell gliding motility, while extensively studied only in *Chlamydomonas* species (Bloodgood, 1990), has been observed in many other flagellated green algae (Ettl, 1970) as well as in *Peranema* (Tamm, 1967; Saito et al., 2003). For a sense of the wide range of protists that exhibit flagella-dependent gliding motility, search with the term "gliding" at http://microscope.mbl.edu. Flagella-dependent whole-cell gliding motility in *Chlamydomonas* is mechanistically independent of swimming motility; a wide range of mutant cell lines deficient in bend propagation (and hence swimming motility) because of defects in dynein arms, radial spokes, and central pair components exhibit normal gliding motility.

Sequence of frames from a video taken using DIC microscopy showing the gliding motility of a single cell of the pf18 mutant. Reproduced from Bloodgood (1990) with permission of Plenum Press, New York.

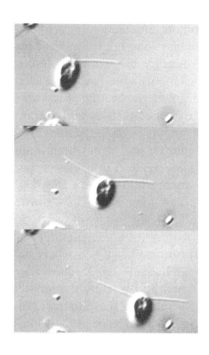

Chlamydomonas cells adhere to a solid substrate via their flagellar surfaces and adopt a characteristic gliding morphology in which the two flagella are straight and are oriented at 180° from one other. Gliding occurs at 1.0–2.0 μm/second in the direction of the long axis of the flagella without any obvious deformation of either flagellum (Figure 11.11; see video "Gliding" at http://www.elsevierdirect.com/companions/9780123708731). A gliding cell can cease movement while maintaining contact with the substrate and resume gliding in the same or the opposite direction. Two observations suggest that gliding motility is powered by the leading flagellum, whose traction with the substrate "drags" the cell body and trailing flagellum along the substrate. First, uniflagellate cells glide with the cell body trailing the flagellum and never stop or reverse direction (Bloodgood, 1981, 1990); this also indicates that the machinery for gliding motility always generates force against the substrate in a direction from flagellar tip to base, suggesting dependence on a minus-end directed motor protein. Second, a wild-type cell, while gliding, will sometimes detach one of its flagella from the substrate and exhibit a single beat (bend propagation). If the leading flagellum loses contact with the substrate, gliding motility temporarily ceases; when a trailing flagellum loses contact with the substrate, gliding motility continues with no interruption. However, a biflagellate cell can cease gliding and resume gliding in the opposite direction. Thus either flagellum can serve as the dominant, force-generating flagellum for gliding. When cells cease gliding while maintaining the gliding configuration (i.e., with the two flagella straight and oriented 180° from one another), it is presumed that both flagella are in functional contact with the substrate and are expressing balanced traction forces. Resumption of gliding is then associated with one flagellum "winning out" over the other in terms of force applied to the substrate.

During gliding, the flagellar surface must exhibit both sensory and motor functions, and there is evidence that both of these are mediated by FMG-1B. Contact with the substrate initiates a transmembrane signaling pathway that activates the gliding motor machinery that applies force to the substrate, thus translocating the flagellum and the entire cell. Important aspects of the mechanism of gliding motility to be addressed include: (1) the mechanism of adhesion of the flagellar surface to the substrate, (2) the signaling mechanism that senses adhesive contact and activates the motor machinery, and (3) the mechanism by which force is transduced to the substrate.

2. Substrate adhesion

Adhesion of the flagellar surface to a solid or semi-solid substrate is necessary for the expression of gliding motility. If a glass surface is derivatized with an immobilized iodination system and cells are incubated with ^{125}I during gliding on this surface, the only flagellar protein that is observed to become iodinated is FMG-1B (Bloodgood and Workman, 1984). This suggests that FMG-1B mediates both sensory (recognition of contact with a substrate, presumably because the substrate acts as a multivalent ligand to crosslink FMG-1B and initiate a signaling cascade) and motor (application of force to the substrate through FMG-1B) interactions between the cell and the substrate.

Adhesion of the flagellar surface to a substrate appears to suppress beating motility. This observation makes sense since beating motility can interfere with gliding motility. This suggests that there may even be an adhesion-dependent signaling mechanism that both suppresses beating motility and activates gliding (Mitchell et al., 2004). Protease treatment of live cells results in loss of adhesion for microspheres, loss of substrate adhesion, and loss of gliding, concomitant with proteolytic cleavage of FMG-1B glycoproteins in the flagellar membrane (Bloodgood and May, 1982). Some gliding-defective mutants exhibit both a loss of adhesion for microspheres and for a solid substrate (Lewin, 1982; Kozminski, 1995). Functional adhesion of flagellar surfaces with the substrate can be quantitated by the percentage of cells in the gliding configuration. This configuration is initially achieved in an interesting manner. Observation of cells in suspension reveals that cells generally initiate contact with a solid substrate via their flagellar tips, which are seen to adhere to and then migrate along the substrate; random migration of the two flagellar tips along the substrate will inevitably result in the cell attaining the gliding configuration with both flagella straight and appearing to be in close apposition to the substrate along their entire length. In reality, interference reflection microscopy shows that the entire length of the flagellum is not equally adherent to the substrate during gliding motility (Figure 3 in Bloodgood, 1990). Unlike the case in *Peranema* (Saito et al., 2003), where gliding velocity varies with flagellar length, *Chlamydomonas* gliding velocity is constant and independent of flagellar length. Indeed, contact of a very small region of the flagellar surface with a solid substrate is sufficient to move the entire cell at the normal gliding velocity; see: http://people.virginia.edu/~rab4m/ RABResearch/Images/MicrosphereMovement/WebPics/pages/Slide3_jpg. htm and video "Cell moving along bead" at http://www.elsevierdirect. com/companions/9780123708731.

As is the case for mammalian whole-cell locomotion (Palecek et al., 1997), it appears that an optimal level of flagellar surface adhesion is necessary for gliding motility. One of the non-gliding mutants of *C. moewusii* isolated by Lewin (1982) exhibits elevated flagellar surface adhesion, based on an increased level of adhesion of polystyrene microspheres (Reinhart and Bloodgood, 1988). Bloodgood (1981) showed that pre-coating the glass surface with 0.1 mg/ml poly-L-lysine enhanced the percentage of gliding cells (with no change in gliding velocity). However, 10 mg/ml poly-L-lysine inhibited gliding completely (presumably by mediating overly tight adhesion between the flagellar surfaces and the glass substrate) although microsphere movements could still be observed on the free surface of these immobilized flagella.

3. Adhesion/crosslinking-induced signaling

The signaling pathway by which *Chlamydomonas* recognizes flagellar contact with a solid substrate and responds by activating cytoskeletal motor proteins to create traction on the substrate involves FMG-1B membrane protein crosslinking, calcium influx, and calcium/calmodulin-activated protein phosphorylation and dephosphorylation. Gliding motility, microsphere movements along the flagellar surface, and crosslinking-induced movement of FMG-1B within the plane of the flagellar membrane all require micromolar concentrations of free calcium in the medium (Bloodgood, 1990; Bloodgood and Salomonsky, 1990). A number of calcium-channel blockers, calmodulin antagonists, and serine–threonine protein kinase inhibitors reversibly inhibit the crosslinking-induced movement of FMG-1B in the flagellar membrane (Bloodgood & Salomonsky, 1990, 1991). The membrane-plus-matrix fraction isolated from flagella exhibits a dramatic but transient increase in protein phosphorylation in response to the addition of micromolar concentrations of calcium (Bloodgood, 1992); indeed, micromolar concentrations of calcium elevated both protein kinase and protein phosphatase activities in the membrane-plus-matrix fraction of the flagella (Bloodgood, 1992), which is known also to contain abundant calmodulin

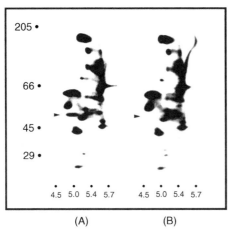

(A)   (B)

FIGURE 11.12 *Phosphoproteins in the membrane-plus-matrix compartment of the flagella of Chlamydomonas. Cells were labeled in vivo with ^{32}P-phosphate, flagella were purified and extracted with non-ionic detergent in the presence of kinase and phosphatase inhibitors, and the proteins in the detergent-soluble membrane-plus-matrix extract were separated by two-dimensional isoelectric focusing/SDS-PAGE. These two images are autoradiograms. (A) Control cells. (B) Cells that were treated with a monoclonal antibody to FMG-1B in order to crosslink the FMG-1B glycoproteins on the flagellar surface and thereby activate the signaling pathway associated with movement of FMG-1B in the plane of the flagellar membrane. Note that the signaling pathway activated by crosslinking of FMG-1B is associated with a dramatic dephosphorylation of a single phosphoprotein (arrowhead) in the membrane-plus-matrix compartment of the flagellum. Molecular mass standards (in kD) are shown on the left. The pH scale is shown at the bottom of the figures. Reproduced from Bloodgood and Salomonsky (1994) with permission of the Rockefeller University Press, New York.*

(Gitelman and Witman, 1980). The flagellar membrane-plus-matrix fraction contains more than 30 phosphoproteins (Bloodgood and Salomonsky, 1994) (Figure 11.12A). Crosslinking of FMG-1B on live cells (prelabeled with ^{32}P-phosphate) using antibodies, the lectin concanavalin A, or polystyrene microspheres induces both increases and decreases in serine/threonine phosphorylation of selected phosphoproteins in the membrane-plus-matrix compartment of the flagellum (Bloodgood & Salomonsky, 1994, 1998). The most dramatic *in vivo* change in protein phosphorylation that results from crosslinking of FMG-1B is the total dephosphorylation of a highly phosphorylated 55- to 60-kD membrane-plus-matrix phosphoprotein (arrowhead in Figure 11.12A), which co-immunoprecipitates with FMG-1B (Bloodgood & Salomonsky, 1994, 1998). This phosphoprotein can be dephosphorylated *in vitro* by calcineurin (a calcium/calmodulin-dependent protein phosphatase of class PP2B) (Bloodgood & Salomonsky, 1994). These observations suggest that the signaling pathway that regulates the movement of FMG-1B within the flagellar membrane involves changes in serine/threonine phosphorylation through changes in the activity of calcium/calmodulin-regulated protein kinases and protein phosphatases. This signaling pathway is incomplete and has yet to be connected to activation of a motor protein responsible for directed movements of FMG-1B (and hence gliding motility).

Evidence for involvement of a calcium/calmodulin-regulated (CaM) kinase in gliding motility comes from a non-gliding mutant (Kozminski, 1995). This mutant has an insertion in a gene coding for a novel serine–threonine protein kinase, which is named gliding-associated kinase (GAK) (Bloodgood and Spano, 2002). The gene sequence is continuous on each side of the vector used for the insertional mutagenesis, suggesting that no deletion occurred during the insertional event. The entire sequence was obtained independently by RT-PCR, from an overlapping set of ESTs, and from the *Chlamydomonas* genome. The sequence has a single open reading frame that encodes a 465-amino-acid (49.5-kD) protein (AAQ23078) containing a complete serine/threonine protein kinase domain that most closely matches that of mammalian Type I CaM kinases (Bloodgood and Spano, 2002). The N-terminal region of this kinase contains a CaM Kinase I calmodulin-binding domain, a proline-rich domain (with PSPPP repeats), an alanine-rich domain containing a number of Ala–Pro repeats, and predicted serine–threonine phosphorylation sites. Some CaM kinases are known to be regulated by both calcium and phosphorylation (Soderling, 1999; Fujisawa, 2001). It is possible that GAK is the FMG-1B-associated protein that is dephosphorylated *in vivo* during FMG-1B crosslinking-induced signaling (Bloodgood and Salomonsky, 1994).

4. Microsphere movements

Force transduction at the flagellar surface can be dramatically visualized by the use of polystyrene microspheres (or other inert marker particles) that adhere to and are transported along the flagellar surface in a bidirectional, saltatory fashion (see videos "Microsphere 1," "Microsphere 2," and "MicrosphereMove" at http://www.elsevierdirect.com/companions/9780123708731). Individual microspheres can traverse the entire length of the flagellum at a constant velocity of 1.5–2.0 μm/second or can stop and start and reverse direction at any point along the flagellum. During a saltation or excursion, microsphere velocity is constant and is very similar to the velocity of gliding motility, although the latter is unidirectional with regard to a single flagellum. Microspheres never appear to move circumferentially on the flagellar surface or to spiral along the flagellar surface, suggesting that they are following paths defined by the outer doublet microtubules, tracks that are perhaps indicated by the linear arrays of intramembrane particles in the flagellar plasma membrane (Weiss et al., 1977; Figure 11.4). The velocity of microsphere movement is load independent; objects of very different size and shape (e.g., a non-motile *E. coli* mutant, a 1.0-μm diameter microsphere, and a 0.2-μm diameter microsphere) are all moved at the same velocity (see video "Bacteria moving on flagella" at http://www.elsevierdirect.com/companions/9780123708731) (Bloodgood, 1977). An unusual observation demonstrates that the force exerted on a single microsphere is sufficient to translocate an entire cell (http://people.virginia.edu/~rab4m/RABResearch/Images/MicrosphereMovement/WebPics/pages/Slide3_jpg.htm; also video "Cell moving along bead" at http://www.elsevierdirect.com/companions/9780123708731). In this movie, a cell attaches by its flagellar surface to a substrate-immobilized microsphere and the flagellum repeatedly moves back and forth relative to the immobilized microsphere at the same velocity as a single microsphere moves along a flagellum (or a cell exhibits gliding motility along a planar substrate). Multiple microspheres can be observed moving along a single flagellum; the behavior of each microsphere is independent of all others. Microspheres moving on the free surface of the leading flagellum of a gliding cell can reverse direction while the cell continues to glide in the same direction. These observations suggest that there are many independent motility domains within the flagellar membrane. It is worth noting that it is much easier to quantitate changes in microsphere adhesion and movement than to quantitate changes in gliding motility in a population of cells (Bloodgood, 1995). Collectively, gliding motility and microsphere movement have been referred to as "flagellar surface motility," although they may be distinct processes (see below).

If polystyrene microspheres are derivatized with an immobilized iodination system and are allowed to attach to and be moved along the flagellar surface in the presence of ^{125}I, the only flagellar protein that is iodinated is FMG-1B (Bloodgood and Workman, 1984). Attachment of microspheres to the flagellar surface results in clustering of FMG-1B at the sites of contact (Bloodgood and Salomonsky, 1998). Calmodulin antagonists (such as trifluoperazine) inhibit both antibody-induced movements of FMG-1B (Bloodgood and Salomonsky, 1990) and microsphere movements (Goodenough, 1993). Lidocaine, which can act as a calcium-channel blocker, also inhibits both antibody-induced movements of FMG-1B (Bloodgood and Salomonsky, 1990) and microsphere movements (Snell et al., 1982). Concanavalin-A-induced immobilization of FMG-1B in a mutant of *Chlamydomonas* inhibits polystyrene-microsphere movement and gliding motility (Bloodgood and Salomonsky, 1989; section VI.C.6). Proteolysis of live cells results in reversible loss of adhesion of polystyrene microspheres to the flagellar surface coincident with proteolysis of the FMG-1B glycoprotein (Bloodgood and May, 1982). Certain non-gliding mutants lose the ability to bind microspheres as well as to adhere to planar substrates (Reinhart and Bloodgood, 1988; Kozminski, 1995). Taken together, these observations strongly suggest that crosslinking-induced movements of FMG-1B in the *Chlamydomonas* flagellar membrane are responsible for inducing the movements of polystyrene microspheres.

Microsphere movements and gliding motility in *Chlamydomonas* occur at similar velocities and both require the presence of micromolar concentrations of free calcium in the medium. Saito et al. (2003) point out that, in a variety of gliding organisms with rather different velocities of gliding motility, microsphere movements tend to exhibit a velocity similar to that of gliding. While there are a number of reasons to assume, from the observations cited in this chapter, that microsphere movements are a manifestation of the mechanism responsible for whole-cell gliding motility, there are some confounding observations in the literature. There are two sets of non-gliding mutants that have been isolated in *Chlamydomonas*. Lewin (1982) used ultraviolet mutagenesis and a colony morphology assay to isolate a series of non-gliding mutants in *C. moewusii*; all of these mutants fail to exhibit microsphere movement (Reinhart and Bloodgood, 1988). On the other hand, Kozminski (1995) used insertional mutagenesis to isolate a set of non-gliding mutants in *C. reinhardtii*, which he divided into several phenotypic categories. While some of the non-gliding mutants were defective in microsphere movement, others were not, raising some question about the relationship of microsphere movement and gliding motility. One phenomenological difference between microsphere movement and gliding is that a flagellum is only capable of pulling a cell (and not pushing a cell) during gliding motility (implying the involvement of a minus-end directed microtubule-associated motor protein) whereas microsphere movement is bidirectional on every flagellum (suggesting the involvement of both plus-end and minus-end directed microtubule-associated motor proteins) (see video "Gliding and microsphere" at http://www.elsevierdirect.com/companions/9780123708731).

5. Force transduction and the role of motor proteins

The mechanism for force transduction underlying gliding motility is not well understood, but it is presumed that gliding motility is dependent on cytoskeletal motor proteins. Only limited, and often indirect, information is available about the involvement of specific motor proteins in whole-cell gliding, microsphere movements along the flagellar surface, and the directed movements of FMG-1B and other glycoproteins within the flagellar membrane. Kozminski et al. (1995) demonstrated that polystyrene-microsphere movements on the flagellar surface and IFT-particle movements are both inhibited in the kinesin-2 temperature-sensitive mutant *fla10* at the non-permissive temperature. McCommis and Dentler (personal communication) have observed an inhibition of gliding motility in the *fla10* mutant at the non-permissive temperature. Huang et al. (2005) demonstrated that the movement of concanavalin-A-induced patches (primarily composed of FMG-1B) within the flagellar membrane is inhibited in the *fla10* mutant at non-permissive temperature. The kinesin-2 motor protein that is defective in *fla10* powers anterograde IFT

(Walther et al., 1994; see Chapter 4). Taken at face value, the observations cited suggest that IFT, gliding motility, microsphere movement, and directed movements of the flagellar membrane glycoproteins (including FMG-1B) are all dependent, directly or indirectly, on the kinesin-2 motor protein.

This does not mean that these processes all have a similar mechanism (and indeed arguments against this conclusion have been raised elsewhere in this chapter). Microsphere movement and gliding motility are flagellar surface phenomena that are clearly associated with the FMG-1B flagellar membrane glycoprotein and appear to be regulated processes that require the crosslinking-induced, directed movement of FMG-1B glycoproteins within the flagellar membrane. Microsphere movements are saltatory and individual microspheres can reverse direction anywhere along the length of the flagellum. On the other hand, the hallmark of IFT is the constitutive, non-saltatory movement of protein particle arrays bidirectionally within the matrix compartment located between the axoneme and the flagellar membrane. IFT particles usually change direction only upon reaching one end of the flagellum. While IFT particles are closely apposed to the inside surface of the flagellar membrane, there is no evidence that they mechanically interact with the FMG-1B flagellar membrane glycoproteins or any other flagellar membrane component. Qin et al. (2004) failed to find evidence that FMG-1B was a cargo for IFT; Pan and Snell (2002) observed sexual agglutinins moving from the gametic cell body to the flagella in the absence of IFT (*fla10* mutant at the non-permissive temperature). Non-gliding mutants of *Chlamydomonas* exhibit IFT movements. The velocities of IFT-particle movements differ from the velocities of polystyrene-microsphere movements (Kozminski et al., 1993), and light microscope observations suggest that polystyrene-microsphere movements do not correlate with IFT-particle movements (see video "Gliding and microsphere" at http://www.elsevierdirect.com/companions/9780123708731) (Kozminski et al., 1998; Dentler, 2005). In contrast to gliding motility, microsphere movements, and crosslinking-induced movements of FMG-1B, IFT-particle movements do not require micromolar levels of free calcium in the medium (Kozminski et al., 1993).

Even though kinesin-2, a plus-end directed microtubule motor, is the only biological motor protein that has been thus far associated with force transduction at the flagellar surface, it is important to point out that, based on the current model for the mechanism of gliding motility in *Chlamydomonas* (see section VI.C.7), gliding motility should be directly dependent upon on a minus-end directed microtubule motor. Dyneins and some kinesins are minus-end directed microtubule motors. Although retrograde IFT is associated with a specific dynein (cytoplasmic dynein 1b), mutants in a subunit of dynein 1b do not appear to inhibit microsphere movement (Pazour et al., 1998); however, these mutants have short flagella, making it very difficult to assay gliding motility.

Another possible candidate for a minus-end directed motor protein that could be involved in gliding motility (as well as the retrograde movements of polystyrene microspheres and crosslinking-induced movements of flagellar membrane proteins in the retrograde direction) is the *Chlamydomonas* homologue of the kinesin-like calmodulin-binding protein (KCBP; ABF50981) (Dymek et al., 2006). In flowering plants, KCBP is a minus-end directed motor protein that binds calmodulin and is negatively regulated by calcium. Although the *Chlamydomonas* homologue is found in the flagellum and binds to axonemal microtubules in an ATP-dependent manner, Dymek et al. (2006) were not able to demonstrate calmodulin-binding or calcium regulation. These observations, coupled with the fact that plant KCBP is negatively regulated by calcium and calmodulin while surface motility in the *Chlamydomonas* flagellum appears to be positively regulated by calcium and calmodulin, do not suggest that KCBP is a strong candidate for the retrograde motor responsible for gliding motility or for retrograde movement of microspheres or flagellar membrane glycoproteins.

Even though gliding motility must be driven by a minus-end directed motor that would move the FMG-1B glycoprotein along the axonemal outer doublet microtubules in the direction of the basal body, it is not hard to rationalize the observed requirement for the plus-end directed microtubule-motor protein kinesin-2. If gliding is driven by the action of minus-end directed motor proteins that move substrate-attached FMG-1B glycoproteins along the axoneme towards the base of the flagellum, the minus-end directed motor proteins and the flagellar FMG-1B glycoproteins would accumulate at the base of the flagellum, were there not a plus-end directed mechanism for redistributing them along the length of the flagellum.

6. Role of directed glycoprotein movements in whole-cell gliding motility

As already described, FMG-1B is the membrane protein that contacts the substrate during gliding motility (Bloodgood and Workman, 1984) and FMG-1B can be actively moved through the flagellar membrane in a regulated manner that is induced by FMG-1B crosslinking (section VI.B). An interesting glycosylation mutant demonstrates that the active movement of concanavalin-A-binding glycoproteins within the flagellar membrane is essential for whole-cell gliding motility (Bloodgood and Salomonsky, 1989). *Chlamydomonas* cells were mutagenized, grown for several generations, and labeled with fluorescent monoclonal antibodies to a carbohydrate epitope on FMG-1B. Cells that had totally lost the ability to bind the antibody were then isolated using flow cytometry. These same cell lines were observed to exhibit a dramatic increase in the binding of concanavalin A to the flagellar surface. The loss of the carbohydrate epitope recognized by the monoclonal antibody coupled with the increase in concanavalin A binding suggested a glycosylation mutant (possibly a mutation in a glycosyl transferase or a mannosidase) that results in the N-linked carbohydrates on the FMG-1B glycoproteins (and possibly on other flagellar membrane glycoproteins) being switched from a complex carbohydrate to a high mannose structure. The same concentration of concanavalin A that induced glycoprotein movements in wild-type cells resulted in such extensive crosslinking in the mutant cells that the concanavalin-A-binding glycoproteins were frozen in position and unable to move within the flagellar membrane. The mutant cells were unable to exhibit microsphere movements or whole-cell gliding motility in the presence of concanavalin A concentrations that allowed microsphere movement and gliding motility in wild-type cells (Bloodgood and Salomonsky, 1989). If the concanavalin A concentration was greatly reduced such that a wild-type amount of concanavalin-A-binding and crosslinking occurred on the mutant cells, then the mutant cell lines exhibited microsphere movement and gliding motility. These observations suggest that both microsphere movement and gliding motility are dependent upon the crosslinking-induced movements of flagellar membrane glycoproteins. Because FMG-1B is, by far, the predominant concanavalin-A-binding glycoprotein in the flagellar membrane, this experiment suggests that FMG-1B movement is required for microsphere movement and gliding motility.

7. A model for gliding motility

The observational and experimental data described in sections VI.B and VI.C of this chapter suggest the following working model for whole-cell gliding motility in *Chlamydomonas*. When a region of the flagellar surface contacts a solid substrate, the substrate serves as a multivalent ligand that locally crosslinks and clusters FMG-1B glycoproteins within the flagellar membrane. This induces adhesion to the substrate as well as an influx of calcium into the flagellum (through an as yet uncharacterized calcium channel). The rise in intraflagellar calcium concentration activates calcium/calmodulin-activated protein kinases and/or phosphatases that change the level of phosphorylation of certain flagellar phosphoproteins. In particular, the dephosphorylation of an approximately 55- to 60-kD phosphoprotein associated with the cytoplasmic domain of FMG-1B is one of the earliest steps in this kinase/phosphatase cascade that ultimately activates a minus-end directed microtubule-associated motor protein that walks along an outer doublet microtubule, moving a locally crosslinked cluster of FMG-1B molecules towards the base of the flagellum. Because this cluster of FMG-1B glycoproteins is adherent to the substrate, the translocation of the cluster relative to the doublet microtubule by the motor protein complex translocates the axoneme (and the entire flagellum and the entire cell) in the opposite direction relative to the substrate.

While this scenario is consistent with most of the observations, there are many remaining mysteries. What is the primary, minus-end directed motor protein that is responsible for gliding and what are the kinases and phosphatases that regulate movement of FMG-1B by the motor? What is the function of gliding motility? Since *Chlamydomonas* is a photosynthetic organism, it is assumed that most cellular movements have evolved to allow the cell to seek out the intensity of light that is optimal for photosynthesis. If so, this leaves us with a mystery as to how gliding motility can be regulated by light direction or intensity in order to allow *Chlamydomonas* cells living within soil (for instance) to position themselves in the optimum light intensity. All currently understood photoresponses in *Chlamydomonas* require swimming motility through liquid medium allowing a light antenna (the eyespot) to continuously scan the environment through 360° (see Chapter 13). Clearly, photoresponses coupled to gliding motility would have to utilize a very different mechanism, both in terms of sensory input (light reception) and motor output (changes in gliding direction) than that used by swimming cells. Huang et al. (2004) have shown that a blue-light receptor molecule (phototropin) is located in the flagellum (as well as in the cell body) of *Chlamydomonas*. Although phototropin plays important roles in the sexual cycle of *Chlamydomonas*, including flagellar adhesion during mating (Pan et al., 1997), the function of the flagellar population of phototropin has not been clearly defined because of the inability to selectively eliminate either the cell body phototropin or the flagellar phototropin (both the product of a single phototropin gene). Under conditions where gliding would occur in nature, it is also possible that gliding could move the cell up or down an oxygen or other chemical gradient. Although most of the length of the flagellum appears not to change shape during gliding motility, the tip of the leading flagellum will occasionally wiggle. On some occasions, the course of the whole-cell movement will follow a curved path defined by a bend in the tip of the flagellum (Bloodgood, 1981); this might be one means for regulating the direction of gliding motility. It would also be of interest to know whether external stimuli (light, carbon sources such as acetate, ionic environment) can cause the cell to reverse the direction of gliding by 180° by changing the dominant flagellum, perhaps through changes in relative adhesion of the flagella with the substrate.

At face value, flagella-based gliding motility (and indeed all gliding motility in prokaryotes and eukaryotic protists) is rather different from metazoan cell motility, which tends to utilize an actin cytoskeleton and integrins in the plasma membrane to mechanically link the cytoskeleton to the substrate (Ridley et al., 2003; Vicente-Manzanares et al., 2005). Integrins are absent from protists (including *Chlamydomonas*), and their functions must be served by other proteins. In many ways, the function of FMG-1B in *Chlamydomonas* gliding motility is very analogous to the function of integrins in mammalian cell motility. Integrins mediate adhesion of the cell surface to the substrate. Crosslinking of integrins results in transmission of a signal into the cell which activates cytoplasmic signaling pathways (involving changes in protein phosphorylation). This results in force being applied, by the cytoskeleton and cytoskeleton-associated motor proteins, to the integrins and hence to the substrate at points of adhesive contact, resulting in traction. The cell moves relative to the contact points (called focal contacts or focal adhesions in mammalian cells). FMG-1B in the flagellar membrane serves the same sensory and motor functions for gliding motility as integrins do for mammalian cell migration; the main differences are that the interaction between the cell surface and the substrate is relegated to a specific plasma-membrane domain (the flagellar membrane), and that it is the microtubule cytoskeleton that is involved in force generation instead of the actin cytoskeleton. It has been shown that this same mammalian concept of focal adhesion complexes (albeit using different proteins) can explain one form of gliding motility in the bacterium *Myxococcus* (Kearns, 2007; Mignot et al., 2007).

D. Cell–cell adhesion and flagellar dynamics during mating

Chlamydomonas gametes exhibit normal gliding and microsphere movement. In addition, gametic flagella exhibit regulated and directed movements of flagellar membrane proteins specific for gametic flagellar membranes (see section VI.B). The initial interaction of gametes of opposite type is through the flagellar surfaces (see section IV.A.7 and Chapter 12). Flagellar adhesion, flagellar glycoprotein dynamics, flagellar reorientation, and flagellar signaling pathways play critical roles in the mating process and are likely to utilize many of the same mechanisms seen in vegetative flagella.

E. Regulation of the intraflagellar ionic environment and flagellar beating motility

One of the important roles of the flagellar membrane is to regulate the ionic environment within the flagellum. One theme of this chapter (and of this entire volume) is the recognition that numerous flagellar functions are regulated by calcium; these include: (1) axonemal waveform (see Chapter 7), (2) cellular photoresponses that result in a calcium action potential in the flagellar membrane, allowing an influx of calcium which alters waveform, sometimes selectively within one flagellum (see Chapter 13), (3) directed movement of FMG-1B and other flagellar membrane proteins, (4) microsphere movements, (5) whole-cell gliding motility, (6) events in mating (see Chapter 12) and (7) flagellar excision (see Chapter 3). While the calcium regulating some of these events may be derived from intracellular calcium stores, evidence strongly suggests that some of these events are regulated by calcium influx from the extracellular medium. A variety of experimental evidence, as well as the *Chlamydomonas* flagellar proteome (Pazour et al., 2005), suggests that the *Chlamydomonas* flagellar membrane contains calcium ATPases and cation channels, including a polycystin-2 homologue (see section IV.A.6); more work is needed to characterize these membrane pumps and channels.

VII. SUMMARY

The *Chlamydomonas* flagellar surface is very dynamic. It appears that the directed movements of flagellar membrane proteins over long distances are responsible for most of the dynamic properties of the flagellar surface, both those common to the vegetative and gametic flagellar surface, such as gliding motility, and those unique to the gametic flagella, such as the early events of mating. These directed protein movements appear to be highly regulated, and not constitutive, although the details of their regulation are still incomplete. It is adhesion-induced crosslinking of flagellar membrane glycoproteins that initiates transmembrane signaling pathways that engage a poorly understood clutch that links an intraflagellar, force-transducing cytoskeletal machinery to specific flagellar glycoproteins, whose directed movements within the flagellar membrane are then manifest as "capping" or "tipping" of flagellar membrane glycoproteins, microsphere movements, whole-cell gliding motility, and the reorientation of flagellar surfaces (during the early events of mating).

The earliest observed and most obvious function of cilia and flagella is the creation and propagation of bends; once it was shown that this form of motility could be reactivated with ATP using permeabilized cilia and flagella (Hoffmann-Berling, 1954) and subsequently using purified axonemes (Gibbons and Gibbons, 1972; Allen and Borisy, 1974), the attention of cell biologists became focused on the axoneme; indeed, the non-ionic-detergent-soluble components of the cilium or flagellum were often sent directly down the laboratory sink! Because of this, the study of the flagellar surface has lacked the attention it deserves, but this is changing. A greater appreciation of the importance of non-motile cilia and of the sensory roles that both motile and non-motile cilia and flagella play (Pazour and Witman, 2003) and of the dynamic nature of the flagellar surface is being accompanied by increased attention to the composition and functions of the ciliary and flagellar membranes. For many of the same reasons that *Chlamydomonas* has proven to be the best model system for the study of axonemal function and for the study of flagellar assembly and length regulation, it is also the best system for studying the many important properties of the ciliary/flagellar membrane.

REFERENCES

Absalon, S., Blisnick, T., Kohl, L., Toutirais, G., Doré, G., Julkowska, D., Tavenet, A., and Bastin, P. (2008). Intraflagellar transport and functional analysis of genes required for flagellum formation in trypanosomes. *Mol. Biol. Cell.* **19**, 929–944.

Adair, W.S., Hwang, C., and Goodenough, U.W. (1983). Identification and visualization of the sexual agglutinin from the mating-type plus flagellar membrane of *Chlamydomonas*. *Cell* **33**, 183–193.

Allen, C. and Borisy, G.G. (1974). Structural polarity and directional growth of microtubules of *Chlamydomonas* flagella. *J. Mol. Biol.* **90**, 381–402.

Andersen, R.A., Barr, D.J.S., Lynn, D.H., Melkonian, M., Moestrup, O., and Sleigh, M.A. (1991). Terminology and nomenclature of the cytoskeletal elements associated with the flagellar/ciliary apparatus in protists. *Protoplasma* **164**, 1–8.

Arisz, S.A., van Himbergen, J.A.J., Musgrave, A., van den Ende, H., and Munnik, T. (2000). Polar glycerolipids of Chlamydomonas moewusii. *Phytochemistry* **53**, 265–270.

Bae, Y.K., Qin, H., Knobel, K.M., Hu, J., Rosenbaum, J.L., and Barr, M.M. (2006). General and cell-type specific mechanisms target TRPP2/PKD-2 to cilia. *Development* **133**, 3859–3870.

Beck, C. and Uhl, R. (1994). On the localization of voltage-sensitive calcium channels in the flagella of *Chlamydomonas reinhardtii*. *J. Cell Biol.* **125**, 1119–1125.

Bergman, K., Goodenough, U.W., Goodenough, D.A., Jawitz, J., and Martin, H. (1975). Gametic differentiation in *Chlamydomonas reinhardtii*. II: Flagellar membranes and the agglutination reaction. *J. Cell Biol.* **67**, 606–622.

Bernstein, M. (1995). Isolation of *Chlamydomonas* mastigonemes. *Meth. Cell Biol.* **47**, 425–430.

Bernstein, M. and Rosenbaum, J.L. (1993). Transport to the cell surface of *Chlamydomonas*: mastigonemes as a marker for the flagellar surface. In: *Molecular Mechanisms of Membrane Traffic*, Vol. H 74 (D.J. Morre, K.E. Howell, and J.M. Bergeron, Eds.), pp. 179–180. NATO ASI Series, Springer-Verlag, Berlin and Heidelberg.

Bloodgood, R.A. (1977). Rapid motility occurring in association with the *Chlamydomonas* flagellar membrane. *J. Cell Biol.* **75**, 983–989.

Bloodgood, R.A. (1981). Flagella-dependent gliding motility in *Chlamydomonas*. *Protoplasma* **106**, 183–192.

Bloodgood, R.A. (1984). Preferential turnover of membrane proteins in the intact *Chlamydomonas* flagellum. *Exp. Cell Res.* **150**, 488–493.

Bloodgood, R.A. (1987). Glycoprotein dynamics in the *Chlamydomonas* flagellar membrane. In: *Advances in Cell Biology*, Vol. 1 (K.R. Miller, Ed.), pp. 97–130. JAI Press, New York.

Bloodgood, R.A. (1988). Gliding motility and the dynamics of flagellar membrane glycoproteins in *Chlamydomonas reinhardtii*. *J. Protozool.* **35**, 552–558.

Bloodgood, R.A. (1990). Gliding motility and flagellar dynamics in *Chlamydomonas*. In: *Ciliary and Flagellar Membranes* (R.A. Bloodgood, Ed.), pp. 91–128. Plenum Press, New York.

Bloodgood, R.A. (1992). Calcium-regulated phosphorylation of proteins in the membrane–matrix compartment of the *Chlamydomonas* flagellum. *Exp. Cell Res.* **198**, 228–236.

Bloodgood, R.A. (1995). Flagellar surface motility: Gliding and microsphere movements. *Meth. Cell Biol.* **47**, 273–279.

Bloodgood, R.A. (2000). Protein targeting to flagella of trypanosomatid protozoa. *Cell Biology Inter.* **24**, 857–862.

Bloodgood, R.A. and Levin, E.N. (1983). Transient increase in calcium efflux accompanies fertilization in *Chlamydomonas*. *J. Cell Biol.* **97**, 397–404.

Bloodgood, R.A. and May, G.S. (1982). Functional modification of the *Chlamydomonas* flagellar surface. *J. Cell Biol.* **93**, 88–96.

Bloodgood, R.A. and Salomonsky, N.L. (1989). Use of a novel *Chlamydomonas* mutant to demonstrate that flagellar glycoprotein movements are necessary for the expression of gliding motility. *Cell Motil. Cytoskeleton* **13**, 1–8.

Bloodgood, R.A. and Salomonsky, N.L. (1990). Calcium influx regulates antibody-induced glycoprotein movements within the *Chlamydomonas* flagellar membrane. *J. Cell Sci.* **96**, 27–33.

Bloodgood, R.A. and Salomonsky, N.L. (1991). Regulation of flagellar glycoprotein movements by protein phosphorylation. *Eur. J. Cell Biol.* **54**, 85–89.

Bloodgood, R.A. and Salomonsky, N.L. (1994). The transmembrane signaling pathway involved in directed movements of *Chlamydomonas* flagellar membrane glycoproteins involves the dephosphorylation of a 60 kDa phosphoprotein that binds to the major flagellar membrane glycoprotein. *J. Cell Biol.* **127**, 803–811.

Bloodgood, R.A. and Salomonsky, N.L. (1995). Phosphorylation of *Chlamydomonas* flagellar proteins. *Meth. Cell Biol.* **47**, 121–127.

Bloodgood, R.A. and Salomonsky, N.L. (1998). Microsphere attachment induces glycoprotein redistribution and transmembrane signaling in the *Chlamydomonas* flagellum. *Protoplasma* **202**, 76–83.

Bloodgood, R.A. and Spano, A.J. (2000). Cloning and characterization of genes encoding the gliding-associated flagellar membrane glycoprotein of *Chlamydomonas reinhardtii*. *Mol. Biol. Cell* **11**, 539a.

Bloodgood, R.A. and Spano, A.J. (2002). Gliding motility in *Chlamydomonas reinhardtii* requires a calcium dependent protein kinase. *Mol. Biol. Cell* **13**, 191a.

Bloodgood, R.A. and Workman, L.J. (1984). A flagellar surface glycoprotein mediating cell–substrate interaction in *Chlamydomonas*. *Cell Motil.* **4**, 77–87.

Bloodgood, R.A., Leffler, E.M., and Bojczuk, A.T. (1979). Reversible inhibition of *Chlamydomonas* flagellar surface motility. *J. Cell Biol.* **82**, 664–674.

Bloodgood, R.A., Woodward, M.P., and Salomonsky, N.L. (1986). Redistribution and shedding of flagellar membrane glycoproteins visualized using an anti-carbohydrate monoclonal antibody and concanavalin A. *J. Cell Biol.* **102**, 1797–1812.

Bloodgood, R.A., Woodward, M.P., and Young, W.W. Jr., (1995). Unusual distribution of a glycolipid antigen in the flagella of *Chlamydomonas*. *Protoplasma* **185**, 123–130.

Bouck, G.B. (1971). The structure, origin, isolation and composition of the tubular mastigonemes of the *Ochromonas* flagellum. *J. Cell Biol.* **50**, 362–384.

Bouck, G.B. (1972). Architecture and assembly of mastigonemes. In: *Advances in Cell and Molecular Biology*, Vol. 2 (E.J. DuPraw, Ed.), pp. 237–271. Academic Press, New York.

Bouck, G.B., Rogalski, A., and Valaitis, A. (1978). Surface organization and composition of *Euglena* II. Flagellar Mastigonemes. *J. Cell Biol.* **77**, 805–826.

Bouck, G.B., Rosier, T.K., and Levasseur, P.J. (1990). *Euglena gracilis*: A model for flagellar surface assembly, with reference to other cells that bear mastigonemes and scales. In: *Ciliary and Flagellar Membranes* (R.A. Bloodgood, Ed.), pp. 65–90. Plenum Press, New York.

Brederoo, J., de Wildt, P., Popp-Snijders, C., Irvine, R.F., Musgrave, A., and van den Ende, H. (1991). Polyphosphoinositol lipids in *Chlamydomonas eugametos* gametes. *Planta* **184**, 175–181.

Brown, R.M., Jr., Johnson, S.C., and Bold, H.C. (1968). Electron microscopy and phase contrast microscopy of sexual reproduction in *Chlamydomonas moewusii*. *J. Phycol.* **4**, 100–125.

Chen, L.L., Pousada, M., and Haines, T.H. (1976). The flagellar membrane of *Ochromonas danica*. Lipid Composition. *J. Biol. Chem.* **251**, 1835–1842.

Chen, S.-J. and Bouck, G.B. (1984). Endogenous glycosyltransferases glucosylate lipids in flagella of *Euglena*. *J. Cell Biol.* **98**, 1825–1835.

Chinskey, N.D., Ulland, M.M.D., and Sloboda, R.D. (2006). Protein phosphorylation and intraflagellar transport in *Chlamydomonas*. *Mol. Biol. Cell* **17**, 1613. (CD-ROM)

Cole, D.G., Diener, D.R., Himelblau, A.L., Beech, P.L., Fuster, J.C., and Rosenbaum, J.L. (1998). *Chlamydomonas* kinesin-II-dependent intraflagellar transport (IFT): IFT particles contain proteins required for ciliary assembly in *Caenorhabditis elegans* sensory neurons. *J. Cell Biol.* **141**, 993–1008.

Davidge, J.A., Chambers, E., Dickinson, H.A., Towers, K., Ginger, M.L., McKean, P.G., and Gull, K. (2006). Trypanosome IFT mutants provide insight into the motor location for mobility of the flagella connector and flagellar membrane formation. *J. Cell Sci.* **119**, 3935–3943.

Demets, R., Tomson, A.M., Himan, W.L., Stegwee, D., and van den Ende, H. (1988). Cell–cell adhesion in conjugating *Chlamydomonas* gametes: a self-enhancing process. *Protoplasma* **145**, 27–36.

Denning, G.M. and Fulton, A.B. (1989). Purification and characterization of clathrin-coated vesicles from *Chlamydomonas*. *J. Protozool.* **36**, 334–340.

Dentler, W.L. (1980). Microtubule–membrane interactions in cilia. I: Isolation and characterization of *Tetrahymena* ciliary membranes. *J. Cell Biol.* **84**, 364–380.

Dentler, W.L. (1990). Linkages between microtubules and membranes in cilia and flagella. In: *Ciliary and Flagellar Membranes* (R.A. Bloodgood, Ed.), pp. 31–64. Plenum Press, New York.

Dentler, W. (2005). Intraflagellar transport (IFT) during assembly and disassembly of *Chlamydomonas* flagella. *J. Cell Biol.* **170**, 649–659.

Dentler, W.L. (2006). Importance of Golgi for the regulation of flagellar length in *Chlamydomonas*. *Mol. Biol. Cell* **17**. abstract 1596. (CD-ROM)

Dolle, R. (1988). Isolation of plasma membrane and binding of the calcium antagonist nimodipine in *Chlamydomonas reinhardtii*. *Physiol. Plant.* **73**, 7–14.

Dymek, E.E., Goduti, D., Kramer, T., and Smith, E.F. (2006). A kinesin-like calmodulin-binding protein in *Chlamydomonas*: Evidence for a role in cell division and flagellar functions. *J. Cell Sci.* **119**, 3107–3116.

Eichenberger, W. (1976). Lipids of *Chlamydomonas reinhardtii* under different growth conditions. *Phytochemistry* **15**, 459–463.

Ettl, H. (1970). Die Gattung *Chloromonas gobi* emend. *Wille. Beihefte zur Nova Hedwigia* **34**, 1–283.

Ferris, P.J., Woessner, J.P., Waffenschmidt, S., Kilz, S., Drees, J., and Goodenough, U.W. (2001). Glycosylated polyproline II rods with kinks as a structural motif in plant hydroxyproline-rich glycoproteins. *Biochemistry* **40**, 2978–2987.

Ferris, P.J., Waffenschmidt, S., Umen, J.G., Lin, H., Lee, J.-H., Ishida, K., Kubo, T., Lau, J., and Goodenough, U.W. (2005). Plus and minus sexual agglutinins from *Chlamydomonas reinhardtii*. *Plant Cell* **17**, 597–615.

Follit, J.A., Tuft, R.A., Fogarty, K.E., and Pazour, G.J. (2006). The Intraflagellar protein IFT20 is associated with the Golgi complex and is required for cilia assembly. *Mol. Biol. Cell* **17**, 3781–3792.

Fujisawa, H. (2001). Regulation of the activities of multifunctional Ca^{2+}/calmodulin -dependent protein kinases. *J. Biochem.* **129**, 193–199.

Gealt, M.A., Adler, J.H., and Nes, W.R. (1981). The sterols and fatty acids from purified flagella of *Chlamydomonas reinhardtii*. *Lipids* **16**, 133–136.

Gibbons, B.H. and Gibbons, I.R. (1972). Flagellar movement and adenosine triphosphatase activity in sea urchin sperm extracted with Triton X-100. *J. Cell Biol.* **54**, 75–97.

Gilula, N.B. and Satir, P. (1972). The ciliary necklace: a ciliary membrane specialization. *J. Cell Biol.* **53**, 494–509.

Giroud, C., Gerber, A., and Eichenberger, W. (1988). Lipids of *Chlamydomonas reinhardtii*. Analysis of molecular species and intracellular site(s) of biosynthesis. *Plant Cell Physiol.* **29**, 587–595.

Gitelman, S.E. and Witman, G.B. (1980). Purification of calmodulin from *Chlamydomonas*: calmodulin occurs in cell bodies and flagella. *J. Cell Biol.* **87**, 764–770.

Goodenough, U.W. (1989). Cyclic AMP enhances the sexual agglutinability of *Chlamydomonas* flagella. *J. Cell Biol.* **109**, 247–252.

Goodenough, U.W. (1993). Tipping of flagellar agglutinins by gametes of *Chlamydomonas reinhardtii*. *Cell Motil. Cytoskeleton* **25**, 179–189.

Goodenough, U.W. and Heuser, J.E. (1999). Deep-etch analysis of adhesion complexes between gametic flagellar membranes of *Chlamydomonas reinhardtii* (Chlorophyceae). *J. Phycol.* **35**, 756–767.

Goodenough, U.W. and Jurivich, D. (1978). Tipping and mating-structure activation induced in *Chlamydomonas* gametes by flagellar membrane antisera. *J. Cell Biol.* **79**, 680–693.

Goodenough, U.W., Adair, W.S., Collin-Osdoby, P., and Heuser, J.E. (1985). Structure of the *Chlamydomonas* agglutinin and related flagellar surface proteins *in vitro* and *in situ*. *J. Cell Biol.* **101**, 924–941.

Goodenough, U.W., Shames, B., Small, L., Saito, T., Crain, R.C., Sanders, M.A., and Salisbury, J.L. (1993). The role of calcium in the *Chlamydomonas reinhardtii* mating reaction. *J. Cell Biol.* **121**, 365–374.

Grenier, G., Guyon, D., Roche, O., Dubertret, G., and Tremolieres, A. (1991). Modification of the membrane fatty acid composition of *Chlamydomonas reinhardtii* cultured in the presence of liposomes. *Plant Physiol. Biochem.* **29**, 429–440.

Harz, H. and Hegemann, P. (1991). Rhodopsin-regulated calcium currents in *Chlamydomonas*. *Nature* **351**, 489–491.

Heintzelman, M.B. (2003). Gliding motility: The molecules behind the motion. *Curr. Biol.* **13**, R57–R59.

Heintzelman, M.B. (2006). Cellular and molecular mechanics of gliding locomotion in eukaryotes. *Int. Rev. Cytol.* **251**, 79–129.

Hoffman, J.L. and Goodenough, U.W. (1980). Experimental dissection of flagellar surface motility in *Chlamydomonas*. *J. Cell Biol.* **86**, 656–665.

Hoffmann-Berling, H. (1954). Adenosintriphosphat als Betriebsstoff von Zellbewegungen. *Biochim. Biophys. Acta* **14**, 182–194.

Holland, E.M., Braun, F.-J., Nonnengäßer, C., Harz, H., and Hegemann, P. (1996). The nature of rhodopsin triggered photocurrents in *Chlamydomonas*. I: Kinetics and influence of divalent ions. *Biophys. J.* **70**, 924–931.

Holland, E.M., Harz, H., Uhl, R., and Hegemann, P. (1997). Control of phobic behavioral responses by rhodopsin-induced photocurrents in *Chlamydomonas*. *Biophys. J.* **73**, 1395–1401.

Holwill, M.E.J. and Sleigh, M.A. (1967). Propulsion in hispid flagella. *J. Exp. Biol.* **47**, 267–276.

Homan, W., Sigon, C., van den Briel, W., Wagter, R., de Nobel, H., Mesland, D., Musgrave, A., and van den Ende, H. (1987). Transport of membrane receptors and the mechanics of sexual cell fusion in *Chlamydomonas eugametos*. *FEBS Lett.* **215**, 323–326.

Homan, W., Musgrave, A., de Nobel, H., Wagter, R., de Wit, D., Kolk, A., and van den Ende, H. (1988). Monoclonal antibodies directed against the sexual binding site of *Chlamydomonas eugametos* gametes. *J. Cell Biol.* **107**, 177–189.

Huang, K., Kunkel, T., and Beck, C.F. (2004). Localization of the blue-light receptor phototropin to the flagella of the green alga *Chlamydomonas reinhardtii*. *Mol. Biol. Cell* **15**, 3605–3614.

Huang, K., Diener, D.R., Karki, R., Pedersen, L.B., Geimer, S., and Rosenbaum, J.L. (2005). The cilium/flagellum: A secretory organelle? *Mol. Biol. Cell* **16**, 411a–412a. (CD-ROM)

Huang, K., Diener, D.R., Mitchell, A., Pazour, G.J., Witman, G.B., and Rosenbaum, J.L. (2007). Function and dynamics of PKD2 in *Chlamydomonas reinhardtii* flagella. *J. Cell Biol.* **179**, 501–514.

Hunnicutt, G.R., Kosfiszer, M.G., and Snell, W.J. (1990). Cell body and flagellar agglutinins in *Chlamydomonas reinhardtii*: The cell body plasma membrane is a reservoir for agglutinins whose migration to the flagella is regulated by a functional barrier. *J. Cell Biol.* **111**, 1605–1616.

Iomini, C., Li, L., Mo, W., Dutcher, S.K., and Piperno, G. (2006). Two flagellar genes, AGG2 and AGG3, mediate orientation to light in *Chlamydomonas*. *Curr. Biol.* **16**, 1147–1153.

Jahn, T.L., Landman, M.D., and Fonseca, J.R. (1964). The mechanism of locomotion of flagellates. II: Function of the mastigonemes of *Ochromonas*. *J. Protozool.* **11**, 291–296.

Janero, D.R. and Barrnett, R. (1981a). Cellular and thylakoid-membrane glycolipids of *Chlamydomonas reinhardtii* 137+. *J. Lipid Res.* **22**, 1119–1125.

Janero, D.R. and Barrnett, R. (1981b). Cellular and thylakoid-membrane phospholipids of *Chlamydomonas reinhardtii* 137+. *J. Lipid Res.* **22**, 1126–1130.

Janero, D.R. and Barrnett, R. (1982). Membrane biogenesis in *Chlamydomonas reinhardtii* 137+. *Exp. Cell Res.* **138**, 451–454.

Jekely, G. and Arendt, D. (2006). Evolution of intraflagellar transport from coated vesicles and autogenous origin of the eukaryotic cilium. *BioEssays* **28**, 191–198.

Kalshoven, H.W., Musgrave, A., and van den Ende, H. (1990). Mating receptor complex in the flagellar membrane of *Chlamydomonas* gametes. *Sex. Plant Reprod.* **3**, 77–87.

Kaneshiro, E.S. (1990). Lipids of ciliary and flagellar membranes. In: *Ciliary and Flagellar Membranes* (R.A. Bloodgood, Ed.), pp. 241–265. Plenum Press, New York.

Kateriya, S., Nagel, G., Bamberg, E., and Hegemann, P. (2004). "Vision" in single-celled algae. *News Physiol. Sci.* **19**, 133–137.

Kearns, D.B. (2007). Bright insight into bacterial gliding. *Science* **315**, 773–774.

Kooijman, R., deWildt, P., Homan, W.L., Musgrave, A., and van den Ende, H. (1988). Light affects flagellar agglutinability in *Chlamydomonas eugametos* by modification of the agglutinin molecules. *Plant Physiol.* **86**, 216–223.

Kooijman, R., de Wildt, P., Beumer, S., van der Vliet, G., Homan, W., Kalshoven, H., Musgrave, A., and van den Ende, H. (1989). Wheat germ agglutinin induces mating reactions in *Chlamydomonas eugametos* by crosslinking agglutinin-associated glycoproteins in the flagellar membrane. *J. Cell Biol.* **109**, 1677–1687.

Kottgen, M. and Walz, G. (2005). Subcellular localization and trafficking of polycystins. *Pflugers Arch-European J. Physiol.* **451**, 286–293.

Kozminski, K.G. (1995). Beat-independent flagellar motilities in *Chlamydomonas* and an analysis of the function of alpha-tubulin acetylation. Ph.D. Dissertation. Yale University, New Haven, CT, 151 pp.

Kozminski, K.G., Johnson, K.A., Forscher, P., and Rosenbaum, J.L. (1993). A motility in the eukaryotic flagellum unrelated to flagellar beating. *Proc. Natl. Acad. Sci. USA* **90**, 5519–5523.

Kozminski, K.G., Beech, P.L., and Rosenbaum, J.L. (1995). The *Chlamydomonas* kinesin-like protein FLA10 is involved in motility associated with the flagellar membrane. *J. Cell Biol.* **131**, 1517–1527.

Kozminski, K.G., Forscher, P., and Rosenbaum, J.L. (1998). Three motilities in *Chlamydomonas* unrelated to flagellar beating. *Cell Motil. Cytoskeleton Video Supplement* **5**.

Lewin, R.A. (1952). Studies on the flagella of algae. I: General observations of *Chlamydomonas moewusii* Gerloff. *Biol. Bull.* **103**, 74–79.

Lewin, R.A. (1982). A new kind of motility mutant (non-gliding) in *Chlamydomonas*. *Experientia* **38**, 348–349.

Li, J.B., Gerdes, J.M., Haycraft, C.J., Fan, Y., Teslovich, T.M., May-Simera, H., Li, H., Blacque, O.E., Li, L., Leitch, C.C., Lewis, R.A., Green, J.S., Parfrey, P.S., Leroux, M.R., Davidson, W.S., Beales, P.L., Guay-Woodford, L.M., Yoder, B.K., Stormo, G.D., Katsanis, N., and Dutcher, S.K. (2004). Comparative genomics identifies a flagellar and basal body proteome that includes the BBS5 human disease gene. *Cell* **117**, 541–552.

Markey, D.R. and Bouck, G.B. (1977). Mastigoneme attachment in *Ochromonas*. *J. Ult. Res.* **59**, 173–177.

Marshall, W.F. (2004). Cellular length control systems. *Ann. Rev. Cell Dev. Biol.* **20**, 677–693.

Matsuda, Y., Sakamoto, K., Kiuchi, N., Mizuochi, T., Tsubo, Y., and Kobata, A. (1982). Two Tunicamycin-sensitive components involved in agglutination and fusion of *Chlamydomonas* gametes. *Arch. Microbiol.* **131**, 87–90.

McFadden, G.I. and Wetherbee, R. (1985). Flagellar regeneration and associated scale deposition in *Pyramimonas gelidicola* (Prasinophyceae, Chlorophyta). *Protoplasma* **128**, 31–37.

McLean, R.J., Laurendi, C.J., and Brown, R.M. Jr. (1974). The relationship of gamone to the mating reaction in *Chlamydomonas moewusii*. *Proc. Nat. Acad. Sci. USA* **71**, 2610–2613.

McLean, R.J., Katz, K.R., Sedita, N.J., Menoff, A.L., Laurendi, C.J., and Brown, R.M. (1981). Dynamics of concanavalin A binding sites on *Chlamydomonas moewusii* flagellar membranes. *Ber. Deutsch. Bot. Ges.* **94**, 387–400.

Melkonian, M. (1982). The functional analysis of the flagellar apparatus in green algae. In: *Prokaryotic and Eukaryotic Flagella. Symp. Soc. Exp. Biol.* **35**, 589–606.

Mesland, D.A.M. (1976). Mating in *Chlamydomonas eugametos*. A scanning electron microscopical study. *Arch. Microbiol.* **109**, 31–35.

Mesland, D.A.M., Hoffman, J.L., and Goodenough, E.W. (1980). Flagellar tip activation stimulated by membrane adhesions in *Chlamydomonas* gametes. *J. Cell Biol.* **84**, 599–617.

Mignot, T., Shaevitz, J.W., Hartzell, P.L., and Zusman, D.R. (2007). Evidence that focal adhesion complexes power bacterial gliding motility. *Science* **315**, 853–856.

Mitchell, B.F., Grulich, L.E., and Mader, M.M. (2004). Flagellar quiescence in *Chlamydomonas*: Characterization and defective quiescence in cells carrying *sup-pf-1* and *sup-pf-2* outer dynein arm mutations. *Cell Motil. Cytoskeleton* **57**, 186–196.

Monk, B.C., Adair, W.S., Cohen, R.A., and Goodenough, U.W. (1983). Topography of *Chlamydomonas*: Fine structure and polypeptide components of the gametic flagellar membrane surface and the cell wall. *Planta* **158**, 517–533.

Moore, T.S., Du, Z., and Chen, Z. (2001). Membrane lipid biosynthesis in *Chlamydomonas reinhardtii. In vitro* biosynthesis of diacylglyceryltrimethylhomoserine. *Plant Physiol.* **125**, 423–429.

Musgrave, A., Homan, W., van den Briel, W., Lelie, N., Schol, D., Ero, L., and van den Ende, H. (1979). Membrane glycoproteins of *Chlamydomonas eugametos* flagella. *Planta* **145**, 417–425.

Musgrave, A., de Wildt, P., Broekman, R., and van den Ende, H. (1983). The cell wall of Chlamydomonas eugametos. *Planta* **158**, 82–89.

Musgrave, A., DeWildt, P., van Etten, I., Pijst, H., Schholma, C., Kooyman, R., Homan, W., and van den Ende, H. (1986). Evidence for a functional membrane barrier in the transition zone between the flagellum and cell body of *Chlamydomonas eugametos* gametes. *Planta* **167**, 544–553.

Musgrave, A., Schuring, F., Munnik, T., and Visser, K. (1993). Inositol 1,4,5-trisphosphate as a fertilization signal in plants: Test case *Chlamydomonas eugametos. Planta* **191**, 280–284.

Nakamura, S., Tanaka, G., Maeda, T., Kamiya, R., Matsunaga, T., and Nikaido, O. (1996). Assembly and function of *Chlamydomonas* mastigonemes as probed with a monoclonal antibody. *J. Cell Sci.* **109**, 57–62.

Nasser, M.I. and Landfear, S.M. (2004). Sequences required for the flagellar targeting of an integral membrane protein. *Mol. Biochem. Parasitol.* **135**, 89–100.

Overath, P., Stierhof, Y.-D., and Wiese, M. (1997). Endocytosis and secretion in trypanosomatid parasites-tumultuous traffic in a pocket. *Trends Cell Biol.* **7**, 27–33.

Palecek, S.P., Loftus, J.C., Ginsberg, M.H., Lauffenburger, D.A., and Horwitz, A.F. (1997). Integrin-ligand binding properties govern cell migration speed through cell-substratum adhesiveness. *Nature* **385**, 537–540.

Pan, J. and Snell, W.J. (2000). Signal transduction during fertilization in the unicellular alga, *Chlamydomonas. Curr. Opinion Microbiol.* **3**, 596–602.

Pan, J. and Snell, W.J. (2002). Kinesin-II is required for flagellar sensory transduction during fertilization in *Chlamydomonas*. *Mol. Biol. Cell* **13**, 1417–1426.

Pan, J., Haring, M.A., and Beck, C.F. (1997). Characterization of blue light signal transduction chains that control development and maintenance of sexual competence in *Chlamydomonas reinhardtii*. *Plant Physiol.* **115**, 1241–1249.

Pasquale, S.M. and Goodenough, U.W. (1987). Cyclic AMP functions as a primary sexual signal in gametes of *Chlamydomonas reinhardtii*. *J. Cell Biol.* **105**, 2279–2292.

Pazour, G.J. and Witman, G.B. (2003). The vertebrate primary cilium is a sensory organelle. *Curr. Opin. Cell Biol.* **15**, 105–110.

Pazour, G.J., Sineshchekov, O.A., and Witman, G.B. (1995). Mutational analysis of the phototransduction pathway of *Chlamydomonas reinhardtii*. *J. Cell Biol.* **131**, 427–440.

Pazour, G.J., Wilkerson, C.G., and Witman, G.B. (1998). A dynein light chain Is essential for the retrograde particle movement of intraflagellar transport (IFT). *J. Cell Biol.* **141**, 979–992.

Pazour, G.J., Agrin, N., Leszyk, J., and Witman, G.B. (2005). Proteomic analysis of a eukaryotic cilium. *J. Cell Biol.* **170**, 103–113.

Ponting, C.P. and Aravind, L. (1997). PAS: a multifunctional domain family comes to light. *Curr. Biol.* **7**, R674–R677.

Qin, H., Diener, D.F., Geimer, S., Cole, D.G., and Rosenbaum, J.L. (2004). Intraflagellar transport (IFT) cargo: IFT transports flagellar precursors to the tip and turnover products to the cell body. *J. Cell Biol.* **164**, 255–266.

Qin, H., Burnette, D.T., Bae, Y.K., Forscher, P., Barr, M.M., and Rosenbaum, J.L. (2005). Intraflagellar transport is required for the vectorial movement of TRPV channels in the ciliary membrane. *Curr. Biol.* **15**, 1695–1699.

Quarmby, L.M., Yueh, Y.G., Cheshire, J.L., Keller, L.R., Snell, W.J., and Crain, R.C. (1992). Inositol phospholipid metabolism may trigger flagellar excision in *Chlamydomonas reinhardtii*. *J. Cell Biol.* **116**, 737–744.

Ray, D.L. and Gibor, A. (1982). Tunicamycin-sensitive glycoproteins involved in the mating of *Chlamydomonas reinhardtii*. *Exp. Cell Res.* **141**, 245–252.

Reinhart, F.D. and Bloodgood, R.A. (1988). Gliding defective mutant cell lines of *Chlamydomonas moewusii* exhibit alterations in a 240 kDa surface-exposed flagellar glycoprotein. *Protoplasma* **144**, 110–118.

Reiter, J.F. and Mostov, K. (2006). Vesicle transport, cilium formation and membrane specialization: The origins of a sensory organelle. *Proc. Natl. Acad. Sci. USA* **103**, 18383–18384.

Remillard, S.P. and Witman, G.B. (1982). Synthesis, transport and utilization of specific flagellar proteins during flagellar regeneration in *Chlamydomonas*. *J. Cell Biol.* **93**, 615–631.

Ridley, A.J., Schwartz, M.A., Burridge, K., Firtelk, R.A., Ginsberg, M.H., Borisy, G., Parsons, J.T., and Horwitz, A.R. (2003). Cell migration: Integrating signals from front to back. *Science* **302**, 1704–1709.

Ringo, D.L. (1967). Flagellar motion and fine structure of the flagellar apparatus in *Chlamydomonas*. *J. Cell Biol.* **33**, 543–571.

Robenek, H. and Melkonian, M. (1981). Sterol-deficient domains correlate with intramembrane particle arrays in the plasma membrane of *Chlamydomonas reinhardtii*. *Eur. J. Cell Biol.* **25**, 258–264.

Roberts, K., Phillips, J., Shaw, P., Grief, C., and Smith, E. (1985). An immuno-logical approach to the plant cell wall. In: *Biochemistry of Plant Cell Walls* (C.T. Brett and J.R. Hillman, Eds.), pp. 125–154. Cambridge University Press, London.

Rogalski, A.A. and Bouck, G.B. (1982). Flagellar surface antigens in *Euglena*: Immunological evidence for an external glycoprotein pool and its transfer to the regenerating flagellum. *J. Cell Biol.* **93**, 758–766.

Rosenbaum, J.L. and Witman, G.B. (2002). Intraflagellar transport. *Nat. Rev. Mol. Cell Biol.* **3**, 813–825.

Saito, A., Suetomo, Y., Arikawa, M., Omura, G., Mostafa Kamal Khan, S.M., Kakuta, S., Suzaki, E., Kataoka, K., and Suzaki, T. (2003). Gliding movement in *Peranema trichophorum* is powered by flagellar surface motility. *Cell Motil. Cytoskeleton* **55**, 244–253.

Saito, T., Tsubo, Y., and Matsuda, Y. (1985). Synthesis and turnover of cell body agglutinin as a pool of flagellar surface agglutinin in *Chlamydomonas reinhardtii* gamete. *Arch. Microbiol.* **142**, 207–210.

Shin-ichiro, Y., Engerer, J., Fuchs, E., Haas, A.K., and Barr, F.A. (2007). Functional dissection of Rab GTPases involved in primary cilium formation. *J. Cell Biol.* **178**, 363–369.

Smith, E., Roberts, K., Hutchings, A., and Galfre, G. (1984). Monoclonal antibodies to the major structural glycoprotein of the *Chlamydomonas* cell wall. *Planta* **161**, 330–338.

Snell, W.J. (1976). Mating in *Chlamydomonas*: A system for the study of specific cell adhesion. I: Ultrastructural and electrophoretic analysis of the flagellar surface components involved in adhesion. *J. Cell Biol.* **68**, 48–69.

Snell, W.J. and Moore, W.S. (1980). Aggregation-dependent turnover of flagellar adhesion molecules in *Chlamydomonas* gametes. *J. Cell Biol.* **84**, 203–210.

Snell, W.J., Buchanan, M., and Clausell, A. (1982). Lidocaine reversibly inhibits fertilization in *Chlamydomonas*: A possible role for calcium in sexual signaling. *J. Cell Biol.* **94**, 607–612.

Soderling, T.R. (1999). The Ca^{2+}-calmodulin-dependent protein kinase cascade. *Trends Biochem. Sci.* **24**, 232–236.

Song, L. and Dentler, W.L. (2001). Flagellar protein dynamics in *Chlamydomonas*. *J. Biol. Chem.* **276**, 29754–29763.

Stephens, R.E. and Good, M.J. (1990). Filipin-sterol complexes in molluscan gill ciliated cell membranes: intercalation into ciliary necklaces and induction of gap junctional particle arrays. *Cell Tissue Struct.* **262**, 301–306.

Tamm, S.L. (1967). Flagellar development in the protozoan *Peranema trichophorum*. *J. Exp. Zool.* **164**, 163–186.

Tomson, A.M., Demets, R., van Spronsen, E.A., Brakenhoff, G.J., Stegwee, D., and van den Ende, H. (1990). Turnover and transport of agglutinins in conjugating *Chlamydomonas* gametes. *Protoplasma* **155**, 200–209.

Ulehla, V.V. (1911). Ultramikroskopische Studien uber Geisselbewegung. *Biol. Zentr.* **31**, 689–705.

Veira, O.V., Gaus, K., Verkade, P., Fullekrug, J., Vaz, W.L.C., and Simons, K. (2006). FAPP2, cilium formation and compartmentalization of the apical membrane in polarized Madin-Darby canine kidney (MDCK) cells. *Proc. Natl. Acad. Sci. USA* **103**, 18556–18561.

Versluis, M., Schuring, F., Klis, F.M., van Egmond, P., and van den Ende, H. (1992). Sexual agglutination in *Chlamydomonas eugametos* is mediated by a single pair of hydroxyproline-rich glycoproteins. *FEBS Microbiol. Lett.* **97**, 101–105.

Versluis, M., Klis, F.M., van Egmond, P., and van den Ende, H. (1993). The sexual agglutinins in *Chlamydomonas eugametos* are sulphated glycoproteins. *J. Gen. Microbiol.* **139**, 763–767.

Vicente-Manzanares, M., Webb, D.J., and Horwitz, A.R. (2005). Cell migration at a glance. *J. Cell Sci.* **118**, 4719–4917.

Wagner, V., Gessner, G., Heiland, I., Kaminski, M., Hawat, S., Scheffler, K., and Mittag, M. (2006). Analysis of the phosphoproteome of *Chlamydomonas reinhardtii* provides new insights into various cellular pathways. *Eukaryotic Cell* **5**, 457–468.

Walther, Z., Vashishtha, M., and Hall, J.L. (1994). The *Chlamydomonas* FLA10 gene encodes a novel kinesin-homologous protein. *J. Cell Biol.* **126**, 175–188.

Wang, Q. and Snell, W.J. (2003). Flagellar adhesion between mating type plus and mating type minus gametes activates a flagellar-tyrosine kinase during fertilization in *Chlamydomonas*. *J. Biol. Chem.* **278**, 32936–32942.

Wang, Q., Pan, J., and Snell, W.J. (2006). Intraflagellar transport particles participate directly in cilium-generated signaling in *Chlamydomonas*. *Cell* **125**, 549–562.

Weiss, R.L., Goodenough, D.A., and Goodenough, U.W. (1977). Membrane particle arrays associated with the basal body and with contractile vacuole secretion in *Chlamydomonas*. *J. Cell Biol.* **72**, 133–143.

Wiese, L. and Mayer, R.A. (1982). Unilateral tunicamycin sensitivity of gametogenesis in dioecious isogamous *Chlamydomonas* species. *Gamete Res.* **5**, 1–9.

Witman, G.B. (1986). Isolation of *Chlamydomonas* flagella and flagellar axonemes. *Methods Enzymol.* **134**, 280–290.

Witman, G.B., Carlson, K., Berliner, J., and Rosenbaum, J.L. (1972). *Chlamydomonas* flagella. I. Isolation and electrophoretic analysis of microtubules, matrix, membranes and mastigonemes. *J. Cell Biol.* **54**, 507–539.

Yang, W., Mason, C.B., Pollock, S.V., Lavezzi, T., Moroney, J.V., and Moore, T.S. (2004). Membrane lipid biosynthesis in *Chlamydomonas reinhardtii*: Expression and characterization of CTP:phosphoethanolamine citidylyltransferase. *Biochem. J.* **382**, 51–57.

Yoshimura, K. (1996). A novel type of mechanoreception by the flagella of *Chlamydomonas*. *J. Exp. Biol.* **199**, 295–302.

CHAPTER 12

Flagellar Adhesion, Flagellar-Generated Signaling, and Gamete Fusion during Mating

William J. Snell[1] and Ursula W. Goodenough[2]

[1]Department of Cell Biology, University of Texas Southwestern
Medical Center, Dallas, Texas, USA

[2]Department of Biology, Washington University, St. Louis, Missouri, USA

CHAPTER CONTENTS

I. INTRODUCTION

Although the widely studied flagella of *Chlamydomonas* are critical for cell motility, the organism uses its flagella for more than just finding optimal light conditions for photosynthesis. During the sexual phase of the life cycle the flagella retain their normal morphology and motility even as they are transformed to carry out specialized sensory functions related to mating. Both mating type *plus* and mating type *minus* gametes express adhesion/signaling molecules, the *plus* and *minus* agglutinins, on their respective flagellar surfaces. The agglutinins are multifunctional and serve as both cell–cell adhesion molecules and sensory transduction proteins. In their adhesion role, the interacting agglutinins on flagella tether cells together similarly to integrins and their ligands; in their signaling role, the interacting agglutinins activate signaling pathways similarly to the odorant

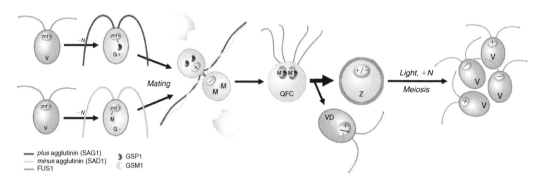

FIGURE 12.1 *Life cycle of* Chlamydomonas reinhardtii. *Haploid vegetative (V) cells of two mating types (mt+ and mt−) divide mitotically. When exogenous nitrogen becomes limiting, they differentiate into gametes (G+ and G−), expressing mating type-specific gametic traits. When gametes are mixed, the plus (SAG1) and minus (SAD1) agglutinins displayed on their flagellar surfaces mediate the initial adhesion reaction; adhesion generates a rise in intracellular cAMP which triggers gamete cell wall release and mating-structure activation; the FUS1 protein on the plus mating structure (cf. Figure 12.5) interacts with partner protein(s) on the minus mating structure to trigger cell fusion and the formation of the binucleate QFC. Homeoproteins GSP1 and GSM1, pre-synthesized in plus and minus gametes respectively, interact to activate transcription of zygote-specific genes. Nuclei fuse, flagella are resorbed, and a thick cell wall is assembled around the zygote (Z). In the laboratory, zygotes subjected to 5 days of dormancy in the dark and returned to light in N-containing media undergo meiosis to release four haploid meiotic products that resume vegetative growth. Occasional QFCs forego the meiotic pathway and instead resume vegetative growth as +/− vegetative diploids (VD). (From Goodenough et al., 2007.)*

receptors in the ciliary membrane of olfactory epithelial cells in vertebrates, to chemosensory receptors in the sensory cilia of *C. elegans*, to mechanosensory molecules of primary cilia on kidney epithelial cells, and to receptors on vertebrate primary cilia that function in essential developmental signaling pathways (Pan et al., 2005; Goodenough et al., 2007).

Gametogenesis is induced in the laboratory by transferring vegetatively growing *plus* and *minus* cells into medium that does not contain a nitrogen source (Figure 12.1) (detailed protocols for gamete induction are provided in Volume 1, Chapter 8). Gametes, which are similar in appearance to their vegetative precursors except for their smaller size, express several gamete-specific and mating type-specific proteins. Mating type (*MT*) is under the control of the *MT* locus, a complex region on linkage group VI (Ferris and Goodenough, 1994; Ferris et al., 2002) described in detail in Volume 1, Chapter 5. Cells that carry the *MT+* locus undergo gametogenesis to become *plus* gametes and cells that carry the *MT−* locus become *minus* gametes (see below).

When *plus* and *minus* gametes are mixed together, the cells initially adhere to each other via their flagellar agglutinins (Figure 12.1). Once an agglutinin molecule binds its ligand (the agglutinin molecule on the flagellum of a gamete of the opposite mating type), the agglutinins undergo complex inter-molecular structural interactions that presumably are essential to form the extremely strong adhesive bonds that tether the two gametes together as they bounce to and fro in the high-energy dance of fertilization. At the same time, the agglutinin interactions trigger a complex signaling pathway (Figure 12.2) that brings about cellular changes required to prepare the gametes for fusion; some of these are illustrated in Figures 12.1 and 12.5. A gamete-specific flagellar adenylyl cyclase (AC) is activated via a

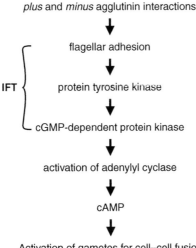

Flagellar adhesion-induced signal transduction

plus and *minus* agglutinin interactions

flagellar adhesion

IFT

protein tyrosine kinase

cGMP-dependent protein kinase

activation of adenylyl cyclase

cAMP

Activation of gametes for cell–cell fusion

FIGURE 12.2

Outline of events triggered by agglutinin interactions during flagellar adhesion and flagellum-generated signaling.

protein kinase- and kinesin-2-dependent pathway (Zhang et al., 1991; Saito et al., 1993; Pan and Snell, 2002). The almost immediate (30–90 seconds), 10–20-fold increase in intracellular cyclic AMP (cAMP) induced by flagellar adhesion (Pijst et al., 1984; Pasquale and Goodenough, 1987) generates several cellular responses that prepare the gametes for cell fusion. These responses include recruitment of additional agglutinin molecules from the cell body plasma membrane onto the flagellar surface; movement of flagellar adhesion sites to the flagellar tip ("tipping"); elongation (1–2 μm) of the A-microtubules of the axonemal doublets at the flagellar tip; swelling of the flagellar tip accompanied by accumulation of electron-dense material underneath the membrane; secretion of a serine protease that activates a metalloprotease that removes the extracellular matrix in preparation for cell fusion; activation of microvillus-like fusion organelles, called mating structures, between the two flagella at the apical end of each cell; and redistribution of a *plus*-specific cell adhesion protein, FUS1, from its location as an apical patch on the unactivated *plus* mating structure to the entire surface of the elongating mating structure. Most or all of the responses can be induced by addition of dibutyryl cAMP (db-cAMP) to gametes of a single mating type (Pasquale and Goodenough, 1987). Addition of db-cAMP does not induce these responses in vegetative cells, indicating that the events downstream of cAMP production also are specific to gametes.

When the *plus* and *minus* gametes begin their interactions, gametes often adhere to several other gametes, forming large aggregates of adhering cells. Pairs of *plus* and *minus* gametes form within the aggregates, and their flagellar motility and constant movement bring the mating structures into close physical proximity. Almost as soon as the activated mating structures are extended from the surfaces of *plus* and *minus* gametes, the tips of these organelles adhere and fuse to establish cytoplasmic continuity between the two gametes. The tube-like connection rapidly shortens and expands as the cells merge into a quadriflagellate cell (QFC) (Figure 12.1). This entire process – from the initial flagellar adhesion after gametes of opposite mating types are mixed together, through cell fusion – is extremely fast, and QFCs can be detected as soon as 15 seconds after mixing.

Cell fusion itself generates intracellular signals that have both short-term and long-term effects. In the short term, and as with fertilization in other species, the fused gametes disengage from other gametes within seconds after fusion occurs as the flagella lose their adhesiveness. Soon thereafter, the beat of all four flagella becomes coordinated and the new zygote becomes motile. Both gametes contain distinct sex-specific homeodomain transcription factors, which heterodimerize once the cytoplasms of the two gametes coalesce in the zygote (see below and Figure 12.1). The newly formed complexes migrate to the two nuclei to activate gene transcription and initiate a new developmental pathway in the zygote. Within a few hours a new, highly impermeable cell wall is formed, and under the proper regimen of light, germination ensues within a few days and yields two *plus* and two *minus* daughter cells, the haploid cell products of meiosis (Figure 12.1). Zygote differentiation and maturation are described in Volume 1, Chapter 5.

The ease of observing and quantifying *Chlamydomonas* mating has made it possible to generate sterile mutants blocked at key steps in fertilization (Goodenough et al., 1995), and the availability of mating mutants, along with the ease of isolating biochemical quantities of the organelles involved in adhesion and fusion (Wilson et al., 1997), have generated a depth of knowledge about fertilization in *Chlamydomonas* that is unparalleled in multicellular organisms. In this chapter we describe our current understanding of the cellular and molecular events that occur during the mating reaction in *Chlamydomonas reinhardtii*, where most is known. Parallel and complementary studies with *C. eugametos* are reviewed in Gloeckner and Beck (1995) and van den Ende (1992). Table 12.1 summarizes the properties of some of the proteins, discussed in more detail below, that are involved in flagellar adhesion and gamete fusion.

Table 12.1 Proteins involved in flagellar adhesion and gamete fusion

Protein	Location	Gene; mutants	Accession No.	Linkage group	Mating phenotype	References
Plus agglutinin	Flagellar membrane	*SAG1*; *sag1-1* to *sag1-6*	AAS07044	VIII	Flagellar sexual adhesion	Goodenough et al. (1978), Ferris et al. (2005), Lee et al. (2007)
Plus agglutinin (O-glycosylation)	Flagellar membrane	*SAG2/GAG1*; *gag1, imp8*	–	–	Flagellar sexual adhesion (O-glycosylation)	Goodenough et al. (1978), Vallon and Wollman (1995)
Minus agglutinin	Flagellar membrane	*SAD1*; *sad1-1, sad1-2*	AAS07042	VI (*MT* locus)	Flagellar sexual adhesion	Hwang et al. (1981), Ferris et al. (2005), Lee et al. (2007)
Gametolysin	Secreted (in response to cAMP)	*MMP1*	BAB68382	XIX	Cell-wall loss	Kubo et al. (2001)
FUS1	Plus mating structure	*FUS1*; *fus1-1*	AAC49416	VI (*MT+* locus)	Mating-structure adhesion	Ferris et al. (1996), Misamore et al. (2003)
MID	Nucleus (predicted)	*MID*; *mid-1, mid-2*	AAC49753	VI (*MT−* locus)	*Minus* sexual differentiation	Ferris and Goodenough (1997), Lin and Goodenough (2007)
MTD1		*MTD1*; RNA interference knockdowns	AAL14635	VI (*MT−* locus)	*Minus* sexual differentiation	Lin and Goodenough, (2007)
GSP1	Cytoplasm → nucleus	*GSP1*	AAL14119	II	Zygote gene expression	Zhao et al. (2001), Lee et al. (2008)
GSM1	Cytoplasm → nucleus	*GSM1*	ABJ15867	VIII	Zygote gene expression	Lee et al. (2008)

II. FLAGELLAR AGGLUTINATION

A. The flagellar agglutinins are large extrinsic glycoproteins

Mating initiates with the adhesion of *plus* and *minus* agglutinins. Bloodgood provides detailed coverage of many features of these proteins in Chapter 11. Presented here is additional information related to their structure, evolution, and relationship to downstream signaling events.

In pioneering work from the Roberts lab (Roberts et al., 1985), it was shown that the *Chlamydomonas* cell wall is a highly ordered self-assemblage of hydroxyproline-rich glycoproteins (HRGPs), a protein family that is also prominent in the cell walls of land plants (Cassab, 1998). When an extensive effort to identify the sexual agglutinins, pioneered by Adair, was eventually successful (Adair et al., 1983), the agglutinins proved to be HRGPs as well (Cooper et al., 1983). They have the signature HRGP conformation of a "shaft" domain carrying hydroxyproline and serine motifs in a polyproline II configuration, and globular "head" domains at the N- and C-termini (Goodenough et al., 1985; Ferris et al., 2005) (Figure 12.3), where the N-termini associate with the flagellar membrane. It was quickly recognized that features enabling HRGP wall assembly might correspond to features enabling sexual adhesion (Cooper et al., 1983; Adair, 1985). Subsequent studies of cell-wall assembly established that wall HRGPs form head-to-shaft and shaft-to-shaft interactions (Goodenough et al., 1985; Goodenough and Heuser, 1988), and this is presumed to be the case as well for the agglutinins. Images of agglutinin–agglutinin interactions, however, show only dense meshworks of fibrous proteins (Goodenough and Heuser, 1999), reminiscent of cell-wall assemblages, that have thus far yielded no information on which protein domains are involved.

Figure 12.3 illustrates an interesting feature of the *plus* agglutinin: when the protein adsorbs to mica it occasionally denatures, revealing the "head-loop" configuration adopted by the underlying shaft (Figure 12.3D and E). Such images are interpreted to indicate that the head assembles around this loop, in the fashion of a lollipop surrounding its stick.

Most mutant strains incapable of sexual agglutination map to two loci: *plus* mutations reside in the *SAG1* locus on linkage group VIII, and *minus* mutations reside in the *SAD1* locus on linkage group VI in the *MT* locus (Ferris et al., 2005). Interestingly, a functional *SAD1* gene occupies a syntenic position in both the *MT+* and *MT−* loci (Ferris et al., 2002). The *plus* copy is not expressed in *plus* gametes because expression of all *minus*-specific genes requires the presence of the MID protein, encoded exclusively in the *MT−* locus (Ferris and Goodenough, 1997). When *plus* cells carry a *MID* transgene, however, they agglutinate normally as *minus*, indicating that the *MT+* copy is fully functional (Ferris and Goodenough, 1997). It is not known whether the *plus* copy of *SAD1* represents an aberrant translocation

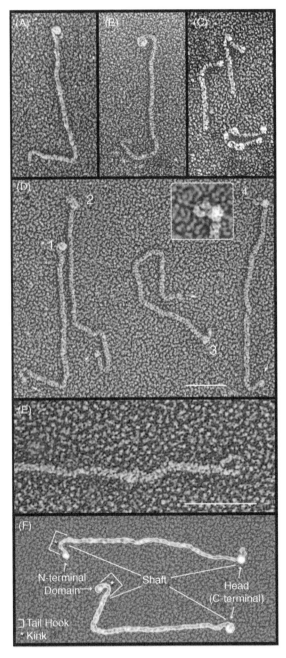

FIGURE 12.3 *Morphology of agglutinins and related cell-wall proteins. (A)* **Plus** *agglutinin. Curved shaft fibril protrudes beneath the large globular head. (B)* **Minus** *agglutinin. (C) GP1 (upper pair of molecules) and GP2 (lower) cell-wall HRGPs, displaying the same head-shaft organization. The dominant repeat motif of the agglutinin shaft, pro-pro-ser-pro-X, is shared by GP1, and their shafts are similar in diameter. The kink in GP1 is curved (left) or acute (right). (D)* **Plus** *agglutinins. Arrows indicate globular domains at the proximal shaft termini. Head 1 is globular; heads 2 and 3 are bilobed; head 4 displays a protruding curved fibril, shown at higher magnification in the inset. (E)* **Plus** *shaft showing head loop free of its globular head domain. (F) Diagram of the agglutinins. All images generated by quick-freeze deep-etch electron microscopy. Bar = 50 nm. (From Ferris et al., 2005.)*

or whether it plays some unknown role in the life cycle. Mutations in the *SAG2* locus, which encodes a gene involved in O-glycosylation, render *plus* but not *minus* cells non-agglutinable (Vallon and Wollman, 1995).

The *SAD1* gene was encountered during a chromosomal walk through the *MT* loci (Ferris and Goodenough, 1994), and the *SAG1* gene was identified by insertional mutagenesis (Ferris et al., 2005). The Goodenough lab has also cloned and sequenced their orthologues in the closely related (<10 million-years-ago divergence) species *C. incerta*. All four proteins are enormous (Figure 12.3): they contain at least 3300 amino acids, plus a heavy endowment of sugar, and have shaft lengths ranging from 225 to 325 nm (Ferris et al., 2005). All four display the same overall pattern of domain organization, indicating that they originated from a common ancestral gene, but the *plus* and *minus* proteins are, with the exception noted below, totally different in sequence, indicating that this common ancestry is ancient.

When the two SAG1 orthologues or two SAD1 orthologues are compared, >30% amino-acid divergence is encountered in the head domains (Lee, Waffenschmidt, and Goodenough, manuscript in preparation), another example of the rapid between-species evolution encountered in numerous sex-related proteins (Swanson and Vacquier, 2002). Even more dramatic, the central regions of the two SAG1 shafts are completely different in sequence from one another, and the central regions of the two SAD1 shafts are completely different in sequence from one another, a divergence that apparently occurred at about the time that *C. reinhardtii* and *C. incerta* last shared a common ancestor (Lee et al., 2007). These observations reinforce the central tenet that the agglutinins mediate species-specific mating interactions, but such dramatic sequence differences discourage hope of identifying interaction domains via inter-species comparisons. The one commonality is that all four head domains contain two highly conserved regions predicted to form hydrophobic α-helices (Lee et al., 2007, and Lee, Waffenschmidt, and Goodenough, manuscript in preparation); these are interpreted to form

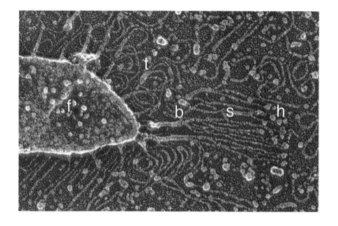

FIGURE 12.4

Agglutinins associated with the flagellar membrane surface (f), visualized by quick-freeze deep-etch electron microscopy after adsorption to a mica flake. s, agglutinin shaft; h, agglutinin head; b, shaft bundle associated with flagellar membrane; t, tiered array of agglutinins. (Goodenough, U. and Heuser, J. E. unpublished.)

robust interior folds that are able to sustain the numerous amino-acid substitutions incurred elsewhere in the heads.

Figure 12.4 shows a quick-freeze deep-etch electron microscopic image
of *plus* agglutinins associated with the flagellar surface of an unmated
gamete after adsorption to a mica flake. Extended single agglutinins can be
identified (see also Figure 11.5), with shafts terminating in distal heads.
These self-associate in proximal bundles of ~8 proteins as they approach
the flagellar surface. In fact, the straight extended configuration is the
exception, and possibly induced by mica adsorption; usually, the agglutinins curve back on themselves so that both the N- and C-terminal globular
domains associate with the flagellar surface, forming staggered tiered arrays
(see also Goodenough et al., 1985). The proteins are presumably organized
in such bundled tiered configurations when they are loaded onto flagellar membranes. Interestingly, the tiered configuration has the outcome
that the central shaft domains are prominently displayed on the unmated
gametic flagellum; these are the same domains that have diverged completely in sequence when *C. reinhardtii* and *C. incerta* proteins of the same
mating type are compared (Lee et al., 2007).

These observations suggest a model wherein initial *plus/minus* contacts
are mediated by the exposed central shaft domains, inducing the release of
heads from their tiered self-membrane associations and allowing the heads
to extend out and mediate the assembly of the extended fibrous system
that interconnects adhered flagella. Supporting this model is the observation that when gametes are incubated in db-cAMP, mimicking the endogenous mating signal, the flagellar agglutinins are uniformly in the extended
configuration (Goodenough and Heuser, 1999).

Early success with agglutinin purification came with the discovery
that they could be released from the flagella by exposing gametes to 5 mM
EDTA, indicating that they are peripheral and not transmembrane proteins
(Adair et al., 1982). This inference was confirmed by the absence of any
candidate transmembrane or fatty-acid-attachment motifs at the N-terminal
globular ends of any of the four sequenced agglutinins (Ferris et al., 2005,
and Lee, Waffenschmidt and Goodenough, manuscript in preparation).
Therefore, the agglutinins are presumably anchored to the flagellum via
additional transmembrane proteins, and adhesion-induced signaling is
presumably transmitted to the flagellum, and then the cell, via these
transmembrane proteins, which await identification.

B. Control of flagellar agglutinability

Flagellar agglutinability is under complex control:

1. Purified agglutinins of opposite type give no evidence of interacting
   with or neutralizing one another in solution. By contrast, if
   agglutinins of one type are affixed to a glass slide, living gametes

adhere to them avidly (providing a bioassay for agglutinin purification) (Adair et al., 1982, 1983) and, in *C. eugametos*, charcoal particles adsorbed with purified agglutinins of opposite type aggregate with each other (Versluis et al., 1992).

2. When isolated *plus* and *minus* gametic flagella are mixed, with or without prior fixation with glutaraldehyde, they aggregate into large clumps that are stable for days at 4°C (Goodenough, 1986). By contrast, when live gametes interact with isolated flagella of opposite type, the flagella, but not the cells, lose their agglutinability within 10–20 minutes (Snell and Roseman, 1979). Similarly, when live gametes interact with purified agglutinins of opposite type affixed to a glass slide, the affixed agglutinins are rapidly inactivated (Adair et al., 1982).

3. Live gametes lose their agglutinability under two conditions: (a) when cells fuse to form binucleate QFCs, their flagella disadhere; and (b) when mutants unable to fuse (Goodenough et al., 1982) are mixed with opposite type, adhesion continues for hours, but they disadhere in the presence of the protein-synthesis inhibitor cycloheximide (Snell and Moore, 1980), in the presence of the protein-glycosylation inhibitor tunicamycin (Matsuda et al., 1981; Snell, 1981), or in the presence of inhibitors of proline hydroxylation (Cooper et al., 1983).

4. When adhering gametes are visualized by the quick-freeze deep-etch technique, agglutinin-laden membrane vesicles are observed blebbing from the flagellar tips (Goodenough and Heuser, 1999).

5. Incubation of gametes with low concentrations of trypsin for 5 minutes destroys their agglutinability (Snell, 1976), presumably by releasing the agglutinins in an active form from the flagella, because isolated agglutinins are resistant to such trypsin treatment (Collin-Osdoby et al. 1984; Collin-Osdoby and Adair, 1985).

6. Addition to *plus* gametes of an adhesion-blocking monoclonal antibody inactivates the flagellar agglutinins, but not the cell body agglutinins. Activation of the non-adhesive gametes by addition of db-cAMP in the presence of protein synthesis inhibitors rapidly leads to re-appearance of active flagellar agglutinins (Hunnicutt et al., 1990).

These observations generate the following understandings. (1) Agglutinins require association with a surface (flagellar membrane, glass, charcoal) for adhesive activity. (2) Agglutinins are inactivated by adhesion when one partner is a living cell. (3) When both partners are living cells, agglutinins are either inactivated during adhesion *in situ* or else they are shed. It is not yet know if agglutinin shedding occurs solely by tip vesiculation or if individual agglutinins also are shed during adhesion. Agglutinins are replaced, first from a pool of pre-synthesized agglutinins (Saito et al.,

1985; Goodenough, 1989) resident on the plasma-membrane surface (Hunnicutt et al., 1990), and then by *de novo* synthesis. (4) Upon gamete fusion, the zygote flagella disadhere from each other, even though the flagellar agglutinins remain on the flagella as assessed by a dried spot assay. Within 45 minutes after fusion, the agglutinins no longer are detectable (Hunnicutt and Snell, 1991). (5) Agglutinin movement between the cell body and the flagella is controlled by a functional barrier that can be regulated by gamete activation (Hunnicutt et al., 1990) (see also Chapter 11). Intriguingly, the Hedgehog pathway in vertebrates depends on a functional barrier to movement of the transmembrane protein Smoothened between the cell body and the primary cilium (Rohatgi and Scott, 2007).

A number of studies have demonstrated that the agglutination reaction is sensitive *in vivo* and/or *in vitro* to a variety of manipulations, including relatively high ionic strength (100 mM KCl) (Goodenough, 1986) and various proteases (Collin-Osdoby et al., 1984; Collin-Osdoby and Adair, 1985). The *plus* agglutinin is far more sensitive to dithiothreitol exposure than the *minus* (Goodenough, 1986), and disulfide reduction causes the *plus* but not the *minus* agglutinin heads to denature (Goodenough et al., 1985). Unfortunately, none of these observations have yielded much insight on the molecular mechanisms of agglutinin interactions.

C. Flagellar tips

When flagella grow out from basal bodies, axonemal assembly occurs at the growing tip – the location of the "plus" ends of the microtubule doublets. Studies of the intraflagellar transport (IFT) process, which delivers components for axonemal outgrowth, are increasingly focused on the flagellar tip as the locus for both component "drop off" and the switch from an anterograde (kinesin-driven) to a retrograde (dynein-driven) transport pattern (see Chapter 4).

The flagellar tip also features prominently in gametic flagellar interactions. (1) When gametes adhere, initial contacts can occur anywhere along the flagellar surface, but adhesive foci soon localize at the tips ("tipping") (Goodenough and Jurivich, 1978) and, as noted above, membrane vesicles bearing enmeshed agglutinins also bleb from the tips (Goodenough and Heuser, 1999). (2) Adhering flagella undergo two internal morphological changes at their tips – a selective elongation of the A microtubules and the accumulation of dense material in this elongated domain – collectively referred to as flagellar tip activation (Mesland et al., 1980). Flagellar tip activation is also induced in non-adhering gametes whose intracellular levels of cAMP are experimentally elevated (Pasquale and Goodenough, 1987). (3) When gametes are adhering and/or their intracellular levels of cAMP are elevated, tip-directed membrane motility is activated. This has been observed under two conditions: (a) polystyrene beads that attach to the membrane surface ordinarily move up and down the flagellum in a saltatory fashion (see Chapter 11), but during mating they rapidly accumulate at the tips, resuming bidirectional transport only after the cells fuse (Hoffman and Goodenough, 1980); and (b) isolated gametic flagella, when presented to living partners of opposite type, similarly accumulate at the flagellar tips with cAMP elevation (Goodenough, 1993).

Prominent unanswered questions include the following: (1) Does the dense material that accumulates at the flagellar tips correspond to IFT particles, or to some gamete-specific components, or both? (2) What is the relationship between the saltatory bead-translocation system, the continuous IFT system, and tipping? (3) Is tipping a prelude to gamete signal transduction, the consequence of signal transduction, or an integral feature of signal transduction? These matters will be considered again once the signal-transduction system is described.

III. FLAGELLAR-GENERATED SIGNALING

A. Flagellar adhesion activates a cAMP-dependent signaling pathway leading to gamete activation

In addition to bringing cells of opposite sex into close contact, agglutinin interactions and flagellar adhesion also initiate a cAMP-dependent signaling pathway that leads to gamete activation (Figure 12.2). Although the key observations that led to the discovery that cAMP is an essential second messenger during mating were made some time ago (Pasquale and Goodenough, 1987; Zhang et al., 1991; Saito et al., 1993), the AC that produces the cAMP has not yet been identified. Surprisingly, the flagellar AC appears not to be regulated by G-proteins (Pasquale and Goodenough, 1987; Zhang et al., 1991; Saito et al., 1993), but is regulated by protein phosphorylation and de-phosphorylation (Zhang et al., 1991; Saito et al., 1993). Presumably, this AC will be molecularly distinct from the vegetative flagellar AC, since the gametic flagellar AC exhibits regulatory properties that are significantly different from those of the vegetative flagellar enzyme (Saito et al., 1993; Zhang and Snell, 1993).

The cell body also contains an AC activity that is about 2-fold lower in specific activity than the flagellar AC. The flagellar AC has sufficient activity to generate the nearly 20-fold increase in cellular cAMP that occurs during mating, but the total activity of the cell body cyclase is greater than the total flagellar AC activity, and the cell body form, like the flagellar form, is activated nearly 2-fold during adhesion (Zhang et al., 1991; Saito et al., 1993). Interestingly, the flagellar form, but not the cell body form, of the enzyme is activated 10–70-fold by prior treatment at elevated temperatures. One model for generation of cAMP during fertilization is that the cAMP made in the flagella during adhesion directly or indirectly activates the cell body form. The *Chlamydomonas* genome contains genes predicted to encode at least three putative ACs (EDP05013, EDP06317, EDO95660)

(Wang and Snell, unpublished observations). Future molecular characterization of these genes, along with identification of the gene disrupted in the *imp3* signaling mutant that is defective in activation of AC (Goodenough et al., 1976; Saito et al., 1993) should provide a better understanding of the mechanisms of their regulation.

B. Activation of a protein tyrosine kinase

One of the earliest biochemically detectable events triggered by agglutinin interactions is activation of a protein tyrosine kinase (PTK) within the flagella. Flagella isolated from gametes undergoing flagellar adhesion possess a PTK activity that is absent from flagella isolated from non-adhering gametes (Wang and Snell, 2003). Flagella from non-adhering gametes and from gametes that had been mixed together for 3–5 minutes and were undergoing extensive flagellar adhesion were assayed for PTK activity using immunoblotting with an anti-phosphotyrosine antibody to detect tyrosine-phosphorylated proteins. To optimize detection, detergent extracts of each of the flagellar samples were incubated with ATP in the presence of phosphatase inhibitors. The immunoblots showed that a 105-kD protein was tyrosine-phosphorylated only in the adhering sample. Experiments with an exogenous PTK substrate confirmed the adhesion-induced activation of a flagellar PTK. Several independent approaches indicated that PTK activation was not dependent on increases in cAMP or cell fusion and was induced by flagellar adhesion *per se* (Wang and Snell, 2003). First, gametes of a fusion-defective mutant, *fus1*, which are capable of flagellar adhesion but not gamete fusion, showed adhesion-dependent activation of the PTK. Second, when gametes underwent flagellar adhesion in the presence of H-8, an inhibitor of cAMP-dependent protein kinases, gamete activation was blocked, but the PTK still was activated. Finally, the PTK was not activated in gametes incubated in db-cAMP, and inclusion of cAMP in the *in vitro* assay for PTK activity in flagella isolated from gametes of a single mating type did not lead to activation of the PTK. The *Chlamydomonas* genome is predicted to encode three putative PTKs (EDP08371, EDP04963, and EDP01808). In future experiments it will be important to identify the flagellar-adhesion activated PTK and determine whether its activation indced is upstream of activation of the flagellar AC (Figure 12.2) or resides in a parallel signaling pathway.

C. The substrate of the adhesion-induced PTK is a cGMP-dependent protein kinase

Wang et al. (2006) used immunoprecipitation and mass spectrometry to identify the 105-kD substrate of the PTK as a cGMP-dependent protein kinase (PKG; AAT81143). The *Chlamydomonas* PKG has three predicted cGMP-binding sites and a serine/threonine protein kinase domain. Unlike cAMP-dependent protein kinases, both the regulatory domains and the catalytic domain are present in the same protein in PKGs. Fractionation of gametic cells showed that approximately 50% of the total cellular PKG was in the flagella, which represent only $\sim$3% of total cellular protein. Related experiments showed that the protein kinase activity of the PKG was activated during flagellar adhesion. Although RNA interference experiments showed that the PKG was in the signal-transduction pathway, it will be important to identify the mechanisms of its regulation and its downstream targets. One model is that cGMP binding to the PKG, along with phosphorylation by the adhesion-activated PTK, regulate the activity of the PKG. Such a dual mechanism for regulation of a PKG might be required in this highly motile, phototactic organism. It is known that cGMP and PKG regulate ciliary motility in rat tracheal epithelial cells, bovine airway epithelial cells, and *Paramecium* (see references in Wang et al., 2006). If *Chlamydomonas* also uses cGMP to regulate motility, it might be necessary in fertilization to have the additional layer of PKG regulation that could be provided through tyrosine phosphorylation by the adhesion-activated PTK. We should note, however, that cellular levels of cGMP increase $\sim$2-fold during flagellar adhesion. Thus, another model for this pathway is that agglutinin interactions trigger activation both of the PTK and a guanylyl cyclase. Activation of either one alone would be insufficient to trigger the signaling pathway, and the PKG would serve as an enforcer in the pathway to ensure that only agglutinin interactions push the cells down the risky pathway of shedding their protective cell walls and getting ready to fuse.

D. IFT and flagella adhesion-activated signaling

Studies on IFT (see Chapter 4) have generated fresh insights into ciliary and flagellar function. Increasing numbers of important functions are being ascribed to this motility process in many organisms. In most if not all instances, however, demonstration that a component of the IFT machinery is essential for a particular cell or tissue function documents that cilia/flagella are essential for that function. Such a result does not indicate that the function directly requires IFT, because without IFT, cilia are not formed. *Chlamydomonas* is one of the few systems that offer a way to overcome this ciliary version of the uncertainty principle and thereby study the role of IFT in an existing flagellum. In the temperature-sensitive kinesin-2 (the anterograde IFT motor) mutant *fla10-1*, IFT becomes undetectable within 30–40 minutes after transfer to the restrictive temperature, yet the flagella remain fully motile and of nearly full length for $\sim$1 hour (Lux and Dutcher, 1991; Kozminski et al., 1995). Thus, this mutant provides a unique system for studying an intact flagellum in which IFT is non-functional. It has been possible to take advantage of this mutant to examine the role of IFT in signal transduction induced by flagellar adhesion.

E. The IFT machinery is required to couple agglutinin interactions to activation of adenylyl cyclase

In studies using *fla10-1* cells, a confluence has emerged between IFT and flagellum-generated signaling during fertilization in *Chlamydomonas*. Piperno et al. (1996) first noticed that *fla10-1* gametes lost the ability to form zygotes soon after being shifted to 32°C – well before flagella were lost – documenting that at least one of the many steps in fertilization required kinesin-2. Subsequently, Pan and Snell (2002) identified the key step in fertilization that was blocked in *fla10-1* gametes. They found that kinesin-2 was required to couple agglutinin interactions to increases in cAMP. Results using several approaches supported this idea. First, microscopic examination and quantitative flagellar adhesion assays demonstrated that flagellar adhesion of *fla10-1* gametes after 40 minutes at 32°C was indistinguishable from that of wild-type gametes at both temperatures and of *fla10-1* gametes at 21°C; yet the *fla10*, 32°C gamete samples were incapable of cell–cell fusion. Second, incubation of the adhering but non-fusing 32°C *fla10-1* cells in db-cAMP rescued their ability to undergo cell fusion, indicating that they retained the ability to respond to cAMP. Finally, direct assays of cAMP showed that *fla10-1* gamete samples at 32°C did not undergo the substantial flagellar adhesion-dependent increase in cAMP that was observed in experiments with wild-type gametes at both temperatures and with *fla10-1* gamete samples at 21°C. These experiments documented that kinesin-2 is essential for sensory transduction in an otherwise intact and functional flagellum.

Pan and Snell (2002) considered several possibilities that could explain the failure of signaling in the *fla10-1* mutant gametes. One was that kinesin-2 and IFT are essential for maintaining proper levels in the flagella of a protein required for coupling agglutinin interactions to the increases in cAMP. For example, the presence in the flagella of sufficient levels of a coupling protein might require the continuous action of kinesin-2. Alternatively, kinesin-2 might participate directly in sensory transduction by moving agglutinins or agglutinin complexes within the flagella after the initial interactions between *plus* and *minus* agglutinins occur. For example, the initial interaction of a *plus* agglutinin molecule with its ligand (a *minus* agglutinin molecule) might induce conformational changes that allow the agglutinin-containing complexes in both cells to link to the kinesin-2/IFT system to form signaling scaffolds. Such linking might be essential for coupling to AC. A related possibility was that agglutinins must be moved to the flagellar tips in order to signal maximally (Mesland et al., 1980; Goodenough, 1993; Piperno et al., 1996; Wang et al., 2006). As described below, subsequent experiments indicated that the IFT machinery participates directly in the signaling pathway.

F. An adhesion-dependent change in cGMP-dependent protein kinase properties requires IFT

Additional studies dissecting the role of IFT in regulation of the signaling pathway focused on the adhesion-regulated PTK and PKG described above. First, Wang and Snell (2003) reported that the adhesion-regulated PTK activity was not detected in *fla10-1* gametes undergoing flagellar adhesion at the restrictive temperature, indicating that this early step in the pathway required IFT. More importantly, the PKG underwent a change in its biochemical properties during adhesion, and the change required the IFT machinery. The PKG in flagella isolated from non-adhering flagella was associated with the flagellar membrane, remained with the flagellum after disruption of the flagella by freezing and thawing, and was sedimented by low-speed centrifugation. On the other hand, in flagella isolated from adhering gametes, much of the PKG was released from the flagella by freeze/thaw, remained in the low-speed supernatant, but was sedimented at high speed, suggesting that the PKG was part of a larger complex. The PKG did not undergo this adhesion-dependent change in properties in flagella of *fla10-1* gametes at the restrictive temperature (Wang et al., 2006).

Analysis of the IFT machinery in wild-type gametes showed that the IFT subcomplexes also underwent changes in properties during adhesion. Similar to the PKG, IFT subcomplexes in non-adhering flagella either remained associated with flagella after freeze/thaw, or were released by freeze/thaw but remained in the supernatant after high-speed centrifugation. These results, along with analysis by velocity sedimentation, were consistent with previous biochemical characterization of IFT subcomplexes (Cole et al., 1998), which indicated that they were individual complexes of about 17S. On the other hand, a substantial portion of IFT subcomplexes in flagella isolated from adhering gametes were released by freeze/thaw and could not be sedimented at low speed, but could be sedimented at high speed. In addition, the IFT subcomplexes behaved as larger assemblies upon velocity sedimentation analysis on sucrose gradients. These assemblies were disrupted by detergent, and no difference in the properties of the IFT subcomplexes in non-adhering and adhering flagella were observed if the samples were prepared in detergent. Thus, flagellar adhesion and signaling were associated with alterations in the biochemical properties of the PKG and the IFT subcomplexes, including their association with the flagellar membrane. These results suggested that the IFT subcomplexes participate directly in the signaling pathway.

G. A direct role for IFT in flagellum-generated signaling

The presence of the larger assemblies of IFT particles is open to at least two interpretations. One is that adhesion and agglutinin interactions lead to a change in the IFT subcomplexes that cause them newly to associate with each other and with the flagellar membrane. A second is that pre-existing assemblies of IFT subcomplexes, which have been shown by electron microscopy to be present in cilia and flagella, are somehow stabilized during adhesion. Although the aggregations (formerly called "rafts") of IFT subcomplexes are easily observed in electron microscopy, apparently they are held together by low affinity interactions (at least in vegetative cells and non-adhering gametes). Future studies to characterize the molecular composition of the larger assemblies should provide new insights into the role of IFT subcomplexes in signaling.

H. IFT-independent movement of agglutinins from the cell body to the flagella

A second observation raises the possibility of a kinesin-2-independent flagellar translocation system. As indicated above, even though the rate and extent of flagellar adhesion are normal in *fla10-1* gametes at 32°C, the cells fail to fuse, presumably because they do not increase their levels of cAMP. An additional consequence of the failure to increase cAMP is that the cells do not recruit additional agglutinins from the cell body to the flagella (Goodenough, 1989; Hunnicutt et al., 1990). Since, as noted above, the agglutinin molecules undergo rapid turnover during adhesion (Snell and Moore, 1980; Goodenough, 1989), the gametes disadhere from each other and become freely swimming single cells again unless agglutinins are replenished. Importantly, addition of db-cAMP to the disadhered gametes brought about rapid restoration of flagellar adhesion. Thus, the mechanism for agglutinin translocation from the cell body to the flagella is intact in the *fla10-1* mutants at 32°C (Pan and Snell, 2002). As noted above, *Chlamydomonas* flagella exhibit a second type of motility, called flagellar surface motility, which is visualized experimentally as the rapid movement of latex microspheres along the surface of the flagella (Bloodgood, 1995; and see Chapter 11). An appealing explanation for these observations, therefore, is that movement of membrane proteins from the cell body to flagella depends on this or another motility mechanism, which perhaps is powered by as yet unidentified motors.

IV. GAMETE FUSION

A. Overview

One of the most significant downstream events of flagellar adhesion-induced signaling is activation of plasma membrane sites on both gametes that are specialized for cell–cell fusion. During gamete activation, these membrane sites, called mating structures, undergo dramatic changes in

FIGURE 12.5

Mating structures of plus *gametes. (A) TEM of* plus *mating structure in unactivated (upper left), partially activated (lower left), and fully activated (right) forms (Wilson, 1996) with their membrane zone (mz), doublet zone (dz), and actin filaments (af). The small arrowheads in the lower left image point to membrane vesicles within the growing fertilization tubule.*
(B) Immunofluorescence image of fertilization tubules on activated plus *gametes stained for actin (green; red, chlorophylll autofluorescence).*
(Wilson, N. and Snell, W. J. unpublished.)

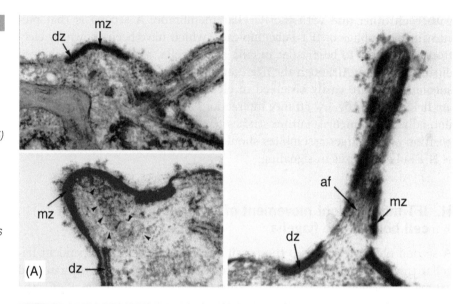

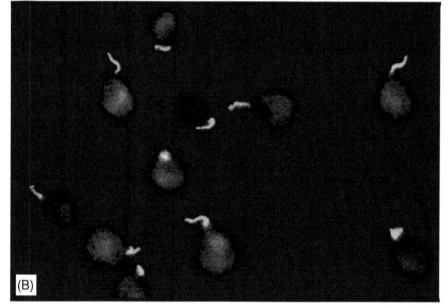

morphology and properties that enlarge them and render them competent to fuse (Figure 12.5). Almost immediately after activation of the mating structures, they fuse to establish cytoplasmic continuity between the two gametes, and the two cells merge into a QFC.

Cell fusion itself generates intracellular signals that have both short- and long-term effects. In the short term, and as with fertilization in other species, the fused gametes are no longer capable of interactions with other gametes. Fusion also activates gene transcription (Ferris and Goodenough,

1987) and initiates a new developmental pathway in the zygote, leading ultimately to meiosis to yield two *plus* and two *minus* haploid daughter (vegetative) cells (Figure 12.1). Intriguingly, *Chlamydomonas* has exploited the confluence of the previously separate cytoplasms of the two gametes to regulate zygote development. A nuclease activity (Umen and Goodenough, 2001; Nishimura et al., 2002) and a homeodomain protein (GSP1) (Zhao et al., 2001) are present only in the *plus* gamete, but the target DNA for the nuclease and the binding partner for the homeodomain protein (GSM1; Lee et al. 2008) are present only in the *minus* gamete (see Figure 12.1). After fusion, the nuclease selectively enters the *minus* chloroplast as a key step in uniparental inheritance of chloroplast traits (Armbrust et al., 1993; Nishimura et al., 2002; see also Volume 1, Chapter 7), and the homeodomain protein complex activates transcription of zygote-specific genes (see Volume 1, Chapter 5).

B. The *Chlamydomonas* mating structures

In thin sections, both *minus* and *plus* mating structures in unactivated gametes appear as small protrusions or buds at the apical ends of the cells (Goodenough and Weiss, 1975), where they associate with the two-membered microtubule "rootlet" of the basal body complex (Goodenough and Weiss, 1978, and see Chapter 2). The unactivated *plus* structure (see panel A in Figure 12.5) is about 0.3 μm in diameter and about 0.2 μm in length. The unactivated *minus* structure is somewhat smaller (not shown). The plasma membranes in the buds of cells of both mating types are lined on their cytoplasmic face with electron-dense material that is called the "membrane zone." In addition, the *plus* mating structure has a "doublet zone" beneath the membrane zone. In three dimensions, this doublet zone appears as a thin, slightly lopsided doughnut. The *minus* mating structure does not have a doublet zone. Additional differences between the mating types become apparent in the activated mating structures (Goodenough and Weiss, 1975; Goodenough et al., 1980, 1982; Detmers et al., 1983).

C. The activated *minus* mating structure

The activated *minus* structure appears as a small, wide dome, as if the plasma membrane had ballooned out at the apex of the original bud to about twice the height of the unactivated structure. The base of the dome retains its membrane zone, but the cytoplasmic face of the membrane at the apex of the dome is uncoated and the cytoplasm in the dome is relatively free of organelles (Goodenough and Weiss, 1975; Weiss et al., 1977; Goodenough and Weiss, 1978). No actin filaments have been detected in activated *minus* mating structures by thin-section electron microscopy or by staining with anti-actin antibodies or bodipy phallicidin (Detmers et al., 1985). By deep-etch electron microscopy, the activated *minus* mating structure displays large intramembranous particles not evident in unmated samples (Weiss et al., 1977).

D. The activated *plus* mating structure: the fertilization tubule

The activated *plus* mating structure, called the fertilization tubule, is an actin-filled, microvillus-like organelle that is 2–5 μm in length (Figure 12.5). Panel B of Figure 12.5 shows bodipy phallicidin-stained fertilization tubules on *plus* gametes that were activated with db-cAMP. Wilson et al. (1997) developed methods for isolating fertilization tubules and showed that they retain their ability to adhere to the mating structures of activated *minus* gametes.

E. *fus1*, a fusion-defective *plus* mutant

The *plus* mutant, *fus1* (Goodenough and Weiss, 1975), agglutinates normally with wild-type *minus* gametes, undergoes sexual signaling, makes seemingly normal mating structures, albeit missing the fuzzy coating called fringe (Goodenough et al., 1982), but fails to fuse. The *FUS1* gene is present only in the *MT+* mating-type locus (Ferris et al., 1996). While gametes containing the original *fus1-1* allele show a slight amount of leakiness, thereby permitting genetic analysis, gametes containing other *fus1* alleles are completely blocked in fusion (Ferris et al., 1996). The deduced amino-acid sequence of the FUS1 protein predicts an N-terminal signal peptide, a single transmembrane domain at the C-terminus, a short cytoplasmic tail, and several motifs for glycosylation (Ferris et al., 1996). FUS1 is not a conserved protein in eukaryotes (see below).

Misamore et al. (2003) used an antibody raised against a FUS1 peptide to identify endogenous FUS1 and learn more about its localization and function. The endogenous protein is ~95 kD. Immunofluorescence showed that it is localized to an apical patch at the site of the mating structure on unactivated gametes and re-distributes during gamete activation to become displayed over the entire surface of the microvillus-like activated *plus* mating structure. Results from experiments using trypsin treatment of live cells showed that FUS1 on both unactivated and activated gametes was localized on the external surface of the *plus* mating structure (Misamore et al., 2003). Thus, FUS1 is in the right place at the right time to be intimately involved in gamete fusion. Analysis of the amino-acid sequence of FUS1 showed that although it has no homologues in eukaryotes, it has several domains similar to prokaryotic Ig domains present in bacterial proteins – intimins and adhesins – used by *Escherichia coli* and other pathogenic prokaryotes to bind to and invade their eukaryotic host cells in the gut.

The availability of mutant *minus sad1-2* gametes (*imp12*) that do not express functional flagellar agglutinins (Ferris et al., 2005), and of the ability to activate gametes of a single mating type experimentally by incubating them in db-cAMP, made it possible to assess the function of FUS1 during gamete fusion. When activated wild-type *plus* gametes were mixed with activated *sad1-2 minus* gametes, they adhered to each other via their mating structures, as expected from earlier studies (Pasquale and Goodenough, 1987). On the other hand, activated *fus1 plus* gametes failed to bind via their mating structures to the activated *sad1-2 minus* gametes, thereby demonstrating that FUS1 is essential for the mating structure adhesion that is a prelude to gamete fusion. In addition, unactivated wild-type *plus* gametes whose walls were removed to expose their mating structures failed to bind to the activated *sad1-2 minus* gametes, suggesting that FUS1 must undergo a gamete activation-induced modification before it can function in mating-structure adhesion (Misamore et al., 2003).

V. EVOLUTIONARY PERSPECTIVES

Current phylogenies suggest that the eukaryotes engaged in eight major radiations, with the nature and placement of the original "root" eukaryote, and its relationship to subsequent radiations, a matter of uncertainty (Baldauf, 2003). Since modern members of all eight groups possess mitochondria or mitochondria-related organelles, there is general agreement that the common ancestor of modern eukaryotes possessed mitochondria (Embley and Martin, 2006). All eight groups prove to share a second commonality as well: representatives of each group have been observed to engage in a sexual cycle, either obligately or facultatively (Dacks and Roger, 1999), or else to possess a highly conserved set of genes that are selectively expressed during meiosis (Ramesh et al., 2005). This suggests that the common ancestor to modern eukaryotes was both mitochondrial and sexual, engaging in haploid-diploid transitions that were followed by meiotic reduction. A third commonality is the possession of motile or sensory cilia/flagella (Mitchell, 2004), a trait lost by fungi and angiosperms but displayed by other members of their respective radiations.

Solter and Gibor (1977) were the first to point out that the gametic flagella of *Chlamydomonas* engage in both motility and sensory transduction, and research in the ensuing decades, summarized in this chapter, has provided ample documentation for that inference. Jékely & Arendt (2006) have provocatively suggested that flagella originated as a differentiated sensory membrane patch, endowed with proteins in the COPI-clathrin families that later evolved into IFT components. While they and Mitchell (2004) note that the signals perceived by this patch might have entailed light, chemical, or mechanical cues, they are particularly interested in the possibility that the patch initially participated in mate recognition and sexual signaling during fertilization. They posit that a microtubule-based protrusion of this patch then augmented exposure of such recognition/signaling components, and only later did this protrusion acquire an axoneme, and the capacity for motility, in some unicellular lineages and cell types.

If this scenario is correct, then *Chlamydomonas* may be displaying a truly ancient use of cilia, a use reflected in the role of cilia as signaling organelles during mating reactions of certain ciliated protozoa (Kitamura and Hiwatashi, 1976; Brown et al., 1993; Plumper et al., 1995) and in the role of sperm flagella as responders to chemotaxis signals in multicellular animals (Spehr et al., 2006). It is intriguing to speculate, therefore, that the original "root" eukaryote, when eventually reconstructed from genomic inferences, may in fact prove to be very similar indeed to the modern *Chlamydomonas*. Given that plastids are often lost during evolution (Embley and Martin, 2006), this organism might well even have been green.

REFERENCES

Adair, W.S. (1985). Characterization of *Chlamydomonas* sexual agglutinins. *J. Cell Sci.*, 233–260.

Adair, W.S., Monk, B.C., Cohen, R., Hwang, C., and Goodenough, U.W. (1982). Sexual agglutinins from the *Chlamydomonas* flagellar membrane. Partial purification and characterization. *J. Biol. Chem.* **257**, 4593–4602.

Adair, W.S., Hwang, C., and Goodenough, U.W. (1983). Identification and visualization of the sexual agglutinin from the mating-type plus flagellar membrane of *Chlamydomonas*. *Cell* **33**, 183–193.

Armbrust, E.V., Ferris, P.J., and Goodenough, U.W. (1993). A mating type-linked gene cluster expressed in *Chlamydomonas* zygotes participates in the uniparental inheritance of the chloroplast genome. *Cell* **74**, 801–811.

Baldauf, S.L. (2003). The deep roots of eukaryotes. *Science* **300**, 1703–1706.

Bloodgood, R.A. (1995). Flagellar surface motility: gliding and microsphere movements. *Meth. Cell Biol.* **47**, 273–279.

Brown, F., Tirone, S., and Wolfe, J. (1993). Early encounters of the repetitive kind: a prelude to cell adhesion in conjugating *Tetrahymena thermophila*. *Dev. Dyn.* **196**, 195–204.

Cassab, G.I. (1998). Plant cell wall proteins. *Annu. Rev. Plant Physiol. Plant. Mol. Biol.* **49**, 281–309.

Cole, D.G., Diener, D.R., Himelblau, A.L., Beech, P.L., Fuster, J.C., and Rosenbaum, J.L. (1998). *Chlamydomonas* kinesin-II-dependent intraflagellar transport (IFT): IFT particles contain proteins required for ciliary assembly in *Caenorhabditis elegans* sensory neurons. *J. Cell Biol.* **141**, 993–1008.

Collin-Osdoby, P. and Adair, W.S. (1985). Characterization of the purified *Chlamydomonas* minus agglutinin. *J. Cell Biol.* **101**, 1144–1152.

Collin-Osdoby, P., Adair, W.S., and Goodenough, U.W. (1984). *Chlamydomonas* agglutinin conjugated to agarose beads as an in vitro probe of adhesion. *Exp. Cell Res.* **150**, 282–291.

Cooper, J.B., Adair, W.S., Mecham, R.P., and Heuser, J.E. (1983). *Chlamydomonas* agglutinin is a hydroxyproline-rich glycoprotein. *Proc. Natl. Acad. Sci. U. S. A.* **80**, 5898–5901.

Dacks, J. and Roger, A.J. (1999). The first sexual lineage and the relevance of facultative sex. *J. Mol. Evol.* **48**, 779–783.

Detmers, P.A., Goodenough, U.W., and Condeelis, J. (1983). Elongation of the fertilization tubule in *Chlamydomonas*: new observations on the core microfilaments and the effect of transient intracellular signals on their structural integrity. *J. Cell Biol.* **97**, 522–532.

Detmers, P.A., Carboni, J.M., and Condeelis, J. (1985). Localization of actin in *Chlamydomonas* using antiactin and NBD-phallacidin. *Cell Motil.* **5**, 415–430.

Embley, T.M. and Martin, W. (2006). Eukaryotic evolution, changes and challenges. *Nature* **440**, 623–630.

Ferris, P.J. and Goodenough, U.W. (1987). Transcription of novel genes, including a gene linked to the mating-type locus, induced by *Chlamydomonas* fertilization. *Mol. Cell Biol.* **7**, 2360–2366.

Ferris, P.J. and Goodenough, U.W. (1994). The mating-type locus of *Chlamydomonas reinhardtii* contains highly rearranged DNA sequences. *Cell* **76**, 1135–1145.

Ferris, P.J. and Goodenough, U.W. (1997). Mating type in *Chlamydomonas* is specified by mid, the minus-dominance gene. *Genetics* **146**, 859–869.

Ferris, P.J., Woessner, J.P., and Goodenough, U.W. (1996). A sex recognition glycoprotein is encoded by the *plus* mating-type gene *fus1* of *Chlamydomonas reinhardtii*. *Mol. Biol. Cell* **7**, 1235–1248.

Ferris, P.J., Armbrust, E.V., and Goodenough, U.W. (2002). Genetic structure of the mating-type locus of *Chlamydomonas reinhardtii*. *Genetics* **160**, 181–200.

Ferris, P.J., Waffenschmidt, S., Umen, J.G., Lin, H., Lee, J.H., Ishida, K., Kubo, T., Lau, J., and Goodenough, U.W. (2005). Plus and minus sexual agglutinins from *Chlamydomonas reinhardtii*. *Plant Cell* **17**, 597–615.

Gloeckner, G. and Beck, C.F. (1995). Genes involved in light control of sexual differentiation in *Chlamydomonas reinhardtii*. *Genetics* **141**, 937–943.

Goodenough, U.W. (1986). Experimental analysis of the adhesion reaction between isolated *Chlamydomonas* flagella. *Exp. Cell Res.* **166**, 237–246.

Goodenough, U.W. (1989). Cyclic AMP enhances the sexual agglutinability of *Chlamydomonas* flagella. *J. Cell Biol.* **109**, 247–252.

Goodenough, U.W. (1993). Tipping of flagellar agglutinins by gametes of *Chlamydomonas reinhardtii*. *Cell Motil. Cytoskeleton* **25**, 179–189.

Goodenough, U.W. and Heuser, J.E. (1988). Molecular organization of cell-wall crystals from *Chlamydomonas reinhardtii* and *Volvox carteri*. *J. Cell Sci.* **90**, 717–733.

Goodenough, U.W. and Heuser, J. (1999). Deep-etch analysis of adhesion complexes between gametic flagellar membranes of *Chlamydomonas reinhardtii* (Chlorophyceae). *J. Phycol.* **35**, 756–767.

Goodenough, U.W. and Jurivich, D. (1978). Tipping and mating-structure activation induced in *Chlamydomonas* gametes by flagellar membrane antisera. *J. Cell Biol.* **79**, 680–693.

Goodenough, U.W. and Weiss, R.L. (1975). Gametic differentiation in *Chlamydomonas reinhardtii*. III. Cell wall lysis and microfilament-associated mating structure activation in wild-type and mutant strains. *J. Cell Biol.* **67**, 623–637.

Goodenough, U.W. and Weiss, R.L. (1978). Interrelationships between microtubules, a striated fiber, and the gametic mating structure of *Chlamydomonas reinhardi*. *J. Cell Biol.* **76**, 430–438.

Goodenough, U.W., Hwang, C., and Martin, H. (1976). Isolation and genetic analysis of mutant strains of *Chlamydomonas reinhardi* defective in gametic differentiation. *Genetics* **82**, 169–186.

Goodenough, U.W., Hwang, C., and Warren, A.J. (1978). Sex-limited expression of gene loci controlling flagellar membrane agglutination in the *Chlamydomonas* mating reaction. *Genetics* **89**, 235–243.

Goodenough, U.W., Adair, W.S., Caligor, E., Forest, C.L., Hoffman, J.L., Mesland, D.A., and Spath, S. (1980). Membrane–membrane and membrane–ligand interactions in *Chlamydomonas* mating. *Soc. Gen. Physiol. Ser.* **34**, 131–152.

Goodenough, U.W., Detmers, P.A., and Hwang, C. (1982). Activation for cell fusion in *Chlamydomonas*: analysis of wild-type gametes and nonfusing mutants. *J. Cell Biol.* **92**, 378–386.

Goodenough, U.W., Adair, W.S., Collin-Osdoby, P., and Heuser, J.E. (1985). Structure of the *Chlamydomonas* agglutinin and related flagellar surface proteins *in vitro* and *in situ*. *J. Cell Biol.* **101**, 924–941.

Goodenough, U.W., Armbrust, E., and Campbell, Ferris,P. (1995). Molecular genetics of sexuality in *Chlamydomonas*. *Annu. Rev. Plant Physiol.* **46**, 21–44.

Goodenough, U., Lin, H., and Lee, J.H. (2007). Sex determination in *Chlamydomonas*. *Semin. Cell Dev. Biol.* **18**, 350–361.

Hoffman, J.L. and Goodenough, U.W. (1980). Experimental dissection of flagellar surface motility in *Chlamydomonas*. *J. Cell Biol.* **86**, 656–665.

Hunnicutt, G.R. and Snell, W.J. (1991). Rapid and slow mechanisms for loss of cell adhesiveness during fertilization in *Chlamydomonas*. *Dev. Biol.* **147**, 216–224.

Hunnicutt, G.R., Kosfiszer, M.G., and Snell, W.J. (1990). Cell body and flagellar agglutinins in *Chlamydomonas reinhardtii*: the cell body plasma membrane is a reservoir for agglutinins whose migration to the flagella is regulated by a functional barrier. *J. Cell Biol.* **111**, 1605–1616.

Hwang, C., Monk, B.C., and Goodenough, U.W. (1981). Linkage of mutations affecting minus flagellar membrane agglutinability to the *mt−* mating type locus of *Chlamydomonas*. *Genetics* **99**, 41–47.

Jékely, G. and Arendt, D. (2006). Evolution of intraflagellar transport from coated vesicles and autogenous origin of the eukaryotic cilium. *Bioessays* **28**, 191–198.

Kitamura, A. and Hiwatashi, K. (1976). Mating-reactive membrane vesicles from cilia of *Paramecium caudatum*. *J. Cell Biol.* **69**, 736–740.

Kozminski, K.G., Beech, P.L., and Rosenbaum, J.L. (1995). The *Chlamydomonas* kinesin-like protein FLA10 is involved in motility associated with the flagellar membrane. *J. Cell Biol.* **131**, 1517–1527.

Kubo, T., Saito, T., Fukuzawa, H., and Matsuda, Y. (2001). Two tandemly-located matrix metalloprotease genes with different expression patterns in the *Chlamydomonas* sexual cell cycle. *Curr. Genet.* **40**, 136–143.

Lee, J.H., Waffenschmidt, S., Small, L., and Goodenough, U. (2007). Between-species analysis of short-repeat modules in cell wall and sex-related hydroxyproline-rich glycoproteins of *Chlamydomonas*. *Plant Physiol.* **144**, 1813–1826.

Lee, J-H., Lin, H., Joo, S., and Goodenough, U. (2008) Early sexual origins of homeoprotein heterodimerization and evolution of the plant KNOX/BELL family. *Cell* **133**, 829–840.

Lin, H. and Goodenough, U.W. (2007). Gametogenesis in the *Chlamydomonas reinhardtii minus* mating type is controlled by two genes, *MID* and *MTD1*. *Genetics* **176**, 913–925.

Lux, F.G., III and Dutcher, S.K. (1991). Genetic interactions at the FLA10 locus: suppressors and synthetic phenotypes that affect the cell cycle and flagellar function in *Chlamydomonas reinhardtii*. *Genetics* **128**, 549–561.

Matsuda, Y., Sakamoto, K., Misuochi, T., Kobata, A., Tamura, G., and Tsubo, Y. (1981). Mating type specific inhibition of gametic differentiation of *Chlamydomonas reinhardtii* by tunicamycin. *Plant. Cell Physiol.* **22**, 1607–1611.

Mesland, D.A., Hoffman, J.L., Caligor, E., and Goodenough, U.W. (1980). Flagellar tip activation stimulated by membrane adhesions in *Chlamydomonas* gametes. *J. Cell Biol.* **84**, 599–617.

Misamore, M.J., Gupta, S., and Snell, W.J. (2003). The *Chlamydomonas* Fus1 protein is present on the mating type plus fusion organelle and required for a critical membrane adhesion event during fusion with minus gametes. *Mol. Biol. Cell* **14**, 2530–2542.

Mitchell, D.R. (2004). Speculations on the evolution of 9 + 2 organelles and the role of central pair microtubules. *Biol. Cell* **96**, 691–696.

Nishimura, Y., Misumi, O., Kato, K., Inada, N., Higashiyama, T., Momoyama, Y., and Kuroiwa, T. (2002). An mt^+ gamete-specific nuclease that targets mt^- chloroplasts during sexual reproduction in *C. reinhardtii*. *Genes. Dev.* **16**, 1116–1128.

Pan, J. and Snell, W.J. (2002). Kinesin-II is required for flagellar sensory transduction during fertilization in *Chlamydomonas*. *Mol. Biol. Cell* **13**, 1417–1426.

Pan, J., Wang, Q., and Snell, W.J. (2005). Cilium-generated signaling and cilia-related disorders. *Lab. Invest.* **85**, 452–463.

Pasquale, S.M. and Goodenough, U.W. (1987). Cyclic AMP functions as a primary sexual signal in gametes of *Chlamydomonas reinhardtii*. *J. Cell Biol.* **105**, 2279–2292.

Pijst, H.L.A., van Driel, R., Janssens, P.M.W., Musgrave, A., and van den Ende, H. (1984). Cyclic AMP is involved in sexual reproduction of *Chlamydomonas eugametos*. *FEBS Lett.* **174**, 132–136.

Piperno, G., Mead, K., and Henderson, S. (1996). Inner dynein arms but not outer dynein arms require the activity of kinesin homologue protein KHP1(FLA10) to reach the distal part of flagella in *Chlamydomonas*. *J. Cell Biol.* **133**, 371–379.

Plumper, E., Freiburg, M., and Heckmann, K. (1995). Conjugation in the ciliate *Euplotes octocarinatus*: comparison of ciliary and cell body-associated glycoconjugates of non-mating-competent, mating-competent, and conjugating cells. *Exp. Cell Res.* **217**, 490–496.

Ramesh, M.A., Malik, S.B., and Logsdon, J.M., Jr. (2005). A phylogenomic inventory of meiotic genes; evidence for sex in *Giardia* and an early eukaryotic origin of meiosis. *Curr. Biol.* **15**, 185–191.

Roberts, K., Grief, C., Hills, G.J., and Shaw, P.J. (1985). Cell wall glycoproteins: structure and function. *J. Cell Sci.*, Suppl. 2 105–127.

Rohatgi, R. and Scott, M.P. (2007). Patching the gaps in Hedgehog signalling. *Nat. Cell Biol.* **9**, 1005–1009.

Saito, T., Tsubo, Y., and Matsuda, Y. (1985). Synthesis and turnover of cell body agglutinin as a pool of flagellar surface agglutinin in *Chlamydomonas reinhardtii* gamete. *Arch. Microbiol.* **142**, 207–210.

Saito, T., Small, L., and Goodenough, U.W. (1993). Activation of adenylyl cyclase in *Chlamydomonas reinhardtii* by adhesion and by heat. *J. Cell Biol.* **122**, 137–147.

Snell, W.J. (1976). Mating in *Chlamydomonas*: a system for the study of specific cell adhesion. II. A radioactive flagella-binding assay for quantitation of adhesion. *J. Cell Biol.* **68**, 70–79.

Snell, W.J. (1981). Flagellar adhesion and deadhesion in *Chlamydomonas* gametes: effects of tunicamycin and observations on flagellar tip morphology. *J. Supramol. Struct. Cell. Biochem.* **16**, 371–376.

Snell, W.J. and Moore, W.S. (1980). Aggregation-dependent turnover of flagellar adhesion molecules in *Chlamydomonas* gametes. *J. Cell Biol.* **84**, 203–210.

Snell, W.J. and Roseman, S. (1979). Kinetics of adhesion and de-adhesion of *Chlamydomonas* gametes. *J. Biol. Chem.* **254**, 10820–10829.

Solter, K.M. and Gibor, A. (1977). Evidence for role of flagella as sensory transducers in mating of *Chlamydomonas reinhardi*. *Nature* **265**, 444–445.

Spehr, M., Schwane, K., Riffell, J.A., Zimmer, R.K., and Hatt, H. (2006). Odorant receptors and olfactory-like signaling mechanisms in mammalian sperm. *Mol. Cell Endocrinol.* **250**, 128–136.

Swanson, W.J. and Vacquier, V.D. (2002). The rapid evolution of reproductive proteins. *Nat. Rev. Genet.* **3**, 137–144.

Umen, J.G. and Goodenough, U.W. (2001). Chloroplast DNA methylation and inheritance in *Chlamydomonas*. *Genes Dev.* **15**, 2585–2597.

Vallon, O. and Wollman, F.A. (1995). Mutations affecting O-glycosylation in *Chlamydomonas reinhardtii* cause delayed cell wall degradation and sex-limited sterility. *Plant Physiol.* **108**, 703–712.

van den Ende, H. (1992). Sexual signalling in Chlamydomonas. In: *Cellular Recognition* (J.A. Callow and J.R. Green, Eds.), pp. 1–19. Cambridge University Press, Cambridge.

Versluis, M., Schuring, F., Klis, F.M., Vanegmond, P., and van den Ende, H. (1992). Sexual agglutination in *Chlamydomonas eugametos* is mediated by a single pair of hydroxyproline-rich glycoproteins. *FEMS Microbiol. Lett.* **97**, 101–105.

Wang, Q. and Snell, W.J. (2003). Flagellar adhesion between mating type plus and mating type minus gametes activates a flagellar protein-tyrosine kinase during fertilization in *Chlamydomonas*. *J. Biol. Chem.* **278**, 32936–32942.

Wang, Q., Pan, J., and Snell, W.J. (2006). Intraflagellar transport particles participate directly in cilium-generated signaling in *Chlamydomonas*. *Cell* **125**, 549–562.

Weiss, R.L., Goodenough, D.A., and Goodenough, U.W. (1977). Membrane differentiations at sites specialized for cell fusion. *J. Cell Biol.* **72**, 144–160.

Wilson, N.F. (1996). The Chlamydomonas mt+ fertilization tubule: a model system for studying the role of cell fusion organelles in gametic cell fusion. Ph. D. Thesis. University of Texas Southwestern Medical Center, Dallas, Texas.

Wilson, N.F., Foglesong, M.J., and Snell, W.J. (1997). The *Chlamydomonas* mating type plus fertilization tubule, a prototypic cell fusion organelle: isolation, characterization, and in vitro adhesion to mating type minus gametes. *J. Cell Biol.* **137**, 1537–1553.

Zhang, Y. and Snell, W.J. (1993). Differential regulation of adenylylcyclases in vegetative and gametic flagella of *Chlamydomonas*. *J. Biol. Chem.* **268**, 1786–1791.

Zhang, Y.H., Ross, E.M., and Snell, W.J. (1991). ATP-dependent regulation of flagellar adenylyl cyclase in gametes of *Chlamydomonas reinhardtii*. *J. Biol. Chem.* **266**, 22954–22959.

Zhao, H., Lu, M., Singh, R., and Snell, W.J. (2001). Ectopic expression of a *Chlamydomonas* mt + -specific homeodomain protein in mt- gametes initiates zygote development without gamete fusion. *Genes Dev.* **15**, 2767–2777.

CHAPTER 13

Sensory Photoreceptors and Light Control of Flagellar Activity

Peter Hegemann and Peter Berthold
Institute for Biology, Experimental Biophysics
Humboldt University, Berlin, Germany

CHAPTER CONTENTS

I. BEHAVIORAL LIGHT RESPONSES

A. Definition and strains

Chlamydomonas displays distinct behavioral responses to light. The most basic responses are directional changes of swimming upon absorption of single photons (Hegemann and Marwan, 1988). At low light, these infrequent events are of purely statistical nature. But, since the *Chlamydomonas* eye is a directional device (see section II), absorption of light is more likely if the eye is facing toward the light source. This bias leads to net movement towards the light source; this movement is traditionally named phototaxis. At high light intensities, when light absorption occurs frequently, responses to individual photon absorptions can no longer be distinguished and the cells swim more or less directly to the light or away from it (positive or negative phototaxis). Phototaxis of microalgae at high light has been observed for at least two centuries (Trevianus, 1817; Buder, 1917). *Chlamydomonas* cells displaying strong negative phototaxis are shown in the video "Chlamy_phototaxis_and_photoshock_Witman_lab" at http://www.elsevierdirect.com/companions/9780123708731.

The switch between positive and negative phototaxis is not fully understood, but the ion composition of the medium and efficiency of photosynthesis are the major factors that affect it (Takahashi et al., 1992; Takahashi and Watanabe, 1993; Hoops et al., 1999). Strains like 137c mt^+ are negatively phototactic to blue/green actinic light in the presence of red background light ($\lambda > 620$ nm), whereas they swim towards blue/green light without background illumination (Takahashi and Watanabe, 1993). In most strains, clean positive phototaxis is visible only when respiration and photosynthesis are rather limited. Both respiration and photosynthesis influence the swimming speed and consequently the phototactic rate (net swimming rate towards the light), a phenomenon named photokinesis (Buder, 1917)

In addition to positive and negative phototaxis, *Chlamydomonas* cells undergo a stop response, also termed photoshock or photophobic response, when suddenly exposed to very bright light. In this response, the cells transiently switch from normal forward swimming to backwards swimming (Hegemann and Bruck, 1989). Photoshock is demonstrated in the videos "Chlamy_photoshock_response_Hegemann_lab" and "Chlamy_phototaxis_and_photoshock_Witman_lab" at http://www.elsevierdirect.com/companions/9780123708731.

Table 13.1 lists the principal *C. reinhardtii* strains used for analysis of photoresponses. Two derivatives of the wild-type strain 137c have been used for many behavioral studies during the past three decades: strain 806 isolated by R.D. Smyth, and strain 495 isolated by A.S. Chunaev, St. Petersburg. Cells of these strains swim fast ($V > 120$ μm/second), have a large eye, and a high phototactic sensitivity (defined below). For complementation experiments with retinal or retinal analogues, the most widely

Table 13.1	Principal strains used for analysis of photoresponses		
Strain number[a]	**Other names**	**Properties**	**References**
CC-125	Strain 137c *mt*[+]	Most frequently used wild-type strain	see Volume 1, Chapter 1
CC-124	Strain 137c *mt*[−]	Contains *agg1* mutation that modifies phototactic accumulation	see Volume 1, Chapters 1 and 4; also Zacks and Spudich (1994)
CC-2682	FN68	Contains *fn68* mutation, an allele at the *PSY1* locus; blocked in carotenoid synthesis, very light-sensitive	Foster et al. (1984), Lawson and Satir (1994), McCarthy et al. (2004)
CC-2359		Contains *lts1-30* mutation, an allele at the *PSY1* locus; blocked in carotenoid synthesis, very light-sensitive	Iroshnikova et al. (1982), Lawson and Satir (1994), McCarthy et al. (2004)
	Smyth strain 806	Swims fast, has a large eye, and high phototactic sensitivity	Isolated by R.D. Smyth; Hegemann et al. (1988)
	Chunaev strain 495	Swims fast, has a large eye, and high phototactic sensitivity	Sineshchekov et al. (1992)
CC-851	CW2	Contains *cw2* mutation, lacks cell wall	Davies and Plaskitt (1971), Hyams and Davies (1972)

[a]*Chlamydomonas Resource Collection; see http://www.chlamycollection.org*

used strains are the carotenoid- and chlorophyll-deficient mutants *fn68* (also known as *car-1* or *F1*, or as the FN68 strain), isolated by W.-Y. Wang, and *lts1-30*, isolated by A.S. Chunaev. These are allelic mutations at the *LTS1* locus, corresponding to the *PSY1* gene encoding phytoene synthase (McCarthy et al., 2004). Both *fn68* and *lts1-30* are almost white, have shorter flagella than wild-type cells, and swim with a speed below $80 \mu m$/second. The cell wall-less strain CW2 has been used for some types of experiments; the lack of a cell wall allows single cell recording and facilitates nuclear transformation by cloned DNA. In earlier studies many different wild-type isolates were used as well.

B. Methods for behavioral analysis

1. Studies on cell populations

The dish test, "reinvented" by Foster (Foster et al., 1984), is the most simple but efficient assay for testing phototaxis without specialized equipment (Figure 13.1). Cells with a density of 1×10^7 per ml are poured into a small Petri dish with a diameter of 3.5 cm. Cells are illuminated from one side using a slide projector equipped with monochromatic interference filters. After 10 minutes, the distance the cells have traveled towards or away from the light is determined and plotted versus the log light intensity (photon irradiance, expressed in W m^{-2} or photons m^{-2} s^{-1}). Extrapolation of the linear part of the resulting stimulus-response curve (dose-response curve) to zero response gives the threshold light intensity. The reciprocal of the

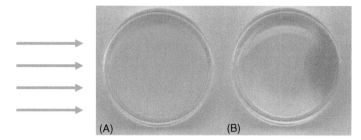

FIGURE 13.1 *Negative phototaxis of* Chlamydomonas *strain CW2 gametes measured in a 3.5-cm petri dish. Photographs were taken immediately after white light was switched on (A) and after 6 minutes of light (B). Arrows indicate the light direction.*

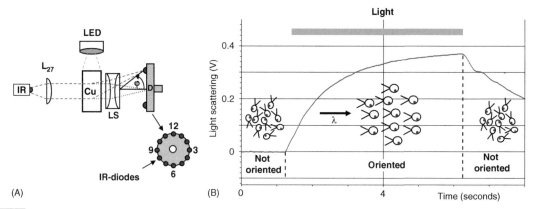

FIGURE 13.2 *(A) Schematic representation of the light scattering system. IR infrared emitter: λ = 850 nm monitoring light, L27 achromatic collector lens f = 27 mm, Cu cuvette, LED light-emitting diode for actinic light, LS achromatic collector lens-system f = 35 mm, φ angle of scattered light, D detector ring of 12 infrared-sensitive photodiodes. (Adapted from Schaller et al., 1997.) (B) A light scattering signal from a cw15 culture before, during, and after a uni-lateral light pulse.*

threshold, i.e., the sensitivity of the cell at a given wavelength, is plotted on a log scale versus the wavelength (or the energy) resulting in a threshold action spectrum that reflects the absorption of the photoreceptor mediating the observed response (Foster et al., 1984; Hegemann et al., 1988).

An alternative approach to behavioral studies of populations of cells uses light scattering. Cells in a cuvette are illuminated with infrared light and the transmitted light is focused beyond the cuvette to form a single spot. In contrast, the light scattered by the cells is focused to form rings surrounding the central spot and is detected by infrared photodiodes (Figure 13.2A). Reorientation of asymmetric cells such as *Chlamydomonas* can be monitored with high time resolution and high signal-to-noise ratio (Figure 13.2B). Such an assay has been used for recording directional changes, stop responses, and phototaxis from *Chlamydomonas* cell populations. The recordings have shown that the responses of *Chlamydomonas* are multi-phasic. The cells show only smooth positive or negative phototaxis at low light. At higher light, very often an initial positive phototaxis is followed by negative phototaxis. During the stop response, the flagella switch from a breast-stroke swimming style, in which asymmetrical bends are generated, to an undulatory type of waveform, in which symmetrical bends are formed and propagated, causing the cells to swim backwards. This typically persists for 300–500 milliseconds. The switch is caused by a massive voltage-activated Ca^{2+} influx into the flagella. Whereas the switch from forward to backward swimming is precisely synchronized, the switch from backward to forward swimming is not. After the stop response, the cells go through an adaptation phase during which they perform many turns before clear phototaxis is resumed. After light-off, the cells perform step-down responses consisting of large loops and extra large beats of one of the flagella before the flagella beat again more or less in synchrony (Schmidt and Eckert, 1976; Rüffer and Nultsch, 1990, 1995; Holland et al., 1997). Light scattering is a sensitive means of monitoring behavior of a population of cells, e.g., a mutant strain or transformant under various light and ionic conditions. However, it gives no information about the beating mode or the signaling defect itself. For extended phototaxis measurements, a modified set-up is recommended. The scattered image of the measuring cuvette is focused on a quadrant detector in such a way that the movement of cells from one side to the other is recorded with high resolution. Nultsch et al. (1971) described an apparatus for automatic recording of phototaxis, but this system was based on absorption and required high cell densities for a good signal.

2. Individual cells

The largest contribution to our understanding of *Chlamydomonas* flagellar movement and beating changes upon light stimulation as the basis of phototaxis came from the observations of Ursula Rüffer using high-speed cinematography (Rüffer and Nultsch, 1985–1998). Recording at a frequency of up to 500 frames/second, she monitored the flagellar beating of both freely swimming cells and cells held with a micropipette. She observed that the *trans*-flagellum beats with a 30% higher frequency than the *cis*-flagellum; this was shown by Takada and Kamiya (1997) to be caused by differences in the outer dynein arms or the outer dynein arm docking complex (see section IV.D.4.a in Chapter 6). When Rüffer stimulated the cells with green light pulses of 1-Hz frequency, the response consisted of inverse shifts of the frontal flagellar beat amplitude in response to light step-up (light intensity increase) and step-down (light intensity decrease) (Figure 13.3). When the eyespot of a positively phototactic cell comes to face the light source as a result of the normal rotation of the cell body that occurs during forward swimming, light absorption causes a decrease in the front amplitude of the *cis*-flagellum (the flagellum nearest

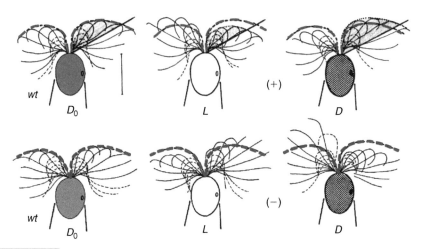

FIGURE 13.3 *Front amplitude changes by 1 Hz step-up and step-down light stimulation. (Upper diagrams) In a positively phototactic cell, the front amplitude of the* trans-*flagellum is increased in the light (L) and that of the* cis-*flagellum reduced. (Lower diagrams) In contrast, in a negatively phototactic cell, the front amplitude of the* cis-*flagellum is increased and that of the* trans-*flagellum is decreased. Front amplitudes in dark-adapted cells, D_0, are indicated by orange-dashed lines. (Modified from Rüffer and Nultsch, 1991.)*

the eyespot) and an increase in the front amplitude of the *trans*-flagellum. This causes the cell to turn toward the light. The opposite occurs in negatively phototactic cells.

After about half a rotation, when the eyespot faces away from the light, the cell undergoes a light-down response resulting in a turn in the opposite direction with respect to the cell, but in the same direction with respect to space since the cell has rotated by 180°. The responses of pipette-held cells monitored upon light step-up and step-down come close to those of freely swimming cells where the light intensity changes smoothly and in a quasi-sinusoidal fashion during cell rotation, and were interpreted as the basis for phototaxis. However, the results describe only high-intensity responses since the visual system is always light-adapted. The rhodopsins absorb red photons of the intensive monitoring light with significant efficiency. Nevertheless, the beating differences of the two flagella were lost in the motile but non-phototactic *ptx1* mutant (Horst and Witman, 1993; Rüffer and Nultsch, 1997), providing convincing support for the interpretation.

In subsequent studies, the beat frequency and stroke velocity of the flagella of pipette-held cells were measured by monitoring beating via two light guides that recorded the infrared light from two 1-μm spots, from a light-guide bundle recording beating from many 1-μm spots, or by projecting a dark-field image of the cell on a quadrant detector that records the differential response produced in different quadrants. These opto-electrical measurements provided a better sensitivity than high-speed cinematography, and the monitoring light could be restricted to wavelengths of $\lambda > 800$ nm. Finally, more cells could be analyzed per unit time (Holland et al., 1997; Josef et al., 2005a,b). However, these methods provided little additional information about the three-dimensional form of the flagellar beat; for the most part, the results confirmed Rüffer's earlier findings. The major advantage is that opto-electrical measurements may be combined with electrical recordings, thus allowing a better understanding of how the observed beating changes are triggered by external light stimuli and subsequent ion channel activities (see section III).

Individual cells also can be studied using motion analysis systems. Video data are collected at a rate of 15–60 frames/second and digitized by the computer. For each cell, the speed, the rate of change of direction, and the mean direction of traveling may be calculated (Hegemann and Bruck, 1989; Lawson et al., 1991; Zacks et al., 1993; Zacks and Spudich, 1994). Motion analysis is powerful in the sense that it monitors a two-dimensional projection of the swimming. It allows monitoring of individual directional changes after a low-intensity flash, and a photophobic response upon flashing with bright light. Upon step-up stimulation, it records the twiddling and dancing during the adaptation period that follows a phobic response, and the large loops *Chlamydomonas* cells perform after light step-down (Hegemann and Bruck, 1989). Finally, data may be collected on large numbers of cells and pooled to give an idea of the variability of the responses within the population. Motion analysis has been widely used to study the behavior of individual white cells after reconstitution with retinal or retinal compounds (Lawson et al., 1991; Zacks et al., 1993; Zacks and Spudich, 1994), or of mutant cells with phototactic defects (Pazour et al., 1995). However, responses in dim light are not readily detected by motion analysis. The signal from a population in which cells respond differently or only a small percentage of cells respond is weak. In this respect, population methods like the dish test or light scattering that monitor the net movement of a population over time are more sensitive. However, without the information from single cells, population assays are difficult to interpret.

II. THE *CHLAMYDOMONAS* EYE

The eyespot (termed stigma in the older literature) of *Chlamydomonas* is ~1 μm in size. In bright-field microscopy, it is seen peripherally near the cell's equator as a conspicuous, single orange-red spot (Figure 13.4A). The ultrastructure of the functional eyespot involves local specializations of membranes from different compartments (Melkonian and Robenek, 1984). In *C. reinhardtii*, the eyespot is usually composed of two highly ordered layers of carotenoid-rich lipid globules inside the chloroplast (also see Volume 1, Chapter 2). The globules are hexagonally closely packed and have a diameter of 80–130 nm. Each layer is subtended by a thylakoid. Additionally, the

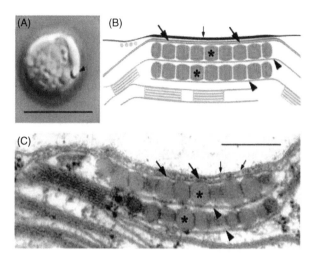

FIGURE 13.4 *(A) Image of a* Chlamydomonas *cell with the carotenoid-enriched eyespot structure (arrowhead); bar = 10 µm. (B) Schematic drawing of the eyespot apparatus. Asterisks indicate the carotenoid-rich eyespot globule layers inside the chloroplast, which are associated with the thylakoids (arrowheads). The plasma membrane (small arrow) is attached to the chloroplast envelope (large arrows) in the region overlying the eyespot globule layers. (C) Transmission electron micrograph of the eyespot apparatus of* Chlamydomonas. *Arrows and asterisks same as in B, bar = 300 µm. (Schmidt et al., 2006.)*

outermost globule layer is attached to specialized areas of the two chloroplast envelope membranes and the adjacent plasma membrane (Figure 13.4B, C). Freeze-fracturing revealed that the plasma membrane and the outer chloroplast envelope membrane above the eyespot globules are rich in intramembrane particles most likely representing membrane proteins or membrane protein complexes (Melkonian and Robenek, 1984). Several mutations affecting assembly of the eye and its positioning have been functionally described (Hartshorne, 1953; Morel-Laurens and Feinleib, 1983; Kreimer et al., 1992; Pazour et al., 1995; Lamb et al., 1999; Nakamura et al., 2001; Roberts et al., 2001; Dieckmann, 2003). In some cases the molecular origin of the defect remains unknown, but genes identified by several mutations encode proteins related to disulfide isomerases (Roberts et al., 2001), plastid-localized kinases, or membrane-associated proteins (T.M. Mittelmeier and C.L. Dieckmann, personal communication). The eye structure in the mutant strains (*fn68, lts1-30*) used in complementation experiments with retinal and retinal analogues have different defects (see Table 13.1).

Together with the photoreceptor molecules and the downstream amplification and/or adaptation system, the pigmented droplets form the functional eye. During rotation, the eye scans the environment for light intensity and quality. The location of the eye is of fundamental importance for receiving an interpretable light signal. During rotation, the eye is about 45° ahead of the flagellar beat plane (Rüffer and Nultsch, 1985), corresponding to about 40 milliseconds during 2-Hz cell rotation.

Modulation of direction of the light incidence with respect to the photoreceptor molecules does not automatically provide a strong modulation of the received light signal. Photon absorption by the photoreceptor molecules varies with the light incidence only if the cell produces a significant contrast. In fully transparent cells with randomly distributed photoreceptors, the contrast is low, the intensity is homogeneous during rotation, and the direction of the incident light cannot be detected. This is the case in bacteria, which are trapped in the light but do not perform real phototaxis in the sense of orienting towards a light source (Buder, 1917). In microalgae the photoreceptor is clearly localized (see below) and an associated optical system provides the directivity. Due to the small eye size, in most microalgae a lens system would provide only refraction and scattering and would not be beneficial. Ken Foster and Robert Smyth elegantly explained the directivity of the eye by proposing that the eyespot is a quarter-wave stack antenna – i.e., the light waves are reflected by the surfaces of the carotenoid vesicles and undergo constructive interference if the difference of the light path of two light waves reflected by adjacent layers is a multiple of $\lambda/2$, producing waves of interference maxima and minima (Land, 1978; Foster and Smyth, 1980). Since the brightest interference maximum is expected to coincide with the plasma membrane overlying the eyespot, the plasma membrane is the ideal location for the photoreceptor (Foster and Smyth, 1980; Melkonian and Robenek, 1984; Kreimer and Melkonian, 1990; Harz et al., 1992). Such a location provides a plausible means of communication with the flagella, since the plasma membrane is continuous with the flagellar membrane. Microspectrometric studies have confirmed and quantified the eyespot reflectivity. Surprisingly, the reflectivity maximum turned out to be in the range of 540 nm, which is clearly red-shifted relative to the phototaxis maximum (500 nm) (Schaller and Uhl, 1997; Schaller et al., 1997).

Reflection and interference increase with the number of layers. Some algae, such as *C. eugametos*, *Tetraselmis*, *Haematococcus* or gametes of the seawater algae *Acetabularia* (order Dasyclades) and *Ulva* (order Ulvales) have only one or two layers. These eyes operate as quarter-wave plates with relatively weak reflection and front-to-back contrast. Eyes of *C. reinhardtii* most frequently have two layers whereas *Hafniomonas reticulata* may have up to eight. Consequently, the contrast between light incident directly upon the eyespot and that reaching the eyespot after traveling through the cell is higher than in species with single-layered eyespots. The half-beam width of the antenna decreases with the number of carotenoid layers, from 130° in a single-layered eye towards 60° in a multilayered system. This spatial resolution is further improved by the concave curvature of the eye and the orientation of the receptor molecules within the eyespot membrane. Nevertheless, algal eyes are designed for tracking diffuse light direction instead of point light sources.

Mutants with unpigmented eyes that also lack chlorophyll and thus are almost completely white (*fn68* or *lts1*) become phototactic after retinal complementation. Therefore, scattering also generates enough contrast for provisional orientation. On the other hand, the phototactic rate towards the light is low in eyeless mutants (Foster and Smyth, 1980), demonstrating that the optical system is necessary for a good performance.

Since reflection and especially interference depend on the color of the light, algal eyes are optimized for a certain color range smaller than the range of a lens system. For wavelengths shorter than the optimum, interference occurs at a tilted angle, whereas for longer wavelengths interference is just reduced. In other words, optimal light incidence is rare as in any other visual system. The perceived light signal is color-modulated during cell rotation, and the sensitivity is shifted to the blue when the cell is approaching the light source (Hegemann and Harz, 1998).

In a proteomic analysis of the isolated eyespot of *Chlamydomonas*, 202 different proteins were identified by at least two different peptides (Schmidt et al., 2006). These included proteins known to be important for eyespot development (EYE2, MIN1), proteins involved in photoreception (CR1, CR2, ChR1, ChR2, and phototropin; see sections IV and V), protein kinases and phosphatases, and calcium-sensing and calcium-binding proteins.

In the colonial alga *Volvox carteri*, the 2000–4000 *Chlamydomonas*-like somatic cells are embedded in the cell matrix at the surface of the spheroid. Somatic cells possess eyes and flagella and are responsible for guiding the colony to places that are optimal for photosynthetic growth. These somatic cells show clear differences from *Chlamydomonas*. One obvious difference is that the flagella beat in parallel and cannot switch from the forward to the backward beating mode (Hoops et al., 1999). The cells in the front part of the colony contain bigger eyes with higher light sensitivity than the rear cells (Kirk, 1998). The 16 larger, eyeless, and flagella-less reproductive cells (gonidia) are located inside the sphere and have no contact with the extracellular medium.

III. PHOTOCURRENTS AS RAPID LIGHT SIGNALING ELEMENTS

A. Recording techniques

1. Individual cells

The analysis of electrical processes in chlorophycean algae, i.e., rapid charge movement across the plasma membrane or the flagellar membrane, became possible by application of a suction pipette technique, first in a pioneering study on *Haematococcus pluvialis* (Litvin et al., 1978) and only many years later on cell-wall-deficient *Chlamydomonas* cells (Harz and Hegemann, 1991) and individual cells of a *Volvox* dissolver mutant (Braun and Hegemann, 1999) (Figure 13.5). Suction pipettes have been widely

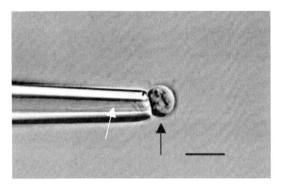

FIGURE 13.5   *A* Chlamydomonas *cell sucked into a glass electrode for photocurrent measurements. White arrow, flagella inside pipette; black arrow, eyespot outside the pipette; bar = 8 μm. (Harz and Hegemann, 1991.)*

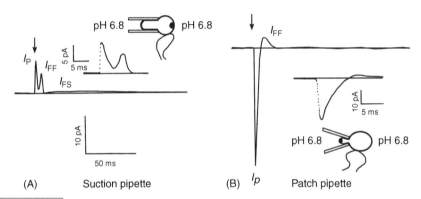

FIGURE 13.6   *Photoreceptor currents recorded indirectly using suction pipettes (A) and directly from the eye using patch pipettes (B). Arrows indicate a light flash. I_P, photoreceptor current. I_{FF}, fast flagellar current, I_{FS}, slow flagellar current. (From Ehlenbeck et al., 2002 with modification.)*

used for recording photocurrents from animal photoreceptor cells (Baylor et al., 1979). Photocurrents may be measured under three different configurations: first, eyespot and flagella outside the pipette (Figure 13.6A); second, eyespot in the pipette (Figure 13.6B), and third, flagella in the pipette. The last configuration is not recommended since flagellar currents rapidly decline due to mechanical inactivation of the flagella. The different configurations allow assignment of the currents to their source regions – i.e., the eyespot or flagella. Finally, photoreceptor currents may be recorded directly from the eye when only the eye is sucked into a patch pipette with a steep cone angle (Ehlenbeck et al., 2002).

2. Cell populations

For recording photocurrents from many cells simultaneously, cells are placed in a $2 \times 0.8 \times 2$ cm cuvette and excited by a unilateral flash. The

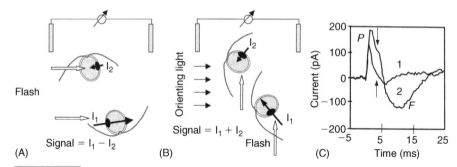

FIGURE 13.7 *Measurements of photocurrents from a cell population under non-preoriented conditions (unilateral or UL mode, A) and after preorientation with low background light (PO mode, B). (C) Photoelectric signals recorded from strain 494 (mt+) under UL (1) and PO (2) conditions. P and F are photoreceptor current and flagellar current, respectively. (Adapted from Sineshchekov et al., 1994 and Sineshchekov and Spudich, 2005.)*

electrical current is measured parallel to the light beam (unilateral mode; Figure 13.7A,C). The method is based on a differential response, i.e., the cells whose eyespots face toward the light show stronger photocurrents compared to cells whose eyespots face away from the light. With increasing flash intensity the amplitude goes through a maximum. At very high flash energies the current amplitude decreases due to excitation of all cells in the cuvette, thereby randomizing the response. In this respect the population assay is similar to a light scattering population assay. In a second configuration, cells are pre-oriented by background light parallel to the plane of the electrodes and then stimulated by a flash perpendicular to the orienting beam (Figure 13.7B,C). The response should be larger because all cells contribute to the response; on the other hand, the background light necessary for significant pre-orientation desensitizes the cells (Sineshchekov et al., 1992, 1994). If the current is recorded from more than 10^7 cells, the current amplitude may reach up to 1 nA, which is less than 0.01% of the total current. Kinetic analysis of photoreceptor currents is difficult with this assay because cells with different sensitivity and orientation to the light are averaged, and the fraction of the total current that is recorded is so small. The great advantage of the population assay over single cell experiments is that photocurrents may be recorded from cells with a wild-type cell wall. Thus, the population assay is preferable for screening electrical responses in different strains and transformants (Pazour et al., 1995; Matsuda et al., 1998).

B. Ion channel activation

Light excitation of the *Chlamydomonas* eye results in a cascade of electrical events. Photocurrents are a composition of several individual components that have been attributed to at least four conductances. The initial major current component is localized in the eyespot region and reflects cation influx across the localized region of the plasma membrane that overlies the eyespot. It was named photoreceptor current I_P (or PC). The amplitude is graded with the flash energy over a wide range in all three Chlorophycean species examined, namely *Chlamydomonas*, *Volvox*, and *Haematococcus* (Litvin et al., 1978; Harz and Hegemann, 1991; Braun and Hegemann, 1999). At saturating flash energies, the *Chlamydomonas* photoreceptor current is activated with virtually no delay ($\tau < 50$ microseconds) and the current rises single exponentially (Holland et al., 1996). It peaks within 1 millisecond, and decays within 20 milliseconds after the flash (Figure 13.6). The amplitude of I_P may reach 40 pA in large cells, reflecting about 50% of the total current. Under these conditions ~10^6 charges enter the eyespot area, transiently depolarizing the cell by ~80 mV (Harz et al., 1992). At lower flash energies, when fewer photoreceptor molecules are activated, the amplitude of the current is smaller and the current kinetics – mainly the decay – are much slower. The kinetics of the photocurrent do not reflect the kinetics of the photoreceptor itself, which in flash experiments should be independent of the light intensity. At low flash intensities, the current integral is proportional to the amount of light that is applied, whereas at higher flash energies the current integral is almost constant. This is because the current decay is a voltage-dependent process: larger ion fluxes result in a greater depolarization of the cell, causing acceleration of the decay and reduction of the total inward current. All properties of the current, namely high light saturation, extremely fast rise, and voltage-dependent decay can be explained by a rhodopsin-ion channel complex that is activated directly by light, a hypothesis that was only slowly accepted (see section IV.C.2).

The ions that carry the photoreceptor current have been a matter of debate since the first discovery of photocurrents in *Haematococcus* (Litvin et al., 1978). At neutral pH the photoreceptor current – at least in *Chlamydomonas* – is mostly carried by Ca^{2+}, but H^+ and K^+ also contribute when the driving force is high enough, i.e., at low pH or high extracellular K^+ (Holland et al., 1996; Nonnengasser et al., 1996; Ehlenbeck et al., 2002). While it is clear now that H^+ is the most readily conducted ion, Ca^{2+} contributes the most current under nearly all physiological conditions, when proton concentrations are relatively low.

Upon step-up stimulation, photoreceptor currents show a transient phase that decays to a stationary level, I_S, of only a few per cent of the maximum. Such adaptation makes sense for an alga that during the day may be continuously exposed to high light intensities. Despite our knowledge from *in vitro* studies about partial photoreceptor inactivation in constant light (see section IV.C.2), the more extreme adaptation of the photoreceptor current *in vivo* is still quite unclear. After a light flash, K^+ efflux repolarizes the cell towards the K^+ equilibrium potential and in continuous light is in equilibrium with the local Ca^{2+}/H^+ influx. Upon repetitive flashing or sinoidal light stimulation, repetitive photoreceptor currents were observed

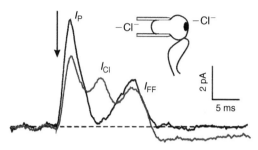

FIGURE 13.8 *Flash-induced photocurrents from strain CW2 cells under "Eyespot/Flagella-Out" configuration at pH 7 and 100 µM Ca²⁺ in presence (black line) and absence (red trace) of Cl⁻. The time of flashing is indicated by the arrow. (Ehlenbeck, 2002.)*

but only when the frequencies were kept between 0.5 and 5 Hz, indicating that photocurrent kinetics and adaptation are aligned with the rotation frequency of the cell (≈ 2 Hz) (Govorunova et al., 1997; Yoshimura and Kamiya, 2001). The currents resulting from sigmoidal stimulation deserve more attention because modulated currents constitute the basis for phototaxis at high light levels.

When the integral of the photoreceptor current exceeds a critical level in unicellular algae, flagellar currents are triggered. Since unicellular flagellates are nearly isopotential, the flagellar channels sense the primary depolarization originating from the eye. In *Chlamydomonas* and *Haematococcus* a fast action potential-like flagellar current I_{FF} and a subsequent slow flagellar current I_{FS} with extremely small amplitude are observed (Figures 13.6 and 13.8). The fast flagellar current is the trigger for switching from forward to backward swimming during the phobic response, as shown by simultaneous detection of photocurrents and flagellar beating (Holland et al., 1997). Thus, no signal amplification and no transmitter are required for flagellar current activation leading to the photophobic response. However, I_{FF} currents are not detected after low-intensity flashes, so the nature of the signaling process between eyespot and flagella under these conditions is less clear. Since depolarization is in the range of only 1 mV at 1% rhodopsin bleaching, modulation of the flagellar beat frequency and beating plane might involve intracellular signaling from the eye to the flagellar base (Sineshchekov and Spudich, 2005).

Since freshwater algae have to cope with low extracellular ionic conditions, they have developed special mechanisms for surviving at low ionic strength. At low extracellular Cl⁻, a Cl⁻ efflux I_{Cl} can be observed between I_P and I_{FF} (Figure 13.8). Since photocurrents are conventionally recorded at high Cl⁻ concentration (mM range), which inhibits the Cl⁻ efflux, this current is hidden in most experiments.

Chlamydomonas also performs striking behavioral responses after light off. Off-responses depend on the duration, color, and intensity of the adaptation light (Schmidt and Eckert, 1976; Hegemann and Bruck, 1989).

The underlying electrical and molecular processes are almost completely unknown.

C. Downstream flagellar processes

To assign flagellar proteins to specific behavioral responses is not an easy task, because for proper orientation of the cell in the light the complete three-dimensional bending pattern of both flagella, including their differences, must be intact. What is known is that intraflagellar Ca^{2+} is the transmitter that causes transient flagellar bending changes that finally lead to direction changes and phototaxis, as well as the switch to backward swimming during the phobic response (Schmidt and Eckert, 1976). Moreover, Ca^{2+} determines the open time of the flagellar channels and the duration of the phobic response (Schmidt and Eckert, 1976). When Ca^{2+} is replaced by Ba^{2+}, the channels remain open for long time periods, resulting in a phobic response of many seconds before the cells tend to deflagellate (Holland et al., 1997). Although flagellar Ca^{2+} channels are quite evenly distributed along both flagella (Beck and Uhl, 1994), *cis*- and *trans*-flagella react in a slightly different manner. The *cis*-flagellum is more sensitive. The *cis*-flagellar current occurs about 1–2 milliseconds earlier, and the undulatory pattern of flagellar beating associated with the phobic response lasts tens of milliseconds longer in the *cis*-flagellum compared to the *trans*-flagellum (Beck and Uhl, 1994; Holland et al., 1997; Rüffer and Nultsch, 1995), resulting in cell rotation at the end of a phobic response (Hegemann and Bruck, 1989). The Ca^{2+} sensitivity of the inner flagellar machinery – the axoneme – has been tested on isolated flagellar apparatuses, isolated demembranated axonemes, and demembranated axonemes still attached to the cell ("demembranated cell models"); in these experiments, the beating of the demembranated flagella was reactivated by addition of ATP (Hyams and Borisy, 1978; Bessen et al., 1980; Kamiya and Witman, 1984). Observations on the demembranated cell models revealed that both axonemes beat equally (in the forward swimming mode) only at 10^{-8} M Ca^{2+}. The experiments allowed two clear conclusions. First, Ca^{2+} binds directly to axonemal proteins. Second, the internal resting Ca^{2+} concentration in living cells is around 10^{-8} M. For selective inactivation of the *cis* axoneme as it occurs transiently during a directional change of dark-adapted cells after stimulation with flashes of low intensity, Ca^{2+} has to rise up to 10^{-7}–10^{-6} M. Interestingly, when Ca^{2+} drops to 10^{-9} M, the *trans*-flagellum inactivates just as it does during a step-down response. For a switch of the bending mode to undulation as it occurs during a phobic response, the Ca^{2+} concentration has to rise up to 10^{-4} M.

Several mutants were described in the past that are defective in phototaxis or phobic responses. Among the flagellar proteins that are discussed as being involved in photoresponses are inner and outer dynein arms

(Mitchell and Rosenbaum, 1985), radial spoke proteins (Yang et al., 2006), kinases that phosphorylate the dyneins asymmetrically (King and Dutcher, 1997), a flavodoxin (Iomini et al., 2006), and many more. In most cases the mode of involvement beyond motility changes is not clear. Axonemal components that may be involved in behavioral light responses are discussed further in Chapters 6–9.

IV. RHODOPSINS AS PHOTORECEPTORS FOR BEHAVIORAL RESPONSES

A. Rhodopsin-mediated responses

1. Action spectra

The first plant rhodopsin to be discovered was in *Chlamydomonas*. In contrast to phytochrome and phototropin (see section V.C.2), rhodopsin undergoes a fast photocycle and is most appropriate for motile organisms like microalgae or gametes of macroalgae. It is unclear, however, how widely rhodopsin is distributed in the plant kingdom.

The optical properties of the eye and additional shading pigments contribute to phototaxis action spectra and thus must be carefully considered. Moreover, since phototaxis is measured in continuous light, several parameters such as adaptation phenomena, photochromic (i.e., light-induced shifts in absorption) properties of the photoreceptors, and screening pigments distort the action spectra, especially when measurements are carried out at high light intensities. Consequently, phototaxis action spectra constructed from recordings at low irradiance (threshold spectra) reveal the nature of the principal photoreceptor in these algae in a much better way than high-intensity spectra. Low-intensity phototaxis action spectra for Chlorophyceae are rhodopsin-shaped with maxima between 460 and 560 nm. From such spectra published earlier in the literature, Foster and colleagues concluded correctly that the photoreceptor for phototaxis in *Chlamydomonas* and other chlorophycean algae is rhodopsin (Foster and Smyth, 1980). The claim was later supported by reconstituting phototaxis in blind cells (the *fn68* mutant) by complementation with retinal (Foster et al., 1984). The *fn68* cells are defective in the normal dark pathway for synthesis of carotenoids and retinal, so dark-grown cells are non-phototactic, although they contain the apoprotein opsin. A great advantage of *Chlamydomonas* is that non-photosynthetic cells can easily be maintained in darkness by growth on acetate (cells lacking pigment must be maintained in the dark because they are easily photodamaged). The experiments were carried out using gametes, and the resulting threshold action spectra were clean rhodopsin-like spectra with maxima at 496 nm. In contrast, previously published high-intensity action spectra recorded from vegetative wild-type cells (strain SAG 11–32; see Volume 1, Table 1.3 for strain identification) had a shoulder at 460 nm (Nultsch et al., 1971), which led to speculation that there might be multiple photoreceptor subspecies.

Although they cannot be cultured in the light, cells of the *fn68* mutant do acquire phototactic competence when exposed to light for brief periods. Foster et al. (1988) found that retinal was synthesized in these cells by a light-dependent pathway whose action spectrum was rhodopsin-shaped, suggesting that the small residual amount of functional rhodopsin in the dark-grown cells triggers retinal synthesis and the full reconstitution of the total rhodopsin. In other words, rhodopsin is controlling its own chromophore production. Incubating the cells with a retinal analogue prior to light exposure increased the amount of functional rhodopsin in darkness, and the action spectrum for light induction then shifted to that expected for the analogue.

In *Volvox* cultures synchronized by a light-dark cycle, juvenile colonies containing presumptive somatic and reproductive cells are produced during the dark, but the cells do not fully differentiate until the light comes back on. The pattern of protein synthesis changes rapidly when the light comes on, and the action spectrum is clearly rhodopsin-shaped with a maximum around 510 nm. It is conceivable that the rhodopsin-mediated Ca^{2+}-influx is the trigger for rapid translation of RNA that has accumulated during the dark period (Kirk and Kirk, 1985). This is one of the rare reports about the contribution of rhodopsin to developmental processes.

It also should be noted that rhythmicity is under light control. Under a low light regime, the circadian rhythm is definitely controlled by different light qualities. In dark-adapted cells, the action spectrum for eliciting a 4-hour phase shift of daily rhythmicity shows two primary peaks, at 520 nm and 660 nm (Kondo et al., 1991). The 520 nm peak is likely to be due to rhodopsin whereas the photoreceptor that accounts for the 660 nm peak is unknown. The spectrum is quite different in light-adapted cells, where the maximal activity appears at 440 and 660 nm (Johnson et al., 1991). In this case, phase resetting appears to be mediated by components of the photosynthetic apparatus. This conclusion is supported by application of compounds that inhibit photosynthetic electron transport and phase shifting. Contributions of a separate blue light receptor could not be demonstrated in these studies. The mechanism by which photoreceptor activation resets the clock ("Zeitgeber") is unknown. See Volume 1, Chapter 3 for additional discussion of circadian rhythms in *Chlamydomonas*.

2. The rhodopsin chromophore

All information about the rhodopsin chromophore(s) has been deduced from complementing blind, retinal-deficient cells with retinoids or various retinal analogues. Conclusions from these data should be restricted to those rhodopsins that are involved in movement responses and should not be extrapolated to those involved in other cellular processes. In complemented cells, phototaxis and photophobic responses have been tested using a variety of methods. The major conclusions are summarized by the following: *Chlamydomonas* rhodopsins serving as photoreceptors for phototaxis contain a planar all-*trans*,6-*s*-*trans* retinal chromophore like all other archaeal rhodopsins, but in contrast to animal rhodopsin, which uses *11-cis* retinal with the ring in the non-planar 6-*s*-*cis* position. The ring accelerates reconstitution, but a chromophore with only three conjugated double bonds plus methyl groups is sufficient for full photosensitivity, which is blue-shifted. During illumination, the chromophore undergoes a *13-trans* to *cis* isomerization. The latter is converted into a conformational change of the protein via the 13-methyl group (Yan et al., 1990; Beckmann and Hegemann, 1991; Derguini et al., 1991; Hegemann et al., 1991; Lawson et al., 1991; Zacks et al., 1993; Sakamoto et al. 1998). Zacks and colleagues reported that retinal regeneration of photophobic responses but not of phototaxis is inhibited by a five-membered ring *13-trans*-locked retinal, which they regarded as evidence for differences in the respective photoreceptor chromophores (Zacks et al., 1993).

Most of the researchers agree with these conclusions. However, K.W. Foster's group has, over the years, promoted a quite different idea about rhodopsin activation. The group reported extended phototaxis measurements carried out with the *fn68* mutant using the dish test. They found that short-chained retinal analogues with only one double bond (hexenal) were able to reconstitute phototaxis. Moreover, retinal analogues that were not able to isomerize in the light were reported to be functional (Foster et al., 1988, 1989, 1991). These findings not only contrast with our knowledge about animal rhodopsins, but also are inconsistent with the current and detailed hypotheses about archaeal rhodopsins. No other research group working in the field has been able to confirm these results using other methods, including light scattering, motion analysis, or electrical recording (Hegemann et al., 1991; Zacks et al., 1993; Sineshchekov et al., 1994). But, strictly speaking, the conclusions have never been disproved. Given that channelrhodopsins can transport charge down the electrochemical gradient (see section IV.C), light-induced dipoles within a non-isomerizable chromophore might be sufficient to drive some charge movement, but certainly not the full current.

B. Animal-type rhodopsins

The first rhodopsin-related protein in *Chlamydomonas* was biochemically identified by labeling the protein in the retinal-deficient *fn68* cells using ^{3}H-retinal. Later, this protein was purified from strain 806 cells, a strain that is ideal for biochemical studies due to its thin but stable cell wall (Beckmann and Hegemann, 1991). The purified "chlamyrhodopsin" (CR) was a mixture of two equally sized proteins, CR1 and CR2, which are both encoded by the *COP1* gene. The mRNA is alternatively spliced (Z48968; AF295371), encoding two variants with different hypothetical retinal-binding sites. CR1 is a result of further RNA processing and present at 50 times lower concentration compared to CR2 (Fuhrmann et al., 2003). A rhodopsin homologous to CR2 was found in *Volvox* (volvoxrhodopsin-1, VR1; accession Y11204), whereas the alternatively spliced product was not seen (Ebnet et al., 1999). CR1, CR2, and VR1 are highly charged 27-kD proteins. The seven-transmembrane segments typical for the rhodopsin superfamily cannot be clearly assigned from hydropathy plots. Thus, their topography is still unclear. CR1, CR2, and VR1 are animal-type rhodopsins specifically related to *Calliphora* opsin (23% identity, CR1/2) or *Drosophila* rhodopsin 6 (DROME rh6, 21%, VR1). Both CR1 and CR2 are strongly enriched in the eyespot, although they also occur in the thylakoid membrane (Deininger et al., 2000).

In *Chlamydomonas*, behavioral responses are not affected by reduction in the amount of CR1 and CR2 in the cell. Antisense transformants with low CR1/CR2 content show perfect phototaxis and phobic responses. In contrast, phototaxis is disabled in VR-reduced *Volvox* cells that are obligatorily phototrophic. Based on these data it was concluded that CR1/2 play an indirect role in behavioral responses, and the original hypothesis that these proteins serve as primary photoreceptors for phototaxis and photoshock had to be revised (Fuhrmann et al., 2001).

C. Channelrhodopsins

1. Structural properties

After the *Chlamydomonas* genome project was started and the sequences of >100,000 cDNA fragments became available, two more rhodopsin cDNAs, *COP3* and *COP4*, were identified (AF385748, AF461397). The two encoded proteins are unusually large rhodopsins with molecular weights around 76 kD. Due to their *in vitro* and *in vivo* function (explained below), they were independently named channelrhodopsins (ChR1 and ChR2) (Hegemann et al., 2001; Nagel et al., 2002), *Chlamydomonas* sensory opsins (CSOA and CSOB) (Sineshchekov et al., 2002), or archaeal-type *Chlamydomonas* opsins (ACOP1 and ACOP2) (Suzuki et al., 2003). Since "channelrhodopsin" is most widely accepted now, this name will be used in the following sections. The proteins consist of a seven-transmembrane region and a long C-terminal extension. The seven-transmembrane segments are related to microbial rhodopsins found in archaea, eubacteria, and fungi. Amino acids that form the H^+-conducting network in bacteriorhodopsin are conserved within the seven-transmembrane region but the amino acids (Glu and Asp) that connect the H^+-network to the bulk phases are replaced by His and Ser. Therefore, it is likely that this conserved H^+-network serves as a switch rather than functioning directly as a H^+-transport pathway.

Channelrhodopsins exhibit some striking features that distinguish them from other sensory rhodopsins. They contain six glutamates in helix two and many cysteines and serines. This higher polarity is partially compensated for by an excess of aromatic amino acids (Hegemann et al., 2001; Sineshchekov et al., 2002; Suzuki et al., 2003).

2. Transport activity

To study the transport activity of ChR1 and ChR2, both proteins were expressed in *Xenopus laevis* oocytes. Oocytes are an extremely useful model system in which *Chlamydomonas* RNAs are well expressed. ChR1 caused a large H^+ conductance in the light, which was negative at least at pH 6 – quite independent of any ion, namely Na^+, K^+, Mg^{2+}, Ca^{2+}, and Cl^-, in the medium (Ringer's solution). Acidification of the medium greatly enhanced the current. Moreover, size and direction followed the applied proton-motive force, proving that ChR1 is a light-gated proton channel (Nagel et al., 2002). However, it is clear that ChR1 also conducts K^+, and Ca^{2+}, explaining the *in vivo* function at high pH (Berthold et al., 2008). ChR2 is also a proton channel, although in early experiments conductances for cations like Na^+, K^+, and even Ca^{2+} were observed (Nagel et al., 2003). Both channels are impermeable to Mg^{2+}. In summary, both proteins are proton channels with some conductance for other cations, which to a variable degree compete with H^+.

The onset of the currents of both channelrhodopsins begins within 20 microseconds after a flash, supporting the hypothesis that there is a direct link between sensor and channel. These proteins are the first members of a new rhodopsin family with ion channel properties and, conversely, the first ion channels with an intrinsic light-triggered switch. ChRs show partial inactivation in continuous light, i.e., light adaptation, which is fast in ChR1 and slower in ChR2. ChR2 inactivation is promoted by high external pH and low (i.e., less negative) membrane voltage. In darkness, the proteins recover to the more active dark-adapted state. The ion conductance of both light-gated channels is extremely small. As already mentioned, upon a single flash 10^6 charges enter the *Chlamydomonas* eye. The number of rhodopsins as estimated from retinal extraction and opsin immunoblotting is 10^4–10^5 per eye (Beckmann and Hegemann, 1991; Govorunova et al., 2004), which is in the range of the intramembrane particles counted in the plasma membrane overlying the eyespot (Melkonian and Robenek, 1984). If this range is correct, each rhodopsin conducts only 10–100 charges per absorbed photon (Harz et al., 1992).

3. Spectroscopy

ChRs can be expressed in *Pichia pastoris* (a yeast) and COS-1 cells (green monkey kidney cells) and functionally purified in sufficient amounts to allow the recording of absorption spectra. The spectra of ChR2 from

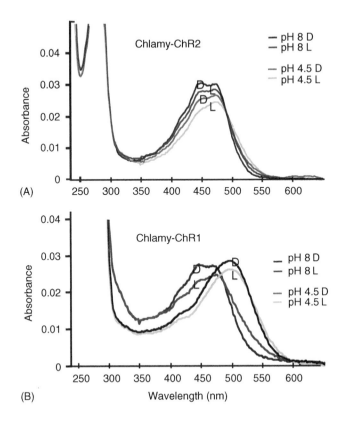

(A)

(B)

Wavelength (nm)

FIGURE 13.9

Absorption spectra of purified dark (D) and light (L)-adapted ChR2 (A) and ChR1 (B) at pH 8 and pH 4.5. Light-adapted ChRs were illuminated for 10 seconds with a 456-nm light-emitting diode before spectra were recorded. (Spectra are a courtesy of Oliver Ernst and Kyoko Tsunoda, Charite Berlin.)

Chlamydomonas and *Volvox* are very similar and exhibit absorption maxima at 470 nm ($ChR2_{470}$) (Figure 13.9). The spectra are vibrationally fine-structured, indicating a rigid retinal chromophore. Upon acidification the absorption slightly shifts asymmetrically in favor of the long-wavelength absorption maximum. The acidic form was named $ChR2_{480}$. Excitation of $ChR2_{470}$ with laser flashes causes immediate absorbance changes ($\tau \leq 100$ ns) and the appearance of a red-shifted photoproduct. This early photoproduct P_{500} deprotonates within a few microseconds to form P_{390}, the latter reacting further to generate P_{510} with $\tau = 200$ microseconds. P_{510} is considered to be the main conducting state consistent with the time of channel activation *in vivo* (Ernst et al., 2008, Bamann et al., 2008). Under more acidic conditions P_{480} undergoes a photocycle with two red-shifted intermediates, P_{515} and P_{530}, but without visible deprotonated species.

Recombinant ChR1 was also purified from COS cells, but the amount was not sufficient to fully characterize the photocycles. *Chlamydomonas* ChR1 also absorbs maximally at 470 nm at neutral pH but is in equilibrium with a species, P_{495}, absorbing maximally at 495 nm. Thus, the photochromic shift is larger in ChR1 compared with ChR2. This was surprising since all low intensity action spectra peak near 500 nm. Since ChR1 is the dominant photoreceptor in *Chlamydomonas*, $ChR1_{495}$ is likely to be the dominant isoform *in vivo*. The dilemma of photochromism can be explained if a negative potential drives the equilibrium from $ChR1_{470}$ towards $ChR1_{495}$ and the maximum is shifted to $ChR1_{470}$ only upon depolarization of the cell or acidification of the medium. The conclusion from these studies is that not only ChR2 but also $ChR1_{470}$ may contribute to the short wavelength shoulder of basically all high intensity phototaxis action spectra in Volvocales (Halldal, 1958; Nultsch et al., 1971). Moreover, since the reflectivity of the eye shifts to a longer wavelength when the angle of light incidence is small, it is conceivable that the ChR1 photochromism contributes to the phototaxis sign inversion (positive vs. negative), which has a strong wavelength dependence. The excitation maximum of *Volvox* ChR1 is shifted even further to the red, with the absorption maximum of the acidic form at 538 nm (Zhang et al., 2008). This accounts for the long wavelength peak (~530 nm) of the phototaxis action spectra (Schletz, 1976; Sakaguchi and Iwasa, 1979). The long-wavelength absorption of *Volvox* ChR1 makes it highly attractive for neuroscience applications (Zhang et al., 2008).

4. Cellular function

In the first study of channelrhodopsin function *in vivo*, the amounts of ChR1 and ChR2 were reduced independently by employing an antisense approach (Sineshchekov et al., 2002). Cells with reduced ChR2 content exhibited photocurrents with slightly reduced size and fast kinetics in population experiments. However, in transformants with reduced ChR1, much reduced photocurrents were observed. Compared to those of wild-type cells, these photocurrents were significantly delayed upon stimulation with flashes of low energy. The authors concluded that ChR1 mediates fast responses and is tightly associated with a secondary channel or is itself a channel. In contrast, it was concluded that ChR2 was responsible for low intensity responses, in particular phototaxis, and was signaling via biochemical steps to a remote ion channel (Sineshchekov et al., 2002; Sineshchekov and Spudich, 2005). However, the latter conclusion seems to conflict with several other observations. First, there is an extremely low abundance of any cation channel-related protein in the eyespot proteome (Schmidt et al., 2006). Second, both ChR1 and ChR2 are activated in *Xenopus* oocytes without delay. Third, a photoreceptor that triggers phototaxis at low light intensity must be present in high amount, whereas a photoreceptor for high light intensity responses might be present at lower levels. Finally, action spectra of phototaxis, photophobic responses, and photocurrents measured at low light intensities or from dark-adapted cells show maxima at 500 nm, indicating that ChR1 is the dominant photoreceptor (Foster et al., 1984; Harz and Hegemann, 1991; Ehlenbeck et al., 2002). Nevertheless, this first study on antisense transformants clearly demonstrated that both ChR1 and ChR2 contribute to the photocurrents. Later, an antisense approach was used to knockdown ChR1 in the cell wall-deficient strain CW2, the gametes of which contain only very small amounts of ChR2. In the knockdown transformant with almost no ChR1 photocurrents, phobic response and phototaxis completely disappeared, indicating that ChR1 is able to trigger both behavioral responses. The recently described ChR1 cation conductance is sufficient to explain photocurrents at pH 9 and the behavior of *Chlamydomonas* at alkaline pH in general (Berthold et al., 2008). Considering that the driving force for Ca^{2+} is much higher than the one for K^+, and that the extracellular concentration of Ca^{2+} is usually higher than that of H^+, Ca^{2+} dominates the photocurrents under most conditions. The additive value of ChR2 and the reason for the different ratios of ChR1 and ChR2 in different strains is still unclear. The strong voltage and pH-dependent photochromism of ChR1 might contribute to the variability of the switch between positive and negative phototaxis under different physiological conditions.

5. Application of channelrhodopsin as a light sensor in neuronal cells

Temporally precise, noninvasive control of ion-channel activities in neuronal cells or tissues is a long-term goal in neuroscience. Several groups have reported expression of ChR2 in neuronal cells and neuronal tissues for exactly such a purpose. They reported reliable control of neuronal spiking as well as excitatory and inhibitory synaptic transmission simply by illuminating the ChR2-expressing cells with light (Boyden et al., 2005; Li et al., 2005; Ishizuka et al., 2006; Zhang and Oertner, 2007). ChR2 also was expressed in chicken embryos and *Caenorhabditis elegans*, where it triggers rapid behavioral responses after light excitation (Li et al., 2005; Nagel et al., 2005). Expression of channelrhodopsin in surviving inner retinal neurons is a potential strategy for partial restoration of vision after rod and cone degeneration. For example, Bi et al. (2006) expressed ChR2 in inner retinal neurons of a mouse with photoreceptor degeneration and restored the ability of the retina to detect light signals and transmission of the light signal to the visual cortex.

Considering the small number of ions that channelrhodopsins carry in *Chlamydomonas*, it might be surprising that electrophysiologists are using channelrhodopsins quite successfully. However, most physiological experiments on *Chlamydomonas* were carried out under low ionic conditions, i.e., pH 7 and $300\,\mu M$ Ca^{2+} plus $300\,\mu M$ K^+, whereas neurophysiologists use much higher ion concentrations, in the range of $100\,mM$. Under these conditions, the current rises at least 10 fold, up to 100–1000 charges per ChR, and, at pH 6, when H influx contributes significantly, values up to 500–5000 charges per excited rhodopsin may be reached. The decay time of the conducting state is in the range of 10–100 milliseconds depending on the conditions. Factors that influence the cycling time are the internal and external pH and the membrane voltage. At high light intensities, each rhodopsin is excited 10–100 times per second, producing a current of 10^3–10^4

charges per second at -100 mV and neutral pH. The level of depolarization achieved with a light pulse is limited by the expression level that is reached in a mammalian cell. However, less than 1000 rhodopsins per cell should be sufficient to depolarize cells fast enough for triggering voltage-gated Na^+ or K^+ channels and evoking action potentials by a 10–100-millisecond light pulse.

D. Enzyme rhodopsins

At least three more opsins are predicted to be encoded by the *Chlamydomonas* genome. These were named COP5 (accession AAQ16277), COP6 (EDO99289), and COP7. All three predicted proteins contain seven-transmembrane regions with homology to microbial-type rhodopsins. Surprisingly, all three sequences contain domains that are known from other signaling systems. These include histidinyl kinase, phosphor acceptor regulator, and guanylyl cyclase domains (Kateriya et al., 2004). Such elements are common in two-component signaling systems, a canonical mechanism that mediates diverse biological responses in many taxa, including chemotaxis in *Escherichia coli*, "phototaxis" in halobacteria, and cytokinin-mediated transcriptional control in *Arabidopsis* or maize (Sheen, 2002). If COP5–COP7 are all expressed and functional as expected from their sequences, they constitute a new class of sensory rhodopsins with C-terminal enzymatic activities (enzyme rhodopsins) and represent the first two-component systems identified in microalgae. It is not unlikely that some of these enzyme rhodopsins are involved in retinal biosynthesis or in modulation of behavioral responses at low light.

V. BLUE-LIGHT SENSITIVE FLAVIN-BASED PHOTORECEPTORS

A. Physiological perspectives

Beyond its functions as energy source and informer for cellular orientation, light also plays a pivotal role as a developmental regulator for green algae. In contrast to higher plants, there is little evidence for red-light sensitive receptors, such as phytochrome, in *Chlamydomonas*. Moreover, no genes with significant homology to phytochrome were found in the genome (Mittag et al., 2005). However, several processes in green algae, and in *Chlamydomonas* in particular, are clearly regulated by blue light. First, gametogenesis, or more precisely the transition from pre-gametes to gametes, requires blue light (Sager and Granick, 1954) (also see Volume 1, Chapter 5). Second, the circadian (daily) rhythm is an endogenous biological program that is entrained by blue light in addition to green light (see Section IV.A.1). Several cellular processes are regulated in synchrony with the clock indirectly by blue light. Among these are phototaxis (measured as photoaccumulation), chemotaxis (chemoaccumulation), adhesion to surfaces, and cell division (Bruce, 1970; Straley and Bruce, 1979; Mergenhagen, 1984; Byrne et al., 1992). However, because the action spectra in the older literature are poor, the contribution of a *bona fide* blue-light receptor needs to be confirmed. Third, *Chlamydomonas* is able to sense the season of the year by measuring the duration of the day in the natural environment, a phenomenon called photoperiodic time measurement. For example, the germination efficiency of *Chlamydomonas* zygospores is enhanced in long-day light/dark cycles (Suzuki and Johnson, 2002). Again, blue light is most efficient in this response (Gloeckner and Beck, 1995; Huang and Beck, 2003).

B. The blue-light photoreceptor proteins

A cryptochrome-like protein, CPH1 ("*Chlamydomonas* photolyase homologue 1", accession AAC37438), has been identified as a potential photoreceptor. The protein is light-labile and almost completely disappears in continuous light. The function of the protein has not been studied in detail. It has been suggested that CPH1 is involved in regulation of transcription of selected genes and/or might function as a "Zeitgeber" for the synchronization of rhythmicity with the actual light-dark cycle (Reisdorph and Small, 2004; Mittag et al., 2005). Since illumination of several hours is required for efficient gametogenesis, CPH1 probably could not function as a photoreceptor for gametogenesis (see next paragraph). A search of the *Chlamydomonas* genome using *Arabidopsis*, *D. melanogaster*, or mammalian Cry proteins as templates resulted in the discovery of a second gene encoding a cryptochrome-related protein, which has been annotated as UVR3 (EDO99639). Thus, *Chlamydomonas* is likely to express two cryptochromes, one of which is more closely related to *Arabidopsis* Cry (CPH1) and the other to animal Cry (UVR2).

The photoreceptor for blue-light induced gametogenesis is phototropin, a homologue of those photoreceptors (phot-proteins) that mediate phototropism and chloroplast relocation in higher plants (Briggs and Christie, 2002). Phototropins contain two blue-light sensitive domains LOV1 and LOV2 and a C-terminal serine-threonine kinase. The three domains are conserved in all phototropins. *Arabidopsis* possesses two phototropins (Phot1 and Phot2), whereas *Chlamydomonas* has only one (PHOT, accession CAC94941), which appears to be unusually small (81 kD instead of the typical 120 kD). Knock-down strains with reduced phototropin levels were partially impaired in three steps of the life cycle: in gametogenesis, maintenance of mating ability, and germination of zygotes. These observations suggested that phototropin is a more general light sensor for controlling the life cycle in *Chlamydomonas* than originally thought (Huang and

Beck, 2003). Phototropin was immunolocalized as a soluble protein in the cytoplasm and as an insoluble protein in the flagella, where it is associated with the outer doublet microtubules. Huang and Beck proposed that phototropin is a cargo for intraflagellar transport (see Chapter 4). In the *fla10* mutant, in which the anterograde intraflagellar transport motor kinesin-2 is defective, transport of phototropin to the flagella is inhibited and the cells are not competent for mating. It was suggested that phototropin controls activation of the agglutinins, which mediate flagellar interaction between sexual mating partners (Huang et al., 2004; also see Chapter 12). Finally, phototropin-reduced strains showed impaired expression of genes encoding light-harvesting complex apoproteins and proteins involved in chlorophyll and carotenoid biosynthesis (Im et al., 2006).

C. *In vitro* studies on CPH1 and phototropin

1. Cryptochromes

The N-terminal 500 amino acids of CPH1, comprising the light-sensor domain, were expressed in high yield in *E. coli* as a yellow protein that bears exclusively flavin adenine dinucleotide in its oxidized state (Immeln et al., 2007). Illumination with blue light induces formation of a neutral flavin radical with absorption at 580 nm even in the absence of an electron donor. The protein is autophosphorylated in blue light, which might have relevance for adaptation or photo-degradation *in vivo*. In contrast to plant cryptochromes, for which protein quantity is a limiting factor, *Chlamydomonas* CPH1 can be obtained in sufficient amounts for further biophysical characterization.

2. Phototropin

The general reaction mechanism for light-activation of phototropins has been most widely studied for the *Chlamydomonas* LOV1-domain. The photoreaction of all phototropins is the formation of a thioadduct between the light-sensing flavin mononucleotide (FMN)-cofactor and a nearby cysteine. The reaction occurs within a few microseconds through a triplet-excited state. Structures of the dark state (Figure 13.10) and the light-activated signaling state are available (Fedorov et al., 2003). A variety of mutated proteins were analyzed, mainly by application of time-resolved ultraviolet/visible and fluorescence spectroscopy (Holzer et al., 2002; Kottke et al., 2003) as well as Fourier transform infrared spectroscopy (Ataka et al., 2003), leading to a detailed model of the reaction cycle (Kottke et al., 2003). The most likely reaction mechanism is the one originally proposed by Kay et al. (2003). It begins during the excited triplet state with the abstraction of an electron from the sulfur of the reactive cysteine to the FMN and is followed by a H^+ transfer from the sulfur to N5 of the isoalloxacine. The two neutral radicals FMNH• and Cys• recombine to form the adduct

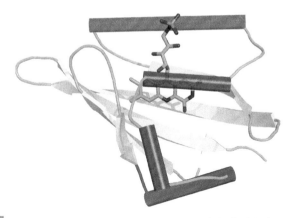

FIGURE 13.10 *Structure of a blue-light photoreceptor domain (LOV1) of the* Chlamydomonas *blue-light receptor phototropin; α-helical segments are red, β-sheets are yellow, the bound FMN is green. (Modified from Fedorov et al., 2003.)*

FMNH(4a)-Cys. As a consequence of the thioadduct formation, small conformational changes were observed that ought to trigger activation of the kinase. The detailed mechanism of kinase activation is not known, and it is unclear why phototropin needs two photoreceptor domains. Finally, the substrate of the kinase needs to be discovered before the cellular function of phototropin will be completely understood.

VI. SUMMARY AND FUTURE PROSPECTS

The study of sensory photoreceptors in *Chlamydomonas* is especially appealing because the reactions can be triggered both *in vivo* and *in vitro* with unparalleled precision. Seven rhodopsins, two cryptochromes, and one phototropin have been identified as sensory photoreceptors in *Chlamydomonas*. Most unusual are the channelrhodopsins, which are light-gated ion channels serving as photoreceptors for photophobic responses and phototaxis. They carry a photoreceptor current that causes depolarization of the whole cell. The depolarization is sensed by voltage-gated Ca^{2+} channels in the flagellar membrane, which then mediate a massive Ca^{2+} influx that causes transient changes in flagellar beating. Three "enzyme rhodopsins" have not been studied *in vitro* yet, but are expected to control slow developmental processes that include gene expression. The two cryptochromes (one plant and one animal type) seem to adjust synchrony with the daily or "circadian" rhythm, but in both cases more studies are necessary to understand their biological function. Phototropin (Phot) controls the sex life of this alga, i.e., gametogenesis and zoospore release from zygotes. The phototropin LOV-domains that occur in both the light- and dark-adapted states have been determined at atomic resolution. However, the immediate signaling partners await discovery. A dissection of the different photoreceptor pathways will almost certainly require the generation and use of mutants with defined photoreceptor deletions – a goal that should be readily achieved in *Chlamydomonas*.

ACKNOWLEDGMENTS

We thank Christoph Beck, Carol Dieckmann, Elizabeth Harris, Georg Kreimer, Telsa Mittelmeier, and George Witman for constructive criticism and stimulating discussions. The work of the authors was generously supported over more than 20 years by the Deutsche Forschungsgemeinschaft, which is gratefully acknowledged here.

REFERENCES

Ataka, K., Hegemann, P., and Heberle, J. (2003). Vibrational spectroscopy of an algal Phot-LOV1 domain probes the molecular changes associated with blue-light reception. *Biophys. J.* **84**, 466–474.

Bamann, C., Kirsch, T., Nagel, G., and Bamberg, E. (2008). Spectral characteristics of the photocycle of channelrhodopsin-2 and its implication for channel function. *J. Mol. Biol.* **18**, 686–694.

Baylor, D.A., Lamb, T.D., and Yau, K.W. (1979). The membrane current of single rod outer segments. *J. Physiol. (Lond)* **288**, 589–611.

Beck, C. and Uhl, R. (1994). On the localization of voltage-sensitive calcium channels in the flagella of *Chlamydomonas reinhardtii*. *J. Cell Biol.* **125**, 1119–1125.

Beckmann, M. and Hegemann, P. (1991). *In vitro* identification of rhodopsin in the green alga *Chlamydomonas*. *Biochemistry* **30**, 3692–3697.

Berthold, P., Tsunoda, S.P., Ernst, O.P., Mages, W., Gradmann, D., and Hegemann, P. (2008). Channelrhodopsin-1 initiates phototaxis and photophobic responses in *Chlamydomonas* by immediate, light-induced depolarization. *Plant Cell* **20**, 1665–1677.

Bessen, M., Fay, R.B., and Witman, G.B. (1980). Calcium control of waveform in isolated flagellar axonemes of *Chlamydomonas*. *J. Cell Biol.* **86**, 446–455.

Bi, A.D., Cui, J.J., Ma, Y.P., Olshevskaya, E., Pu, M.L., Dizhoor, A.M., and Pan, Z.H. (2006). Ectopic expression of a microbial-type rhodopsin restores visual responses in mice with photoreceptor degeneration. *Neuron* **50**, 23–33.

Boyden, E.S., Zhang, F., Bamberg, E., Nagel, G., and Deisseroth, K. (2005). Millisecond-timescale, genetically targeted optical control of neural activity. *Nat. Neurosci.* **8**, 1263–1268.

Braun, F.J. and Hegemann, P. (1999). Two light-activated conductances in the eye of the green alga *Volvox carteri*. *Biophys. J.* **76**, 1668–1678.

Briggs, W.R. and Christie, J.M. (2002). Phototropins 1 and 2: versatile plant blue-light receptors. *Trends Plant Sci.* **7**, 204–210.

Bruce, V.G. (1970). The biological clock in *Chlamydomonas reinhardi*. *J. Protozool.* **17**, 328–334.

Buder, J. (1917). Zur Kenntnis der phototaktischen Richtungsbewegung. *Jahrbuch f. wiss. Botanik* **58**, 106–220.

Byrne, T.E., Wells, M.R., and Johnson, C.H. (1992). Circadian rhythms of chemotaxis to ammonium and of methylammonium uptake in *Chlamydomonas*. *Plant Physiol.* **98**, 879–886.

Davies, D.R. and Plaskitt, A. (1971). Genetical and structural analyses of cell-wall formation in *Chlamydomonas reinhardi*. *Genet. Res.* **17**, 33–43.

Deininger, W., Fuhrmann, M., and Hegemann, P. (2000). Opsin evolution: out of wild green yonder? *Trends Genet.* **16**, 158–159.

Derguini, F., Mazur, P., Nakanishi, K., Starace, D.M., Saranak, J., and Foster, K.W. (1991). All-trans retinal is the chromophore bound to the photoreceptor of the alga *Chlamydomonas reinhardtii*. *Photochem. Photobiol.* **54**, 1017–1022.

Dieckmann, C.L. (2003). Eyespot placement and assembly in the green alga *Chlamydomonas*. *Bioessays* **25**, 410–416.

Ebnet, E., Fischer, M., Deininger, W., and Hegemann, P. (1999). Volvoxrhodopsin, a light-regulated sensory photoreceptor of the spheroidal green alga *Volvox carteri*. *Plant Cell b*, 1473–1484.

Ehlenbeck, S. (2002). Licht-induzierte H^+ und Cl^--Ströme in *Chlamydomonas reinhardtii*. Dissertation. Universität Regensburg.

Ehlenbeck, S., Gradmann, D., Braun, F.J., and Hegemann, P. (2002). Evidence for a light-induced H^+ conductance in the eye of the green alga *Chlamydomonas reinhardtii*. *Biophys. J.* **82**, 740–751.

Ernst, O., Sanchez Murcia, P.A., Daldrop, P., Tsunoda, S.P., Kateriya, S., and Hegemann, P. (2008). Photoactivation of channelrhodopsin. *J. Biol. Chem.* **283**, 1637–1643.

Fedorov, R., Schlichting, I., Hartmann, E., Domratcheva, T., Fuhrmann, M., and Hegemann, P. (2003). Crystal structures and molecular mechanism of a light-induced signaling switch: The Phot-LOV1 domain from *Chlamydomonas reinhardtii*. *Biophys. J.* **84**, 2474–2482.

Foster, K.W. and Smyth, R.D. (1980). Light antennas in phototactic algae. *Microbiol. Rev.* **44**, 572–630.

Foster, K.W., Saranak, J., Patel, N., Zarilli, G., Okabe, M., Kline, T., and Nakanishi, K. (1984). A rhodopsin is the functional photoreceptor for phototaxis in the unicellular eukaryote *Chlamydomonas*. *Nature* **311**, 756–769.

Foster, K.W., Saranak, J., and Zarrilli, G. (1988). Autoregulation of rhodopsin synthesis in *Chlamydomonas reinhardtii*. *Proc. Natl. Acad. Sci. USA* **85**, 6379–6383.

Foster, K.W., Saranak, J., Derguini, F., Zarrilli, G.R., Johnson, R., Okabe, M., and Nakanishi, K. (1989). Activation of *Chlamydomonas* rhodopsin *in vivo* does not require isomerization of retinal. *Biochemistry* **28**, 819–824.

Foster, K.W., Saranak, J., and Dowben, P.A. (1991). Spectral sensitivity, structure and activation of eukaryotic rhodopsins. Activation spectroscopy of rhodopsin analogs in *Chlamydomonas*. *J. Photochem. Photobiol. B* **8**, 385–408.

Fuhrmann, M., Stahlberg, A., Govorunova, E., Rank, S., and Hegemann, P. (2001). The abundant retinal protein of the *Chlamydomonas* eye is not the photoreceptor for phototaxis and photophobic responses. *J. Cell Sci.* **114**, 3857–3863.

Fuhrmann, M., Deininger, W., Kateriya, S., and Hegemann, P. (2003). Rhodopsin-related proteins, cop1, cop2 and chop1, in *Chlamydomonas reinhardtii*. In: *Photoreceptors and Light Signalling* (A. Batschauer, Ed.), pp. 124–135. Royal Society of Chemistry, Cambridge, UK.

Gloeckner, G. and Beck, C.F. (1995). Genes involved in light control of sexual differentiation in *Chlamydomonas reinhardtii*. *Genetics* **141**, 937–943.

Govorunova, E.G., Sineshchekov, O.A., and Hegemann, P. (1997). Desensitization and dark recovery of the photoreceptor current in *Chlamydomonas reinhardtii*. *Plant Physiol.* **115**, 633–642.

Govorunova, E.G., Jung, K.H., Sineshchekov, O.A., and Spudich, J.L. (2004). *Chlamydomonas* sensory rhodopsins A and B: cellular content and role in photophobic responses. *Biophys. J.* **86**, 2342–2349.

Halldal, P. (1958). Action spectra of phototaxis and related problems in Volvocales, *Ulva* gametes and Dinophyceae. *Physiol. Plant* **11**, 118–153.

Hartshorne, J.N. (1953). The function of the eyespot in *Chlamydomonas*. *New Phytol.* **52**, 292–297.

Harz, H. and Hegemann, P. (1991). Rhodopsin-regulated calcium currents in *Chlamydomonas*. *Nature* **351**, 489–491.

Harz, H., Nonnengasser, C., and Hegemann, P. (1992). The photoreceptor current of the green alga *Chlamydomonas*. *Phil. Trans. Roy. Soc. Lond. Ser. B Biol. Sci.* **338**, 39–52.

Hegemann, P. and Bruck, B. (1989). Light-induced stop response in *Chlamydomonas reinhardtii*: occurrence and adaptation phenomena. *Cell Motil. Cytoskeleton* **14**, 501–515.

Hegemann, P. and Harz, H. (1998). How microalgae see the light. In: *Microbial Responses to Light and Time. Society for General Microbiology Symposium* (M.X. Caddick, S. Baumberg, D.A. Hodgson, and M.K. Phillip-Jones, Eds.), pp. 95–105. Cambridge University Press, Cambridge.

Hegemann, P. and Marwan, W. (1988). Single photons are sufficient to trigger movement responses in *Chlamydomonas reinhardtii*. *Photochem. Photobiol.* **48**, 99–106.

Hegemann, P., Hegemann, U., and Foster, K.W. (1988). Reversible bleaching of *Chlamydomonas reinhardtii* rhodopsin *in vivo*. *Photochem. Photobiol.* **48**, 123–128.

Hegemann, P., Gärtner, W., and Uhl, R. (1991). All-trans retinal constitutes the functional chromophore in *Chlamydomonas* rhodopsin. *Biophys. J.* **60**, 1477–1489.

Hegemann, P., Fuhrmann, M., and Kateriya, S. (2001). Algal sensory photoreceptors. *J. Phycol.* **37**, 668–676.

Holland, E.M., Braun, F.J., Nonnengasser, C., Harz, H., and Hegemann, P. (1996). Nature of rhodopsin-triggered photocurrents in *Chlamydomonas*. 1. Kinetics and influence of divalent ions. *Biophys. J.* **70**, 924–931.

Holland, E.M., Harz, H., Uhl, R., and Hegemann, P. (1997). Control of phobic behavioral responses by rhodopsin-induced photocurrents in *Chlamydomonas*. *Biophys. J.* **73**, 1395–1401.

Holzer, W., Penzkofer, A., Fuhrmann, M., and Hegemann, P. (2002). Spectroscopic characterization of flavin mononucleotide bound to the LOV1 domain of Phot1 from *Chlamydomonas reinhardtii*. *Photochem. Photobiol.* **75**, 479–487.

Hoops, H.J., Brighton, M.C., Stickles, S.M., and Clement, P.R. (1999). A test of two possible mechanisms for phototactic steering in *Volvox carteri* (Chlorophyceae). *J. Phycol.* **35**, 539–547.

Horst, C.J. and Witman, G.B. (1993). *ptx1*, a nonphototactic mutant of *Chlamydomonas*, lacks control of flagellar dominance. *J. Cell Biol.* **120**, 733–741.

Huang, K.Y. and Beck, C.F. (2003). Photoropin is the blue-light receptor that controls multiple steps in the sexual life cycle of the green alga *Chlamydomonas reinhardtii*. *Proc. Natl. Acad. Sci. USA* **100**, 6269–6274.

Huang, K.Y., Kunkel, T., and Beck, C.F. (2004). Localization of the blue-light receptor phototropin to the flagella of the green alga *Chlamydomonas reinhardtii*. *Mol. Biol. Cell.* **15**, 3605–3614.

Hyams, J.S. and Borisy, G.G. (1978). Isolated flagellar apparatus of *Chlamydomonas*: characterization of forward swimming and alteration of waveform and reversal of motion by calcium ions *in vitro*. *J. Cell Sci.* **33**, 235–253.

Hyams, J. and Davies, D.R. (1972). The induction and characterisation of cell wall mutants of *Chlamydomonas reinhardi*. *Mutation Res.* **14**, 381–389.

Im, C.S., Eberhard, S., Huang, K.Y., Beck, C.F., and Grossman, A.R. (2006). Phototropin involvement in the expression of genes encoding chlorophyll and carotenoid biosynthesis enzymes and LHC apoproteins in *Chlamydomonas reinhardtii*. *Plant J.* **48**, 1–16.

Immeln, D., Schlesinger, R., Heberle, J., and Kottke, T. (2007). Blue light induces radical formation and autophosphorylation in the light-sensitive domain of *Chlamydomonas cryptochrome*. *J. Biol. Chem.* **282**, 21720–21728.

Iomini, C., Li, L.Y., Mo, W.J., Dutcher, S.K., and Piperno, G. (2006). Two flagellar genes, *AGG2* and *AGG3*, mediate orientation to light in *Chlamydomonas*. *Curr. Biol.* **16**, 1147–1153.

Iroshnikova, G.A., Rakhimberdieva, M.G., and Karapetyan, N.V. (1982). Study of pigmentation-modifying mutations in strains of *Chlamydomonas reinhardii* of different ploidy. III. Characteristics of disturbances of the photosynthetic apparatus in the presence of mutations in the *lts1* locus. *Sov. Genet.* **18**, 1350–1356.

Ishizuka, T., Kakuda, M., Araki, R., and Yawo, H. (2006). Kinetic evaluation of photosensitivity in genetically engineered neurons expressing green algae light-gated channels. *Neurosci. Res.* **54**, 85–94.

Johnson, C.H., Kondo, T., and Hastings, J.W. (1991). Action spectrum for resetting the circadian phototaxis rhythm in the *CW15* strain of *Chlamydomonas*. II. Illuminated cells. *Plant Physiol.* **97**, 1122–1129.

Josef, K., Saranak, J., and Foster, K.W. (2005a). An electro-optic monitor of the behavior of *Chlamydomonas reinhardtii* cilia. *Cell Motil. Cytoskeleton* **61**, 83–96.

Josef, K., Saranak, J., and Foster, K.W. (2005b). Ciliary behavior of a negatively phototactic *Chlamydomonas reinhardtii*. *Cell Motil. Cytoskeleton* **61**, 97–111.

Kamiya, R. and Witman, G.B. (1984). Submicromolar levels of calcium control the balance of beating between the two flagella in demembranated models of *Chlamydomonas*. *J. Cell Biol.* **98**, 97–107.

Kateriya, S., Nagel, G., Barnberg, E., and Hegemann, P. (2004). "Vision" in single-celled algae. *News Physiol. Sci.* **19**, 133–137.

Kay, C.W.M., Schleicher, E., Kuppig, A., Hofner, H., Rudiger, W., Schleicher, M., Fischer, M., Bacher, A., Weber, S., and Richter, G. (2003). Blue light perception in plants. Detection and characterization of a light-induced neutral flavin radical in a C450A mutant of phototropin. *J. Biol. Chem.* **278**, 10973–10982.

King, S.J. and Dutcher, S.K. (1997). Phosphoregulation of an inner dynein arm complex in *Chlamydomonas reinhardtii* is altered in phototactic mutant strains. *J. Cell Biol.* **136**, 177–191.

Kirk, D.L. (1998). *Volvox: Molecular Genetic Origins of Multicellularity and Cellular Differentiation*. Cambridge University Press.

Kirk, M.M. and Kirk, D.L. (1985). Translational regulation of protein synthesis, in response to light, at a critical stage of Volvox development. *Cell* **41**, 419–428.

Kondo, T., Johnson, C.H., and Hastings, J.W. (1991). Action spectrum for resetting the circadian phototaxis rhythm in the *CW15* strain of *Chlamydomonas*. 1. Cells in darkness. *Plant Physiol.* **95**, 197–205.

Kottke, T., Heberle, J., Hehn, D., Dick, B., and Hegemann, P. (2003). Phot-LOV1: Photocycle of a blue-light receptor domain from the green alga *Chlamydomonas reinhardtii*. *Biophys. J.* **84**, 1192–1201.

Kreimer, G. and Melkonian, M. (1990). Reflection confocal laser scanning microscopy of eyespots in flagellated green algae. *Eur. J. Cell Biol.* **53**, 101–111.

Kreimer, G., Overlander, C., Sineshchekov, O.A., Stolzis, H., Nultsch, W., and Melkonian, M. (1992). Functional analysis of the eyespot in *Chlamydomonas reinhardtii* mutant *ey* 627, *mt*-. *Planta* **188**, 513–521.

Lamb, M.R., Dutcher, S.K., Worley, C.K., and Dieckmann, C.L. (1999). Eyespot-assembly mutants in *Chlamydomonas reinhardtii*. *Genetics* **153**, 721–729.

Land, M.F. (1978). Animal eyes with mirror optics. *Sci. Am.* **239**, 126–134.

Lawson, M.A. and Satir, P. (1994). Characterization of the eyespot regions of blind *Chlamydomonas* mutants after restoration of photophobic responses. *J. Euk. Microbiol.* **41**, 593–601.

Lawson, M.A., Zacks, D.N., Derguini, F., Nakanishi, K., and Spudich, J.L. (1991). Retinal analog restoration of photophobic responses in a blind *Chlamydomonas reinhardtii* mutant. Evidence for an archaebacterial like chromophore in a eukaryotic rhodopsin. *Biophys. J.* **60**, 1490–1498.

Li, X., Gutierrez, D.V., Hanson, M.G., Han, J., Mark, M.D., Chiel, H., Hegemann, P., Landmesser, L.T., and Herlitze, S. (2005). Fast noninvasive activation and inhibition of neural and network activity by vertebrate rhodopsin and green algae channelrhodopsin. *Proc. Natl. Acad. Sci. USA* **102**, 17816–17821.

Litvin, F.F., Sineshchekov, O.a., and Sineshchekov, V.a. (1978). Photoreceptor electric potential in the phototaxis of the alga *Haematococcus pluvialis*. *Nature* **271**, 476–478.

Matsuda, A., Yoshimura, K., Sineshchekov, O.A., Hirono, M., and Kamiya, R. (1998). Isolation and characterization of novel *Chlamydomonas* mutants that display phototaxis but not photophobic response. *Cell Motil. Cytoskeleton* **41**, 353–362.

McCarthy, S.S., Kobayashi, M.C., and Niyogi, K.K. (2004). White mutants of *Chlamydomonas reinhardtii* are defective in phytoene synthase. *Genetics* **168**, 1249–1257.

Melkonian, M. and Robenek, H. (1984). The eyespot apparatus of flagellated green algae: a critical review. *Prog. Phycol. Res.* **3**, 193–268.

Mergenhagen, D. (1984). Circadian clock: genetic characterization of a short period mutant of *Chlamydomonas reinhardii*. *Eur. J. Cell Biol.* **33**, 13–18.

Mitchell, D.R. and Rosenbaum, J.L. (1985). A motile *Chlamydomonas* flagellar mutant that lacks outer dynein arms. *J. Cell Biol.* **100**, 1228–1234.

Mittag, M., Kiaulehn, S., and Johnson, C.H. (2005). The circadian clock in *Chlamydomonas reinhardtii*. What is it for? What is it similar to?. *Plant Physiol.* **137**, 399–409.

Morel-Laurens, N.M.L. and Feinleib, M.E. (1983). Photomovement in an eyeless mutant of *Chlamydomonas*. *Photochem. Photobiol.* **37**, 189–194.

Nagel, G., Ollig, D., Fuhrmann, M., Kateriya, S., Mustl, A.M., Bamberg, E., and Hegemann, P. (2002). Channelrhodopsin-1: a light-gated proton channel in green algae. *Science* **296**, 2395–2398.

Nagel, G., Szellas, T., Huhn, W., Kateriya, S., Adeishvili, N., Berthold, P., Ollig, D., Hegemann, P., and Bamberg, E. (2003). Channelrhodopsin-2, a directly light-gated cation-selective membrane channel. *Proc. Natl. Acad. Sci. USA* **100**, 13940–13945.

Nagel, G., Brauner, M., Liewald, J.F., Adeishvili, N., Bamberg, E., and Gottschalk, A. (2005). Light activation of channelrhodopsin-2 in excitable cells of Caenorhabditis elegans triggers rapid behavioral responses. *Curr. Biol.* **15**, 2279–2284.

Nakamura, S., Ogihara, H., Jinbo, K., Tateishi, M., Takahashi, T., Yoshimura, K., Kubota, M., Watanabe, M., and Nakamura, S. (2001). *Chlamydomonas reinhardtii* Dangeard (Chlamydomonadales, Chlorophyceae) mutant with multiple eyespots. *Phycol. Res.* **49**, 115–121.

Nonnengasser, C., Holland, E.M., Harz, H., and Hegemann, P. (1996). The nature of rhodopsin-triggered photocurrents in *Chlamydomonas*. 2. Influence of monovalent ions. *Biophys. J.* **70**, 932–938.

Nultsch, W., Throm, G., and Rimscha, I.V. (1971). Phototaktische Untersuchungen an *Chlamydomonas reinhardii* Dangeard in homokontinuierlicher Kultur. *Arch. Mikrobiol.* **80**, 351–369.

Pazour, G., Sineschekov, O., and Witman, G. (1995). Mutational analysis of the phototransduction pathway of *Chlamydomonas reinhardtii*. *J. Cell Biol.* **131**, 427–440.

Reisdorph, N.A. and Small, G.D. (2004). The *CPH1* gene of *Chlamydomonas reinhardtii* encodes two forms of cryptochrome whose levels are controlled by light-induced proteolysis. *Plant Physiol.* **134**, 1546–1554.

Roberts, D.G.W., Lamb, M.R., and Dieckmann, C.L. (2001). Characterization of the EYE2 gene required for eyespot assembly in *Chlamydomonas reinhardtii*. *Genetics* **158**, 1037–1049.

Rüffer, U. and Nultsch, W. (1985). High speed cinematographic analysis of the movement of *Chlamydomonas*. *Cell Motil. Cytoskeleton* **5**, 251–263.

Rüffer, U. and Nultsch, W. (1987). Comparison of the beating of cis- and trans-flagella of *Chlamydomonas* held on micropipettes. *Cell Motil. Cytoskeleton* **7**, 87–93.

Rüffer, U. and Nultsch, W. (1990). Flagellar photoresponses of *Chlamydomonas* cells held on micropipettes: I. Change in flagellar beat frequency. *Cell Motil. Cytoskeleton* **15**, 162–167.

Rüffer, U. and Nultsch, W. (1991). Flagellar photoresponses of *Chlamydomonas* cells held on micropipettes: II. Change in flagellar beat pattern. *Cell Motil. Cytoskeleton* **18**, 269–278.

Rüffer, U. and Nultsch, W. (1995). Flagellar photoresponses of *Chlamydomonas* cells held on micropipettes. III. Shock responses. *Bot. Acta* **108**, 255–265.

Rüffer, U. and Nultsch, W. (1997). Flagellar photoresponses of *ptx1*, a nonphototactic mutant of *Chlamydomonas*. *Cell Motil. Cytoskeleton* **37**, 111–119.

Rüffer, U. and Nultsch, W. (1998). Flagellar coordination in *Chlamydomonas* cells held on micropipettes. *Cell Motil. Cytoskeleton* **41**, 297–307.

Sager, R. and Granick, S. (1954). Nutritional control of sexuality in *Chlamydomonas reinhardi*. *J. Gen. Physiol.* **37**, 729–742.

Sakaguchi, H. and Iwasa, K. (1979). Two photophobic responses in Volvox carteri. *Plant Cell Physiol.* **20**, 909–916.

Sakamoto, M., Wada, A., Akai, A., Ito, M., Goshima, T., and Takahashi, T. (1998). Evidence for the archaebacterial-type conformation about the bond between the β-Ionone ring and the polyene chain of the chromophore retinal in chlamyrhodopsin. *FEBS Lett.* **434**, 335–338.

Schaller, K. and Uhl, R. (1997). A microspectrophotometric study of the shielding properties of eyespot and cell body in *Chlamydomonas*. *Biophys. J.* **73**, 1573–1578.

Schaller, K., David, R., and Uhl, R. (1997). How *Chlamydomonas* keeps track of the light once it has reached the right phototactic orientation. *Biophys. J.* **73**, 1562–1572.

Schletz, K. (1976). Phototaxis in Volvox. Pigments involved in perception of light direction. *Z. Pflanzenphysiol* **77**, 189–211.

Schmidt, J.A. and Eckert, R. (1976). Calcium couples flagellar reversal to photostimulation in *Chlamydomonas reinhardtii*. *Nature* **262**, 713–715.

Schmidt, M., Gessner, G., Matthias, L., Heiland, I., Wagner, V., Kaminski, M., Geimer, S., Eitzinger, N., Reissenweber, T., Voytsekh, O., Fiedler, M., Mittag, M., and Kreimer, G. (2006). Proteomic analysis of the eyespot of *Chlamydomonas reinhardtii* provides novel insights into its components and tactic movements. *Plant Cell* **18**, 1908–1930.

Sheen, J. (2002). Phosphorelay and transcription control in cytokinin signal transduction. *Science* **296**, 1650–1652. erratum Science 298, 1172

Sineshchekov, O.A. and Spudich, J.L. (2005). Sensory rhodopsin signaling in green flagellate algae. In: *Handbook of Photosensory Receptors* (W.R. Briggs and J.L. Spudich, Eds.), pp. 25–42. Wiley-VCH Verlag GmbH & Co KGaA.

Sineshchekov, O.A., Govorunova, E.G., Der, A., Keszthelyi, L., and Nultsch, W. (1992). Photoelectric responses in phototactic flagellated algae measured in cell-suspension. *J. Photochem. Photobiol. B Biol.* **13**, 119–134.

Sineshchekov, O.A., Govorunova, E.G., Der, A., Keszthelyi, L., and Nultsch, W. (1994). Photoinduced electric currents in carotenoid-deficient *Chlamydomonas* mutants reconstituted with retinal and its analogs. *Biophys. J.* **66**, 2073–2084.

Sineshchekov, O.A., Jung, K.H., and Spudich, J.L. (2002). Two rhodopsins mediate phototaxis to low- and high-intensity light in *Chlamydomonas reinhardtii*. *Proc. Natl. Acad. Sci. USA* **99**, 8689–8694.

Straley, S.C. and Bruce, V.G. (1979). Stickiness to glass. Circadian changes in the cell surface of *Chlamydomonas reinhardi*. *Plant Physiol.* **63**, 1175–1181.

Suzuki, L. and Johnson, C.H. (2002). Photoperiodic control of germination in the unicell *Chlamydomonas*. *Naturwissenschaften* **89**, 214–220.

Suzuki, T., Yamasaki, K., Fujita, S., Oda, K., Iseki, M., Yoshida, K., Watanabe, M., Daiyasu, H., Toh, H., Asamizu, E., Tabata, S., Miura, K., Fukuzawa, H., Nakamura, S., and Takahashi, T. (2003). Archaeal-type rhodopsins in *Chlamydomonas*: model structure and intracellular localization. *Biochem. Biophys. Res. Commun.* **301**, 711–717.

Takada, S. and Kamiya, R. (1997). Beat frequency difference between the two flagella of *Chlamydomonas* depends on the attachment site of outer dynein arms on the outer-doublet microtubules. *Cell Motil. Cytoskeleton* **36**, 68–75.

Takahashi, T. and Watanabe, M. (1993). Photosynthesis modulates the sign of phototaxis of wild-type *Chlamydomonas reinhardtii*. Effects of red background illumination and 3-(3′,4′-dichlorophenyl)-1,1-dimethylurea. *FEBS Lett.* **336**, 516–520.

Takahashi, T., Kubota, M., Watanabe, M., Yoshihara, K., Derguini, F., and Nakanishi, K. (1992). Diversion of the sign of phototaxis in a *Chlamydomonas reinhardtii* mutant incorporated with retinal and its analogs. *FEBS Lett.* **314**, 275–279.

Trevianus, L.G. (1817). Vermischte Schriften. II, 84.

Yan, B., Takahashi, T., Johnson, R., Derguini, F., Nakanishi, K., and Spudich, J.L. (1990). All-trans/13-cis isomerization of retinal is required for phototaxis signaling by sensory rhodopsins in *Halobacterium halobium*. *Biophys. J.* **57**, 807–814.

Yang, P., Diener, D.R., Yang, C., Kohno, T., Pazour, G.J., Dienes, J.M., Agrin, N.S., King, S.M., Sale, W.S., Kamiya, R., Rosenbaum, J.L., and Witman, G.B. (2006). Radial spoke proteins of *Chlamydomonas* flagella. *J. Cell Sci.* **119**, 1165–1174.

Yoshimura, K. and Kamiya, R. (2001). The sensitivity of *Chlamydomonas* photoreceptor is optimized for the frequency of cell body rotation. *Plant Cell Physiol.* **42**, 665–672.

Zacks, D.N. and Spudich, J.L. (1994). Gain setting in *Chlamydomonas reinhardtii*: mechanism of phototaxis and the role of the photophobic response. *Cell Motil. Cytoskeleton* **29**, 225–230.

Zacks, D.N., Derguini, F., Nakanishi, K., and Spudich, J.L. (1993). Comparative study of phototactic and photophobic receptor chromophore properties in *Chlamydomonas reinhardtii*. *Biophys. J.* **65**, 508–518.

Zhang, F., Prigge, M., Beyrière, F., Tsunoda, S.P., Mattis, J., Yizhar, O., Hegemann, P., and Deisseroth, K. (2008). Red-shifted optogenetic excitation: a tool for fast neural control derived from *Volvox carteri*. *Nat. Neurosci.* **11**, 631–633.

Zhang, Y.-P. and Oertner, T.G. (2007). Optical induction of synaptic plasticity using a light-sensitive channel. *Nat. Meth.* **4**, 139–141.

CHAPTER 14

Mitosis and Cytokinesis

Wallace F. Marshall
Department of Biochemistry and Biophysics, UCSF,
San Francisco, California, USA

CHAPTER CONTENTS

I. INTRODUCTION

A. Why study *Chlamydomonas* cell division?

The mechanisms of cell division in *Chlamydomonas* are of inherent interest for several reasons. First, the extensive work on centrioles and basal bodies in *Chlamydomonas* (see Chapter 2) makes this organism an excellent model system for exploring the role of centrioles in cell division. Second, the fact that the *Chlamydomonas* cell cycle can be synchronized with light/dark cycles provides an important technical advantage to the analysis of cell division. Furthermore, the fact that cell size control is an experimentally tractable phenomenon in *Chlamydomonas* (Umen and Goodenough, 2001) makes this alga an excellent system to study the interplay between cell size and division. Finally, the fact that cell division in *Chlamydomonas* combines features of both plant and animal cells presents a unique opportunity to consider the evolution of mitosis and cytokinesis.

Studies of cell division in *Chlamydomonas* basically employ two types of approaches, often in combination: microscopy and genetics. In this chapter we will survey the existing literature on *Chlamydomonas* cell division, with a focus on the mechanisms of mitosis and cytokinesis. The reader will notice that the powerful genomic and genetic approaches now available, combined with the relatively primitive state of our current knowledge, means that there is a wealth of opportunities for future study in this area.

B. Overview of the events of cell division in *Chlamydomonas*

We start with a timeline of division events in *Chlamydomonas*. The first visible sign that a *Chlamydomonas* cell is getting ready to divide is that its flagella resorb. This resorption takes place over roughly 20 minutes. While resorption of cilia and flagella prior to cell division is a near-universal feature of eukaryotic cells, the functional link between flagellar resorption and cell cycle progression is poorly understood. The mechanism of resorption is also not known although it has been suggested to involve both disassembly of the axoneme at its distal end as well as severing of the axoneme at the transition zone (see Chapters 2 and 3). Within a few minutes of completion of flagellar resorption, the cortex of the cell undergoes a 90-degree rotation. This cortical rotation is sensitive to light, so that it can be inhibited to varying extents during transmitted light imaging; however, this appears to have no effect on mitosis or cytokinesis (Holmes and Dutcher, 1989). The driving force for this process is not known. It is interesting to consider whether this rotation might have mechanistic features in common with the cortical rotation that takes place in fertilized *Xenopus* eggs (Gerhart et al., 1989).

Within minutes after flagellar resorption and cortical rotation, the cleavage furrow begins to form, usually at one end of the cell, and the process completes within 20 minutes to yield two daughter cells. One complicating feature of division in *Chlamydomonas* is that cells can undergo more than one round of rapid division to produce as many as eight progeny cells without an intervening period of growth. This is presumably an adaptation to allow the cells to take maximal advantage of available light, by spending all day in a state of pure growth and then dividing several times at night when the sun has gone down. The number of divisions appears to be regulated by the cell size control machinery, and is adjusted based on the size of the cells at the commitment point (Umen and Goodenough, 2001). This unusual pattern of cell division is also observed in apicomplexan parasites (Morrissette and Sibley, 2002).

The divisions take place within the remnant of the mother cell wall. When the newly formed daughter cells are ready to hatch, they need to escape from the mother wall, and they do this with the assistance of an enzymatic activity secreted by the daughter cells (Matsuda et al., 1995; and see Volume 1, Chapter 2). Cells with defects in flagella or flagellar motility often show delays in hatching and tend to form large clumps of unhatched cells, suggesting that the newly formed flagella have a role in the processing, secretion or display of the cell-wall lytic enzyme. One might also speculate that the mechanical activity of the flagella might be involved in escape from the mother cell wall, much as a chick uses its beak to escape from an egg; however, the fact that some paralyzed flagella (*pf*) mutants do not show a hatching defect argues against a critical role for flagellar motility in the hatching process.

In the remainder of this chapter, we consider the details of mitosis and cytokinesis that lead to the generation of a new set of daughter cells. More general aspects of cell cycle regulation and cell division are discussed in Volume 1, Chapter 3.

II. MITOSIS

A. Chromatin and nuclear envelope

Mitosis in *Chlamydomonas* is accompanied by condensation of the chromosomes, making it a more typical mitosis than, for example, that seen in budding yeast. The molecular players involved in chromosome condensation have not been studied in *Chlamydomonas*, although the genome clearly contains the components of the condensin and cohesin complexes. The histone composition of *Chlamydomonas* chromatin is generally similar to that of other eukaryotes (Morris et al., 1990). Mitotic *Chlamydomonas* chromatin is highly enriched in phosphorylated histone H3 (Figure 14.1). This post-translational modification of histones is carried out by a number of different kinases, including Aurora family kinases, and is a marker for mitotic chromatin in both animal and plant cells (Prigent and Dimitrov, 2003; Kawabe et al., 2005).

Although it has not been quantified, the degree of condensation seen in *Chlamydomonas* mitotic chromosomes is visually similar to that seen in metazoans. Individual chromosomes can be clearly seen provided the imaging is performed using high quality optics and deconvolution, and that care is taken to preserve the three-dimensional nature of the cell and avoid making flattened spread preparations.

In contrast to mammalian cells, the nuclear envelope of *Chlamydomonas* does not break down during mitosis. Instead, the envelope becomes fenestrated at the poles, presumably to allow the basal bodies to interact with the spindle poles (Johnson and Porter, 1968).

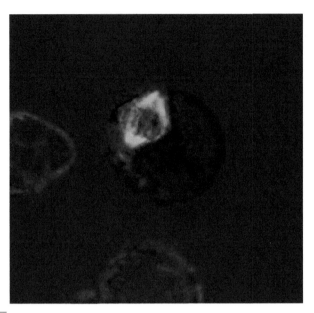

FIGURE 14.1 *Metaphase spindle of* Chlamydomonas. *(Green) alpha tubulin immunofluorescence, (Red) phospho-histone H3 specific antibody immunofluorescence, (Blue) DAPI DNA stain. Only mitotic cells give a positive signal with the phospho-H3 antibody. DAPI signal outside of spindle represents DNA nucleoids of chloroplasts and mitochondria. Image shown is a projection of a deconvolved three-dimensional data set acquired using a DeltaVision system with a 100× NA 1.4 objective lens. Courtesy of Lani Keller.*

B. Mitotic spindle and cytoskeleton

The mitotic spindle of *Chlamydomonas* is readily observed by staining cells with commercially available antibodies specific for alpha tubulin (Figure 14.1). Imaging of *Chlamydomonas* spindles tends to show a more or less classical spindle appearance, although sometimes the spindle looks slightly bent or crescent shaped. The spindle poles are close to the cell surface, probably because the basal bodies do not fully detach from the cortex when they move to the spindle poles (Cavalier-Smith, 1974).

It is unclear whether the *Chlamydomonas* spindle contains astral microtubules in the usual sense. Certainly, they are not readily seen in fluorescence images of spindles, but this is often the case in other cell types as well, even when astral microtubules are known to be present. The problem is that astral microtubules are relatively labile and often do not survive fixation (W. Theurkauf, personal communication). We therefore reserve judgment as to whether or not there are true astral microtubules in *Chlamydomonas*, on the grounds of insufficient evidence. On the other hand, it is clear that a different set of microtubules, the rootlets, plays an important role in spindle positioning and may therefore perform an equivalent function to that normally performed by astral microtubules in metazoans.

The rootlet microtubules consist of two two-membered rootlets, containing two parallel microtubules, and two four-membered rootlets, containing four parallel microtubules (see Chapter 2). The rootlet microtubules are heavily acetylated, making them easy to visualize using antibodies specific for this modification. We will discuss below the potential role of rootlets in guiding cytokinesis, but in mitosis the function of rootlets is much less clear. One possibility is that they determine spindle orientation. When the spindle poles are widely separated, the basal body pair at each pole tends to be closely associated with a microtubule rootlet. This could indicate that the basal bodies migrate along opposing rootlets in order to separate during early mitosis, but it could also indicate that the basal bodies themselves determine the formation of rootlets. Either way, the apparently tight association of centrioles with the rootlets, which are in turn associated with the cell cortex, suggests that rootlets could play a function analogous to astral microtubules in animals, by linking the spindle poles to defined sites on the cell surface. Consistent with this view, *bld2* cells, which have centrioles that are reduced to a short cylinder of nine singlet microtubules (St. Clair and Goodenough, 1975) due to a defect in the centriole-specific protein epsilon tubulin (Dutcher et al., 2002), show disorganized or frayed rootlets and also have randomized spindle orientations (Ehler et al., 1995). The latter observations were initially interpreted as reflecting a role for centrioles in spindle positioning on the grounds that 99% of *bld2* cells lacked centrioles as assayed by electron microscopy (Ehler et al., 1995). However, in immunofluorescence microscopy studies, all *bld2* cells showed two foci when labeled with antibodies to POC1 (EDP07349), a protein recruited early to procentrioles (Marshall, unpublished results). Thus, all *bld2* cells have centrioles, although in a reduced form easily missed by electron microscopy. It therefore seems likely that the spindle orientation defects in these cells are a consequence of the disorganized rootlets. Additional functions of the rootlet microtubules during cytokinesis will be discussed below.

There are no known mutations which specifically ablate the rootlets, nor have mutants been described in which centrioles and rootlets become dissociated from one another. Without such mutations, it is impossible to directly test a causal role for rootlets in mitosis; thus, our understanding of rootlet function is currently restricted to inferences based on spatial correlation.

C. Mitotic spindle poles

The organization of the mitotic spindle pole in *Chlamydomonas* differs from that of animal cells in terms of the location of the centrioles. In contrast to animal cells where centrioles are buried in the center of the mitotic spindle pole, electron microscopic studies in *Chlamydomonas* have clearly shown that the basal bodies are not located directly at the spindle poles

(Johnson and Porter, 1968; Coss, 1974). Instead, the basal bodies maintain an attachment to the poles via centrin-containing fibers known as nucleus-basal body connectors (NBBCs) (Wright et al., 1985). These fibers, defined based on immunofluorescence staining with anti-centrin antibodies, appear similar to fibers known as "rhizoplasts" (Kater, 1929), and indeed centrin was first identified as a protein obtained from isolated rhizoplasts of the green alga *Tetraselmis* (Salisbury et al., 1984). Yet another term sometimes used in the literature to describe this structure is "striated rootlet." The NBBC is dispensable for mitosis, as mutants lacking this structure can perform normal mitosis (Wright et al., 1989).

The situation has been clarified by electron microscopy studies, conducted in the related unicellular green alga *Spermatozopsis* (Lechtreck and Grunow, 1999), in which images of different stages of early mitosis indicate that spindle microtubules are initially nucleated near the base of the basal body, which migrates to the nuclear envelope early in mitosis. As the nuclear envelope becomes fenestrated, the microtubule cloud surrounding the basal body is deposited on the nuclear surface, and the basal body then separates from the pole but remains tethered to it by distinct fibers, presumably composed of centrin. It is likely that a similar situation will hold in *Chlamydomonas*.

Mutants in the centrin protein VFL2 show a loss of basal bodies at the spindle poles. *vfl2* mutants also show an increased rate of mitotic chromosome loss (Zamora and Marshall, 2005). This may indicate that spindle poles lacking basal bodies function less effectively in mitotic chromosome segregation; alternatively, this result might reflect some basal body-independent function for centrin.

If basal bodies are not actually within the spindle pole, this leaves us to ask what composes the spindle pole itself. Gamma-tubulin is present in *Chlamydomonas* and localizes to the basal bodies and can also be seen at the poles of the spindle (Dibbayawan et al., 1995; Silflow et al., 1999), suggesting that the basic microtubule-nucleating function of the spindle pole in *Chlamydomonas* will be similar to that of other organisms. The *Chlamydomonas* genome encodes clear homologues of the gamma-tubulin ring complex proteins GCP2 (EDP00586), GCP3 (EDO98538), and GCP4 (EDP09233), suggesting that the basic molecular organization of the gamma-tubulin ring complex is conserved in *Chlamydomonas*. It is more difficult to say whether or not the overall organization of the centrosome is conserved with animal cells, since the main pericentriolar proteins such as pericentrin contain sufficient coiled-coil and other simple sequence to make it very hard to identify homologues in other eukaryotes.

D. Genetic analysis of mitosis in *Chlamydomonas*

Since the great strength of *Chlamydomonas* lies in its utility for classical genetics, we will consider what is known about genes involved in mitosis in this organism. In contrast to studies of flagella/basal bodies or photosynthesis, there have been relatively few systematic genetic studies of mitosis in *Chlamydomonas*. Instead, many of the known genetic results are based on analyses of genes first identified in screens for flagellar phenotypes. This is therefore an area with tremendous room for future exploration.

One type of screen that has directly identified spindle-related genes has been screens for resistance to microtubule-depolymerizing agents. For example, the *APM1* and *APM2* loci (James et al., 1988) were identified based on resistance to the compound amiprophos-methyl, which inhibits tubulin synthesis (Collis and Weeks, 1978). In many cases the resistance mutations occur in genes encoding alpha- or beta-tubulin and confer increased stability to microtubules (Bolduc et al., 1988; Schibler and Huang, 1991; James et al., 1993). Another way to identify mutations affecting microtubules is to screen visually for defects in cell shape. Using this approach, Horst et al. (1999) identified a mutation called *cmu1-1* in which cytoplasmic microtubules are disorganized, being unusually long and curling around the cell. In such mutants, there appears to be a defect in spindle positioning during mitosis as judged by the fact that some daughter cells fail to inherit nuclei while others inherit two nuclei. This mutation may point the way to an analysis of spindle orientation during mitosis in *Chlamydomonas*.

Several genetic studies have implicated FLA10 kinesin, a component of the heterotrimeric kinesin-2 that drives intraflagellar transport (see Chapter 4), in mitosis or cell division. FLA10 localizes to the basal bodies during mitosis; this discovery immediately raised the possibility that it might be involved in spindle pole function (Vashishtha et al., 1996). Such an involvement was supported by genetic interaction studies with other mutants affecting mitosis. An allele of *APM1*, *apm1-122*, shows a synthetic growth-arrest phenotype in combination with *fla10* mutations (Lux and Dutcher, 1991). This result raises the possibility that the *APM1* gene product might possess a microtubule dynamics-modifying activity that acts through the spindle pole, perhaps by affecting microtubule nucleation, although to date the *APM1* gene has not been cloned. A second type of evidence implicating FLA10 in mitosis comes from chromosome loss analysis. While it is true that a null allele at the *FLA10* locus is completely viable, indicating that FLA10 is not strictly required for mitosis (Matsuura et al., 2002), other *fla10* alleles have been shown by sensitive assays to confer increased rates of chromosome loss (Miller et al., 2005). Extensive genetic analysis suggests that this chromosome loss phenotype is due to a dominant gain-of-function mutation in *FLA10* and thus probably does not reflect a role for normal FLA10 function in the process of mitosis, but this remains a question that calls for further investigation.

The regulation of mitosis in *Chlamydomonas* is still not understood; however, a mutant, *met1*, is known which causes cells to arrest in metaphase (Harper et al., 2004). Cloning MET1 could provide a much-needed foot in the door to understanding the molecular regulatory networks controlling mitosis in this organism. Cell cycle regulation is discussed in more detail in Volume 1, Chapter 3.

III. CYTOKINESIS

A. Comparison with plant and animal cytokinesis

Cytokinetic mechanisms in different phyla can potentially differ in terms of their initiation, progression, and termination (Guertin et al., 2002). Initiation of cytokinesis determines where the division plane will be placed. Once the cleavage plane is determined, the cytoplasm of the two daughter cells must be physically divided. It was once assumed that the progression of cleavage during cytokinesis in plants and animals occurred by radically different mechanisms, with animals relying on contraction of an actin ring to pinch off the two daughter cells and plants building a partition by accumulating vesicles at the site of division. In fact, it is now clear that even in animal cells, vesicle accumulation and ring contraction both take place, with actin ring contraction guiding the early stages of cytokinesis and vesicle fusion driving the final stages (Albertson et al., 2005). Nevertheless, one can say that a contractile actin ring is more characteristic of animal than plant cytokinesis. Another hallmark that distinguishes plant and animal cytokinesis is the presence of organized microtubule arrays. The preprophase band in plants is a ring of parallel microtubules that becomes evident on the cortex of the cell and seems to mark the future division plane. A second microtubule array prominent in plants is the phragmoplast, which contains antiparallel microtubules oriented perpendicular to the cleavage plane. These microtubules recruit vesicles that build a partition between the daughter cells, obviating the need for a constricting actin ring.

Cleavage plane positioning in *Chlamydomonas* appears to be predetermined by the position of the rootlet microtubules (see Chapter 2). Specifically, the four-membered microtubule rootlets foretell the position where the cleavage furrow will form. The position of these rootlets, in turn, is likely dictated by the position of the basal bodies, so that basal body position may be a key factor in ultimately specifying the cleavage plane. Evidence that the rootlets do not merely co-localize with the actin band but play a functional role in its organization came from studies of mutants in which additional rootlets form (Ehler and Dutcher, 1998). Such mutants recruit actin to the additional rootlets and show subsequent defects in cytokinesis. Similarly, a mutation in the centriole-associated epsilon tubulin causes cells to have aberrant organization of rootlet microtubules, and these mutants also show mispositioning of their cleavage furrows (Ehler et al., 1995), further strengthening the correlation between rootlets and cleavage furrow placement. Although at first such a direct coupling of rootlet position with the cleavage plane may seem like a unique feature of *Chlamydomonas*, the fact that the rootlets may be functional analogues of astral microtubules suggests that the situation in *Chlamydomonas* is quite similar to that in animal cells, where non-kinetochore spindle microtubules and/or astral microtubules appear to modulate cortical events during the early stages of cytokinesis in order to determine furrow position (e.g., Canman et al., 2003).

Chlamydomonas cytokinesis resembles animal cell cytokinesis in terms of its actin dynamics, with a clearly demonstrable actin ring forming and then contracting (Harper et al., 1992; Ehler and Dutcher, 1998). In terms of microtubule organization, however, *Chlamydomonas* cytokinesis is strikingly plant-like. Prior to formation of the actin ring, a parallel array of microtubules called the "phycoplast" forms on the cell cortex (Johnson and Porter, 1968; Gaffal, 1988; Holmes and Dutcher, 1989; Gaffal and El-Gammal, 1990; Schibler and Huang, 1991; Ehler and Dutcher, 1998); in its timing and localization, the phycoplast strongly resembles the preprophase band seen in plants. The position of the phycoplast microtubules is foretold by the position of the four-membered rootlet (Gaffal and El-Gammal, 1990; Ehler and Dutcher, 1998), whose centriole-distal portion is termed the "metaphase band" (Doonan and Grief, 1987). Specification of phycoplast microtubule position by the four-membered rootlet may be a mechanism by which the rootlet microtubules influence the position of the cleavage furrow. It is presently unclear whether formation of the phycoplast array truly precedes actin recruitment, since no live-cell studies have been reported in which both have been visualized. However, if the microtubule-based structures are disrupted pharmacologically, both actin deposition and furrow progression are also disrupted (Ehler and Dutcher, 1998), suggesting that the microtubule structures play an important role in regulating cleavage. Interestingly, a kinesin-related calmodulin-binding protein (KCBP; ABF50981) found to localize to the phycoplast in *Chlamydomonas* (Dymek et al., 2006) is homologous to a protein that localizes to the preprophase band in plants (Bowser and Reddy, 1997), further supporting the idea that the two structures are functionally similar.

When the actin ring is finished constricting, a final step of abscission is needed in order to completely separate the two daughter cells from each other. This is the least well-understood step of cytokinesis in all systems, and *Chlamydomonas* is no exception. To my knowledge, there are no published studies of this final step, and it remains unclear how it happens.

Part of what happens in cytokinesis is partitioning of organelles to the two daughter cells. For large organelles such as chloroplasts that do not fragment into large numbers of small vesicles, it is important to understand how the cell manages to distribute them to the daughters. One simplistic model would be that spindle-generated mitotic forces, or else the progression of the cleavage furrow, play a direct role in chloroplast division by pulling the chloroplast apart, but this is unlikely to be the case. In *met1* mutant cells arrested at the nonpermissive temperature, multiple pyrenoids are seen, indicating that mitosis *per se* is not required for chloroplast separation (Harper et al., 2004).

B. Genetic analysis of cytokinesis

Genetic analysis will ultimately be the most productive route to understanding the mechanism of *Chlamydomonas* cytokinesis. In addition to the basal body mutations discussed above that cause furrow placement defects, several other mutants have been described in which cleavage furrow progression is defective. In the *cyt1*, *oca1*, and *oca2* mutants, furrows are initiated but do not complete (Warr, 1968; Hirono and Yoda, 1997). Interestingly, these defects do not result in cell death, but instead produce multinucleated monster cells with multiple pairs of flagella. The fact that such mutations are not lethal provides a unique opportunity to study the correlates of failed cytokinesis without complications due to cell death. Unfortunately, the specific genes defective in these mutants have not yet been identified.

Several RNAi and dominant negative experiments with various genes have revealed cytokinesis defects. RNAi of centrin (Koblenz et al., 2003) leads to production of binucleate cells, suggesting a defect in cytokinesis. However, these are not as severe as the giant monster cells produced by the mutations described above. RNAi of the intraflagellar transport protein IFT27 caused abnormal cytokinesis, including production of large clumped cells similar to standard cytokinesis mutants (Qin et al., 2007). Similar results were seen with RNAi of the basal-body-associated protein DIP13 (Pfannenschmid et al., 2003). Since all three proteins – centrin, IFT27, and DIP13 – localize at basal bodies, these results may provide an entry point to the analysis of basal body function in cytokinesis. It is worth noting, however, that these studies were not originally intended to identify cytokinesis genes, and the results were therefore somewhat fortuitous, suggesting that a more systematic approach might be useful. The *Chlamydomonas* genome is predicted to encode several obvious candidate proteins that are homologues of proteins known to be involved in cytokinesis in other organisms. For example, there is a clear ortholog (EDP07654) of the fission yeast Mob1p protein that regulates initiation of cytokinesis (septation) (Hou et al., 2004). Combining reverse genetic analysis of such candidates by RNAi with forward genetic analysis of mutants obtained in screens might eventually be the key to understanding how *Chlamydomonas* cytokinesis works.

IV. PERSPECTIVES

Chlamydomonas promises to play a unique role in our future understanding of cell division. *Chlamydomonas* provides many of the same genetic advantages as budding yeast for studying basic cell biology, but in

Chlamydomonas the processes of mitosis and cytokinesis are far more similar to those seen in animal cells than are the corresponding processes in yeast. In particular, *Chlamydomonas* allows us to take a yeast-like forward genetic approach to understanding the roles of centrioles in mitosis, the connection between flagellar resorption and cell division, and the interplay of actin and microtubules during cytokinesis. Moreover, the combination of plant- and animal-like features can help to shed light on the evolution of cell division mechanisms. We have only just begun to learn the lessons that this simple organism can teach us about how cells divide.

REFERENCES

Albertson, R., Riggs, B., and Sullivan, W. (2005). Membrane traffic: a driving force in cytokinesis. *Trends Cell Biol.* **15**, 92–101.

Bolduc, C., Lee, V.D., and Huang, B. (1988). Beta-tubulin mutants of the unicellular green alga *Chlamydomonas reinhardtii. Proc. Natl. Acad. Sci. U.S.A.* **85**, 131–135.

Bowser, J. and Reddy, A.S. (1997). Localization of a kinesin-like calmodulin-binding protein in dividing cells of *Arabidopsis* and tobacco. *Plant J.* **12**, 1429–1437.

Canman, J.C., Cameron, L.A., Maddox, P.S., Straight, A., Tirnauer, J.S., Mitchison, T.J., Fang, G., Kapoor, T.M., and Salmon, E.D. (2003). Determining the position of the cell division plane. *Nature* **424**, 1074–1078.

Cavalier-Smith, T. (1974). Basal body and flagellar development during the vegetative cell cycle and the sexual cycle of *Chlamydomonas reinhardtii. J. Cell Sci.* **16**, 529–556.

Collis, P.S. and Weeks, D.P. (1978). Selective inhibition of tubulin synthesis by amiprophos methyl during flagellar regeneration in *Chlamydomonas reinhardtii. Science* **202**, 440–442.

Coss, R.A. (1974). Mitosis in *Chlamydomonas reinhardtii*: basal bodies and the mitotic apparatus. *J. Cell Biol.* **63**, 325–329.

Dibbayawan, T.P., Harper, J.D.I., Elliott, J.E., Gunning, B.E.S., and Marc, J. (1995). A γ-tubulin that associates specifically with centrioles in HeLa cells and the basal body complex in *Chlamydomonas. Cell Biol. Int.* **19**, 559–567.

Doonan, J.H. and Grief, C. (1987). Microtubule cycle in *Chlamydomonas reinhardtii*: an immunofluorescence study. *Cell Motil. Cytoskel.* **7**, 381–392.

Dutcher, S.K., Morrissctte, N.S., Preble, A.M., Rackley, C., and Stanga, J. (2002). ε-tubulin is an essential component of the centriole. *Mol. Biol. Cell* **13**, 3859–3869.

Dymek, E.E., Goduti, D., Kramer, T., and Smith, E.F. (2006). A kinesin-like calmodulin-binding protein in *Chlamydomonas*: evidence for a role in cell division and flagellar functions. *J. Cell Sci.* **119**, 3107–3116.

Ehler, L.L. and Dutcher, S.K. (1998). Pharmacological and genetic evidence for a role of rootlet and phycoplast microtubules in the positioning and assembly of cleavage furrows in *Chlamydomonas reinhardtii. Cell Motil. Cytoskel.* **40**, 193–207.

Ehler, L.L., Holmes, J.A., and Dutcher, S.K. (1995). Loss of spatial control of the mitotic spindle apparatus in a *Chlamydomonas reinhardtii* mutant strain lacking basal bodies. *Genetics* **141**, 945–960.

Gaffal, K.P. (1988). The basal body-root complex of *Chlamydomonas reinhardtii* during mitosis. *Protoplasma* **143**, 118–129.

Gaffal, K.P. and El-Gammal, S. (1990). Elucidation of the enigma of the "metaphase band" of *Chlamydomonas reinhardtii*. *Protoplasma* **156**, 139–148.

Gerhart, J., Danilchik, M., Doniach, T., Roberts, S., Rowning, B., and Stewart, R. (1989). Cortical rotation of the *Xenopus* egg: consequences for the anteroposterior pattern of embryonic dorsal development. *Development* **107**, 37–51.

Guertin, D.A., Trautmann, S., and McCollum, D. (2002). Cytokinesis in eukaryotes. *Microbiol. Mol. Biol. Rev.* **66**, 155–178.

Harper, J.D., McCurdy, D.W., Sanders, M.A., Salisbury, J.L., and John, P.C. (1992). Actin dynamics during the cell cycle in *Chlamydomonas reinhardtii*. *Cell Motil. Cytoskel.* **22**, 117–126.

Harper, J.D., Salisbury, J.L., John, P.C., and Koutoulis, A. (2004). Changes in the centrin and microtubule cytoskeletons after metaphase arrest of the *Chlamydomonas reinhardtii met1* mutant. *Protoplasma* **224**, 159–165.

Hirono, M. and Yoda, A. (1997). Isolation and phenotypic characterization of *Chlamydomonas* mutants defective in cytokinesis. *Cell Struct. Funct.* **22**, 1–5.

Holmes, J.A. and Dutcher, S.K. (1989). Cellular asymmetry in *Chlamydomonas reinhardtii*. *J. Cell Sci.* **94**, 273–285.

Horst, C.J., Fishkind, D.J., Pazour, G.J., and Witman, G.B. (1999). An insertional mutant of *Chlamydomonas reinhardtii* with defective microtubule positioning. *Cell Motil. Cytoskel.* **44**, 143–154.

Hou, M.C., Guertin, D.A., and McCollum, D. (2004). Initiation of cytokinesis is controlled through multiple modes of regulation of the Sid2p-Mob1p kinase complex. *Mol. Cell Biol.* **24**, 3262–3276.

James, S.W., Ranum, L.P., Silflow, C.D., and Lefebvre, P.A. (1988). Mutants resistant to anti-microtubule herbicides map to a locus on the uni linkage group in *Chlamydomonas reinhardtii*. *Genetics* **118**, 141–147.

James, S.W., Silflow, C.D., Stroom, P., and Lefebvre, P.A. (1993). A mutation in the alpha 1 tubulin gene of *Chlamydomonas reinhardtii* confers resistance to anti-microtubule herbicides. *J. Cell Sci.* **106**, 209–218.

Johnson, U.G. and Porter, K.R. (1968). Fine structure of cell division in *Chlamydomonas reinhardtii*. Basal bodies and microtubules. *J. Cell Biol.* **38**, 403–425.

Kater, J.M. (1929). Morphology and division of *Chlamydomonas* with reference to the phylogeny of the flagellate neuromotor system. *Univ. Calif. Pub. Zool.* **33**, 125–169.

Kawabe, A., Matsunaga, S., Nakagawa, K., Kurihara, D., Yoneda, A., Hasezawa, S., Uchiyama, S., and Fukui, K. (2005). Characterization of plant Aurora kinases during mitosis. *Plant Mol. Biol.* **58**, 1–13.

Koblenz, B., Schoppmeier, J., Grunow, A., and Lechtreck, K.F. (2003). Centrin deficiency in *Chlamydomonas* causes defects in basal body replication, segregation, and maturation. *J. Cell Sci.* **116**, 2635–2646.

Lechtreck, K.F. and Grunow, A. (1999). Evidence for a direct role of nascent basal bodies during spindle pole initiation in the green alga *Spermatozopsis similis*. *Protist* **150**, 163–181.

Lux, F.G. and Dutcher, S.K. (1991). Genetic interactions at the FLA10 locus: suppressors and synthetic phenotypes that affect the cell cycle and flagellar function in *Chlamydomonas reinhardtii*. *Genetics* **128**, 549–561.

Matsuda, Y., Koseki, M., Shimada, T., and Saito, T. (1995). Purification and characterization of a vegetative lytic enzyme responsible for liberation of daughter cells during the proliferation of *Chlamydomonas reinhardtii*. *Plant Cell Physiol.* **36**, 681–689.

Matsuura, K., Lefebvre, P.A., Kamiya, R., and Hirono, M. (2002). Kinesin-II is not essential for mitosis and cell growth in *Chlamydomonas*. *Cell Motil. Cytoskel.* **52**, 195–201.

Miller, M.S., Esparza, J.M., Lippa, A.M., Lux, F.G., Cole, D.G., and Dutcher, S.K. (2005). Mutant kinesin-2 motor subunits increase chromosome loss. *Mol. Biol. Cell* **16**, 3810–3820.

Morris, R.L., Keller, L.R., Zweidler, A., and Rizzo, P.J. (1990). Analysis of *Chlamydomonas reinhardtii* histones and chromatin. *J. Protozool.* **37**, 117–123.

Morrissette, N.S. and Sibley, L.D. (2002). Cytoskeleton of apicomplexan parasites. *Microbiol. Mol. Biol. Rev.* **66**, 21–38.

Pfannenschmid, F., Wimmer, V.C., Rios, R.M., Geimer, S., Kröckel, U., Leiherer, A., Haller, K., Nemcová, Y., and Mages, W. (2003). *Chlamydomonas* DIP13 and human NA14: a new class of proteins associated with microtubule structures is involved in cell division. *J. Cell Sci.* **116**, 1449–1462.

Prigent, C. and Dimitrov, S. (2003). Phosphorylation of serine 10 in histone H3, what for? *J. Cell Sci.* **116**, 3677–3685.

Qin, H., Wang, Z., Diener, D., and Rosenbaum, J.L. (2007). Intraflagellar transport protein 27 is a small G protein involved in cell-cycle control. *Curr. Biol.* **17**, 193–202.

St. Clair, H.S. and Goodenough, U.W. (1975). BALD-2: a mutation affecting the formation of doublet and triplet sets of microtubules in *Chlamydomonas reinhardtii*. *J. Cell Biol.* **66**, 480–491.

Salisbury, J.L., Baron, A., Surek, B., and Melkonian, M. (1984). Striated flagellar roots: isolation and partial characterization of a calcium-modulated contractile organelle. *J. Cell Biol.* **99**, 962–970.

Schibler, M.J. and Huang, B. (1991). The colR4 and colR15 beta-tubulin mutations in *Chlamydomonas reinhardtii* confer altered sensitivities to microtubule inhibitors and herbicides by enhancing microtubule stability. *J. Cell Biol.* **113**, 605–614.

Silflow, D.C., Liu, B., LaVoie, M., Richardson, E.A., and Palevitz, B.A. (1999). Gamma-tubulin in *Chlamydomonas*: characterization of the gene and localization of the gene product in cell. *Cell Motil. Cytoskel.* **42**, 285–297.

Umen, J.G. and Goodenough, U.W. (2001). Control of cell division by a retinoblastoma protein homolog in *Chlamydomonas*. *Genes Dev.* **15**, 1652–1661.

Vashishtha, M., Walther, Z., and Hall, J.L. (1996). The kinesin-homologous protein encoded by the *Chlamydomonas FLA10* gene is associated with basal bodies and centrioles. *J. Cell Sci.* **109**, 541–549.

Warr, J.R. (1968). A mutant of *Chlamydomonas reinhardi* with abnormal cell division. *J. Gen. Microbiol.* **52**, 243–251.

Wright, R.L., Salisbury, J., and Jarvik, J.W. (1985). A nucleus-basal body connector in *Chlamydomonas reinhardtii* that may function in basal body localization or segregation. *J. Cell Biol.* **101**, 1903–1912.

Wright, R.L., Adler, S.A., Spanier, J.G., and Jarvik, J.W. (1989). Nucleus-basal body connector in *Chlamydomonas*: evidence for a role in basal body segregation and against essential roles in mitosis or in determining cell polarity. *Cell Motil. Cytoskel.* **14**, 516–526.

Zamora, I. and Marshall, W.F. (2005). A mutation in the centriole-associated protein centrin causes genomic instability via increased chromosome loss in *Chlamydomonas reinhardtii*. *BMC Biol.* **3**, 15.

CHAPTER 15

The *Chlamydomonas* Flagellum as a Model for Human Ciliary Disease

Gregory J. Pazour[1] and George B. Witman[2]
[1]Program in Molecular Medicine, and
[2]Department of Cell Biology, University of Massachusetts
Medical School, Worcester, Massachusetts, USA

CHAPTER CONTENTS

I. INTRODUCTION

Chlamydomonas is among a very small group of model organisms where it is possible to combine biochemical, genetic, and cell biological approaches to investigate the basic biology of ciliary and basal body proteins and functions that, when defective, give rise to human diseases – the so-called "ciliopathies" (Badano et al., 2006b). This is possible because cilia and flagella

445

and their basal bodies have been highly conserved throughout evolution. Structurally, the *Chlamydomonas* flagellum and a human airway cilium are virtually indistinguishable (Figure 15.1A–C). Not surprisingly for an organelle whose structure has been so highly conserved, many *Chlamydomonas* flagellar proteins have clear human homologues. Over half of all proteins in the *Chlamydomonas* flagellar proteome have human homologues with a BLAST E score of ≤1e-10 (Pazour et al., 2005). For some individual complexes, the proportion of conserved subunits is even higher. For example, 14 out of 16 outer dynein arm subunits and all 18 intraflagellar transport (IFT) particle proteins have close human homologues (Pazour et al., 2006; Pazour and Witman, unpublished). *Chlamydomonas* proteins and their human homologues are typically 30–50% identical, but in some cases, e.g., LC8, which is a subunit of the outer arm, the inner arm, and the radial spokes, the identity can be as high as ~90%. Finally, the mechanisms by which cilia and flagella are assembled and function have been highly conserved. Because of this conservation of structure, composition, assembly,

FIGURE 15.1

(A) Substructures of a typical motile eukaryotic flagellum or cilium. All of these substructures are present in both Chlamydomonas *flagella* (B) and human airway cilia (C), which are nearly indistinguishable in their ultrastructure. Arrows indicate one of the outer dynein arms in each. Mutation of the outer dynein arm intermediate chain IC1 results in loss of the outer arm in both Chlamydomonas (D) and humans (E). (From Pazour et al., 2006, with modification. The images of Chlamydomonas *flagella* originally appeared in Wilkerson et al., 1995; the images of human cilia originally appeared in Pennarun et al., 1999 and are used with permission.)

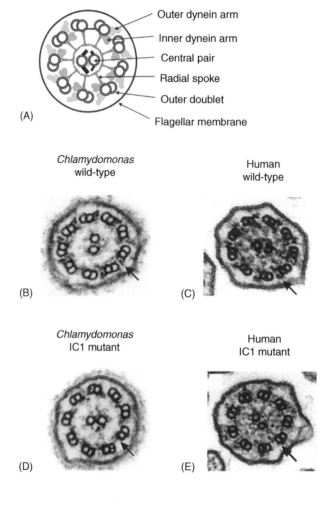

and function, *Chlamydomonas* has long been accepted as the ideal model for diseases such as primary ciliary dyskinesia (PCD) that affect motile cilia and flagella. More recently, with the discovery that non-motile cilia, including vertebrate primary cilia and sensory cilia such as the photoreceptor cell outer segment, use the same assembly process and have many of the same structural proteins and signaling molecules as the *Chlamydomonas* flagellum, the latter has come to be recognized as an excellent model for diseases associated with these cilia as well.

In addition to PCD, diseases and developmental abnormalities caused by defects in motile cilia include situs inversus, hydrocephalus, epilepsy, and male infertility (here treated together with PCD). Diseases associated specifically with motile cilia involve those structures, such as the dynein arms, radial spokes, and central microtubules, that generate movement and are specific to these cilia (Figure 15.1A). Primary cilia and sensory cilia are non-motile and lack dynein arms, radial spokes, and central microtubules; diseases associated with them usually involve defects in the ciliary assembly machinery or ciliary signaling. These diseases include polycystic kidney disease (PKD), retinal degeneration and blindness, and several syndromes that usually involve kidney disease and/or blindness plus other symptoms. Genomic stability is dependent on correct centriole duplication and function (Lingle et al., 1998; Pihan et al., 1998; Doxsey, 2001), and control of the cell cycle may involve the primary cilium (see Chapter 3 and Pan and Snell, 2007), so defects in cilia and basal bodies/centrioles also may lead to cancer.

In this chapter, we review past work on *Chlamydomonas* that has provided important insights into these diseases, and discuss additional diseases for which *Chlamydomonas* has the potential to be very informative in the future. Frequently, the human genes causing a disease have been identified by positional cloning, and the proteins they encode have been shown to be localized to cilia or basal bodies by immunofluorescence microscopy, but the functions of the proteins are understood poorly or not at all. We note when *Chlamydomonas* has apparent orthologues of these proteins. In many of these cases, *Chlamydomonas* offers the best opportunity for a detailed analysis that could lead to an understanding of the basic biology of the human disease protein.

II. PRIMARY CILIARY DYSKINESIA

PCD (also termed "immotile cilia syndrome"; OMIM 242650) is a severe inherited disorder in which specific axonemal structures – most commonly the outer dynein arms, but in some cases the inner dynein arms, the radial spokes, or the central pair of microtubules – are missing (see Afzelius et al., 2001 and Zariwala et al., 2007 for reviews). These defects impair ciliary and flagellar movement, resulting in symptoms that include bronchiectasis (an irreversible and debilitating enlargement of the bronchial tubes caused by recurrent bacterial infection and mucus blockage), chronic bronchitis, and chronic sinusitis due to defects in airway cilia. Male patients are usually infertile due to defects in sperm flagella. About one-half of PCD patients have situs inversus (see section V). The disease is genetically heterogeneous and is likely to be caused by defects in any one of a large number of genes. In most cases, it is inherited in an autosomal recessive manner. The incidence is estimated at 1:20,000–1:60,000 (Zariwala et al., 2007), from which it may be deduced that ~1 in 70–120 people carry a PCD disease mutation.

The connection between cilia and this disorder was first made by Bjorn Afzelius (1976), who found that the airway cilia of several patients with the above clinical manifestations lacked the dynein arms. Previous to this, a large number of *Chlamydomonas* mutants with paralyzed flagella had been isolated (Warr et al., 1966; McVittie, 1972; Lewin, 1974), and data were accumulating that some of these were defective in the central pair of microtubules, the radial spokes, or the outer or inner dynein arms (Warr et al., 1966; Witman et al., 1976, 1978; Piperno et al., 1977; Huang et al., 1979). In 1978, Afzelius visited one of us (GBW) at Princeton University, and we discussed at length how *Chlamydomonas* – with its advantages for genetic and biochemical analyses of the axonemal components – should be an excellent model for elucidating the molecular basis for PCD. Over the next two decades, work in several laboratories focused on identifying the protein subunits making up these structures in *Chlamydomonas* (Piperno et al., 1977; Huang et al., 1979; Adams et al., 1981; Pfister et al., 1982), sequencing the genes and cDNAs that encoded these subunits, including the α, β, and γ heavy chains and the IC1 and IC2 intermediate chains of outer arm dynein (Mitchell and Kang, 1991; Mitchell and Brown, 1994; Wilkerson et al., 1994, 1995; and see Chapter 6), and identifying mutants defective in dynein arm assembly (Kamiya, 1988).

With the development of techniques for insertional mutagenesis in *Chlamydomonas* (Tam and Lefebvre, 1993), it became possible to use reverse genetics to determine which of these subunits were essential for outer arm assembly (reviewed by Pazour and Witman, 2000). This was first applied to the outer arm intermediate chain IC1 (then referred to as IC78), which previously had been shown to be located in an intermediate chain/light chain complex at the base of the dynein (King and Witman, 1990) and to be in direct contact with alpha tubulin in the doublet microtubule (King et al., 1991). Deletion of IC1 resulted in loss of the outer arm, indicating that IC1 was essential for outer arm assembly (Figure 15.1D) (Wilkerson et al., 1995). It was further shown that IC1 was defective in 8 out of 24 insertional mutants selected for slow swimming, which is characteristic of mutants with defects in the dynein arms. The human homologue of IC1 thus became a leading candidate for causing PCD in humans. Acting on this information, Pennarun et al. (1999) in France used conserved sequences of

Chlamydomonas IC1 and its sea urchin homologue (Ogawa et al., 1995) to clone the homologous human sequence *DNAI1*. Sequencing of *DNAI1* from a patient with a PCD phenotype and lacking the outer dynein arms (Figure 15.1E) revealed two loss-of-function mutations in the gene, which is located on chromosome 9p. Remarkably, the structural defect in this patient – loss of the outer arms – is exactly the same as that in the *Chlamydomonas* IC1 deletion mutant (compare Figure 15.1D and E). This was the first human PCD gene to be identified. Approximately 10% of PCD patients are now known to have defects in *DNAI1*; 18 different *DNAI1* mutations have been identified (Zariwala et al., 2007).

Information obtained from *Chlamydomonas* similarly led to the identification of the second PCD disease gene. Omran et al. (2000) used a homozygosity mapping strategy to localize a gene causing PCD and absence of outer dynein arms in a large consanguineous family. The critical disease region on chromosome 5p15-p14 was large (~20 cM) but contained a gene, *DNAH5*, encoding the human homologue of the γ heavy chain of the *Chlamydomonas* outer arm dynein. Because defects in the γ heavy chain cause the slow swimming phenotype of the *Chlamydomonas* mutant *oda2* (Wilkerson et al., 1994; Rupp et al., 1996), *DNAH5* was considered the best candidate for causing PCD in this family. Subsequent sequencing of full-length transcripts of *DNAH5* in PCD patients from 25 families identified four homozygous and six heterozygous mutations in eight of the families; most of these mutations were predicted to result in total or partial loss of the dynein motor domain and microtubule-binding domain (Olbrich et al., 2002). Ultrastructural analysis of patients in six of the families revealed defects in the outer arms. Mutations in *DNAH5* are a common cause of PCD; in 134 PCD families analyzed, 28% had defects in *DNAH5*, and 49% of patients with confirmed defects in the outer arms had mutations in *DNAH5* (Hornef et al., 2006). To date, a total of 66 mutant alleles of *DNAH5* have been identified.

More recently, a gene encoding the β heavy chain of outer arm dynein was shown to be mutated in some PCD patients (Bartoloni et al., 2002; Schwabe et al., 2008). Again, part of the rationale for examining this gene was knowledge that defects in the *Chlamydomonas* β heavy chain (encoded by *ODA4*) cause abnormal motility (Brokaw and Kamiya, 1987; Porter et al., 1994). One of the human mutations identified is predicted to truncate the protein at the beginning of the fourth AAA domain (see Chapter 6), while the others are located closer to the C-terminal end of the molecule. Interestingly, the cilia still beat, albeit with abnormal waveform, and no ultrastructural defects were observed. This precisely parallels results for the *Chlamydomonas oda4-s7* mutant, in which the C-terminal two-thirds of the β heavy chain is truncated (Sakakibara et al., 1993). The remaining N-terminal portion (~160 kD) of the polypeptide is assembled with the other heavy chains into an outer arm that is only barely distinguishable ultrastructurally from the wild-type outer arm, and the mutant cells still swim, but with reduced velocity.

Information originally obtained from *Chlamydomonas* greatly facilitated the identification of the PCD disease genes described above. However, the genes causing the disease in the majority of PCD families are still unknown. To assist in the discovery of new PCD genes, Pazour et al. (2006) systematically searched for human genes encoding homologues of the known *Chlamydomonas* outer arm proteins. This work identified a total of 24 human genes predicted to encode outer dynein arm subunits or associated proteins, plus 12 human genes likely to encode inner dynein arm subunits. When the chromosomal locations of these genes were compared with chromosomal regions suspected of harboring PCD genes on the basis of familial inheritance studies, several of the genes overlapped with suspected PCD disease gene loci (Figure 15.2). Thus, these genes are excellent candidates for screening for disease-causing mutations in PCD patients with outer and/or inner dynein arm defects.

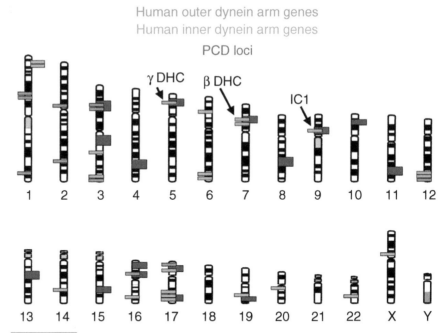

FIGURE 15.2 *Ideograms of human chromosomes showing the locations of known or suspected PCD disease gene loci (orange) based on published studies of familial inheritance. The locations of human genes encoding homologues of* Chlamydomonas *outer dynein arm (green) and inner dynein arm (blue) subunits are shown. The locations of known PCD disease genes encoding the β and γ heavy chains and intermediate chain IC1 of outer arm dynein are labeled. Note that there is overlap between dynein genes and uncharacterized PCD loci on chromosomes 3, 16, and 17. Other suspected PCD disease loci do not overlap with homologues of known dynein genes; disease in these families could be caused by mutations in genes encoding radial spoke and central pair proteins, or as yet uncharacterized ciliary proteins. (Data from Pazour et al., 2006.)*

A large number of *Chlamydomonas* central pair proteins also have been characterized at the molecular level (see Table 8.2). Targeted disruption of the genes encoding Spag6 and Spag16, the mouse homologues of *Chlamydomonas* PF16 and PF20 respectively, caused male infertility due to defects in sperm motility (Sapiro et al., 2002; Zhang et al., 2006). Mice with defects in these proteins generally do not exhibit airway disease, probably because the mouse is less susceptible than humans to airway infection due to impaired ciliary function. Defects in Spag6 and Hydin, another central pair protein initially characterized in *Chlamydomonas* (Lechtreck and Witman, 2007), produce hydrocephalus (see next section) (Sapiro et al., 2002; Lechtreck et al., 2008). A mutation affecting human SPAG16 has been identified, but, at least in the heterozygous state, does not cause PCD or male infertility (Zhang et al., 2007).

To date, no PCD genes have been found that cause loss of the radial spokes. This is likely to be due in part to a paucity of studies of PCD families with defects in these structures, and in part to a lack of candidate PCD genes encoding radial spoke proteins. As part of a proteomic analysis of the *Chlamydomonas* radial spoke, 14 human genes were identified that are likely to encode radial spoke proteins (Yang et al., 2006, and see Table 7.2); these are excellent candidates for screening in PCD patients with radial spoke defects.

III. HYDROCEPHALUS

Hydrocephalus (OMIM 236600) is the abnormal accumulation of cerebrospinal fluid in the ventricles of the brain, which leads to enlargement of the skull. The most severely affected individuals die perinatally; mental retardation is common in survivors. Most cases of human hydrocephalus are of genetic origin, and the incidence is estimated at 0.12–2.5 per 1000 live and stillbirths (Schrander-Stumpel and Fryns, 1998). In the mouse, mutations in the *Hydin* gene cause hydrocephalus. In humans, a mutation causing hydrocephalus maps to within 1.2 Mb of the *HYDIN* gene (Callen et al., 1990), so this gene is a strong candidate for causing the disorder in humans.

In the mouse, Hydin is expressed in the ciliated epithelial cells lining the ventricles, as well as in the ciliated epithelia of bronchi and oviducts (Davy and Robinson, 2003). *Chlamydomonas* has a close homologue (EDP09735; BLAST E = 0.0) of Hydin, and numerous peptides derived from this protein were found in the axonemal fractions of the flagellar proteome (Pazour et al., 2005), indicating that it is a flagellar protein. Lechtreck and Witman (2007) prepared antibodies to *Chlamydomonas* hydin and showed that the protein is associated with the C2 microtubule of the central pair apparatus (see Chapter 8). RNAi knockdown of hydin in *Chlamydomonas* resulted in specific loss of the C2b projection from the central microtubules, indicating that hydin is an essential component of this structure (Figure 15.3A–D).

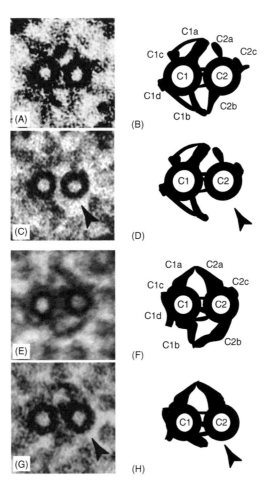

FIGURE 15.3 *Electron micrographs and diagrams of cross-sections of the central pair apparatus from wild-type* Chlamydomonas *(A, B), a* Chlamydomonas *cell in which hydin was knocked down by RNAi (C, D), and brain ependymal cilia from a wild-type mouse (E, F) and a Hydin mutant mouse (G, H). Note that the C2b projection is missing (arrowheads) in both the* Chlamydomonas *hydin RNAi flagellum (C, D) and the mouse Hydin mutant cilium (G, H). C1 and C2, central microtubule 1 and 2, respectively; the central pair projections are labeled according to Mitchell and Sale (1999) (and see Figure 8.2). (The images and diagrams of* Chlamydomonas *central pairs are from Lechtreck and Witman, 2007; the images and diagrams of mouse central pairs are from Lechtreck et al., 2008 and are used with permission.)*

Interestingly, the hydin RNAi cells exhibited a phenotype not previously reported: the flagella, which occasionally beat, were generally arrested at the beginning of the effective stroke and at the beginning of the recovery stroke. As these are the times during the beat cycle when dynein arm activity in one-half of the axoneme is turned off and dynein arm activity in the other half of the axoneme turned on, it was proposed that hydin had a role in switching the dynein arms on and off at the appropriate time in the beat cycle. It was further hypothesized that mutations in mammalian hydin would cause malfunctioning of the cilia due to a defect in the central pair.

The *Chlamydomonas* flagellum differs from metazoan cilia in one important aspect: its central pair rotates within the ring of doublets once per beat cycle, whereas the central pair of metazoans does not rotate (see Chapter 8). To determine if the findings on *Chlamydomonas* hydin were applicable to mammalian cilia, Lechtreck et al. (2008) compared the cilia of brain epithelial cells and tracheal cells of wild-type and Hydin mutant mice and observed that the mutant cilia specifically lacked a projection of the central pair (Figure 15.3E–H). Although the structure of the *Chlamydomonas* and mouse central pair differ in such details as the length of the major projections, conserved markers indicated that the missing structure is the C2b projection – the same structure missing in the *Chlamydomonas* hydin RNAi strains. The cilia of the Hydin mutant mouse are still motile; however, high-speed video analysis revealed that they have attenuated waveforms and tend to stall at the beginning of the effective and recovery strokes – a motility defect closely related to that observed in the *Chlamydomonas* hydin RNAi strains. As a result, the cilia fail to generate flow of cerebrospinal fluid, and this in turn leads to hydrocephalus, probably through a failure to keep open the narrow aqueduct connecting the third and fourth ventricles (Ibanez-Tallon et al., 2004). The cilia of the airway were similarly affected, but this did not lead to obvious respiratory distress. These observations indicated that, despite the differences in central pair rotation, the composition, structure, and function of a specific component of the axonemal central pair have been remarkably conserved between *Chlamydomonas* and mammals.

The above studies on *Chlamydomonas* provided the first explanation for the role of Hydin in cilia and why mutations in it might cause hydrocephalus. Two considerations indicate that mutations in other proteins that affect ciliary motility also are likely to cause hydrocephalus in humans. First, in the mouse, defects in the outer arm dynein heavy chain Mdnah5, the central pair protein Spag6, and the IFT-particle protein Ift88 all cause hydrocephalus (Ibanez-Tallon et al., 2002; Sapiro et al., 2002; Banizs et al., 2005). Second, human patients with PCD are about 80 times more likely than the general population to develop hydrocephalus as a result of aqueductal closure (Ibanez-Tallon et al., 2004). Why more patients with PCD do not develop hydrocephalus is not clear, but it appears that defects in motile cilia in humans are more likely to result in airway disease, whereas the same or similar defects in mice are more likely to cause hydrocephalus.

IV. JUVENILE MYOCLONIC EPILEPSY

Juvenile myoclonic epilepsy (JME; OMIM 606904) is a common form of epilepsy in which individuals have afebrile seizures and myoclonic jerks, with onset usually in adolescence. Suzuki et al. (2004) examined the candidate gene *EFHC1* in 44 families with JME and detected several point mutations that cosegregated with the disease in six of the families, indicating that the mutations in *EFHC1* likely were causing JME in these families. Immunohistological analysis indicated that the Efhc1 protein was widely expressed in adult mouse brain and largely overlapped with the R-type voltage-dependent Ca^{2+} channel $Ca_v2.3$, which was examined in part because EFHC1 has EF-hands, and Ca^{2+} channels previously had been implicated in epilepsy. Overexpression of Efhc1 in mouse hippocampal primary neurons induced apoptosis; this effect was reduced when the overexpressed proteins carried the point mutations that had been identified or when the cells were treated with agonists of $Ca_v2.3$. The investigators hypothesized that the mutations in EFHC1 compromise the apoptotic activity of EFHC1 through $Ca_v2.3$ and thereby prevent elimination of unwanted neurons during development of the central nervous system. As a result, the patient has an increased density of neurons and hyperactivatable neuronal circuits.

This hypothesis might have been much different if Suzuki and collaborators had taken into consideration that human EFHC1 is very likely the orthologue of *Chlamydomonas* RIB72 (EDP06952), which had been cloned previously by Patel-King et al. (2002) and independently by Ikeda et al. (2003). *Chlamydomonas* RIB72 is 38% identical to human EFHC1; a BLAST comparison of the two sequences yields an expect value of 2e-96, and each protein returns the other in a reciprocal BLAST search. Both proteins have three DM10 repeats, of unknown function, plus two EF-hands at the C-terminus (see King, 2006 for review). Immunofluorescence microscopy revealed that RIB72 is located along the entire length of the flagellum but not in the transition zone or basal body (Patel-King et al., 2002; Ikeda et al., 2003). In fractionation experiments, at least 87% of the RIB72 was associated with the three-protofilament ribbons that are derived from the outer doublet microtubules when axonemes are treated with Sarkosyl or urea. Immuno-electron microscopy confirmed localization of the protein to the outer doublet microtubules and specifically to the three-protofilament ribbons (Ikeda et al., 2003). Therefore, RIB72 is an outer doublet protein in *Chlamydomonas*. The function of RIB72 is not completely clear, but it has been proposed to have a role in the assembly and function of the axoneme and possibly to have a relationship to the interdoublet links that hold the outer doublets in their ninefold configuration (Ikeda et al., 2003).

Because the *Chlamydomonas* studies showed that RIB72 was a flagellar component, Ikeda et al. (2005) reexamined the tissue distribution and localization of Efhc1 in the mouse. Western blot analysis using a new anti-Efhc1 polyclonal antibody revealed that Efhc1 was abundant in ciliated tissues such as testis, trachea, and oviduct, and present at much lower levels in the brain, but not detectable in tissues with no cilia or only primary cilia. Immunofluorescence microscopy showed that Efhc1 was located in trachea cilia and sperm flagella, but not in primary cilia. These results were consistent with the axonemal location of RIB72 in *Chlamydomonas*. Homologues of RIB72 are absent from *Caenorhabditis elegans*, which has non-motile sensory cilia but lacks motile cilia; because of this and the absence of an immunofluorescence signal from mouse primary cilia, Ikeda et al. suggested that the protein functions only in motile cilia and flagella. These results cast doubt on the conclusions of Suzuki et al. (2004), and suggested that JME may involve ciliary dysfunction.

In a subsequent study, Suzuki et al. (2008) reported that the antibody used in their 2004 investigation was non-specific, and described the generation of a new monoclonal antibody to mouse Efhc1. Using the new antibody, they carried out immunohistochemical localization in 1-month-old mouse brain and found that Efhc1 was located in cilia and cytoplasm of the ependymal cells lining the ventricles of the brain; no signal was detected in any other part of the adult brain. The ventricular staining was absent from Efhc1-null mutant mice, indicating that it was specific for Efhc1. In agreement with Ikeda et al. (2005) and King (2006), they concluded that JME caused by *EFHC1* mutations is likely to be a ciliopathy.

Questions still remain about whether the *EFHC1* mutations affect the motility of ependymal cilia, and whether these mutations cause JME through an effect on the guidance of migrating neurons, which is dependent on the flow of cerebrospinal fluid (Sawamoto et al., 2006), or some other mechanism. Nevertheless, the history of discovery and characterization of EFHC1/RIB72 is an outstanding illustration of the value of *Chlamydomonas* for understanding human disease. In this case, knowledge of RIB72's location in the *Chlamydomonas* axoneme guided experiments that revealed a similar location of the mammalian homologue in motile cilia of the brain, resulting in reevaluation of the function of EFHC1 and revealing a previously unsuspected connection between cilia and JME.

V. SITUS INVERSUS

As mentioned in section II, about half of PCD patients have situs inversus, a condition in which the normal left–right positioning of the internal organs is reversed; in combination with bronchiectasis and chronic sinusitis, this is known as Kartagener syndrome (OMIM 244400) (Afzelius et al., 2001). At the time when Afzelius discovered that PCD was caused by defects in the airway cilia (Afzelius, 1976), the reason why some of these patients had situs inversus was not clear. However, Afzelius speculated that the beating of cilia on the embryonic epithelia must somehow determine the direction of rotation of the viscera during development. Because some of the patients with dynein arm defects had normal left–right asymmetry, Afzelius proposed that in the absence of this beating, chance alone determines if the visceral asymmetry will be normal or reversed. However, it took more than two decades and the discovery and characterization of IFT in *Chlamydomonas* before this hypothesis could be proven.

As discussed in detail in Chapter 4, IFT involves the movement of particles from the base to the tip of the flagellum and then back to the base. The IFT particles carry flagellar precursors as cargo to the tip of the flagellum, where the precursors are assembled into the axoneme. The motor for anterograde IFT is kinesin-2; retrograde IFT is powered by cytoplasmic dynein 1b. The IFT motor and particle proteins are highly conserved in almost all ciliated organisms, and IFT is essential for the assembly of most cilia and flagella in these organisms.

Early studies on the temperature-sensitive *Chlamydomonas* mutant *fla10* showed that shifting the cells to restrictive temperature resulted in flagellar resorption, and that cells deflagellated following a 30-minute incubation at the restrictive temperature were unable to regenerate their flagella (Huang et al., 1977; *fla10-1* was termed *dd-a-224* at that time). With the discoveries that *FLA10* encoded a flagellar kinesin-like protein (Walther et al., 1994) and that this was the motor for anterograde IFT (Kozminski et al., 1995), it became apparent that FLA10 and IFT were essential for the assembly and maintenance of *Chlamydomonas* flagella. FLA10 was subsequently recognized to be a member of the kinesin-2 family of heterotrimeric kinesins; each kinesin-2 molecule contains two different motor subunits and a non-motor kinesin accessory protein (KAP) (Scholey, 1996).

Because IFT proteins are so highly conserved, the above discoveries in *Chlamydomonas* immediately suggested that knockout of these proteins in mammals would block ciliary assembly and thus allow investigation of the function of IFT and cilia in higher organisms. Mammalian genomes encode three kinesin-2 motor subunits: Kif3A, Kif3B, and Kif3C; Kif3A is thought to interact with either Kif3C or Kif3B and KAP to form the holoenzyme. To determine the function of the kinesin-2 complex in the mouse, Nonaka et al. (1998) generated a null mutant for Kif3B. The mutant embryos died before 12.5 days post-conception, and examination of embryos at 9.5 days post-conception revealed that they had randomized left–right asymmetry. Left–right symmetry of the early mouse embryo is thought to be broken by the action of the embryonic node, which is a small pit on day 7.5 embryos. Cells of the node of the embryo previously had been shown to each have a single cilium, although whether or not these cilia were motile was in dispute (Sulik et al., 1994; Bellomo et al., 1996). When Nonaka and colleagues examined the node of the mutant embryos, the nodal cells were found to lack these cilia. Observation of the nodal cilia of the wild-type embryo confirmed that they were indeed motile, and that this motility generated a leftward flow across the surface of the node. They proposed that the leftward flow initiated a signaling cascade that generated left–right asymmetry, thus providing the first plausible explanation for the connection between cilia and situs inversus reported by Afzelius in 1976. Knockout of the Kif3A subunit in the mouse similarly resulted in loss of nodal cilia and randomized establishment of left–right asymmetry (Marszalek et al., 1999).

Clearly, the knowledge of IFT and FLA10 in *Chlamydomonas* were critical for both the conception of these important experiments and the interpretation of the results.

VI. POLYCYSTIC KIDNEY DISEASE

One of the most important and unexpected contributions of *Chlamydomonas* to an understanding of a specific human disease was the discovery that defects in cilia cause PKD (Pazour et al., 2000). This finding was a direct consequence of research on the basic biology of IFT. Cole et al. (1998) showed that the *Chlamydomonas* IFT particle contained at least 16 subunits arranged in two complexes, complex A and complex B (see Chapter 4). In the course of cloning and sequencing the genes encoding these subunits, it was observed that one of them, *IFT88*, encoded a subunit that was a close homologue of a mouse protein, Tg737, which was defective in a mouse model for PKD (Pazour et al., 2000).

PKD is the most common life-threatening genetic disease in humans, affecting 12.5 million people worldwide (Granthum et al., 1996). In humans, there are two main types of non-syndromic PKD, called autosomal dominant PKD and autosomal recessive PKD. Autosomal dominant PKD (OMIM 173900) is by far the most common form of inherited cystic kidney disease and affects between 1 in 400 and 1 in 1000 adults worldwide. The disease is inherited in a dominant manner, but the genes that cause the disease are thought to act recessively at the cellular level, similar to the way that tumor suppressors behave. An affected individual inherits a mutated copy of the gene and then acquires mutations in the good copy of the gene in the epithelial cells of the nephrons. The cells that acquire second hits proliferate and fail to differentiate fully, leading to cystic kidney disease. Autosomal recessive PKD (OMIM 263200) is a disease primarily of newborns and children, and has an incidence of about 1 in 20,000 live births. About 30% of affected individuals die shortly after birth from respiratory insufficiency due to incomplete development of the lungs. Children surviving the neonate period typically develop hypertension, and 20–45% progress to end-stage renal disease by the age of 15 years. In both cases, the disease results from excessive proliferation of the epithelial cells lining the ducts and tubules of the kidney, leading to massive enlargement of the kidneys.

In the mouse, defects in Ift88/Tg737 were believed to cause the autosomal recessive form of the disease, but it was not known why. The predominant hypothesis was that *Tg737* was a tumor suppressor gene (Richards et al., 1996; Isfort et al., 1997). To learn more about the function of IFT88, Pazour et al. (2000) identified a *Chlamydomonas* insertional mutant in which the *IFT88* gene was disrupted. The *ift88* mutant cells grew and divided normally, indicating that IFT88 was not essential in

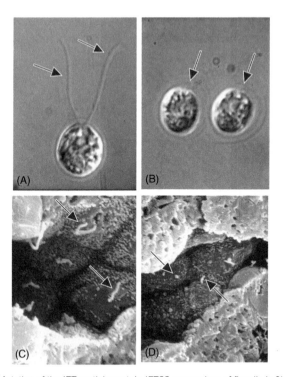

FIGURE 15.4 *Mutation of the IFT-particle protein IFT88 causes loss of flagella in* Chlamydomonas *and loss of primary cilia in mouse kidney cells. Flagella, seen in wild-type* Chlamydomonas *(A, arrows), are completely missing in the* ift88 *mutant cells (B, arrows). Similarly, cilia, present on the epithelial cells of the collecting ducts of a wild-type mouse kidney (C, arrows), occur only as very short stubs on the collecting ducts of an* ift88/Tg737 *mutant mouse kidney (D, arrows). (From Pazour et al., 2000 with modification.)*

Chlamydomonas. However, the cells completely lacked flagella (compare Figure 15.4A and B). Electron microscopy showed that they had normal basal bodies and flagellar transition regions (see Chapter 2), but failed to assemble the flagellar axoneme beyond the transition region. This was the first demonstration that disruption of an IFT-particle protein could block flagellar assembly.

The finding that IFT88 was essential for flagellar assembly in *Chlamydomonas* indicated that there might be a connection between Ift88/Tg737, ciliary assembly, and PKD in mammals. The kidney does in fact have cilia: each of the epithelial cells lining the kidney collecting ducts and tubules has a single non-motile primary cilium that extends into the lumen of the duct. To determine if Ift88/Tg737 is essential for assembly of these kidney cilia, Pazour et al. (2000) compared wild-type and *Tg737* mutant mouse kidneys by scanning electron microscopy. As expected, the epithelial cells of the wild-type kidney had long, well-formed cilia; in contrast, the cilia in the mutant were just short stubs (compare Figure 15.4C and D). Therefore, Ift88 is important for assembly of kidney primary cilia in the mouse, just as it is important for assembly of flagella in *Chlamydomonas*. These results indicated that the basic defect in these mutant mice is an inability to assemble primary cilia due to a malfunction of IFT. As a result, the mice develop PKD.

Prior to this research, it was not known what function, if any, the primary cilium had (Alberts et al., 1994). The results of Pazour et al. (2000) indicated that the primary cilium must have an important function, although it was not yet clear what that function was. While the above work was being carried out, Barr and Sternberg (1999) reported that the *C. elegans* homologues of the PKD proteins polycystin-1 and polycystin-2 were specifically located in sensory neurons, with high levels in both the cell body and ciliated endings. Almost all cases of human autosomal dominant PKD are caused by defects in one of these two proteins. Polycystin-1 is a large transmembrane protein with receptor-like properties. Polycystin-2 is a Ca^{2+}-selective cation channel that is believed to interact with polycystin-1. Together, these two proteins are proposed to form a receptor/channel complex that acts in an early step of a signal transduction pathway that controls cell proliferation and differentiation and, when defective, leads to PKD (reviewed in Igarashi and Somlo, 2002). To determine if there is a similar connection between polycystin-2 and the mammalian primary cilium, Pazour et al. (2002b) used immunofluorescence microscopy to examine the distribution of polycystin-2 in ciliated cultures of human and mouse kidney cells and mouse kidney sections. Although the antibody labeled internal membranes, the strongest labeling, and the only labeling on the surface of the cell, was in the cilium. These results demonstrated that polycystin-2 is displayed specifically on the primary cilium. Shortly thereafter, polycystin-1 was shown to have a similar localization (Yoder et al., 2002; Nauli et al., 2003). Therefore, polycystin-1 and polycystin-2 are ciliary proteins. This, together with other results, indicated that the primary cilium is a sensory organelle, receiving signals from the environment and relaying them to the cell body (Pazour and Witman, 2003). When this signaling is disrupted, PKD results. Thus, PKD can be caused by defects in either the ciliary membrane proteins polycystin-1 or polycystin-2, or by an inability to assemble the entire cilium, as in the case of the *Tg737* mouse. Most known cystic kidney disease proteins have now been shown to be associated with the cilium or its basal body, indicating that PKD is truly a disease of the cilium (Pazour, 2004).

Although the *Tg737* mutant mouse has defects in its primary cilia, its airway cilia are relatively unaffected (Vucica and Pazour, unpublished results) and it does not develop obvious respiratory disease. This indicates that cilia in some tissues are more susceptible to certain mutations than cilia in other tissues. This same tissue-to-tissue variability undoubtedly accounts for some of the phenotypic differences seen in human syndromes caused by defects in cilia and basal body proteins (see next section).

With its advantages for biochemistry and physiology, *Chlamydomonas* has the potential to teach us much more about the proteins involved in PKD.

Chlamydomonas has a close homologue of mammalian polycystin-2. This protein, called PKD2 (ABR14113), is present in the flagellum (Pazour et al., 2005, and see section IV.A.6 of Chapter 11). Like mammalian polycystin-2, PKD2 is a membrane protein with six transmembrane domains (Huang et al., 2007). However, a pool of the protein is anchored to the axoneme (Pazour et al., 2005; Huang et al., 2007). Moreover, the 210-kD full-length protein is cleaved into 120- and 90-kD fragments, and only the two fragments are present in the flagellum. The fragments also are present in cell bodies, including cell bodies of *bld1* cells, which have only short flagellar stubs due to a defect in the IFT-particle protein IFT52. Therefore, cleavage must occur in the cell body. Both cleavage products become anchored to the axoneme after transport into the flagellum. It will be of interest to determine the significance of cleavage and anchorage to the axoneme, and whether similar processing occurs with mammalian polycystin-2. It should be noted that another member of the TRP family of channels, to which mammalian polycystin-2 and *Chlamydomonas* PKD2 belong, also undergoes cleavage (cited in Huang et al., 2007). As in PKD2, cleavage occurs in the large extracellular loop. Therefore, cleavage may be a mechanism for regulating TRP channel activity that is conserved in *Chlamydomonas*.

Interestingly, flagellar levels of PKD2 are elevated fourfold in *Chlamydomonas* gametes, and RNAi knockdown of PKD2 blocks mating (Huang et al., 2007). Agglutination of RNAi mating-type *plus* gametes with wild-type mating-type *minus* gametes was normal; mating appeared to be blocked at a step between flagellar adhesion and phosphorylation of cGMP-dependent protein kinase (see Chapter 12 for a full discussion of the steps involved in mating). There is evidence that the early steps of mating are dependent upon an influx of extracellular Ca^{2+}, and it is possible that PKD2 is responsible for this influx. As noted above, mammalian polycystin-2 is believed to interact with polycystin-1 to generate a receptor/channel complex. However, the *Chlamydomonas* genome does not appear to encode a homologue of polycystin-1, so PKD2 may interact with some other receptor, such as the sexual agglutinins or transmembrane proteins that bind the sexual agglutinins (see Chapter 12).

Chlamydomonas also has homologues of mammalian fibrocystin/polyductin. In humans, fibrocystin is encoded by *PKHD1*, and defects in this gene account for at least 85% of cases of autosomal recessive PKD (Bergmann et al., 2005). Mammalian fibrocystin is a large ~450-kD protein that is predicted to be almost entirely extracellular with a small cytoplasmic C-terminal tail (Harris and Rossetti, 2004). Fibrocystin has been localized to cilia and basal bodies in mammalian cells (Ward et al., 2003; Wang et al., 2004; Zhang et al., 2004), but its function is almost completely unknown. The *Chlamydomonas* genome encodes two fibrocystin-like molecules, EDP03875 (BLAST E = 1e-35) and EDP05658 (BLAST E = 2e-81), but only the former, here termed fibrocystin, is likely to have a transmembrane domain near the C-terminal end (based on genomic sequence and ESTs). Peptides derived from fibrocystin were found in the *Chlamydomonas* flagellar proteome, strongly suggesting that this protein is the functional homologue of mammalian fibrocystin. Interestingly, these peptides were found in both the axoneme and membrane-plus-matrix fractions, suggesting that fibrocystin, like PKD2, is a membrane protein anchored to the axoneme. Surprisingly, homologues of fibrocystin are not present in *C. elegans* or *Drosophila*. Therefore, *Chlamydomonas* may be the best genetic model for elucidating the function of this important ciliary protein.

VII. SYNDROMES INVOLVING PRIMARY CILIA

Primary cilia are present on numerous cell types and tissues throughout the human body, and human photoreceptor outer segments are modified primary cilia. Therefore, it would be expected that defects that involve fundamental processes in cilia would be syndromic – i.e., affect multiple tissues. Indeed, the *Tg737* mutant mouse, which is defective in IFT (see previous section), has multiple abnormalities, including polycystic kidneys, liver cysts and fibrosis, pancreatic cysts, retinal degeneration, male infertility, hydrocephalus, skeletal defects, and polydactyly (reviewed in Pazour and Rosenbaum, 2002; Lehman et al., 2008). Because of this, it was predicted early on that defects in cilia might be the basis for similar human syndromes (Pazour and Rosenbaum, 2002; Pazour et al., 2002a; Rosenbaum and Witman, 2002), and this has now been confirmed. Most commonly, these disorders involve cystic kidney disease and hepatic fibrosis in combination with retinal degeneration and/or some other clinical feature (Lehman et al., 2008). In some cases, different mutations in the same gene can cause a range of overlapping phenotypes. For example, defects in CEP290/nephrocystin-6 can cause Leber congenital amaurosis (isolated or non-syndromic blindness) (den Hollander et al., 2006), Senior–Loken syndrome (blindness and nephronophthisis, a type of cystic kidney disease) (Helou et al., 2007), Joubert syndrome (blindness, nephronophthisis, and cerebellar defects) (Sayer et al., 2006), Bardet–Biedl syndrome (BBS) (blindness and retinal defects, kidney defects and cysts, liver fibrosis, polydactyly, mental retardation, and mild obesity) (Leitch et al., 2008), and Meckel syndrome (cystic kidneys, hepatic fibrosis, developmental anomalies of the central nervous system, and polydactyly) (Frank et al., 2007), which is perinatally fatal and the most severe of the syndromes caused by mutations in nephrocystin-6/CEP290. This variability may be because some alleles affect cilia in one tissue more than in another, or because some patients carry modifier genes that affect the severity and symptoms of the disorder (Badano et al., 2006a). Many of the genes that cause these disorders encode proteins that have homologues in *Chlamydomonas*. In most cases, little or nothing is known about the functions of these proteins, so *Chlamydomonas* has the potential to contribute much to our understanding of the basic biology of these diseases.

BBS (OMIM 209900) is a relatively rare disorder that typifies the range of symptoms caused by ciliary dysfunction (Koenig, 2003). Human genes encoding 12 BBS proteins have been cloned. The *Chlamydomonas* genome encodes very good matches to BBS1 (EDO97795), BBS2 (EDP08340; incomplete gene model), BBS3/ARL6 (EDO97476), BBS5 (AAS89977), BBS7 (EDP09445), BBS8 (EDO97459), and BBS9 (EDP04216). These all have BLAST E values <1e-50 and identify the human disease protein when reciprocal BLAST searches are performed, indicating that all are likely to be orthologous to the human disease proteins. BBS4 matches *Chlamydomonas* EDP02926 with a BLAST E = 2e-09. In spite of the relatively weak match, reciprocal BLAST searches with the *Chlamydomonas* protein identify mammalian BBS4, suggesting it is orthologous. As a class, these proteins are highly conserved in ciliated organisms but not in non-ciliated organisms, and this fact was exploited to identify the *BBS5* gene (Li et al., 2004 and see Chapter 2). Defects in the BBS proteins are thought to cause dysfunction of cilia or basal bodies. BBS1, 2, 4, 5, 7, and 8 have been localized to basal bodies and cilia in *C. elegans* and/or mammalian cells, and BBS7 and 8 have been visualized moving along sensory cilia in *C. elegans* (Ansley et al., 2003; Blacque et al., 2004; Kim et al., 2004; Kulaga et al., 2004; Li et al., 2004; Mykytyn et al., 2004). Ou et al. (2005) proposed that, in *C. elegans*, BBS7 and 8 stabilized interactions between IFT complex A and complex B. However, in the *Chlamydomonas* flagellar proteome, no BBS proteins were identified, whereas numerous peptides representing all of the complex A and B proteins were identified, so if the BBS proteins are associated with the IFT complexes in *Chlamydomonas* flagella, it must be with only a subset of the complexes. Nachury et al. (2007) found that, in mammalian cells, seven of the conserved BBS proteins (BBS1, 2, 4, 5, 7, 8, and 9) form a stable complex that they termed the BBSome. They presented evidence that BBS1 in the BBSome binds the guanosyl exchange factor (GEF) for the small GTPase Rab8 at the basal body. Rab8, which has a homologue (EDO96660) in *Chlamydomonas*, previously was implicated in fusion of rhodopsin carrier vesicles to the base of the frog photoreceptor connecting cilium (Moritz et al., 2001). Based on this and other data, Nachury et al. proposed that the BBSome functions to recruit membrane vesicles to the cilium. Given the high degree of conservation of the BBSome, *Chlamydomonas* promises to be a useful model for testing this hypothesis.

Nephronophthisis (OMIM 256100), a type of cystic kidney disease characterized by interstitial fibrosis and medullary cyst formation, most frequently occurs in isolation but also is present in patients with Senior–Loken syndrome (OMIM 606996) and Joubert syndrome (OMIM 213300). It is the most common genetic cause of renal failure in children and young adults. At least eight nephronophthisis genes have been identified thus far. These genes encode proteins called nephrocystin-1 (or simply nephrocystin) (Hildebrandt et al., 1997), nephrocystin-2 (which is known in mouse as inversin) (Otto et al., 2003), nephrocystin-3 (Omran et al., 2002), nephrocystin-4 (also known as nephroretinin) (Otto et al., 2002), nephrocystin-5 (Otto et al., 2005), nephrocystin-6/CEP290 (defects in which cause 21% of all cases of Leber congenital amaurosis (den Hollander et al., 2006)) (Sayer et al., 2006), the transcription factor GLIS2 (Attanasio et al., 2007), and RPGRIP1L (Delous et al., 2007). Mutations in the NIMA-related protein kinase NEK8 also may cause nephronophthisis in humans (Otto et al., 2008). All have been localized to cilia or basal bodies in mammals. *Chlamydomonas* has excellent matches to nephrocystin-4, nephrocystin-6/CEP290, and NEK8. A BLAST comparison of *Chlamydomonas* nephrocystin-4 (EDP03489), encoded by *NPH4*, and human nephrocystin-4 yields a BLAST E value of 4e-81. It is likely that these proteins are orthologous as reciprocal BLAST searches identify mammalian nephrocystin-4 proteins as top matches and the similarity extends over a significant portion of the proteins. The BLAST E value between *Chlamydomonas* nephrocystin-6 (EDP06028, encoded by *POC3*) and human nephrocystin-6/CEP290 is 1e-25, and again, a reciprocal BLAST search identifies mammalian nephrocystin-6/CEP290 proteins as the top matches. Moreover, both *Chlamydomonas* proteins were found in the *Chlamydomonas* basal body proteome (Keller et al., 2005). The *Chlamydomonas* genome encodes at least 10 NEK homologues (Quarmby and Mahjoub, 2005), two of which were found in the flagellar proteome (Pazour et al., 2005). Mutations in one of these, FA2 (AAL86904), cause deflagellation and cell cycle defects (see Chapter 3 and Mahjoub et al., 2002). The radiation of the NEK family in *Chlamydomonas* makes it hard to establish orthology between the nephronophthisis gene NEK8 and a particular NEK in *Chlamydomonas*. The best match to NEK8 in *Chlamydomonas* is CNK3/NMA3 (EDP08224) (BLAST E = 2e-60) but phylogenetic analysis shows that *Chlamydomonas* CNK3/NMA3, CNK4/NMA4, CNK5/NMA5, and FA2 all group with a number of human NEKs, including NEK8 (Bradley et al., 2004). The *Chlamydomonas* genome also is predicted to encode proteins similar to nephrocystin-2 (BLAST E = 8e-36) and nephrocystin-3 (BLAST E = 2e-23), but the similarities are within ankyrin or TPR repeats, respectively, and reciprocal BLAST searches do not detect the corresponding mammalian nephrocystins as the top matches, so orthology is not certain.

Oral–facial–digital syndrome (OMIM 311200) is a developmental disorder that includes abnormalities of the face, oral cavity, and digits, and is often accompanied by cystic kidneys. The disease manifests itself as an X-linked dominant disorder in females with lethality in males. The only gene known to cause this syndrome encodes OFD1, a coiled-coil protein that localizes to both the basal body of the primary cilium and the nucleus (Ferrante et al., 2001; Romio et al., 2003). The closest match encoded by the *Chlamydomonas* genome is OFD1 (EDP04691). The BLAST E score is 1e-17 and the *Chlamydomonas* protein identifies human OFD1 as the top match in a reciprocal BLAST search, indicating that they are likely to be orthologues. OFD1 was found in the *Chlamydomonas* basal body proteome, where it was termed BUG11 (Keller et al., 2005). Mammalian OFD1 interacts with RuvB-like 1 (Giorgio et al., 2007), which is part of a complex that includes the closely related protein RuvB-like 2 (Kanemaki et al., 1999), and is believed to be involved in chromatin remodeling. The closest *Chlamydomonas* homologues to RuvB-like 1 and RuvB-like 2 are EDP00834 (75% identity, BLAST E = 0.0) and EDP05775 (78% identity, BLAST E = 0.0), respectively. Both *Chlamydomonas* proteins return the corresponding human protein in reciprocal BLAST searches, indicating that the human and *Chlamydomonas* proteins are very likely to be orthologues. Interestingly, mRNAs for *Chlamydomonas* OFD1, RuvB-like 1, and RuvB-like 2 are strongly induced by deflagellation (Stolc et al., 2005); this expression pattern would not be expected for OFD1 if it were strictly a basal body structural component. Because of their expression patterns, it was predicted that *Chlamydomonas* RuvB-like 1 and RuvB-like 2 act in the nucleus to regulate transcription of genes encoding flagellar proteins or to mediate gene regulation in response to ciliary signaling (Stolc et al., 2005). Further investigations of these proteins in *Chlamydomonas* have the potential to determine if OFD1 does indeed function in both the nucleus and basal body, and perhaps to elucidate a novel signaling pathway between the cilium and the nucleus.

Jeune syndrome (OMIM 611263), also known as Jeune asphyxiating thoracic dystrophy, involves cystic lesions of the kidney, liver, and pancreas, retinal degeneration, polydactyly, and skeletal abnormalities, including deformity of the chest. Patients often die in infancy because of a severely constricted thoracic cage and respiratory insufficiency. Jeune syndrome and short-rib polydactyly (OMIM 263530) are likely to be variants of the same disorder (Ho et al., 2000) and are very similar to Ellis–van Creveld syndrome (OMIM 225500). Beales et al. (2007) screened 12 consanguineous families with Jeune syndrome and identified three in which the disease was linked to a large region on chromosome 3q. One of two candidate genes within this region encoded the homologue of the IFT-particle protein known as Che-2 in *C. elegans* and IFT80 in *Chlamydomonas*. Sequencing of this gene in the affected families revealed one single-amino-acid deletion and two missense mutations that segregated with the disease. This is the first human mutation to be identified in an IFT protein. The mutations are likely to cause only a partial loss of function. The specific role of IFT80 in IFT is not known, but null mutations of Ift88 and Ift172 cause embryonic lethality in mice (Murcia et al., 2000; Huangfu et al., 2003), and a similar phenotype might be expected for a null mutation of IFT80 in humans.

VIII. RETINAL DEGENERATION AND BLINDNESS

Vision in vertebrates is intimately connected to cilia because the rod and cone outer segments are developmentally derived from a cilium, and this cilium persists as the connecting cilium, which is the only connection between the photoreceptor cell's inner segment and outer segment. RetNet Retinal Information Network (www.sph.uth.tmc.edu/Retnet/) lists over 200 genes that cause retinal disease. Many human retinal degeneration and blindness disease genes are conserved in *Chlamydomonas*. Some of these cause blindness as part of a syndrome and have been discussed in the previous section. However, others cause blindness without pleiotropic effects. Most interestingly, the tubby-like protein TULP1 is similar to *Chlamydomonas* TLP1 (EDP00750; BLAST E = 3e-62) and TLP2 (EDP04606; BLAST E = 2e-19). Tubby-like proteins were first identified in the mouse as causing obesity and retinal degeneration (Ikeda et al., 2002). Humans with mutations in TULP1 show retinal degeneration (OMIM 602280). Reciprocal BLAST searches with *Chlamydomonas* TLP1 identify human TUB, TULP3, and TULP1 (BLAST E values 2e-66, 5e-65, and 3e-62, respectively) as the top matches; searches with TLP2 also identify human TUB, TULP3, and TULP1 (BLAST E values 8e-21, 2e-20, and 2e-19, respectively). Human mutations in TULP3 have not been described, but mice with mutations show severe neural development defects and die embryonically (Ikeda et al., 2001). Transcripts of *Chlamydomonas TLP2* are induced during flagellar regeneration (Stolc et al., 2005). Also of interest, the human retinitis pigmentosa protein XRP2 is similar to *Chlamydomonas* XRP2 (EDP03211, BLAST E = 7e-41). XRP2 is similar in sequence to a tubulin-specific chaperone cofactor and can stimulate the GTPase activity of tubulin *in vitro* (Bartolini et al., 2002). Reciprocal BLAST searches with the *Chlamydomonas* protein identify the human disease protein as a top match. This protein is not conserved in either *Saccharomyces* or *Arabidopsis*, which lack cilia and flagella, making it likely to play a role in cilia or basal body/centrosome function. Mutations in the human protein HRG4, which is a close homologue of *C. elegans* UNC119, cause rod–cone dystrophy (Kobayashi et al., 2000). The *Chlamydomonas* genome encodes a single homologue of HRG4 (EDO97256, BLAST E = 6e-42), and the *Chlamydomonas* protein returns the human protein in a reciprocal BLAST search. HRG4/UNC119 is present exclusively in ciliated organisms (Li et al., 2004) and was found in the basal body proteome (Keller et al., 2005), indicating that it is likely to be a basal body component.

In addition to its promise for characterizing ciliary and basal body proteins that cause blindness when defective, *Chlamydomonas* has opened the door to a new avenue for potentially restoring vision to patients with retinal degeneration. Many of the genes that cause blindness by disrupting photoreceptor cell functions lead to a progressive loss of the dysfunctional rod and cone cells by apoptosis. Once the photoreceptor cells are lost, there are few possibilities for restoring vision, even by *avant garde* approaches that show promise for restoring function to extant photoreceptor cells or cells and tissues affected by other diseases (Flannery and Greenberg, 2006). For example, replacement of the defective gene by gene therapy is not an option once the photoreceptor cells have died, and techniques for replacement of photoreceptor cells by either adult or embryonic stem cells have not yet been developed. Importantly, however, the neuronal cells of the inner retina may persist for many years after the photoreceptor cells are lost. The discovery of the *Chlamydomonas* photoreceptor channelrhodopsin (Nagel et al., 2002; Sineshchekov et al., 2002; Suzuki et al., 2003) has raised the possibility that vision might be restored by turning these retinal neurons into photoresponsive cells. As discussed in detail in Chapter 13, channelrhodopsin is a cation-selective ion channel that is directly gated by light. Bi et al. (2006) have shown that a cDNA encoding the opsin for channelrhodopsin-2 can be delivered by an adeno-associated virus to retinal ganglion cells in the eye of a mouse that is a model for retinal degeneration. The gene is expressed in these cells, the cells become capable of generating light-evoked currents, and the channelrhodopsin-2-generated light responses are transmitted to the visual cortex. Taking a somewhat different approach, Lagali et al. (2008) used subretinal injection and *in vivo* electroporation to deliver constructs expressing channelrhodopsin-2 to degenerated retinas of the same mouse model; the construct utilized a cell-specific promoter to genetically target ON bipolar cells. Expression was stable for at least 6 months and restored light responsiveness to the retinas. The light-induced signals were relayed to the visual cortex and restored the ability of the mice to follow movement. These results suggest that, with further development and refinement, ectopic expression of channelrhodopsin could be used to restore vision in humans suffering from advanced rod and cone degeneration.

IX. CANCER

Chlamydomonas is not often thought of as a model for cancer, but centrosomes play a role in genetic instability that can lead to cancer (Lingle et al., 1998; Pihan et al., 1998; Doxsey, 2001). A number of cancer-related genes are conserved in *Chlamydomonas*. Most of these also are conserved in organisms without cilia and are involved in processes like DNA damage detection and repair. Some cancer-related genes, however, are conserved in *Chlamydomonas* but not *Saccharomyces* or *Arabidopsis*. DMBT1 ("deleted in malignant brain tumors 1"; NP_015568) is similar to *Chlamydomonas* SRR6 (EDP09946) with a highly significant BLAST E value (1e-81), and a BLAST search of the human genome with *Chlamydomonas* SRR6 returns DMBT1 as the top match. DMBT1 is a member of the scavenger receptor super family and is often deleted in glioblastoma brain tumors (Mollenhauer et al., 1997). DLEC1 ("deleted in lung and esophageal cancer 1"; NP_031363) is similar to the *Chlamydomonas* flagellar protein FAP81 (EDP05051) (BLAST E = 1e-42), and a reciprocal BLAST search of the human genome returns DLEC1 as the best match. The *DLEC1* gene, previously named *DLC1*, is thought to be a negative regulator of cell proliferation (Daigo et al., 1999), as was also thought to be the case for mouse *Ift88/Tg737* (see section VI). The "coiled-coil domain containing 6" protein (also known as H4; NP_005427) is similar to *Chlamydomonas* CCD6 (EDP01714; BLAST E = 5e-25). Inversion of the long arm of human chromosome 10 results in a juxtaposition of the RET proto-oncogene and the gene encoding coiled-coil domain containing 6 protein. This produces a chimeric protein containing the tyrosine kinase domain fused to coiled-coil domain containing 6 protein (Grieco et al., 1990). Coiled-coil domains are abundant in centrosomes (Andersen et al., 2003) and cilia (Pazour et al., 2005), and the fact that this protein is conserved in *Chlamydomonas* and not *Saccharomyces* or *Arabidopsis* suggests that coiled-coil domain containing 6 protein is a centrosomal or ciliary protein and may function to promote cancer by inappropriately targeting the tyrosine kinase activity to the centrosome or cilium. Human tumor suppressing subtransferable candidate 5 (NP_899056) is similar to *Chlamydomonas* SOC1 (EDP00464; BLAST E = 3e-22), and a reciprocal BLAST search of the human genome returns the same human protein. Somatic mutations and loss of heterozygosity have been found in this gene in kidney tumor and lung cancer tissues (Lee et al., 1998).

X. CONCLUSION

Studies in *Chlamydomonas* have illuminated the basic biology of many ciliary and basal body proteins involved in human diseases and developmental disorders, including PCD, hydrocephalus, epilepsy, situs inversus, and PKD. *Chlamydomonas* has close homologues for many other human proteins that are known to cause disease. Some of these have been noted in this chapter, but techniques such as positional cloning are identifying new human disease genes at a rapid pace, and many of these also are likely to have homologues in *Chlamydomonas*. Conversely, as we already have seen for PCD and BBS, knowledge of *Chlamydomonas* genes and proteins can be helpful for narrowing down the likely disease gene within a larger region of a chromosome shown by familial inheritance studies to contain the disease locus. Equally importantly, *Chlamydomonas* offers what arguably may be the best model system for elucidating the functions of those disease proteins that are poorly understood. As the chapters in this volume make abundantly clear, more is known about the structure, composition, assembly, and function of flagella and basal bodies of *Chlamydomonas* than of any other organism. This wealth of information provides a uniquely strong foundation for detailed analyses of the components of these organelles. This background of knowledge promises to act synergistically with the advantages of *Chlamydomonas* for genetic, molecular biological, cell biological, and biochemical analyses of flagella and basal bodies to yield new structural and functional insights that could not be obtained readily from any other model system.

ACKNOWLEDGMENTS

We thank Dr. Karl Lechtreck of the University of Massachusetts Medical School, Worcester, for the preparation of Figure 15.3. This work was supported by grants from the National Institutes of Health (GM60992 to GJP and GM30626 to GBW), and the Robert W. Booth Fund at the Greater Worcester Community Foundation (GBW).

REFERENCES

Adams, G.M., Huang, B., Piperno, G., and Luck, D.J. (1981). Central-pair microtubular complex of *Chlamydomonas* flagella: polypeptide composition as revealed by analysis of mutants. *J. Cell Biol.* **91**, 69–76.

Afzelius, B.A. (1976). A human syndrome caused by immotile cilia. *Science* **193**, 317–319.

Afzelius, B., Mossberg, B., and Bergstrom, S. (2001). Immotile cilia syndrome (primary ciliary dyskinesia), including Kartagener syndrome. In: *The Metabolic and Molecular Bases of Inherited Disease* (C. Scriver, A. Beaudet, W. Sly, and D. Valle, Eds.), pp. 4817–4827. McGraw-Hill, New York.

Alberts, B., Bray, D., Lewis, J., Raff, M., Roberts, K., and Watson, J. (1994). *Molecular Biology of the Cell*. Garland Publishing, New York.

Andersen, J.S., Wilkinson, C.J., Mayor, T., Mortensen, P., Nigg, E.A., and Mann, M. (2003). Proteomic characterization of the human centrosome by protein correlation profiling. *Nature* **426**, 570–574.

Ansley, S.J., Badano, J.L., Blacque, O.E., Hill, J., Hoskins, B.E., Leitch, C.C., Kim, J.C., Ross, A.J., Eichers, E.R., Teslovich, T.M., Mah, A.K., Johnsen, R.C., Cavender, J.C., Lewis, R.A., Leroux, M.R., Beales, P.L., and Katsanis, N. (2003). Basal body dysfunction is a likely cause of pleiotropic Bardet–Biedl syndrome. *Nature* **425**, 628–633.

Attanasio, M., Uhlenhaut, N.H., Sousa, V.H., O'Toole, J.F., Otto, E., Anlag, K., Klugmann, C., Treier, A.C., Helou, J., Sayer, J.A., Seelow, D., Nurnberg, G., Becker, C., Chudley, A.E., Nurnberg, P., Hildebrandt, F., and Treier, M. (2007). Loss of GLIS2 causes nephronophthisis in humans and mice by increased apoptosis and fibrosis. *Nat. Genet.* **39**, 1018–1024.

Badano, J.L., Leitch, C.C., Ansley, S.J., May-Simera, H., Lawson, S., Lewis, R.A., Beales, P.L., Dietz, H.C., Fisher, S., and Katsanis, N. (2006a). Dissection of epistasis in oligogenic Bardet–Biedl syndrome. *Nature* **439**, 326–330.

Badano, J.L., Mitsuma, N., Beales, P.L., and Katsanis, N. (2006b). The ciliopathies: an emerging class of human genetic disorders. *Annu. Rev. Genomics Hum. Genet.* **7**, 125–148.

Banizs, B., Pike, M.M., Millican, C.L., Ferguson, W.B., Komlosi, P., Sheetz, J., Bell, P.D., Schwiebert, E.M., and Yoder, B.K. (2005). Dysfunctional cilia lead to altered ependyma and choroid plexus function, and result in the formation of hydrocephalus. *Development* **132**, 5329–5339.

Barr, M.M. and Sternberg, P.W. (1999). A polycystic kidney-disease gene homologue required for male mating behaviour in *C. elegans*. *Nature* **401**, 386–389.

Bartolini, F., Bhamidipati, A., Thomas, S., Schwahn, U., Lewis, S.A., and Cowan, N.J. (2002). Functional overlap between retinitis pigmentosa 2 protein and the tubulin-specific chaperone cofactor C. *J. Biol. Chem.* **277**, 14629–14634.

Bartoloni, L., Blouin, J.L., Pan, Y., Gehrig, C., Maiti, A.K., Scamuffa, N., Rossier, C., Jorissen, M., Armengot, M., Meeks, M., Mitchison, H.M., Chung, E.M., Delozier-Blanchet, C.D., Craigen, W.J., and Antonarakis, S.E. (2002). Mutations in the *DNAH11* (axonemal heavy chain dynein type 11) gene cause one form of situs inversus totalis and most likely primary ciliary dyskinesia. *Proc. Natl. Acad. Sci. U. S. A.* **99**, 10282–10286.

Beales, P.L., Bland, E., Tobin, J.L., Bacchelli, C., Tuysuz, B., Hill, J., Rix, S., Pearson, C.G., Kai, M., Hartley, J., Johnson, C., Irving, M., Elcioglu, N., Winey, M., Tada, M., and Scambler, P.J. (2007). *IFT80*, which encodes a conserved intraflagellar transport protein, is mutated in Jeune asphyxiating thoracic dystrophy. *Nat. Genet.* **39**, 737–739.

Bellomo, D., Lander, A., Harragan, I., and Brown, N.A. (1996). Cell proliferation in mammalian gastrulation: the ventral node and notochord are relatively quiescent. *Dev. Dyn.* **205**, 471–485.

Bergmann, C., Senderek, J., Windelen, E., Kupper, F., Middeldorf, I., Schneider, F., Dornia, C., Rudnik-Schoneborn, S., Konrad, M., Schmitt, C.P., Seeman, T., Neuhaus, T.J., Vester, U., Kirfel, J., Buttner, R., and Zerres, K. (2005). Clinical consequences of *PKHD1* mutations in 164 patients with autosomal-recessive polycystic kidney disease (ARPKD). *Kidney Int.* **67**, 829–848.

Bi, A., Cui, J., Ma, Y.P., Olshevskaya, E., Pu, M., Dizhoor, A.M., and Pan, Z.H. (2006). Ectopic expression of a microbial-type rhodopsin restores visual responses in mice with photoreceptor degeneration. *Neuron* **50**, 23–33.

Blacque, O.E., Reardon, M.J., Li, C., McCarthy, J., Mahjoub, M.R., Ansley, S.J., Badano, J.L., Mah, A.K., Beales, P.L., Davidson, W.S., Johnsen, R.C., Audeh, M., Plasterk, R.H.A., Maillie, D.L., Katsanis, N., Quarmby, L.M., Wicks, S.R., and Leroux, M.R. (2004). Loss of *C. elegans* BBS-7 and BBS-8 protein function results in cilia defects and compromised intraflagellar transport. *Genes Dev.* **18**, 1630–1642.

Bradley, B.A., Wagner, J.J., and Quarmby, L.M. (2004). Identification and sequence analysis of six new members of the NIMA-related kinase family in *Chlamydomonas*. *J. Eukaryot. Microbiol.* **51**, 66–72.

Brokaw, C.J. and Kamiya, R. (1987). Bending patterns of *Chlamydomonas* flagella. IV. Mutants with defects in inner and outer dynein arms indicate differences in dynein arm function. *Cell Motil. Cytoskel.* **8**, 68–75.

Callen, D.F., Baker, E.G., and Lane, S.A. (1990). Re-evaluation of GM2346 from a del (16)(q22) to t(4;16)(q35;q22.1). *Clin. Genet.* **38**, 466–468.

Cole, D.G., Diener, D.R., Himelblau, A.L., Beech, P.L., Fuster, J.C., and Rosenbaum, J.L. (1998). *Chlamydomonas* kinesin-II-dependent intraflagellar transport (IFT): IFT particles contain proteins required for ciliary assembly in *Caenorhabditis elegans* sensory neurons. *J. Cell Biol.* **141**, 993–1008.

Daigo, Y., Nishiwaki, T., Kawasoe, T., Tamari, M., Tsuchiya, E., and Nakamura, Y. (1999). Molecular cloning of a candidate tumor suppressor gene, *DLC1*, from chromosome 3p21.3. *Cancer Res.* **59**, 1966–1972.

Davy, B.E. and Robinson, M.L. (2003). Congenital hydrocephalus in *hy3* mice is caused by a frameshift mutation in *Hydin*, a large novel gene. *Hum. Mol. Genet.* **12**, 1163–1170.

Delous, M., Baala, L., Salomon, R., Laclef, C., Vierkotten, J., Tory, K., Golzio, C., Lacoste, T., Besse, L., Ozilou, C., Moutkine, I., Hellman, N.E., Anselme, I., Silbermann, F., Vesque, C., Gerhardt, C., Rattenberry, E., Wolf, M.T., Gubler, M.C., Martinovic, J., Encha-Razavi, F., Boddaert, N., Gonzales, M., Macher, M.A., Nivet, H., Champion, G., Bertheleme, J.P., Niaudet, P., McDonald, F., Hildebrandt, F., Johnson, C.A., Vekemans, M., Antignac, C., Ruther, U., Schneider-Maunoury, S., Attie-Bitach, T., and Saunier, S. (2007). The ciliary gene *RPGRIP1L* is mutated in cerebello-oculo-renal syndrome (Joubert syndrome type B) and Meckel syndrome. *Nat. Genet.* **39**, 875–881.

den Hollander, A.I., Koenekoop, R.K., Yzer, S., Lopez, I., Arends, M.L., Voesenek, K.E., Zonneveld, M.N., Strom, T.M., Meitinger, T., Brunner, H.G., Hoyng, C.B., van den Born, L.I., Rohrschneider, K., and Cremers, F.P. (2006). Mutations in the *CEP290* (*NPHP6*) gene are a frequent cause of Leber congenital amaurosis. *Am. J. Hum. Genet.* **79**, 556–561.

Doxsey, S. (2001). Re-evaluating centrosome function. *Nat. Rev. Mol. Cell Biol.* **2**, 688–698.

Ferrante, M.I., Giorgio, G., Feather, S.A., Bulfone, A., Wright, V., Ghiani, M., Selicorni, A., Gammaro, L., Scolari, F., Woolf, A.S., Sylvie, O., Bernard, L., Malcolm, S., Winter, R., Ballabio, A., and Franco, B. (2001). Identification of the gene for oral–facial–digital type I syndrome. *Am. J. Hum. Genet.* **68**, 569–576.

Flannery, J.G. and Greenberg, K.P. (2006). Looking within for vision. *Neuron* **50**, 1–11.

Frank, V., Ortiz Brüchle, N., Mager, S., Frints, S.G., Bohring, A., du Bois, G., Debatin, I., Seidel, H., Senderek, J., Besbas, N., Todt, U., Kubisch, C., Grimm, T., Teksen, F., Balci, S., Zerres, K., and Bergmann, C. (2007). Aberrant splicing is a common mutational mechanism in *MKS1*, a key player in Meckel–Gruber syndrome. *Hum. Mutat.* **28**, 638–639.

Giorgio, G., Alfieri, M., Prattichizzo, C., Zullo, A., Cairo, S., and Franco, B. (2007). Functional characterization of the OFD1 protein reveals a nuclear localization and physical interaction with subunits of a chromatin remodeling complex. *Mol. Biol. Cell* **18**, 4397–4404.

Granthum, J.J., Nair, V., and Winklhofer, F. (1996). Cystic diseases of the kidney. In: *Brenner and Rector's The Kidney* (B.M. Brenner, Ed.), pp. 1699–1730. W.B. Saunders Company, Philadelphia, PA.

Grieco, M., Santoro, M., Berlingieri, M.T., Melillo, R.M., Donghi, R., Bongarzone, I., Pierotti, M.A., Della, P.G., Fusco, A., and Vecchio, G. (1990). PTC is a novel rearranged form of the ret proto-oncogene and is frequently detected *in vivo* in human thyroid papillary carcinomas. *Cell* **60**, 557–563.

Harris, P.C. and Rossetti, S. (2004). Molecular genetics of autosomal recessive polycystic kidney disease. *Mol. Genet. Metab.* **81**, 75–85.

Helou, J., Otto, E.A., Attanasio, M., Allen, S.J., Parisi, M.A., Glass, I., Utsch, B., Hashmi, S., Fazzi, E., Omran, H., O'Toole, J.F., Sayer, J.A., and Hildebrandt, F. (2007). Mutation analysis of *NPHP6/CEP290* in patients with Joubert syndrome and Senior–Loken syndrome. *J. Med. Genet.* **44**, 657–663.

Hildebrandt, F., Otto, E., Rensing, C., Nothwang, H.G., Vollmer, M., Adolphs, J., Hanusch, H., and Brandis, M. (1997). A novel gene encoding an SH3 domain protein is mutated in nephronophthisis type 1. *Nat. Genet.* **17**, 149–153.

Ho, N.C., Francomano, C.A., and van Allen, M. (2000). Jeune asphyxiating thoracic dystrophy and short-rib polydactyly type III (Verma–Naumoff) are variants of the same disorder. *Am J. Med. Genet.* **90**, 310–314.

Hornef, N., Olbrich, H., Horvath, J., Zariwala, M.A., Fliegauf, M., Loges, N.T., Wildhaber, J., Noone, P.G., Kennedy, M., Antonarakis, S.E., Blouin, J.L., Bartoloni, L., Nusslein, T., Ahrens, P., Griese, M., Kuhl, H., Sudbrak, R., Knowles, M.R., Reinhardt, R., and Omran, H. (2006). *DNAH5* mutations are a common cause of primary ciliary dyskinesia with outer dynein arm defects. *Am. J. Respir. Crit. Care Med.* **174**, 120–126.

Huang, B., Rifkin, M., and Luck, D. (1977). Temperature-sensitive mutations affecting flagellar assembly and function in *Chlamydomonas reinhardtii*. *J. Cell Biol.* **72**, 67–85.

Huang, B., Piperno, G., and Luck, D.J.L. (1979). Paralyzed flagella mutants of *Chlamydomonas reinhardtii* defective for axonemal doublet microtubule arms. *J. Biol. Chem.* **254**, 3091–3099.

Huang, K., Diener, D.R., Mitchell, A., Pazour, G.J., Witman, G.B., and Rosenbaum, J.L. (2007). Function and dynamics of PKD2 in *Chlamydomonas reinhardtii* flagella. *J. Cell Biol.* **179**, 501–514.

Huangfu, D., Liu, A., Rakeman, A.S., Murcia, N.S., Niswander, L., and Anderson, K.V. (2003). Hedgehog signalling in the mouse requires intraflagellar transport proteins. *Nature* **426**, 83–87.

Ibanez-Tallon, I., Gorokhova, S., and Heintz, N. (2002). Loss of function of axonemal dynein Mdnah5 causes primary ciliary dyskinesia and hydrocephalus. *Hum. Mol. Genet.* **11**, 715–721.

Ibanez-Tallon, I., Pagenstecher, A., Fliegauf, M., Olbrich, H., Kispert, A., Ketelsen, U.P., North, A., Heintz, N., and Omran, H. (2004). Dysfunction of axonemal dynein heavy chain Mdnah5 inhibits ependymal flow and reveals a novel mechanism for hydrocephalus formation. *Hum. Mol. Genet.* **13**, 2133–2141.

Igarashi, P. and Somlo, S. (2002). Genetics and pathogenesis of polycystic kidney disease. *J. Am. Soc. Nephrol.* **13**, 2384–2398.

Ikeda, A., Ikeda, S., Gridley, T., Nishina, P.M., and Naggert, J.K. (2001). Neural tube defects and neuroepithelial cell death in *Tulp3* knockout mice. *Hum. Mol. Genet.* **10**, 1325–1334.

Ikeda, A., Nishina, P.M., and Naggert, J.K. (2002). The tubby-like proteins, a family with roles in neuronal development and function. *J. Cell Sci.* **115**, 9–14.

Ikeda, K., Brown, J.A., Yagi, T., Norrander, J.M., Hirono, M., Eccleston, E., Kamiya, R., and Linck, R.W. (2003). Rib72, a conserved protein associated with the ribbon compartment of flagellar A-microtubules and potentially involved in the linkage between outer doublet microtubules. *J. Biol. Chem.* **278**, 7725–7734.

Ikeda, T., Ikeda, K., Enomoto, M., Park, M.K., Hirono, M., and Kamiya, R. (2005). The mouse ortholog of EFHC1 implicated in juvenile myoclonic epilepsy is an axonemal protein widely conserved among organisms with motile cilia and flagella. *FEBS Lett.* **579**, 819–822.

Isfort, R.J., Cody, D.B., Doersen, C.-J., Richards, W.G., Yoder, B.K., Wilkinson, J.E., Kier, L.D., Jirtle, R.L., Isenberg, J.S., Klounig, J.E., and Woychik, R.P. (1997). The tetratricopeptide repeat containing Tg737 gene is a liver neoplasia tumor suppressor gene. *Oncogene* **15**, 1797–1803.

Kamiya, R. (1988). Mutations at twelve independent loci result in absence of outer dynein arms in *Chlamydomonas reinhardtii*. *J. Cell Biol.* **107**, 2253–2258.

Kanemaki, M., Kurokawa, Y., Matsu-ura, T., Makino, Y., Masani, A., Okazaki, K., Morishita, T., and Tamura, T.A. (1999). TIP49b, a new RuvB-like DNA helicase, is included in a complex together with another RuvB-like DNA helicase, TIP49a. *J. Biol. Chem.* **274**, 22437–22444.

Keller, L.C., Romijn, E.P., Zamora, I., Yates, J.R.,III., and Marshall, W.F. (2005). Proteomic analysis of isolated *Chlamydomonas* centrioles reveals orthologs of ciliary-disease genes. *Curr. Biol.* **15**, 1090–1098.

Kim, J.C., Badano, J.L., Sibold, S., Esmail, M.A., Hill, J., Hoskins, B.E., Leitch, C.C., Venner, K., Ansley, S.J., Ross, A.J., Leroux, M.R., Katsanis, N., and Beales, P.L. (2004). The Bardet–Biedl protein BBS4 targets cargo to the pericentriolar region and is required for microtubule anchoring and cell cycle progression. *Nat. Genet.* **36**, 462–470.

King, S.M. (2006). Axonemal protofilament ribbons, DM10 domains, and the link to juvenile myoclonic epilepsy. *Cell Motil. Cytoskel.* **63**, 245–253.

King, S.M. and Witman, G.B. (1990). Localization of an intermediate chain of outer arm dynein by immunoelectron microscopy. *J. Biol. Chem.* **265**, 19807–19811.

King, S.M., Wilkerson, C.G., and Witman, G.B. (1991). The M_r 78,000 intermediate chain of *Chlamydomonas* outer arm dynein interacts with α-tubulin *in situ*. *J. Biol. Chem.* **266**, 8401–8407.

Kobayashi, A., Higashide, T., Hamasaki, D., Kubota, S., Sakuma, H., An, W., Fujimaki, T., McLaren, M.J., Weleber, R.G., and Inana, G. (2000). HRG4 (UNC119) mutation found in cone–rod dystrophy causes retinal degeneration in a transgenic model. *Invest Ophthalmol. Vis. Sci.* **41**, 3268–3277.

Koenig, R. (2003). Bardet–Biedl syndrome and Usher syndrome. *Dev. Ophthalmol.* **37**, 126–140.

Kozminski, K.G., Beech, P.L., and Rosenbaum, J.L. (1995). The *Chlamydomonas* kinesin-like protein FLA10 is involved in motility associated with the flagellar membrane. *J. Cell Biol.* **131**, 1517–1527.

Kulaga, H.M., Leitch, C.C., Eichers, E.R., Badano, J.L., Lesemann, A., Hoskins, B.E., Lupski, J.R., Beales, P.L., Reed, R.R., and Katsanis, N. (2004). Loss of BBS proteins causes anosmia in humans and defects in olfactory cilia structure and function in the mouse. *Nat. Genet.* **36**, 994–998.

Lagali, P.S., Balya, D., Awatramani, G.B., Münch, T.A., Kim, D.S., Busskamp, V., Cepko, C.L., and Roska, B. (2008). Light-activated channels targeted to ON bipolar cells restore visual function in retinal degeneration. *Nat. Neurosci.* **11**, 667–675.

Lechtreck, K.F. and Witman, G.B. (2007). *Chlamydomonas reinhardtii* hydin is a central pair protein required for flagellar motility. *J. Cell Biol.* **176**, 473–482.

Lechtreck, K.F., Delmotte, P., Robinson, M.L., Sanderson, M.J., and Witman, G.B. (2008). Mutations in *Hydin* impair ciliary motility in mice. *J. Cell Biol.* **180**, 633–643.

Lee, M.P., Reeves, C., Schmitt, A., Su, K., Connors, T.D., Hu, R.J., Brandenburg, S., Lee, M.J., Miller, G., and Feinberg, A.P. (1998). Somatic mutation of *TSSC5*, a novel imprinted gene from human chromosome 11p15.5. *Cancer Res.* **58**, 4155–4159.

Lehman, J.M., Michaud, E.J., Schoeb, T.R., Aydin-Son, Y., Miller, M., and Yoder, B.K. (2008). The Oak Ridge Polycystic Kidney mouse: Modeling ciliopathies of mice and men. *Dev. Dyn.* **March 25**. [Epub ahead of print]

Leitch, C.C., Zaghloul, N.A., Davis, E.E., Stoetzel, C., Diaz-Font, A., Rix, S., Al Fadhel, M., Lewis, R.A., Eyaid, W., Banin, E., Dollfus, H., Beales, P.L., Badano, J.L., and Katsanis, N. (2008). Hypomorphic mutations in syndromic encephalocele genes are associated with Bardet–Biedl syndrome. *Nat. Genet.* **40**, 443–448.

Lewin, R.A. (1974). Genetic control of flagellar activity in *Chlamydomonas moewusii* (Chlorophyta, Volvocales). *Phycologia* **13**, 45–55.

Li, J.B., Gerdes, J.M., Haycraft, C.J., Fan, Y., Teslovich, T.M., May-Simera, H., Li, H., Blacque, O.E., Li, L., Leitch, C.C., Lewis, R.A., Green, J.S., Parfrey, P.S., Leroux, M.R., Davidson, W.S., Beales, P.L., Guay-Woodford, L.M., Yoder, B.K., Stormo, G.D., Katsanis, N., and Dutcher, S.K. (2004). Comparative genomics identifies a flagellar and basal body proteome that includes the BBS5 human disease gene. *Cell* **117**, 541–552.

Lingle, W.L., Lutz, W.H., Ingle, J.N., Maihle, N.J., and Salisbury, J.L. (1998). Centrosome hypertrophy in human breast tumors: implications for genomic stability and cell polarity. *Proc. Natl. Acad. Sci. U.S.A.* **95**, 2950–2955.

Mahjoub, M.R., Montpetit, B., Zhao, L., Finst, R.J., Goh, B., Kim, A.C., and Quarmby, L.M. (2002). The *FA2* gene of *Chlamydomonas* encodes a NIMA family kinase with roles in cell cycle progression and microtubule severing during deflagellation. *J. Cell Sci.* **115**, 1759–1768.

Marszalek, J.R., Ruiz-Lozano, P., Roberts, E., Chien, K.R., and Goldstein, L.S.B. (1999). Situs inversus and embryonic ciliary morphogenesis defects in mouse mutants lacking the KIF3A subunit of kinesin-II. *Proc. Nat. Acad. Sci. U. S. A.* **96**, 5043–5048.

McVittie, A. (1972). Flagellum mutants of *Chlamydomonas* reinhardtii. *J. Gen. Microbiol.* **71**, 525–540.

Mitchell, D.R. and Brown, K.S. (1994). Sequence analysis of the *Chlamydomonas* alpha and beta dynein heavy chain genes. *J. Cell Sci.* **107**, 635–644.

Mitchell, D.R. and Kang, Y. (1991). Identification of *oda*6 as a *Chlamydomonas* dynein mutant by rescue with the wild-type gene. *J. Cell Biol.* **113**, 835–842.

Mitchell, D.R. and Sale, W.S. (1999). Characterization of a *Chlamydomonas* insertional mutant that disrupts flagellar central pair microtubule-associated structures. *J. Cell Biol.* **144**, 293–304.

Mollenhauer, J., Wiemann, S., Scheurlen, W., Korn, B., Hayashi, Y., Wilgenbus, K.K., von Deimling, A., and Poustka, A. (1997). *DMBT1*, a new member of the SRCR superfamily, on chromosome 10q25.3-26.1 is deleted in malignant brain tumours. *Nat. Genet.* **17**, 32–39.

Moritz, O.L., Tam, B.M., Hurd, L.L., Peranen, J., Deretic, D., and Papermaster, D.S. (2001). Mutant rab8 impairs docking and fusion of rhodopsin-bearing post-Golgi membranes and causes cell death of transgenic *Xenopus* rods. *Mol. Biol. Cell.* **12**, 2341–2351.

Murcia, N.S., Richards, W.G., Yoder, B.K., Mucenski, M.L., Dunlap, J.R., and Woychik, R.P. (2000). The *Oak Ridge Polycystic Kidney (orpk)* disease gene is required for left–right axis determination. *Development.* **127**, 2347–2355.

Mykytyn, K., Mullins, R.F., Andrews, M., Chiang, A.P., Swiderski, R.E., Yang, B., Braun, T., Casavant, T., Stone, E.M., and Sheffield, V.C. (2004). Bardet–Biedl syndrome type 4 (BBS4)-null mice implicate Bbs4 in flagella formation but not global cilia assembly. *Proc. Natl. Acad. Sci. U.S.A.* **101**, 8664–8669.

Nachury, M.V., Loktev, A.V., Zhang, Q., Westlake, C.J., Peranen, J., Merdes, A., Slusarski, D.C., Scheller, R.H., Bazan, J.F., Sheffield, V.C., and Jackson, P.K. (2007). A core complex of BBS proteins cooperates with the GTPase Rab8 to promote ciliary membrane biogenesis. *Cell* **129**, 1201–1213.

Nagel, G., Ollig, D., Fuhrmann, M., Kateriya, S., Musti, A.M., Bamberg, E., and Hegemann, P. (2002). Channelrhodopsin-1: a light-gated proton channel in green algae. *Science* **296**, 2395–2398.

Nauli, S.M., Alenghat, F.J., Luo, Y., Williams, E., Vassilev, P., Li, X., Elia, A.E., Lu, W., Brown, E.M., Quinn, S.J., Ingber, D.E., and Zhou, J. (2003). Polycystins 1 and 2 mediate mechanosensation in the primary cilium of kidney cells. *Nat. Genet.* **33**, 129–137.

Nonaka, S., Tanaka, Y., Okada, Y., Takada, S., Harada, A., Kanai, Y., Kido, M., and Hirokawa, N. (1998). Randomization of left–right asymmetry due to loss of nodal cilia generating leftward flow of extraembryonic fluid in mice lacking KIF3B motor protein. *Cell* **95**, 829–837.

Ogawa, K., Kamiya, R., Wilkerson, C., and Witman, G. (1995). Interspecies conservation of outer arm dynein intermediate chain sequences defines two intermediate chain subclasses. *Mol. Biol. Cell* **6**, 685–696.

Olbrich, H., Haffner, K., Kispert, A., Volkel, A., Volz, A., Sasmaz, G., Reinhardt, R., Hennig, S., Lehrach, H., Konietzko, N., Zariwala, M., Noone, P.G., Knowles, M., Mitchison, H.M., Meeks, M., Chung, E.M., Hildebrandt, F., Sudbrak, R., and Omran, H. (2002). Mutations in *DNAH5* cause primary ciliary dyskinesia and randomization of left–right asymmetry. *Nat. Genet.* **30**, 143–144.

Omran, H., Haffner, K., Volkel, A., Kuehr, J., Ketelsen, U.P., Ross, U.H., Konietzko, N., Wienker, T., Brandis, M., and Hildebrandt, F. (2000). Homozygosity mapping of a gene locus for primary ciliary dyskinesia on chromosome 5p and identification of the heavy dynein chain *DNAH5* as a candidate gene. *Am. J. Respir. Cell Mol. Biol.* **23**, 696–702.

Omran, H., Sasmaz, G., Häffner, K., Volz, A., Olbrich, H., Melkaoui, R., Otto, E., Wienker, T.F., Korinthenberg, R., Brandis, M., Antignac, C., and Hildebrandt, F. (2002). Identification of a gene locus for Senior–Løken syndrome in the region of the nephronophthisis type 3 gene. *J. Am. Soc. Nephrol.* **13**, 75–79.

Otto, E., Hoefele, J., Ruf, R., Mueller, A.M., Hiller, K.S., Wolf, M.T., Schuermann, M.J., Becker, A., Birkenhager, R., Sudbrak, R., Hennies, H.C., Nurnberg, P., and Hildebrandt, F. (2002). A gene mutated in nephronophthisis and retinitis pigmentosa encodes a novel protein, nephroretinin, conserved in evolution. *Am. J. Hum. Genet.* **71**, 1161–1167.

Otto, E.A., Schermer, B., Obara, T., O'Toole, J.F., Hiller, K.S., Mueller, A.M., Ruf, R.G., Hoefele, J., Beekmann, F., Landau, D., Foreman, J.W., Goodship, J.A., Strachan, T., Kispert, A., Wolf, M.T., Gagnadoux, M.F., Nivet, H., Antignac, C., Walz, G., Drummond, I.A., Benzing, T., and Hildebrandt, F. (2003). Mutations in *INVS* encoding inversin cause nephronophthisis type 2, linking renal cystic disease to the function of primary cilia and left–right axis determination. *Nat. Genet.* **34**, 413–420.

Otto, E.A., Loeys, B., Khanna, H., Hellemans, J., Sudbrak, R., Fan, S., Muerb, U., O'Toole, J.F., Helou, J., Attanasio, M., Utsch, B., Sayer, J.A., Lillo, C., Jimeno, D., Coucke, P., De Paepe, A., Reinhardt, R., Klages, S., Tsuda, M., Kawakami, I., Kusakabe, T., Omran, H., Imm, A., Tippens, M., Raymond, P.A., Hill, J., Beales, P., He, S., Kispert, A., Margolis, B., Williams, D.S., Swaroop, A., and Hildebrandt, F. (2005). Nephrocystin-5, a ciliary IQ domain protein, is mutated in Senior–Loken syndrome and interacts with RPGR and calmodulin. *Nat. Genet.* **37**, 282–288.

Otto, E.A., Trapp, M.L., Schultheiss, U.T., Helou, J., Quarmby, L.M., and Hildebrandt, F. (2008). *NEK8* mutations affect ciliary and centrosomal localization and may cause nephronophthisis. *J. Am. Soc. Nephrol.* **19**, 587–592.

Ou, G., Blacque, O.E., Snow, J.J., Leroux, M.R., and Scholey, J.M. (2005). Functional coordination of intraflagellar transport motors. *Nature* **436**, 583–587.

Pan, J. and Snell, W. (2007). The primary cilium: keeper of the key to cell division. *Cell* **129**, 1255–1257.

Patel-King, R.S., Benashski, S.E., and King, S.M. (2002). A bipartite Ca^{2+}-regulated nucleoside-diphosphate kinase system within the *Chlamydomonas* flagellum. The regulatory subunit p72. *J. Biol. Chem.* **277**, 34271–34279.

Pazour, G.J. (2004). Intraflagellar transport and cilia-dependent renal disease: the ciliary hypothesis of polycystic kidney disease. *J. Am. Soc. Nephrol.* **15**, 2528–2536.

Pazour, G.J. and Rosenbaum, J.L. (2002). Intraflagellar transport and cilia-dependent diseases. *Trends Cell Biol.* **12**, 551–555.

Pazour, G.J. and Witman, G.B. (2000). Forward and reverse genetic analysis of microtubule motors in *Chlamydomonas*. *Methods* **22**, 285–298.

Pazour, G.J. and Witman, G.B. (2003). The vertebrate primary cilium is a sensory organelle. *Curr. Opin. Cell Biol.* **15**, 105–110.

Pazour, G.J., Dickert, B.L., Vucica, Y., Seeley, E.S., Rosenbaum, J.L., Witman, G.B., and Cole, D.G. (2000). *Chlamydomonas IFT*88 and its mouse homologue, polycystic kidney disease gene *Tg737*, are required for assembly of cilia and flagella. *J. Cell Biol.* **151**, 709–718.

Pazour, G.J., Baker, S.A., Deane, J.A., Cole, D.G., Dickert, B.L., Rosenbaum, J.L., Witman, G.B., and Besharse, J.C. (2002a). The intraflagellar transport protein, IFT88, is essential for vertebrate photoreceptor assembly and maintenance. *J. Cell Biol.* **157**, 103–113.

Pazour, G.J., San Agustin, J.T., Follit, J.A., Rosenbaum, J.L., and Witman, G.B. (2002b). Polycystin-2 localizes to kidney cilia and the ciliary level is elevated in orpk mice with polycystic kidney disease. *Curr. Biol.* **12**, R378–R380.

Pazour, G.J., Agrin, N., Leszyk, J., and Witman, G.B. (2005). Proteomic analysis of a eukaryotic cilium. *J. Cell Biol.* **170**, 103–113.

Pazour, G.J., Agrin, N., Walker, B.L., and Witman, G.B. (2006). Identification of predicted human outer dynein arm genes: candidates for primary ciliary dyskinesia genes. *J. Med. Genet.* **43**, 62–73.

Pennarun, G., Escudier, E., Chapelin, C., Bridoux, A.-M., Cacheux, V., Roger, G., Clemont, A., Goossens, M., Amselem, S., and Duriez, B. (1999). Loss-of-function mutations in a human gene related to *Chlamydomonas reinhardtii* dynein IC78 result in primary ciliary dyskinesia. *Am. J. Hum. Genet.* **65**, 1508–1519.

Pfister, K.K., Fay, R.B., and Witman, G.B. (1982). Purification and polypeptide composition of dynein ATPases from *Chlamydomonas* flagella. *Cell Motil.* **2**, 525–547.

Pihan, G.A., Purohit, A., Wallace, J., Knecht, H., Woda, B., Quesenberry, P., and Doxsey, S.J. (1998). Centrosome defects and genetic instability in malignant tumors. *Cancer Res.* **58**, 3974–3985.

Piperno, G., Huang, B., and Luck, D. (1977). Two-dimensional analysis of flagellar proteins from wild-type and paralyzed mutants of *Chlamydomonas reinhardtii*. *Proc. Nat. Acad. Sci. U.S.A.* **74**, 1600–1604.

Porter, M.E., Knott, J.A., Gardner, L.C., Mitchell, D.R., and Dutcher, S.K. (1994). Mutations in the *SUP-PF*-1 locus of *Chlamydomonas reinhardtii* identify a regulatory domain in the β-dynein heavy chain. *J. Cell Biol.* **126**, 1495–1507.

Quarmby, L.M. and Mahjoub, M.R. (2005). Caught Nek-ing: cilia and centrioles. *J. Cell Sci.* **118**, 5161–5169.

Richards, W.G., Yoder, B.K., Isfort, R.J., Detilleux, P.G., Foster, C., Neilsen, N., Woychik, R.P., and Wilkinson, J.E. (1996). Oval cell proliferation associated with the murine insertional mutation *TgN737Rpw*. *Am. J. Path.* **149**, 1919–1930.

Romio, L., Wright, V., Price, K., Winyard, P.J., Donnai, D., Porteous, M.E., Franco, B., Giorgio, G., Malcolm, S., Woolf, A.S., and Feather, S.A. (2003). *OFD1*, the gene mutated in oral–facial–digital syndrome type 1, is expressed in the metanephros and in human embryonic renal mesenchymal cells. *J. Am. Soc. Nephrol.* **14**, 680–689.

Rosenbaum, J.L. and Witman, G.B. (2002). Intraflagellar transport. *Nat. Rev. Mol. Cell Biol.* **3**, 813–825.

Rupp, G., O'Toole, E., Gardner, L.C., Mitchell, B.F., and Porter, M.E. (1996). The *sup-pf-2* mutations of *Chlamydomonas* alter the activity of the outer dynein arms by modification of the gamma-dynein heavy chain. *J. Cell Biol.* **135**, 1853–1865.

Sakakibara, H., Takada, S., King, S.M., Witman, G.B., and Kamiya, R. (1993). A *Chlamydomonas* outer arm dynein mutant with a truncated β heavy chain. *J. Cell Biol.* **122**, 653–661.

Sapiro, R., Kostetskii, I., Olds-Clarke, P., Gerton, G.L., Radice, G.L., and Strauss, J.F. III. (2002). Male infertility, impaired sperm motility, and hydrocephalus in mice deficient in sperm-associated antigen 6. *Mol. Cell Biol.* **22**, 6298–6305.

Sawamoto, K., Wichterle, H., Gonzalez-Perez, O., Cholfin, J.A., Yamada, M., Spassky, N., Murcia, N.S., Garcia-Verdugo, J.M., Marin, O., Rubenstein, J.L., Tessier-Lavigne, M., Okano, H., and Alvarez-Buylla, A. (2006). New neurons follow the flow of cerebrospinal fluid in the adult brain. *Science* **311**, 629–632.

Sayer, J.A., Otto, E.A., O'Toole, J.F., Nurnberg, G., Kennedy, M.A., Becker, C., Hennies, H.C., Helou, J., Attanasio, M., Fausett, B.V., Utsch, B., Khanna, H., Liu, Y., Drummond, I., Kawakami, I., Kusakabe, T., Tsuda, M., Ma, L., Lee, H., Larson, R.G., Allen, S.J., Wilkinson, C.J., Nigg, E.A., Shou, C., Lillo, C., Williams, D.S., Hoppe, B., Kemper, M.J., Neuhaus, T., Parisi, M.A., Glass, I.A., Petry, M., Kispert, A., Gloy, J., Ganner, A., Walz, G., Zhu, X., Goldman, D., Nurnberg, P., Swaroop, A., Leroux, M.R., and Hildebrandt, F. (2006). The centrosomal protein nephrocystin-6 is mutated in Joubert syndrome and activates transcription factor ATF4. *Nat. Genet.* **38**, 674–681.

Scholey, J.M. (1996). Kinesin-II, a membrane traffic motor in axons, axonemes, and spindles. *J. Cell Biol.* **133**, 1–4.

Schrander-Stumpel, C. and Fryns, J.P. (1998). Congenital hydrocephalus: nosology and guidelines for clinical approach and genetic counselling. *Eur. J. Pediatr.* **157**, 355–362.

Schwabe, G.C., Hoffmann, K., Loges, N.T., Birker, D., Rossier, C., de Santi, M.M., Olbrich, H., Fliegauf, M., Failly, M., Liebers, U., Collura, M., Gaedicke, G., Mundlos, S., Wahn, U., Blouin, J.L., Niggemann, B., Omran, H., Antonarakis, S.E., and Bartoloni, L. (2008). Primary ciliary dyskinesia associated with normal axoneme ultrastructure is caused by *DNAH11* mutations. *Hum. Mutat.* **29**, 289–298.

Sineshchekov, O.A., Jung, K.H., and Spudich, J.L. (2002). Two rhodopsins mediate phototaxis to low- and high-intensity light in *Chlamydomonas reinhardtii*. *Proc. Natl. Acad. Sci. U.S.A.* **99**, 8689–8694.

Stolc, V., Samanta, M.P., Tongprasit, W., and Marshall, W.F. (2005). Genome-wide transcriptional analysis of flagellar regeneration in *Chlamydomonas reinhardtii* identifies orthologs of ciliary disease genes. *Proc. Natl. Acad. Sci. U.S.A.* **10**, 3703–3707.

Sulik, K., Dehart, D.B., Inagaki, T., Carson, J.L., Vrablic, T., Gesteland, K., and Schoenwolf, G.C. (1994). Morphogenesis of the murine node and notochordal plate. *Dev. Dyn.* **201**, 260–278.

Suzuki, T., Yamasaki, K., Fujita, S., Oda, K., Iseki, M., Yoshida, K., Watanabe, M., Daiyasu, H., Toh, H., Asamizu, E., Tabata, S., Miura, K., Fukuzawa, H., Nakamura, S., and Takahashi, T. (2003). Archaeal-type rhodopsins in *Chlamydomonas*: model structure and intracellular localization. *Biochem. Biophys. Res. Commun.* **301**, 711–717.

Suzuki, T., Delgado-Escueta, A.V., Aguan, K., Alonso, M.E., Shi, J., Hara, Y., Nishida, M., Numata, T., Medina, M.T., Takeuchi, T., Morita, R., Bai, D., Ganesh, S., Sugimoto, Y., Inazawa, J., Bailey, J.N., Ochoa, A., Jara-Prado, A., Rasmussen, A., Ramos-Peek, J., Cordova, S., Rubio-Donnadieu, F., Inoue, Y., Osawa, M., Kaneko, S., Oguni, H., Mori, Y., and Yamakawa, K. (2004). Mutations in *EFHC1* cause juvenile myoclonic epilepsy. *Nat. Genet.* **36**, 842–849.

Suzuki, T., Inoue, I., Yamagata, T., Morita, N., Furuichi, T., and Yamakawa, K. (2008). Sequential expression of Efhc1/myoclonin1 in choroid plexus and ependymal cell cilia. *Biochem. Biophys. Res. Commun.* **367**, 226–233.

Tam, L.-W. and Lefebvre, P.A. (1993). Cloning of flagellar genes in *Chlamydomonas reinhardtii* by DNA insertional mutagenesis. *Genetics* **135**, 375–384.

Walther, Z., Vashishtha, M., and Hall, J.L. (1994). The *Chlamydomonas FLA*10 gene encodes a novel kinesin-homologous protein. *J. Cell Biol.* **126**, 175–188.

Wang, S., Luo, Y., Wilson, P.D., Witman, G.B., and Zhou, J. (2004). The autosomal recessive polycystic kidney disease protein is localized to primary cilia, with concentration in the basal body area. *J. Am. Soc. Nephrol.* **15**, 592–602.

Ward, C.J., Yuan, D., Masyuk, T.V., Wang, X., Punyashthiti, R., Whelan, S., Bacallao, R., Torra, R., LaRusso, N.F., Torres, V.E., and Harris, P.C. (2003). Cellular and subcellular localization of the ARPKD protein; fibrocystin is expressed on primary cilia. *Hum. Mol. Genet.* **12**, 2703–2710.

Warr, J.R., McVittae, A., Randall, J., and Hopkins, J.M. (1966). Genetic control of flagellar structure in *Chlamydomonas reinhardii*. *Genet. Res.* **7**, 335–351.

Wilkerson, C.G., King, S.M., and Witman, G.B. (1994). Molecular analysis of the γ heavy chain of *Chlamydomonas* flagellar outer-arm dynein. *J. Cell Sci.* **107**, 497–506.

Wilkerson, C.G., King, S.M., Koutoulis, A., Pazour, G.J., and Witman, G.B. (1995). The 78,000 M_r intermediate chain of *Chlamydomonas* outer arm dynein is a WD-repeat protein required for arm assembly. *J. Cell Biol.* **129**, 169–178.

Witman, G.B., Fay, R., and Plummer, J. (1976). *Chlamydomonas* mutants: evidence for the roles of specific axonemal components in flagellar movement. In: *Cell Motility* (R.D. Goldman, T.D. Pollard, and J.L. Rosenbaum, Eds.), pp. 969–986. Cold Spring Harbor Laboratory, Cold Spring Harbor, NY.

Witman, G.B., Plummer, J., and Sander, G. (1978). *Chlamydomonas* flagellar mutants lacking radial spokes and central tubules. *J. Cell Biol.* **76**, 729–747.

Yang, P., Diener, D.R., Yang, C., Kohno, T., Pazour, G.J., Dienes, J.M., Agrin, N.S., King, S.M., Sale, W.S., Kamiya, R., Rosenbaum, J.L., and Witman, G.B. (2006). Radial spoke proteins of *Chlamydomonas* flagella. *J. Cell Sci.* **119**, 1165–1174.

Yoder, B.K., Hou, X., and Guay-Woodford, L.M. (2002). The polycystic kidney disease proteins, polycystin-1, polycystin-2, polaris, and cystin, are co-localized in renal cilia. *J. Am. Soc. Nephrol.* **13**, 2508–2516.

Zariwala, M.A., Knowles, M.R., and Omran, H. (2007). Genetic defects in ciliary structure and function. *Annu. Rev. Physiol.* **69**, 423–450.

Zhang, M.Z., Mai, W., Li, C., Cho, S.Y., Hao, C., Moeckel, G., Zhao, R., Kim, I., Wang, J., Xiong, H., Wang, H., Sato, Y., Wu, Y., Nakanuma, Y., Lilova, M., Pei, Y., Harris, R.C., Li, S., Coffey, R.J., Sun, L., Wu, D., Chen, X.Z., Breyer, M.D., Zhao, Z.J., McKanna, J.A., and Wu, G. (2004). PKHD1 protein encoded by the gene for autosomal recessive polycystic kidney disease associates with basal bodies and primary cilia in renal epithelial cells. *Proc. Natl. Acad. Sci. U. S. A.* **101**, 2311–2316.

Zhang, Z., Kostetskii, I., Tang, W., Haig-Ladewig, L., Sapiro, R., Wei, Z., Patel, A.M., Bennett, J., Gerton, G.L., Moss, S.B., Radice, G.L., and Strauss, J.F. III. (2006). Deficiency of SPAG16L causes male infertility associated with impaired sperm motility. *Biol. Reprod.* **74**, 751–759.

Zhang, Z., Zariwala, M.A., Mahadevan, M.M., Caballero-Campo, P., Shen, X., Escudier, E., Duriez, B., Bridoux, A.M., Leigh, M., Gerton, G.L., Kennedy, M., Amselem, S., Knowles, M.R., and Strauss, J.F. III. (2007). A heterozygous mutation disrupting the *SPAG16* gene results in biochemical instability of central apparatus components of the human sperm axoneme. *Biol. Reprod.* **77**, 864–871.

Index

Note: Page numbers referring to figures or tables are in *italic*.

479